T0191998

Statistical Theory of Open Systems

Fundamental Theories of Physics

An International Book Series on The Fundamental Theories of Physics:
Their Clarification, Development and Application

Editor: ALWYN VAN DER MERWE
  University of Denver, U.S.A.

Editorial Advisory Board:

Volume 67

Statistical Theory of Open Systems

Volume 1:
A Unified Approach to Kinetic Description
of Processes in Active Systems

by

Yu. L. Klimontovich

Department of Theoretical Physics,
Moscow State University,
Moscow, Russia

KLUWER ACADEMIC PUBLISHERS
DORDRECHT / BOSTON / LONDON

A C.I.P. Catalogue record for this book is available from the Library of Congress

ISBN 0-7923-3242-3 (PB)

Published by Kluwer Academic Publishers,
P.O. Box 17, 3300 AA Dordrecht, The Netherlands.

Kluwer Academic Publishers incorporates
the publishing programmes of
D. Reidel, Martinus Nijhoff, Dr W. Junk and MTP Press.

Sold and distributed in the U.S.A. and Canada
by Kluwer Academic Publishers,
101 Philip Drive, Norwell, MA 02061, U.S.A.

In all other countries, sold and distributed
by Kluwer Academic Publishers Group,
P.O. Box 322, 3300 AH Dordrecht, The Netherlands.

Manuscript translated by A. Dobroslavsky

Printed on acid-free paper

Printed in the Netherlands

CONTENTS

PREFACE

Let us begin by quoting from the Preface to the author's *Statistical Physics* (Moscow, Nauka 1982; also published in English by Harwood in 1986):

"'My God! Yet another book on statistical physics! There's no room on my bookshelves left!' Such emotions are quite understandable. Before jumping to conclusions, however, it would be worthwhile to read the Introduction and look through the table of contents. Then the reader will find that this book is totally different from the existing courses, fundamental and concise.

... We do not use the conventional division into statistical theories of equilibrium and nonequilibrium states. Rather than that, the theory of nonequilibrium state is the basis and the backbone of the entire course.

... This approach allows us to develop a unified method for statistical description of a very broad class of systems.

... The author certainly does not wish to exaggerate the advantages of the book, considering it as just the first attempt to create a textbook of a new kind."

The next step in this direction was the author's *Turbulent Motion and the Structure of Chaos* (Moscow, Nauka 1990; Kluwer Academic Publishers 1991). This book is subtitled *A New Approach to the Statistical Theory of Open Systems*. Naturally, the "new approach" is not meant to defy the consistent and efficient methods of the conventional statistical theory; it should be regarded as a useful reinforcement of such methods.

The material of this new book has been discussed at many conferences and seminars. It also served as a basis for the author's courses read at Moscow State University and at universities of Germany. The feedback was definitely positive. Of course, there was a number of opponents, none of whom could, however, produce any constructive counter-arguments. Eventually it became clear that the main ideas, methods and results of statistical theory ought to be presented in the form of a monographic textbook which will be helpful not only for the physicists, but also – at least basically – for all specialists who use statistical theory of open systems in their studies.

Let us now explain briefly the title of the present book.

Recall that open systems, unlike the ideal closed (isolated) systems, exchange matter, energy, and (which is very important) information with their environment. Because of this, the evolution of open systems does not necessarily lead towards the equilibrium (and the most chaotic) state. On the contrary, open systems may be involved in processes of self-organization, which result in more complicated and more advanced structures.

All systems that we are going to deal with are macroscopic. This means that they consist of a large number of elements or "particles", which can be atoms and molecules in physics and chemistry, macromolecules, cells and microorganisms in biology, groups of living organisms in sociology, planets and stars in astronomy. In other words, the spectrum of open systems in quite broad.

Such systems are difficult to describe not only because they comprise a very large number of elementary objects. Big problems are also posed by the dynamic instability of motion of separate particles – that is, by the complexity of dynamic motion. This means that the trajectories diverge drastically even with small variations of the initial conditions.

Because of this, small external actions may result in uncontrollable and unpredictable changes in the motion of particles. The only way out is then to rely on the statistical description.

The presence of a large number of elementary objects allows one in many cases to treat the system of particles as a continuous medium. This convention changes dramatically the nature of the system. In order to avoid the potential problems, it is necessary to use a physical definition of continuous medium rather than a formal mathematical definition. This requires a concrete definition of physically infinitesimal time and length scales in terms of characteristic parameters of the system. The corresponding physically infinitesimal volume is the equivalent of a physical "point". Naturally, the definition will depend on the adopted level of description. In this way, a concept of "continuous medium" is introduced into the statistical theory.

We shall see that the number of particles within a "point" in continuous medium can be quite large: in a Boltzmann gas (rarefied gas), for instance, this number is of the order of 10^5. This explains the expedience of smoothing over the states of particles within a physically infinitesimal volume. The judicious choice of the smoothing function allows one to preserve the complexity of motion of individual elements of the medium in the statistical theory.

The Gibbs ensemble for nonequilibrium processes can be defined in terms of the physical definition of continuous medium. The deficiency of description of microstates of separate systems in the ensemble is due to the lack of information on the motion of particles within the "points" of continuous medium.

While the motion of individual particles of the system is described by the reversible equations of motion (the Hamilton equations), the transition to the approximation of continuous medium leads inevitably to irreversible equations, because the information concerning the motion of particles within the "points" is irretrievably lost. It is important that the instability of dynamic motion (that is, the complexity of dynamic motion) plays a constructive role in the development of statistical theory of open systems.

Concern with the structure of "continuous medium" at all levels of description is an important part of the new approach to the statistical theory of open systems. As we shall see, this approach brings us to generalized kinetic equations, which serve as basis for unified description of kinetic, hydrodynamic and diffusion processes without the use of perturbation theory. It also makes it possible to give new treatment to the phenomenon of turbulence, and to propose a unified kinetic description of laminar and turbulent motion.

All this equally applies to both classical and quantum systems. The problem of continuous medium and its structure is inherent in the transition from the idealized reversible equations of quantum mechanics to approximate and irreversible (but realistic) equations of quantum statistics. The use of statistical theory of open systems in quantum mechanics supplies answers to the "eternal questions" — whether quantum mechanical description is complete, and whether there are hidden parameters in quantum mechanics.

Let us once again emphasize that the new approach supplements rather than replaces the traditional ways of describing the nonequilibrium processes. A more or less comprehensive description can only be based on a judicious combination of "the old" and "the new".

It was not easy to select the material to be included in the book. Eventually, it seemed wise to divide the book into two volumes. This first volume presents the basic ideas, methods and results for relatively "simple" systems, avoiding the minute mathematical details which add little to the understanding of the important matters. The interested reader is referred to the sources containing the links which had to be left out.

I am grateful to my teachers, colleagues, and friends for fruitful discussions of numerous problems that are touched upon in the book. To name a few of the many is to do injustice to all.

Yu. Klimontovich
August 1994

CHAPTER 1

INTRODUCTION

As indicated in the Preface, this book purports to give a systematic and comprehensive treatment of the main ideas, methods and results of the modern statistical theory of open systems.

A considerable portion of the book is devoted to the material which has previously been presented in a number of textbooks. Conventional methods of description are critically revised, however, so as to prepare fertile turf for the new approach, the powerful synthesis of "the old" and "the new".

The structure of the course is clear enough from the table of contents. Introduction gives a general outline of our subject. The main attention here will be paid to the less conventional facets of the statistical theory of open systems.

This implies that the Introduction will not touch upon all the problems discussed further on in the book. It is not that these problems are of overwhelming importance: here we just wanted to emphasize the most recent developments.

Observe that our macroscopic open systems exchange not only energy and matter, but also information with surrounding bodies, and that the spectrum of such systems is extremely broad. Because of this, statistical theory of open systems provides a sound scientific basis for many disciplines.

Statistical description of macroscopic open systems is based on the concept of statistical Gibbs ensemble, which is naturally associated with the problem of definition of the structure of "continuous medium".

With these fundamental concepts we are going to begin.

1.1. Structure of "Continuous Medium". Gibbs Ensemble for Nonequilibrium Processes

1.1.1. Gibbs Ensemble. Equilibrium State

Gibbs ensemble is one of the basic concepts of statistical physics. Let us consider the definition of ensemble as formulated by Gibbs himself in his fundamental paper entitled

1

Elementary Principles in Statistical Mechanics, Developed with Especial Reference to the Rational Foundation of Thermodynamics (1902). This definition is given in the Preface to the paper and runs as follows:

"For some purposes, however, it is desirable to take a broader view of the subject. We may imagine a great number of systems of the same nature, but differing in the configurations and velocities which they have at a given instant, and differing not merely infinitesimally, but it may be so as to embrace every conceivable combination of configuration and velocities. And here we may set the problem, not to follow a particular system through its succession of configurations, but to determine how the whole number of systems will be distributed among the various conceivable configurations and velocities at any required time, when the distribution has been given for some one time. The fundamental equation for this inquiry is that which gives the rate of change of the number of systems which fall within any infinitesimal limits of configurations and velocity.

"Such inquiries have been called by Maxwell *statistical*. They belong to a branch of mechanics which owes its origin to the desire to explain the laws of thermodynamics on mechanical principles, and of which Clausius, Maxwell, and Boltzmann are to be regarded as the principal founders. The first inquiries in this field were indeed somewhat narrower in their scope than that which has been mentioned, being applied to the particles of the system rather than to independent systems. Statistical inquiries were next directed to the phases (or conditions with respect to configurations and velocity) which succeed one another in a given system in the course of time. The explicit consideration of a great number of systems and their distribution in phase [space], and of their permanence or alteration of this distribution in the course of time is perhaps first found in Boltzmann's paper on the "Zusammenhang zwischen den Sätzen über das Verhalten mehratomiger Gasmoleküle mit Jacobi's Princip des letzten Multiplicators" (1871).

"But although, as a matter of history, statistical mechanics owes its origin to investigations in thermodynamics, it seems eminently worthy of an independent development, both on account of the elegance and simplicity of its principles, and because it yields new results and places old truths in a new light in departments quite outside of thermodynamics. Moreover, the separate study of this branch of mechanics seems to afford the best foundation for the study of rational thermodynamics and molecular mechanics."

Let X be the complete set of variable which determine the state of the elements of the system. In particular, for a system of structureless atoms X is a $6N$-dimensional vector of coordinates and momenta,

$$X = X(r_1,...,r_N;p_1,...,p_N) \ . \tag{1.1.1}$$

By $f_N(X,t)$ we denote the statistical distribution of the states of the systems of Gibbs ensemble. This distribution is normalized to unity,

$$\int f_N(X,t)dX = 1 . \tag{1.1.2}$$

Statistical distribution will be introduced in Sect. 2.5; here we just note that at equilibrium it does not depend on time, and its concrete form is determined by the Hamilton function $H(X,a)$ of the conservative system in question, where a is the set of external parameters which are controlled for the ensemble of systems.

The most important for practical applications is the case when the system is in a thermostat at the temperature T. Then, as demonstrated by Gibbs (see Sect. 3.2), the statistical distribution is given by

$$f_N(X,a,T) = \exp\frac{F(a,T) - H(X,a)}{\kappa T} , \tag{1.1.3}$$

which is known as the canonical Gibbs distribution, where $F(a,T)$ is the free energy, one of the basic thermodynamic functions.

Distribution (1.1.3) holds for any arbitrary macroscopic system in a thermostat, provided that the motion of individual atoms (elements) of the system is describable by the equations of motion of classical mechanics.

Recall that we are dealing with macroscopic systems. This means that the number of elements (particles) N is very large, and very large is the number of degrees of freedom. For instance, the number of degrees of freedom is $6N$ for a system of N structureless atoms. At the same time, the distribution (1.1.3) contains very few parameters (as compared to the number of degrees of freedom $6N$): the temperature T, and the set of external parameters a, which in thermodynamics are not many.

We see that the equilibrium Gibbs ensemble has very few controllable degrees of freedom, and therefore the uncertainty of definition of microstates of individual subsystems of the ensemble is very high. In other words, the degree of chaoticity is very large.

1.1.2. Gibbs Ensemble. Averaging over Distribution of Initial Values

The above definition of statistical ensemble is of course not suitable for the description of states far from equilibrium, when the degree of uncertainty – chaoticity (we shall give a proper definition of these terms later on) is much lower than at equilibrium. The lowest (zero) chaoticity corresponds to the complete dynamic description of the particles of the system, when the state of the system at any time t is described by $6N$-dimensional vector

$$X = X(X_0, t - t_0) ,$$ (1.1.4)

where X_0 is the set of $6N$ initial (at $t = t_0$) values of all coordinates and momenta of the particles. We see that the complete description of the system demands that the number of parameters should equal the number of the degrees of freedom.

Dynamic description can be associated with the appropriate dynamic (superscript "d") distribution function:

$$f_N^{(d)}(X,t) = \delta(X - X(X_0, t - t_0)) .$$ (1.1.5)

This function is actually used for the transition from the particle variables $X(t)$ to the relevant $6N$-dimensional phase space X. In thermodynamics this is equivalent to the transition from the Lagrangian description of motion of particles of the fluid to the Eulerian description of distribution in the phase space.

Because of the extreme complexity of motion in macroscopic systems, complete dynamic description is not feasible, and we have to go over to statistical description. This brings us back to the definition of the Gibbs ensemble, this time for nonequilibrium states. This definition should reflect the transition from the reversible equations of mechanics to the irreversible equations of the statistical theory. It also should serve as a bridge from complete dynamic description (when the uncertainty of the description can, in principle, be reduced to zero) to the Gibbs ensemble for the state of equilibrium, when the degree of chaoticity is at maximum.

Widely used in the theory of nonequilibrium processes is the following definition of Gibbs ensemble for nonequilibrium states. Assume that the deficiency of description is wholly due to the uncertainty of definition of the initial coordinates and momenta X_0 of the particles. If we denote the relevant distribution function by

$$f_N(X_0, t_0) , \quad \int f_N(X_0, t_0) dX_0 = 1 ,$$ (1.1.6)

then the statistical distribution $f_N(X,t)$ can be derived by averaging the dynamic distribution (1.1.5) over the distribution of initial values (1.1.6):

$$f_N(X,t) = \overline{f_N^{(d)}(X,t)}^{(X_0)} = \int \delta(X - X(X_0, t - t_0)) f_N(X_0, t) dX_0 .$$ (1.1.7)

Hereby we assume that the entire uncertainty is due to insufficient knowledge of the initial values X_0, whereas the time evolution is described by the same equations of motion. In other words, the dynamic equations still remain the reversible equations of Hamiltonian mechanics. Since the initial values X_0 enter equations as parameters, averaging over the initial distribution does not change the equation for the distribution function. In the course of evolution the degree of uncertainty of the states of systems of Gibbs ensemble remains the same, because it is set by the initial distribution $f_N(X_0,t_0)$. There is, however, a crucial problem.

In case of equilibrium processes, the deficiency of description in terms of initial values is, as a rule, much lower than the deficiency which corresponds to the most chaotic equilibrium distribution.

One of the principles of statistical thermodynamics states that the process of time evolution in a "closed"[*] system leads towards equilibrium. Since the equilibrium state is the most chaotic (see Gibbs theorem in Ch. 3), the equation for statistical distribution $f_N(X,t)$ in case of a "closed" system should describe the increase in chaoticity in the course of evolution towards equilibrium.

If, however, we accept the definition of Gibbs ensemble (1.1.7), then the increase in chaoticity in the course of evolution towards equilibrium cannot be described.

Indeed, for the measure of chaos we are going to use the entropy of the system (see Sect. 3.5) which is linked with the statistical distribution $f_N(X,t)$ by integral relationship

$$S(t) = -\kappa \int f_N(X,t)\ln f_N(X,t)dX \ . \tag{1.1.8}$$

This definition of entropy was first given by Boltzmann for an exemplary case of rarefied gas (see Ch. 3 and Ch. 6); then Gibbs used it for an arbitrary equilibrium state, for canonical distribution in particular. Several decades later, Shannon introduced entropy as defined by (1.1.8) for distribution $f_N(X,t)$ of arbitrary set of variables X as measure of chaoticity in the theory of information.

So we use the Boltzmann-Gibbs-Shannon entropy as expressed in terms of statistical distribution to fathom the chaoticity of states of a closed system (which will be defined later on). Then the second law of thermodynamics, which states that entropy of a closed system is a non-decreasing entity, can be expressed as

$$\frac{dS}{dt} \geq 0 \ , \tag{1.1.9}$$

where the "equals" sign corresponds to the state of equilibrium.

However, equation (1.1.7) for statistical distribution is reversible and only describes the "transfer" of degree of chaoticity as determined by the initial distribution $f_N(X_0,t_0)$. Because of this, the entropy (1.1.7) remains constant in the course of time evolution,

---

[*] "Closed" is put in quotes because the increase in entropy occurs under condition of constant *mean* energy rather than constant energy itself; see Sect. 1.6.3 and Ch. 3.

$$\frac{dS}{dt} = 0 \ , \tag{1.1.10}$$

and hence it is impossible to describe evolution towards equilibrium using this definition of Gibbs ensemble.

In spite of this, the classical works of Bogolybov, Born, Green, Kirkwood (see Ch. 6) use the reversible Liouville equation for deriving the basic irreversible kinetic equations of statistical theory of irreversible processes — in particular, the kinetic Boltzmann equation. We shall see that this is actually associated with re-definition (implicit, as a rule) of the Gibbs ensemble.

All known kinetic equations of statistical physics are examples of mechanics of continuous medium. Unlike equations of hydrodynamics and elasticity theory, these equations refer to a continuous medium in a six-dimensional phase space $x = x(r,p)$ of coordinates and momenta, or coordinates and velocities. The principal statistical characteristic is then the distribution function $f(r,p,t)$. Boltzmann equation is an example of kinetic equation for rarefied gas.

We see that the use of kinetic equations implies a transition from a system of particles to the continuous medium approximation. The size of point of continuous medium is determined by the concept of physically infinitesimal volume V_{ph}. The number of particles within physically infinitesimal volume (the number of particles within a "point") is $N_{ph} = nV_{ph}$, where $n = N/V$ is the average number of particles per unit volume.

Obviously, the definition of physically infinitesimal scales (in particular, physically infinitesimal volume) will depend on the accepted level of description. The smallest "point" size will correspond to the kinetic description.

1.1.3. Physically Infinitesimal Scales in Kinetic Theory

By definition of physically infinitesimal scales, the number of particles within V_{ph} is large (that is, $N_{ph} \gg 1$), and the scales of time τ_{ph} and length l_{ph} are small compared to the characteristic scales of time T and length L of the problem:

$$\tau_{ph} \ll T, \ l_{ph} \ll L, \ N_{ph} \gg 1 \ . \tag{1.1.11}$$

Naturally, the definition of physically infinitesimal scales cannot be universal: it will depend on the adopted level of description of nonequilibrium processes (kinetic, hydrodynamic, or diffusion).

Let the system under consideration be a rarefied gas of N atoms which occupies volume V. We use the most simple model which regards atoms as perfectly elastic balls of diameter r_0. The value of r_0 is one of the characteristic parameters of length.

Another characteristic parameter of length is r_{av}, the average distance between atoms. By order of magnitude $r_{av} \approx n^{-\frac{1}{3}}$, because $v = 1/n$ is the volume per one atom.

A third characteristic parameter of length is l, the mean free path length, which by order of magnitude is $l = 1/nr_0^2$. Only two out of the three parameters r_0, r_{av}, l are independent; let these be r_0 and r_{av}. We use them to define a dimensionless parameter of density

$$\varepsilon = nr_0^3 , \qquad\qquad\qquad (1.1.12)$$

which characterizes the ratio of the volume of atom to the volume of space per one atom.

Rarefied gases are those for which the parameter of density ε is small. For instance, air at atmospheric pressure is a rarefied gas, because $\varepsilon \approx 10^{-4}$.

Thus, the following relationships hold for a rarefied gas:

$$r_0 << r_{av} << l , \quad \varepsilon = nr_0^3 << 1 . \qquad\qquad (1.1.13)$$

We can also define the relevant parameters of time (v_T is thermal velocity of atoms):

$$\tau_0 = \frac{r_0}{v_T} , \quad \tau_{av} = \frac{r_{av}}{v_T} , \quad \tau = \frac{l}{v_T} , \quad \text{so that} \quad \tau_0 << \tau_{av} << \tau . \qquad (1.1.14)$$

The time τ is called the mean free path time, or the "collision" time.

We are going to distinguish two types of equilibrium states: total equilibrium and local equilibrium. The time of evolution towards total equilibrium depends on the size of the system and can be relatively large (see Ch. 11). It may be the time of diffusion, which characterizes the process of spatial redistribution of foreign particles in a fluid.

The time of transition to local equilibrium is determined by the mean free path time τ. The relevant length scale is the mean free path l. The transition to local equilibrium is described by the kinetic Boltzmann equation (see Ch. 6).

So, for the characteristic scales in the kinetic theory we may use the values of τ and l in the relationships (1.1.11). In other words, for the kinetic stage of relaxation we set

$$T \Rightarrow \tau , \quad L \Rightarrow l . \qquad\qquad\qquad (1.1.15)$$

Now, how do we choose physically infinitesimal scales τ_{ph}, l_{ph} which satisfy inequalities (1.1.11), (1.1.15) and thus define the structure of "continuous medium"?

At first sight, this attempt to regard a rarefied gas as a continuous medium may seem paradoxical. To show that this is not so, we proceed as follows.

Divide the value of τ, which is the average time lapse between two consecutive collisions of a given particle, into the number of particles within physically infinitesimal volume $N_{ph} = nV_{ph}$. The resulting value is the time between consecutive collisions of one (any one!) particle contained within physically infinitesimal volume V_{ph}; it would be natural to take this for the definition of τ_{ph}. Using the relationship $\tau_{ph} = l_{ph}/v_T$, we obtain two equations,

$$\frac{\tau}{N_{ph}} \sim \frac{\tau}{nl_{ph}^3} = \tau_{ph} \;,\quad \tau_{ph} = \frac{l_{ph}}{v_T} \;. \tag{1.1.16}$$

Hence, making use of the definitions of l and ε, we obtain concrete estimates for physically infinitesimal scales:

$$\tau_{ph} \sim \sqrt{\varepsilon}\,\tau \ll \tau \;,\quad l_{ph} \sim \sqrt{\varepsilon}\,l \ll l \;,\quad N_{ph} \sim \frac{1}{\sqrt{\varepsilon}} \gg 1 \;. \tag{1.1.17}$$

For a rarefied gas, when inequalities (1.1.13) hold, these values satisfy condition (1.1.11).

Introduction of physically infinitesimal scales enables us to make concrete estimates for the degree of uncertainty of definition of microstates of systems of the ensemble. This offers a new opportunity of defining the Gibbs ensemble for nonequilibrium processes. Naturally, smoothing of dynamic characteristics over physically infinitesimal volume (or over the respective time interval τ_{ph}) transforms the initial reversible equations of motion of particles (elements) of the system into irreversible equations of statistical theory. Formulation of these equations constitutes one of the main tasks of this book.

1.1.4. *Physically Infinitesimal Scales in Hydrodynamic Description*

Characteristic parameters τ, l in the kinetic theory belong to the internal parameters (1.1.13), (1.1.14) of the system. In hydrodynamics the situation is different. As a matter of fact, the characteristic length of the problem may be represented by the diameter of the pipe, or by the size of aerofoil. Such parameters may be classified as external, because they are not expressible via the characteristics r_0, r_{av}, l. The relevant relaxation times are now expressed in terms of the external parameter L, and one of the three dissipative coefficients: diffusion D, internal viscous friction ν, or temperature conductivity χ. All three dissipative processes are of diffusion type; accordingly, the "coefficient of diffusion" is represented by one of the coefficients: D, ν, or χ. Then the time of relaxation by diffusion is

$$\tau_D = \frac{L^2}{D} \text{ , where } D = D, \text{ or } \nu, \text{ or } \chi \text{ .} \tag{1.1.18}$$

We mark the hydrodynamic physically infinitesimal scales with superscript "G" (for *Gasdynamic parameter*). Physically infinitesimal time τ_{ph}^G we define by analogy with (1.1.16); now, however, the characteristic time τ is represented by the time of diffusion relaxation τ_D.

For diffusion processes (which include the processes of viscous friction and heat conduction), the linkage between characteristic time and length is given by

$$\tau_{ph}^G = \frac{\left(l_{ph}^G\right)^2}{D} \text{ , } D = D, \text{ or } \nu, \text{ or } \chi \text{ .} \tag{1.1.19}$$

Observe that "traces" of diffusive (hydrodynamic) motion are preserved also within the physically infinitesimal volume V_{ph}^G.

In case of kinetic description, physically infinitesimal time τ_{ph} is defined by the natural equation (1.1.16). For hydrodynamic description we may use a similar definition taking due account of the fact that the characteristic times τ_D, τ_{ph} are proportional to squared lengths L, l_{ph}. Accordingly, we define N_{ph}^G as L^2/l_{ph}^2, and, using (1.1.18), (1.1.19), find that

$$N_{ph}^G \sim \frac{L^2}{\left(l_{ph}^G\right)^2} \sim \frac{\tau_D}{\tau_{ph}} \text{ , } N_{ph}^G \sim n\left(l_{ph}^G\right)^3 \text{ .} \tag{1.1.20}$$

Hence follow the definitions

$$\tau_{ph}^G \sim \frac{\tau_D}{N^{2/5}} \text{ , } l_{ph}^G \sim \frac{L}{N^{1/5}} \text{ , } N_{ph}^G \sim N^{2/5} \text{ ,} \tag{1.1.21}$$

where we have used equation $nL^3 = N$.

We see that, unlike the case of (1.1.17) for the kinetic description, now the physically infinitesimal scales depend on the external parameter L, or, since $nL^3 = N$, on the total number of particles in the volume L^3. This is a clear indication that the definition of physically infinitesimal scales (and the concept of a "point" of continuous medium itself) depends on the adopted level of description.

Since the above definition of Gibbs ensemble for nonequilibrium processes is based on the concept of physically infinitesimal volume, the physical essence of the definition will not be the same for the kinetic and hydrodynamic levels of description.

1.2. Transition from Kinetic to Gasdynamic Description

So, we have defined the concept of "point" for the kinetic and gasdynamic levels of description of nonequilibrium processes. Naturally, the gasdynamic description is rougher than the kinetic, and so the "point" of continuous medium in hydrodynamics is larger — that is, $V_{ph} < V_{ph}^G$.

How then do we proceed from the "precise" description based on the kinetic Boltzmann equation to the "rough" gasdynamic description? This transition can be made in the following manner.

Recall that the characteristic scales in the kinetic and gasdynamic descriptions are, respectively, the mean free path l and the external parameter L. If we introduce a dimensionless parameter called the Knudsen number

$$\text{Kn} = \frac{l}{L} \, , \tag{1.2.1}$$

the transition to equations of gas dynamics amounts to finding an approximate solution of Boltzmann equation using perturbation theory in small Knudsen number (methods of Gilbert, Chapman – Enskog, Grad; see Ch. 11).

To this end, equation for the distribution function $f(r,p,t)$ serves as basis for constructing a closed set of equations in the main gasdynamic functions: density, velocity, and temperature,

$$\rho(r,t) = mn(r,t) \, , \quad u(r,t) \, , \quad T(r,t) \, . \tag{1.2.2}$$

Gasdynamic functions are expressed in terms of the distribution function (see Ch. 5). An exact closed set of equations in functions (1.2.2.), however, cannot be derived from the kinetic Boltzmann equation without simplifying assumptions. Because of this, it is necessary to use perturbation theory in small Knudsen number.

In zero approximation the solution of Boltzmann equation is represented by the so-called local Maxwell distribution

$$nf(r,p,t) = \frac{n(r,t)}{[2\pi m\kappa T(r,t)]^{3/2}} \, \exp \left\{ -\frac{[p - mu(r,t)]^2}{2\kappa T(r,t)} \right\} \, . \tag{1.2.3}$$

We see that the dependence is only explicit on momentum p; implicit dependence on coordinates and time enters via functions $n(r,t)$, $u(r,t)$, $T(r,t)$. Accordingly, for complete solution we need equations for these three functions. They are obtained by substituting (1.2.3) into the Boltzmann equation.

As a result, we get a closed set of equations which is a special case of the complete set of gasdynamic equations: more precisely, equations of gas dynamics neglecting dissipative processes.

In order to obtain a more comprehensive set of hydrodynamic equations, it is natural to use the next approximation of the perturbation theory in small parameter Kn. This feat is feasible, and yields a closed set of gasdynamic equations including dissipative processes due to viscosity and heat conduction. Moreover, it is even possible to calculate the dissipative coefficients as such: kinematic viscosity v, and temperature conductivity χ. This method is also useful for corroboration of gasdynamic equations in more complicated cases, which take account of the internal structure of the atoms (system's elements), and of the chemical transformations of the elements.

We see that the use of kinetic equations for deriving a simpler hydrodynamic description is extremely useful for many practical applications. Some fundamental questions, however, remain unanswered. We are going to discuss them in Ch. 13; at this point we just mark the following.

The first question concerns the definition of heat flux — that is, chaotic motion of gas particles. Traditionally, this definition is given as follows.

As indicated above, the exact equations in functions (1.2.2) (derived without using perturbation theory) are not closed, since they involve higher moments (with respect to velocity) of the distribution function $f(r,p,t)$. For example, equation in energy (or temperature) includes third moment of relative velocity

$$q(r,t) = n\int \delta v \frac{m(\delta v)^2}{2} f dp \; , \quad \delta v = v - u(r,t) \; . \tag{1.2.4}$$

This vector is called the heat flux.

This is not unreasonable, since this vector is determined by the relative kinetic energy, and therefore defines, in a sense, the transfer of chaotic motion. There also is another important point: in zero approximation in Knudsen number (when we are dealing with local equilibrium, and distribution function is given by (1.2.3)) the heat flux vector (1.2.4) is zero. In the first approximation, however, we come to Fourier law

$$q(r,t) = -\lambda \frac{\partial T}{\partial r} \; , \quad \lambda = \rho C_p \chi \; . \tag{1.2.5}$$

This result is, of course, very important. However, the definition of heat flux (1.2.4) is not quite consistent: it has the time-symmetry of a reversible process, by contract to expression (1.2.5).

Moreover, equation (1.2.4) is not a complete characteristic of transfer of chaotic motion. It would be more consistent to define heat flux in terms of entropy gradient. Can this be accomplished within the framework of conventional approach?

These and other difficulties stimulate the quest for alternative approaches to nonequilibrium processes in open systems.

Such difficulties are mostly due to the excessively formal approach to the problem. After all, we are dealing with the transition from the kinetic description of continuous medium, when the concepts of "point" and "instantaneity" are defined by physically infinitesimal scales (1.1.17), to the hydrodynamic description, when these concepts are defined in a quite different way (1.1.21). Because of this, evaluation of the degree of uncertainty of definition of states of individual systems of the Gibbs ensemble is not the same for the kinetic and hydrodynamic descriptions.

It follows that the concept of nonequilibrium Gibbs ensemble must be re-defined when we proceed from the kinetic description to the hydrodynamic description. This leads to re-definition of small dimensionless parameters used by perturbation theory.

In lieu of the Knudsen number it will be natural to use another small parameter which, of course, will not be the same in the kinetic theory and in gas dynamics.

1.3. Unified Description of Kinetic and Hydrodynamic Processes

1.3.1. *Physical Knudsen Number*

The difficulties in applying the perturbation theory in small Knudsen number are also associated with the fact that Kn does not reflect the structure of the "continuous medium". In place of Knudsen number it is expedient therefore to use another dimensionless parameter which is always small as long as we use the approximation of continuous medium.

At the kinetic stage of relaxation, the characteristic parameter of length L is the mean free path l, whereas the physically infinitesimal length l_{ph} is given by (1.1.17). This justifies the following definition of "physical Knudsen number" for the kinetic description:

$$K_{ph} = \frac{l_{ph}}{l} \sim \frac{1}{N_{ph}} \ . \tag{1.3.1}$$

The smallness of this parameter is guaranteed by the fact that $N_{ph} \gg 1$.

In the domain of free molecular flow, when the characteristic length L (for instance, the diameter of the pipe) is much smaller than the mean free path l, the approximation of continuous medium may be used as long as

$$l_{ph} \ll L \ll l .$$ (1.3.2)

In case of hydrodynamic description, the physically infinitesimal scales are defined by (1.1.21), and the physical Knudsen number is

$$K_{ph}^G = \frac{l_{ph}^G}{l} \sim \frac{1}{\left(N_{ph}^G\right)^{1/2}} .$$ (1.3.3)

The smallness of this parameter is ensured by condition $N_{ph}^G \gg 1$.

1.3.2. Reconciliation of Kinetic and Hydrodynamic Definitions of Continuous Medium

Since the structure of continuous medium is considered more coarse-grained in gasdynamic description than in the kinetic description, the gasdynamic physically infinitesimal volume V_{ph}^G is larger than the kinetic physically infinitesimal volume V_{ph}. We can express the relation between these values in terms of density parameter ε and Knudsen number Kn:

$$\frac{V_{ph}}{V_{ph}^G} \sim \varepsilon^{3/10} Kn^{6/5} \le 1 ,$$ (1.3.4)

where the equals sign corresponds to the largest value of Knudsen number (and, consequently, the smallest value of external parameter L_{min}) for which it is still possible to reconcile the hydrodynamic and the kinetic definitions of continuous medium. Using definitions (1.1.17), from (1.3.4) we find that

$$(Kn)_{max} \sim \sqrt{N_{ph}} \; ; \; L_{min} \sim \sqrt{N_{ph}} l_{ph} \sim \frac{l}{\sqrt{N_{ph}}} .$$ (1.3.5)

We see that the minimum length L_{min} (and, therefore, the minimum size of point), for which the traces of diffusive motion are still retained and the gasdynamic description is still feasible, is less than mean free path l and greater than the kinetic physically infinitesimal scale l_{ph}. Between these limits a unified description of kinetic and hydrodynamic processes is possible in a broad range of Knudsen numbers without using perturbation theory.

The relevant characteristic time is defined as

$$\left(\tau_{ph}^{G}\right)_{min} \sim \frac{L_{min}}{D} \sim \sqrt{\varepsilon}\,\tau \sim \tau_{ph} \tag{1.3.6}$$

and is of the order of physically infinitesimal time scale for kinetic description. We shall need this result for the generalized kinetic equation for unified description of kinetic and gasdynamic processes.

We see that a general definition of point of continuous medium is possible. The above-obtained relationships allow making estimates for the number of particles within a "point":

$$N_{min} = nL_{min}^{3} \sim \varepsilon^{-5/4} . \tag{1.3.7}$$

Hence at normal conditions, when $\varepsilon \approx 10^{-4}$, the number of particles within a "point" is $N_{min} \approx 10^{5}$.

1.3.3. Small-Scale and Large-Scale Fluctuations

In addition to the three characteristic parameters of Boltzmann gas (r_0, r_{av}, l), the concept of structure of "continuous medium" allows us to introduce a fourth parameter, physically infinitesimal length l_{ph}. For unified description of kinetic and hydrodynamic processes we replace l_{ph} with L_{min}. The relationship between all these scales is governed by the small dimensionless parameter of density (1.1.12). Also,

$$r_0 \ll r_{av} \ll l_{ph} \ll l , \tag{1.3.8}$$

for length scales, and

$$\tau_0 \ll \tau_{av} \ll \tau_{ph} \ll \tau \tag{1.3.9}$$

for time scales.

The introduction of structural scales for continuous medium allows us to distinguish two classes of fluctuations:

$$\tau_{cor} < \tau_{ph} , \ r_{cor} < l_{ph} ; \qquad\qquad \tau_{cor} > \tau_{ph} , \ r_{cor} > l_{ph} . \tag{1.3.10}$$

If the correlation times satisfy the left-hand inequalities, the fluctuations are small-scale, and large-scale otherwise.

According to this definition, small-scale fluctuations occur inside the point of "continuous medium". The relevant information is therefore lost, which accounts for the dissipation in the kinetic equation. The process of dissipation is only possible when there is interaction between particles. In a rarefied Boltzmann gas this interaction may be regarded as collisions between particles. This is the reason why the dissipative term in the kinetic equation is termed Boltzmann collision integral.

So, one of the sources of dissipation in the kinetic equation is due to the loss of information within the points of "continuous medium", to the damping of small-scale correlations.

What is the role of large-scale fluctuations?

We try to answer this question in Ch. 9. We shall see that the inclusion of large-scale fluctuations results in a double-decked superstructure on top of the building erected by Boltzmann.

The first addition is due to those fluctuations of the distribution function which are a direct consequence of the atomic structure of "continuous medium". Inclusion of such fluctuations is necessary, for instance, for the description of Brownian motion. Fluctuations of this kind may be termed molecular fluctuations; their intensity is determined by the number of particles within the physically infinitesimal volume (the number of particles within a "point")

$$\frac{\delta f}{f} \sim \frac{1}{\sqrt{N_{ph}}} . \tag{1.3.11}$$

Large-scale fluctuations, whose nature is governed by collective interactions, play an important and sometimes the principal role in the states far from equilibrium, such as result from the development of instabilities when, for instance, a laminar flow becomes unstable and the turbulent regime sets in. We call these the turbulent fluctuations, since they determine the nature of turbulent motion.

As indicated above, the inclusion of large-scale fluctuations brings us beyond the scope of Boltzmann theory. There is yet another reason for the revision of Boltzmann theory.

1.3.4. *Equation for Unified Description of Kinetic and Hydrodynamic Processes*

Unlike the initial Hamilton equations for gas particles, the kinetic Boltzmann equation is irreversible. In particular, it describes time evolution towards the equilibrium state. The origin of irreversibility, however, has not been clarified either by Boltzmann or by later students (see Ch. 6).

Whence does the instability come?

We have already emphasized the fundamental importance of dynamic instability of motion, and will discuss it further in Sect. 1.5 and in the forthcoming chapters. Combined with uncontrollable external factors, dynamic instability of motion effectively undermines complete predictability of motion. The minimum characteristic time of dynamic instability of motion is of the order of physically infinitesimal time τ_{ph}, so the effect of dynamic instability may be taken into account by smoothing the reversible dynamic motion over the above-defined physically infinitesimal scales.

Smoothing destroys information about the behavior of particles within the "point", and so the reversible description becomes no longer feasible. Since the structure of "continuous medium" is defined by the relevant infinitesimal scales, it is on these scales that the irreversibility originates.

In Ch. 13 we show that smoothing adds a dissipative term to the kinetic Boltzmann equation:

$$\frac{\partial f}{\partial t} + v\frac{\partial f}{\partial r} + \frac{F}{m}\frac{\partial f}{\partial v} = I_{(v)}(r,p,t) + I_{(r)}(r,p,t) \ . \tag{1.3.12}$$

The left-hand side of this equation describes variation of the distribution function which explicitly depends on the time, on the convective transfer, and on the mean force. If the force is non-dissipative, the time symmetry of the terms on the right-hand side corresponds to a reversible process.

The right-hand side of the kinetic equation contains two dissipative terms. The first of these, $I_{(v)}(r,p,t)$, accounts for dissipation due to the redistribution of particle velocities (hence the subscript "v") — for instance, because of collisions between the particles. In general, $I_{(v)}$ is the Boltzmann collision integral. Owing to the smallness of fluctuations of distribution function (condition (1.3.11)), however, a much simpler expression can be used for this dissipative term (see Ch. 13).

The second dissipative term on the right-hand side of the kinetic equation is due to the redistribution of particle coordinates. This term is due to the dynamic instability of motion, and to the resulting mixing of trajectories in the phase space. Since the minimum characteristic time of development of dynamic instability is of the order of physically infinitesimal time τ_{ph}, the kinetic equation includes an additional dissipative term of the form (see Ch. 13)

$$-\frac{1}{\tau_{ph}}\left(f - \tilde{f}\right) , \tag{1.3.13}$$

where $\tilde{f}$ is the distribution function smoothed over the infinitesimal volume L_{min}^3. The difference $\left(f - \tilde{f}\right)$ can be expanded in the physical Knudsen parameter (cf. (1.3.1),

(1.3.3.))

$$K_{\mathrm{ph}} = \frac{L_{\min}}{L} , \qquad (1.3.14)$$

and so the additional dissipative term has the form

$$I_{(r)}(r,p,t) = \frac{\partial}{\partial r}\left(D\frac{\partial f}{\partial r}\right) - \frac{\partial}{\partial r}\left(\frac{b}{m}Ff\right) , \qquad (1.3.15)$$

where D is one of the three kinetic coefficients: kinematic viscosity v, temperature conductivity χ, or self-diffusion D. For a rarefied gas, all three coefficients are of the same order of magnitude, so we set

$$D = v = \chi . \qquad (1.3.16)$$

Coefficient of spatial diffusion D and mobility b are linked by Einstein formula

$$D = b\frac{\kappa T}{m} , \quad b = \tau . \qquad (1.3.17)$$

Here we have noted that for a rarefied gas the mobility b coincides with the free path time τ.

Generalized kinetic equation (1.3.12) can be employed for a unified description of nonequilibrium processes in both the kinetic and the gasdynamic domains without using perturbation theory in Knudsen number (see Ch. 13). Dissipative effects of viscosity and temperature conduction, as well as dissipation due to self diffusion, are taken into account in a natural way.

The expedience of taking self-diffusion into account has been questioned now and again (see Ch. 13). Let us quote here just one argument in favor of existence of self-diffusion — transport of matter in a one-component system in the presence of density gradient in accordance with Fick law.

Consider a one-component rarefied gas, a Boltzmann gas in the field of gravity. Given self-diffusion and mobility in the field of gravity, the flow of matter $j(r,t)$ is a sum of three components: convection flow ρu, self-diffusion flow, and the flow due to mobility in the field of constant force F. As a result, the continuity equation takes the form

$$\frac{\partial \rho}{\partial t} = -\mathrm{div}\left[\rho u - D\mathrm{grad}\rho + \frac{b}{m}F\rho\right] , \quad F = -\mathrm{grad}U . \qquad (1.3.18)$$

At equilibrium, when $u = 0$, the coefficient of mobility b is linked with coefficient of self-diffusion D by Einstein formula

$$b = m\frac{D}{\kappa T} \ . \tag{1.3.19}$$

Given this relationship, the equation at equilibrium is satisfied by Boltzmann distribution

$$\rho(r) = mN\frac{\exp\left(-\dfrac{U(r)}{\kappa T}\right)}{\int \exp\left(-\dfrac{U(r)}{\kappa T}\right)dr} \ . \tag{1.3.20}$$

We see that actually the existence of Boltzmann distribution proves the existence of self-diffusion. This distribution evolves from the balance between two dissipative counterflows: one due to self-diffusion, and the other due to mobility in the field of external force, provided that the Einstein formula holds.

Obviously, the established Boltzmann distribution satisfies the condition of dynamic equilibrium between the external force and the pressure gradient (the temperature being constant).

1.4. Kinetic Description of Autowave Processes in Active Media. Basic Equation of Theory of Self-Organization (Synergetics)

So far we have been mainly concerned with open systems whose elements are atoms or molecules. More complicated (and often much more interesting) are open systems whose elements are active small macroscopic objects. Such systems are referred to as active media. Of course ,the basic equations in the theory of active media differ from the above-discussed kinetic equations for gases, or equations of gas dynamics. Given, however, an appropriate input of energy, gas and liquid systems may become active. An example is turbulent motion in liquid, which may be regarded as a process in open system composed of active elements (see Ch. 22).

A typical basic equation of statistical theory of open systems (theory of self-organization — synergetics) is a reaction diffusion equation

$$\frac{\partial X(R,t)}{\partial t} = F(X) + D\frac{\partial^2 X}{\partial R^2} \ , \tag{1.4.1}$$

which has been systematically studied by Fisher, Kolmogorov, Petrovsky, Piskunov (see Ch. 18).

Here $X(R,t)$ is a set (vector) of certain characteristics of the system: concentrations of reactants in chemically reacting systems, velocity field in turbulent motion, distributions of charge and current in open systems composed of self-oscillating elements such as Van der Pol oscillators, etc. Nonlinear functions F depend on the structure of elements of the system. For instance, in case of one-component open system of bistable elements F may be defined as

$$F(X) = (a - bX^2)X , \quad b > 0 .$$ (1.4.2)

Then the relevant dynamic equation for an individual element may be written as

$$\frac{dX}{dt} = (a - bX^2)X , \quad b > 0 .$$ (1.4.3)

At $a < 0$ the stationary point is the state of rest; at $a > 0$ we have a bistable state with $X = \pm\sqrt{a/b}$. In this case, $a = 0$ is the point of bifurcation.

Finally, D is the coefficient of spatial diffusion of elements of the open system.

Fisher – Kolmogorov – Petrovsky – Piskunov (FKPP) equation of reaction diffusion type describes a broad class of physical, chemical and biological processes. It is one of the basic equations of the general theory of self-organization which incorporates the theory of autowave processes, the theory of dissipative structures, the theory of nonequilibrium phase transitions. Using the term *synergetics*, coined by Hermann Haken to emphasize the importance of concerted action in the processes of self-organization, one may say that equations of FKPP type belong to the basic equations of synergetics.

An important task of the statistical theory of open systems consists therefore in validating equations which provide the basis for description of various aspects of the theory of self-organization. From the outset we are confronted with a number of very important questions.

First of all, we have to clarify the concept of individual element of open system. Such elements can be represented by structurally distinct small macroscopic elements: individual oscillators or triggers. Sometimes, however, the situation is not so simple.

For instance, what shall we regard as elements of chemically reacting medium which hosts the famous Belousov – Zhabotinsky reaction, a spectacular example of autowave process? After all, the reaction occurs on molecular level. Can such elements be identified in terms of structure of the medium?

This question brings us back to the concept of physically infinitesimal scales. We defined such scales both for the kinetic description (see (1.1.17)), and for the hydrodynamic description of nonequilibrium processes (see (1.1.21)). In the former case the "structural elements" (we put this term in quotes so as to emphasize that these elements differ from, say, distinct Van der Pol oscillators) are entirely defined by the internal parameters of the system.

In the latter case the situation is quite different, because the "structural elements" defined in terms of physically infinitesimal scales depend on the external parameter L (or the total number of particles N).

In this connection we should make an important point.

Dissipation in the kinetic Boltzmann equation is wholly determined by the small-scale fluctuations (correlations). This, however, is not yet sufficient for the transition to hydrodynamic equations, and it was necessary to introduce an additional "collision integral" (1.3.15) which accounts for the dissipative processes due to viscosity, heat conduction and self-diffusion in the hydrodynamic domain. Each of the three dissipative coefficients is related by (1.1.20) to physically infinitesimal scales, and by (1.1.21) to the external scale L (or number $N \approx nL^3$).

In this respect the equations of reaction diffusion type are less informative. Indeed, they take spatial diffusion into account, and therefore reflect the existence of physically infinitesimal scales (1.1.21) which depend on the external scale L. These equations, however, do not keep traces of the internal structure of small macroscopic elements of the open system in question, and thus do not take into account the effects of small-scale correlations. For this reason the equations of FKPP type are not suitable, for instance, for describing the behavior of the system in the neighborhood of points of bifurcation.

To account for the structure of small macroscopic elements of the open system in question, we have to step up to a higher level of description and to use the kinetic equation for the distribution function $f(X,R,t)$, which is the distribution function of the set of characteristic parameters X at different values of R and t. The simplest example of such equation for a medium of bistable elements is (see Ch. 18)

$$\frac{\partial f(X,R,t)}{\partial t} = \frac{\partial}{\partial X}\left(D_{(X)}\frac{\partial f}{\partial X}\right) + \frac{\partial}{\partial X}\left[\left(-a + bX^2\right)Xf\right] + D\frac{\partial^2 f}{\partial R^2} \ . \tag{1.4.4}$$

Here, like in (1.3.12), we take into account the role of two "collision integrals", of which the second is determined by spatial diffusion and is similar to (1.3.15), whereas the first depends on the "atomic structure" of small macroscopic elements of the system. When these elements are structurally distinct (like small Van der Pol oscillators), the coefficient of diffusion D in the additional "collision integral" characterizes the effects of atomic structure of the element perceived as internal noise.

The transition from equation (1.4.4) to FKPP equation is possible in the approximation which neglects this internal noise ($D_{(X)} \equiv 0$), and the distribution function $f(X,R,t)$ only depends on $X(R,t)$:

$$f(X,R,t) = \delta(X - X(R,t)), \text{ and, therefore, } \langle X \rangle = X(R,t) . \tag{1.4.5}$$

Substituting this distribution into the kinetic equation (1.4.4), multiplying by X and carrying out integration, we come to (1.4.1) with $F(X)$ in the form (1.4.2).

Possible is also the other limiting case, when spatial diffusion is a "fast" process, so that a uniform distribution over space is established within $\tau_D < 1/|a|$. Then equation (1.4.4) reduces to an equation for distribution function $f(X,t)$ in internal variable X.

Distribution (1.4.5), which corresponds to FKPP equation, is entirely determined by the first moment, $\langle X \rangle = X(R,t)$. The inclusion of higher moments permits a more detailed description of autowave processes than the one based on FKPP equation. The additional information obtained in this way is especially important for describing the system's behavior in the neighborhood of bifurcation points (see Ch. 18).

1.5. Dynamic and Statistical Description of Complex Motion in Macroscopic Open Systems. Constructive Role of Dynamic Instability of Motion

So, we have made considerable progress in the construction (even if somewhat fragmentary) of the statistical theory of nonequilibrium processes, based on the definition of Gibbs ensemble given in Sect. 1.1. We noted that the uncertainty of definition of microstates in individual systems of the ensemble is due to the lack of complete knowledge of the behavior of particles (elements) of the system within the confines of physically infinitesimal volume. Thus the reversibility of motion is lost at the stage of smoothing of the dynamic characteristics over the small volume V_{ph}.

This method of introducing irreversibility is certainly justified, since the "working equations" in both the kinetic theory and hydrodynamics are the equations of mechanics of continuous media. At the same time, the formality of this approach causes some discontent.

In this connection it is desirable to show that this transition to irreversible equations reflects the complexity of dynamic motion which results from the development of dynamic instability of motion. We raised this point briefly in the beginning of this chapter; now it is time to consider the linkage between the statistical and dynamic descriptions in greater detail. Let us start with some history.

The founding fathers of the modern statistical theory of nonequilibrium processes and the dynamic theory of complex motion are, of course, Ludwig Eduard Boltzmann and Jules Henri Poincaré. Well known is the dramatic rivalry between these two theories. Although the passions have somewhat subsided over the years, these two directions are still being pursued more or less independently from each other. It is clear, however, that a better insight into the problem of transition from the reversible equations of Hamiltonian mechanics to the irreversible equations of statistical theory can only be gained by bringing these two theories together. The expedience of a merger is also evident from the latest developments in the theory of self-organization — synergetics.

Further to the classical papers by Poincaré, we observe two major stages in the advancement of the dynamic theory. The first is associated with the progress in the theory of self-oscillations stimulated by the rapid development of radio engineering. Valuable contributions to this field have been made by Van der Pol, L.I. Mandelstam, A.A. Andronov, A.A. Vitt, L.S. Pontryagin, N.M. Krylov, N.N. Bogolyubov.

The second fruitful period in the dynamic theory is linked with the studies in the theory of turbulence, in particular, with the problem of long-term weather forecast. This period actually started in 1963, with the seminal paper by E. Lorenz. The importance of this paper, however, only became clear much later, when mathematicians D. Ruelle and F. Takens published their results in 1971. A new mathematical representation of complex motion in nonlinear dynamic dissipative systems (the *strange attractor*) was introduced at this stage.

The new concepts of *dynamic chaos* and *deterministic chaos* emerged from the studies of complex disordered motions in relatively simple dynamic systems.

One might ask whether it is physically sensible to speak of "dynamic chaos". This term refers to complex motion which sets in as a result of dynamic instability of motion — exponential divergence of initially close trajectories.

The layman's "chaos" is something uncontrollable, disorderly, inferior. Let us show that dynamic instability in statistical theory may play a constructive role in the derivation of irreversible kinetic equations. First, however, consider a sociological example.

Imagine an international congress on problems of statistical physics. For the initial state we take the state immediately after the closing ceremony and consider two options for further movements of the participants: (1) after leaving the hall the participants continue moving together, and (2) the participants "diverge exponentially" and return to where they live and work; in other words, their motion becomes dynamically unstable. Obviously, the second option is better suited to the purposes of proliferation of new information. We see that in this example the "dynamic instability of motion" plays a constructive role.

And now back to physical systems. Again we consider a rarefied Boltzmann gas. For simplicity, we regard atoms as perfectly elastic balls, which in most cases is a fairly justified assumption.

We know that the motion of atom balls, like the motion of balls in Sinai billiards, is dynamically unstable. So far, however, we did not take this dynamic instability into account when discussing the transition from the reversible equations of mechanics to the irreversible Boltzmann equation. Now let us show that with due regard for dynamic instability we can understand better the origins of irreversibility, and validate the transition from the reversible equations of mechanics to the irreversible equations of statistical theory.

The first steps in this direction have been made by N.S. Krylov. His works allow making estimates which permit us to link the physically infinitesimal time τ_{ph} to the characteristic time of development of dynamic instability of atoms of Boltzmann gas.

The estimated time of development of dynamic instability of motion of a given atom τ_{inst} is of the order of free path time τ. Then the characteristic time of development of dynamic instability of motion per one (any one) particle within physically infinitesimal volume V_{ph} is τ/N_{ph}. As a result, we arrive at the following estimate:

$$\left(\tau_{inst}\right)_{min} \sim \tau_{ph} .\tag{1.5.1}$$

We see that the characteristic time of development of dynamic instability of motion of any particle within physically infinitesimal volume V_{ph} is of the order of physically infinitesimal time τ_{ph}. This adds weight to our definition of τ_{ph}. Important is also the following.

Dynamic instability of motion of atom balls, which manifests itself in the exponential divergence of trajectories and in the resulting mixing, facilitates the transition from the reversible Hamiltonian equations for atoms to the relatively more simple Boltzmann kinetic equation. This confirms the constructive role of dynamic instability of motion of atoms of Boltzmann gas.

This, of course, is just a rough sketch which by no means exhausts the role of dynamic instability of motion in statistical physics. As a matter of fact, dynamic instability occurs not only on the microscopic level (the motion of gas atoms), but also in the evolution of macroscopic characteristics (for instance, velocity and temperature in hydrodynamics), as first demonstrated in the paper by E. Lorenz who used a mathematical model of thermal convection (see Ch. 14).

In connection with the arguments concerning the constructive role of dynamic instability of motion of atoms we may ask whether the dynamic instability of motion of macroscopic characteristics can also be a constructive phenomenon. Does the development of dynamic instability of motion in open systems only lead towards chaos, or it may facilitate the processes of self-organization which result in more advanced structures?

In order to answer these questions we require criteria of relative degree of order of nonequilibrium states in open systems. Using such criteria, we shall demonstrate that processes of self-organization are feasible in the presence of dynamic instability on macroscopic level. More than that, the creation of "new", higher organized, structures starts exactly because the "old" structures become dynamically unstable.

1.6. Criteria of Self-Organization

1.6.1. *Physical Chaos in Open Systems. Controlling Parameters*

"Chaos" and "chaotic motion" are fundamental concepts of physics; nevertheless, they lack a clear-cut definition.

According to Boltzmann, motion is most chaotic in the state of equilibrium. The term "chaotic motion", however, is also applied to states far from equilibrium — for instance, to turbulence. The term "dynamic chaos" is often used as a characteristic of complex motion in dynamic systems. This means that we really need criteria for comparing the degree of order of different states of open systems.

As a first step to reach this end, we introduce the concept of "controlling parameter". Of course, there can be several parameters which control the degree of order in an open system.

Various characteristics may act as controlling parameters. Sometimes it is quite difficult to single out the relevant variables, and mistakes are possible. Because of this, good criteria of the relative degree of order must also be an instrument for checking the correct choice of variables.

Let a be the set of controlling parameters. We consider two states, corresponding to $a = a_0$ and $a = a_0 + \Delta a$; Δa is called the "control". We assume that the state with $a = a_0$ is more chaotic (of course, this has to be verified), and take it for the state of physical chaos. Obviously, this state may be quite far from equilibrium.

1.6.2. *Evolution and Self-Organization*

The concept of evolution is very general. In physics, for example, we consider evolution towards equilibrium in a closed system. At certain conditions, evolution in open systems may lead to different stationary states. We have also mentioned processes of self-organization. In what relation do the concepts of "evolution" and "self-organization" stand to each other?

By processes of self-organization we mean those processes which result in more complicated and more intricate structures. Naturally, we need some criteria to gauge the measure of complication and intricacy.

Is any process of evolution a process of self-organization? No, because neither physical nor, for that matter, biological systems have an intrinsic tendency towards self-organization. We know that evolution may lead to degradation — like the evolution of a physical system towards equilibrium which, according to Boltzmann and Gibbs, is the most chaotic state. Degradation of structures is observed also in biological and sociological systems. So, self-organization is just one of the possible directions of evolution.

Let us now discuss the possibility of quantitative description of the process of degradation associated with the transition to the state of equilibrium.

1.6.3. Processes of Degradation. Boltzmann's H–Theorem

The name of H–theorem (where "H" stands for "heat") was introduced by the British physicist called Burbury in 1894, several years after Ludwig Boltzmann had proved this statement. The name is supposed to emphasize the fact that H–theorem deals with the transition to the equilibrium (thermal) state.

Texts in molecular and statistical physics maintain that Boltzmann's H–theorem holds for closed systems. This assertion needs a certain refinement which will be very important for the formulation of criteria of self-organization, or criteria of the relative degree of order of different states of open systems.

Boltzmann's H–theorem states that the entropy of a closed system increases in the course of transition towards the equilibrium state, and remains constant in the state of equilibrium:

$$\frac{dS}{dt} \geq 0 \ . \tag{1.6.1}$$

The proof (see Sect. 6.4) is based on the definition of entropy similar to (1.1.8), where the distribution function f_N is replaced by function $f(x,t)$ which satisfies Boltzmann's equation.

It is important, however, that it is the mean energy $\langle E \rangle$ of rarefied Boltzmann gas (which reduces to the kinetic energy, since Boltzmann gas is an ideal gas from thermodynamic point of view) that remains constant in the course of evolution:

$$\langle E \rangle = \left\langle \frac{p^2}{2m} \right\rangle = \text{const} \ . \tag{1.6.2}$$

In other words, the conserved quantity is not the energy E (condition of closedness), but rather the mean energy $\langle E \rangle$; therefore, fluctuations of energy are allowed. This points to the internal non-closedness of the Boltzmann gas system, which is due not to the energy exchange with the environment, but to the existence of an infinite (in the thermodynamic limit) buffer of internal degree of freedom. Boltzmann's kinetic equation only contains a tiny fraction of total information about the system.

Explicit inclusion of condition (1.6.2) (condition of internal non-closedness) allows us to formulate Boltzmann's H–theorem in terms of Lyapunov functional. To wit, it is possible to prove that the difference in the entropies of equilibrium (S_0) and nonequilibrium ($S(t)$) states is given by (see Ch. 6)

$$\Lambda_S = S_0 - S(t) = \kappa n \int \ln \frac{f(x,t)}{f_0(x)} f(x,t)dx \geq 0 \ , \tag{1.6.3}$$

where the second inequality follows from the H–theorem (1.6.1):

$$\frac{d\Lambda_S}{dt} = \frac{d}{dt}\left(S_0 - S(t)\right) \leq 0 \ . \tag{1.6.4}$$

Here we denote the difference in the entropies by Λ_S, since, in accordance with (1.6.3) and (1.6.4), it is a Lyapunov functional.

So we have proved the following:
(1) The statement that Boltzmann's H–theorem holds for a closed system is not quite correct, because the constancy of the mean energy indicates that the system in the Boltzmann approximation is internally non-closed, and
(2) Using condition (1.6.2) we can reduce Boltzmann's H–theorem to the statement that there exists a Lyapunov functional Λ_S which is defined as the difference in the entropies $S_0 - S(t)$.

The fact that the functional is defined as the difference in the entropies rather than the difference, for instance, in the free energies, is very advantageous.

First, the entropy can be defined for an arbitrary nonequilibrium state as long as we know the relevant distribution function. Observe that the distribution function can be obtained not only from a mathematical model (for instance, Boltzmann kinetic distribution), but also directly from experimental data: real-time distributions, spectra, spatial distributions. This implies that the entropy can be defined, in principle, for any open system.

Secondly, it is only the entropy (by contrast to the free energy, for example) that possesses all the characteristics of a quantity that can be used as the measure of the relative uncertainty of description of different states in the statistical theory.

On the strength of arguments developed above, we may formulate Boltzmann's H–theorem as follows: as the Boltzmann gas evolves towards the equilibrium state, the degree of chaoticity increases and reaches its maximum at the state of equilibrium.

Note that this statement is valid as long as the mean energy remains constant. For Boltzmann gas, however, (1.6.2) is a direct implication of the kinetic Boltzmann equation rather than an additional condition.

We have formulated Boltzmann's H–theorem for a rarefied Boltzmann gas. Now we are going to lift this restriction.

1.6.4. *Gibbs Theorem*

Boltzmann's H–theorem applies to a nonequilibrium ideal gas. In general, H–theorem for nonideal gas could not be proved. In case of open systems the situation is even more complicated, because in the course of evolution towards equilibrium the value of mean energy or the relevant effective energy (effective Hamilton function) is not conserved. Nevertheless, the Boltzmann – Gibbs – Shannon entropy can be used as the measure of the relative degree of order of different states of open systems in the course of evolution of both in time and in the space of controlling parameters.

First, however, let us discuss a general result obtained by Gibbs without any constraints being imposed on the nature of interaction between the elements of the system in question.

Consider a system characterized by an arbitrary Hamilton function $H(X)$. At equilibrium, the distribution $f_N^{(0)}(X)$ is given by the canonical Gibbs distribution (1.1.3). By $f_N(X,t)$ we denote the distribution function for any arbitrary distribution with one and only (!) restriction: the mean value of the energy (Hamilton function $H(X)$) must coincide with the value corresponding to the equilibrium state,

$$\int H(X) f_N^{(0)}(X) dX = \int H(X) f_N(X,t) dX \ . \tag{1.6.5}$$

Then for the difference in the entropies of the equilibrium S_0 and nonequilibrium $S(t)$ states we get

$$S_0 - S(t) = \kappa \int \ln \frac{f_N(X,t)}{f_N^{(0)}(X)} f_N(X,t) dX \geq 0 \ . \tag{1.6.6}$$

This inequality expresses the statement of the Gibbs theorem: the entropy of the equilibrium state is greater than the entropy of any arbitrary state whose mean energy is equal to the mean energy of the equilibrium state. Unlike condition (1.6.2) for a Boltzmann gas, constraint (1.6.5) is an additional requirement imposed on the class of "arbitrary" distributions.

Gibbs theorem (1.6.6) holds for systems with all kinds of interactions between elements. In this respect it is more universal than inequality (1.6.3). It is not possible to prove the H–theorem in the general case (that is, to show that the entropy increases monotonically as we approach the equilibrium state with Gibbs distribution $f_N^{(0)}$); the proof, however, can be obtained for two very important special cases (see Ch. 6) — in particular, for an ideal gas.

1.6.5. *Examples of H-Theorem for Open Systems*

Now we look at the class of open systems as considered in the theory of Brownian motion. For Brownian motion, both the energy and the mean energy are not conserved in the course of time evolution. Because of this, it is not possible to define a Lyapunov functional as a straightforward difference in the entropies, and one has to use functional defined as differences in the "free energies" for gauging the relative degree of order of different states. To validate the use of Lyapunov functional Λ_S we must define the effective Hamilton function and carry out renormalization (indicated by ~) to the constant value of this function. As a result, we formulate the H–theorem for open systems (Ch. 12):

$$\tilde{\Lambda}_S = S_0 - \tilde{S}(t) \geq 0 \; ,$$

$$\frac{d\tilde{\Lambda}_S}{dt} = \frac{d}{dt}\left(S_0 - \tilde{S}(t)\right) \leq 0 \; , \tag{1.6.7}$$

$$\langle H_{\text{eff}} \rangle = \text{const} \; ,$$

where S_0 is the entropy of the stationary state.

So, the entropy of an open system increases in the course of time evolution towards the equilibrium (or a stationary) state provided that the mean value of the Hamilton function remains constant (additional condition). Since the entropy is the measure of uncertainty of the statistical description, we may state that the degree of chaoticity of open systems increases in the course of time evolution towards a stationary state. In other words, this is a process of degradation.

Now let us consider a situation when self-organization is feasible, and give a possible criterion of self-organization.

1.6.6. *Decrease in Entropy in the Process of Self-Organization. S–Theorem*

So far we have been considering the process of time evolution of an open system towards its stationary state, assuming therewith that the values of the controlling parameters a were fixed.

Of interest is also a different situation when the evolution proceeds via a number of stationary states which each correspond to different values of the controlling parameters. In other words, we may deal with the evolution of stationary states in the space of controlling parameters. Let us discuss a criterion which permits checking whether the process in question is self-organization or degradation.

Let there be just one controlling parameter a. We consider two states, $a = a_0$ and $a = a_0 + \Delta a$. For definiteness, we take the state with $a = a_0$ for the state of physical chaos. Later this assumption will have to be verified. Generally, the state of physical chaos is not necessarily an equilibrium state.

Assume that we have a mathematical model (for instance, the kinetic equation) which provides the basis for finding the distribution function for the state of physical chaos f_0, and the distribution function $f(x, a_0 + \Delta a)$. We represent f_0 in the form of canonical distribution (1.1.3):

$$f_0(x, a) = \exp\frac{F_{\text{eff}} - H_{\text{eff}}(x, a)}{T_{\text{eff}}} , \quad \int f_0(x) dx = 1 , \tag{1.6.8}$$

where we have introduced the relevant effective characteristics. Naturally, the effective Hamilton function is not "energy" in the physical sense, and the macroscopic variable x may have no mechanical analogy.

The mean value of the effective energy $\langle H_{\text{eff}} \rangle$ for the two selected states may be not the same. In order to use the entropy of the two selected states as the measure of the relative degree of order, we must renormalize the distribution (1.6.8) so as to satisfy condition (cf. (1.6.5))

$$\int H_{\text{eff}}(x, a_0) \tilde{f}_0(x, a_0, \Delta a) dx = \int H_{\text{eff}}(x, a_0) f_0(x, a_0 + \Delta a) dx . \tag{1.6.9}$$

The dependence of the renormalized free energy on the effective temperature $\tilde{F}(\tilde{T}_{\text{eff}})$ follows from the condition of normalization. In this way, from (1.6.9) we find

$$\tilde{T}_{\text{eff}} = \tilde{T}_{\text{eff}}(\Delta a) , \quad \tilde{T}\big|_{\Delta a = 0} = \tilde{T}_{\text{eff}} . \tag{1.6.10}$$

Using condition (1.6.9), we obtain the following inequality for the difference in the entropies $\tilde{S}_0 - S$ (cf. Gibbs inequality (1.6.6)):

$$\tilde{S}_0 - S = \kappa \int \ln\frac{f(x, a_0 + \Delta a)}{\tilde{f}_0(x, a_0, \Delta a)} f(x, a_0 + \Delta a) dx \geq 0 . \tag{1.6.11}$$

This inequality defines the relative measure of indeterminacy of the two selected states. However, it says nothing as to which of the two states is the more ordered one. To answer this question we must analyze the solution of equation (1.6.9) which results in the dependence (1.6.10). If we find that

$$\tilde{T}_{eff} > T_{eff} \, , \tag{1.6.12}$$

which means that equation (1.6.9) will be satisfied if we raise the temperature of the state "0" taken a priori for the state of physical chaos, then the state "0" is actually more chaotic. This check validates our choice of the state "0" as the state of higher chaoticity.

So, if conditions (1.6.12), (1.6.11) are satisfied, the transition

$$a_0 \Rightarrow a_0 + \Delta a \tag{1.6.13}$$

is a process of self-organization: movement from the more chaotic to the less chaotic state, whereas the difference in the entropies (1.6.11) serves as the measure of the relative order of the two states.

The combination of (1.6.11), (1.6.12) expresses the contents of the so-called S–theorem (where "S" stands for *self-organization*) formulated by the author (Klimontovich 1983). Criterion based on this statement will be referred to as criterion of self-organization.

1.6.7. Assessment of the Relative Degree of Order Using the S–Theorem on the Basis of Experimental Data

The above-formulated criterion of self-organization requires knowing the structure of the effective Hamilton function H_{eff}. Heretofore we assumed that the effective Hamilton function can be derived from the appropriate mathematical model of the system under consideration. In many cases, however, an adequate mathematical model is not available. Because of this, we face the problem of gaining information about the structure of the effective Hamilton function directly from experimental data. A method of doing this was proposed in (Klimontovich 1988). This method can be used for assessing the relative degree of order of states of open systems of quite different nature, including various physical, chemical and biological systems.

We select a set of time-dependent parameters which characterize (at a certain level of understanding) the open system under consideration. Let it be, for instance, a cardiogram of the patient. The cardiogram changes under the action of controlling parameters (for instance, medication).

Now we consider two different time realizations at two different values of the controlling parameter,

$$x(t,a_0) \, , \ \ x(t,a_0 + \Delta a) \, . \tag{1.6.14}$$

The time realizations must be long enough, so as to enable one to derive distribution functions which are practically stationary:

$$f(x,a_0) \, , \quad f(x,a_0 + \Delta a) \, . \tag{1.6.15}$$

Once again we assume that the state at $a = a_0$ is the state of physical chaos. Using the distribution f_0 we introduce the effective Hamilton function and proceed as described in the preceding section (see also Ch. 12, 15). In this way, the available experimental data can be used for comparing the relative degree of order of states of open system at different values of the controlling parameter.

There are other criteria for assessing the relative degree of order of states of open systems. Of course, it would be very interesting to compare the efficiency of different criteria (see Ch. 21). We shall also see that temporal and spatial spectra can be effectively used in lieu of time realizations.

1.7. Entropy and Entropy Production in Laminar and Turbulent Flows

Let us use the above criterion of relative degree of order of states of open systems for comparing the relative chaoticity of laminar and turbulent flows.

Turbulent motion has been systematically studied for over a hundred years; nevertheless, a still much debated point is whether the transition from laminar to turbulent flow, for instance in a pipe, is a transition from the more ordered to the less ordered state, or a transition from the less ordered state to the more ordered one. The former point of view is almost universally accepted. The opposing view was expressed by the author in (Klimontovich IV), and also in (Prigogine – Stengers 1984; Ebeling – Klimontovich 1984), to be later corroborated by calculations (Klimontovich 1984). The following expression was obtained for the difference in the entropies of laminar and stationary turbulent motion (see Ch. 22):

$$T\left(S_{\text{lam}} - S_{\text{turb}}\right) = \frac{1}{2}\rho\left\langle(\delta u)^2\right\rangle \, . \tag{1.7.1}$$

We see that the entropy of laminar motion is greater than the entropy of averaged turbulent motion. The change in the entropy is determined by the dispersion of the velocity of hydrodynamic flow δu — that is, by the sum of diagonal elements of Reynolds stress tensor. According to S–theorem, the result (1.7.1) indicates that the transition from laminar to turbulent flow is a process of self-organization.

In particular, the higher orderedness of turbulent flow manifests itself in the following. In the laminar flow the transfer of momentum from layer to layer (which causes viscosity) occurs by independent changes in momenta of individual atoms. By contrast, the transfer of momentum in turbulent flow is a collective process. This difference can be expressed in the following words: upon transition to the turbulent regime the individualistic unorganized resistance of laminar flow is replaced by the collective and, therefore, more organized resistance.

Entropy production is another very important characteristic of the degree of order. Entropy production, as we shall see in Ch. 21, is linked with one of the most fundamental characteristics of the degree of chaoticity in the theory of dynamic chaos, the Kolmogorov – Sinai – Krylov metric entropy.

Values of entropy production were compared in (Klimontovich – Engel-Herbert 1984) for two types of flow: (1) steady turbulent flow of Couette and Poiseuille, and (2) unsteady laminar flow at Reynolds numbers above critical ($Re > R_{cr}$). It was found that, under additional provision of constant tension on the walls of the channel, entropy production for turbulent flow (stable at $Re > R_{cr}$) is less than entropy production for laminar flow (unstable at $Re > R_{cr}$) — that is,

$$\sigma_{lam} - \sigma_{turb} > 0. \tag{1.7.2}$$

In other words, entropy production for turbulent flow (new steady regime of motion) is less than entropy production for laminar flow if such flow could have existed at $Re > R_{cr}$.

This particular (but very important) result points to the possible existence of a certain general principle.

1.8. Principle of Minimum Entropy Production in Processes of Self-Organization

Prigogine's "principle of minimum entropy production in stationary state" in nonequilibrium thermodynamics can be explained as follows.

Denote by $\sigma_{station}$ the value of entropy production in stationary state, and by $\sigma(t)$ the value of entropy production in non-stationary (current) state. According to Prigogine's principle,

$$\sigma(t) > \sigma_{station}. \tag{1.8.1}$$

This principle has been proved for linear thermodynamic systems. General proof for nonlinear systems does not exist; moreover, there are examples of violation of this principle (Stratonovich 1985, 1992).

Let us return to inequality (1.7.2), which defines the relationship between two quantities: entropy production in case of unstable laminar flow, and entropy production for averaged steady turbulent flow on condition that the tension on the wall is the same. On the strength of this example we may formulate the following statement which we call *the principle of minimum entropy production in processes of self-organization.*

Namely, consider a process of self-organization which goes via a series of stationary states, corresponding to different values of controlling parameter a. Let a_{cr} be the critical value of controlling parameter at which one in a series of bifurcations (nonequilibrium phase transitions) takes place (Ch. 16). By σ_{stab} we denote the value of entropy production in the new stable state which resulted from this transition. (This value corresponds to σ_{turb} in case of steady turbulent motion.)

At $a > a_{cr}$ the precedent state (the one which existed before the bifurcation) becomes unstable, and by σ_{instab} we denote the value of entropy production in the imaginary state which, because of its instability, is not realized at $a > a_{cr}$. This value corresponds to σ_{lam} in case of imaginary laminar flow extended into the domain of supercritical Reynolds numbers. Like in case of transition from laminar flow to turbulent, the values of σ_{stab} and σ_{instab} must be calculated under additional constraints which depend on the concrete structure of the system under consideration.

The above-formulated principle can be expressed by the following inequality:

$$\sigma_{stab} > \sigma_{instab} , \qquad\qquad\qquad (1.8.2)$$

which states that the system takes the path of decreasing entropy production at nonequilibrium phase transitions which form the process of self-organization.

The problem of obtaining general proof of this principle, like the general proof of Boltzmann's H-theorem, remains open. However, even if it eventually turns out to be of limited applicability, the general statement (1.8.2) based on the concrete result (1.7.2) will stimulate further studies. As Rabindranath Tagore put it, 'if we close the door before Fallacy, how will the Truth come in?'

Of course, in this brief introduction we were able to touch upon just a few problems of the modern statistical theory of open systems. In the forthcoming chapters we will give a more systematic treatment to such problems. From the table of contents the reader will see that many important problems of statistical theory of open systems had to be left out, in spite of the author's earnest effort to make the book as comprehensive as possible.

In particular, we do not discuss the historical views on "chaos" and "order". Suffice it to say that these concepts were very important for ancient philosophers — for instance, for Plato and his followers. Plato's definitions of these fundamental concepts remain essentially valid even today. Indeed, philosophers of antiquity argued that chaos is the state of the system stripped of any and all possibilities of manifesting its properties. From chaos emerges the order of the Universe.

Also, beyond the scope of the book remains the problem of self-organization in cosmology, one of the most thrilling and grandiose problems of today's science. In this field, however, there still are more questions than comprehensible answers. We may only hope that this book will be of some help in solving these fundamental problems.

In particular, such hope is based on the fact that even the philosophers of old had used the fundamental concepts of the modern statistical theory of open systems in their speculations, such as chaos and order, laminar and turbulent motion. The great Lucretius, for example, regarded the Universe as "turbulent order" which emerged from the primal chaos similar to Brownian motion (see Ch. 22).

To some extent we tried to fill these gaps in the Conclusion, and by supplying a concise bibliography. The list of original papers, reviews and monographs will help the reader to find his bearings in the all-important domain of today's science, the Statistical Theory of Open Systems.

CHAPTER 2

DYNAMIC AND STATISTICAL DESCRIPTION OF PROCESSES IN MACROSCOPIC SYSTEMS

Now we embark upon systematic presentation of the basics of statistical theory of open systems. A statistical theory always works with models of some kind. The simplest model object is a system comprised of a large number of identical structureless microscopic elements (atoms). For such a system we may give a concrete definition of Hamilton function and write (quite formally, of course) the appropriate Hamilton equations of motion. The solution of Hamilton equations at specified initial conditions gives us, in principle, complete information about the state of the system in question.

As said in the Introduction, this task cannot be accomplished even with the aid of most powerful computers. Because of this, we inevitably have to switch from the reversible equations of mechanics (or the relevant equations for dynamic distributions in the phase space) to the irreversible equations of statistical theory.

There are other factors (dynamic instability of motion of elementary objects, or existence of uncontrollable external parameters) which call for the transition to irreversible equations. Transition to irreversible equations is based on the concept of the ensemble of identical macroscopic systems. Such ensemble was introduced by Gibbs for the description of equilibrium states. Generalization of this concept for nonequilibrium states of macroscopic systems constitutes one of the central tasks of this chapter.

It does not seem worthwhile to dwell too much on the contents of the chapter. Rather, we should go straight on to the discussion.

2.1. Hamilton Function. Reversible Equations of Motion. Functions of Dynamic Variables

Assume that the system under consideration consists of N identical structureless particles which move according to the laws of classical mechanics. Examples of such systems are one-component gases or simple liquids.

At any time t the state of such system is described by coordinates r_1, r_2, ..., r_N and momenta p_1, p_2, ..., p_N of all particles of the system. In future we are going to use the following shorthand for the $6N$-dimensional vector $X(t)$ and 6-dimensional vector x_i:

$$X(t) = (x_1,...,x_N) , \quad x_i = (r_i,p_i) , \quad i = 1, 2, ..., N . \tag{2.1.1}$$

In order to describe the time evolution of the system of particles, we must express $X(t)$ in terms of its initial value X_0, that is, find the function

$$X(t) = X(x_0,t_0,t) . \tag{2.1.2}$$

For this purpose we construct the relevant Hamilton function and find the equations of motion for our system.

Hamilton function. For the system of N structureless atoms of mass m the Hamilton function is given by

$$H = \sum_{1 \le i \le N} \left(\frac{p_i^2}{2m} + U(r_i,t) \right) + \frac{1}{2} \sum_{\substack{1 \le i,j \le N \\ i \ne j}} \Phi\left(|r_i - r_j|\right) . \tag{2.1.3}$$

The first sum includes the kinetic energy of the particle and its potential energy. The latter may be a function of time. The second (double) summation takes care of interactions between particles. Note that we have made a simplifying assumption that the interaction is described by the double sum of potential energies of pairs of particles. The function $\Phi\left(|r_i - r_j|\right) \equiv \Phi_{ij}$ only depends on the absolute value of the distance between the two particles. In case of condensed media this simplification may be not justified, and one will have to assume a more complicated nature of interactions. Here, however, we use the above expression for the Hamilton function.

The potential of interaction of particles. The concrete form of interaction function $\Phi(|r|)$ can only be calculated on the basis of quantum mechanics. In the classical statistical theory the function $\Phi(|r|)$ is assumed to be known. There is also a number of tentative analytical expressions for the potential energy $\Phi(|r|)$ — for instance, the Liénard–Jones potential.

The model of elastic balls. The atoms are regarded as perfectly elastic balls r_0 in diameter. They interact (collide) when they come within the distance $r = r_0$ between their centers. This fact is expressed as follows:

$$\Phi(|r|) = \begin{cases} 0, & r > r_0 \\ \infty, & r \le r_0 \end{cases} . \tag{2.1.4}$$

Equations of motion — Hamilton equations. In order to find function (2.1.2) which describes the time evolution of all particles of the system given the initial conditions, we must use the relevant equations of motion (Hamilton equations):

$$\frac{dr_i}{dt} = \frac{\partial H(x,t)}{\partial p_i} \ , \ \frac{dp_i}{dt} = -\frac{\partial H(x,t)}{\partial r_i} \ .$$

(2.1.5)

For the Hamilton function we may use expression (2.1.3).
Functions of dynamic variables. Any function $A(X,t)$ we call a function of dynamic variables. The total derivative of such function is

$$\frac{dA}{dt} = \frac{\partial A}{\partial t} + \frac{\partial H}{\partial p_i}\frac{\partial A}{\partial r_i} - \frac{\partial H}{\partial r_i}\frac{\partial A}{\partial p_i} \ .$$

(2.1.6)

If the total derivative is zero, then the function of dynamic variables is the integral of motion.

If we choose the Hamilton function for our function of dynamic variables, the total derivative is

$$\frac{dH}{dt} = \frac{\partial H}{\partial t} \ .$$

(2.1.7)

This means that the Hamilton function is the integral of motion only for a conservative system, when there is no explicit time dependence.

Important examples of functions of dynamic variables are the total momentum and the angular momentum of the particles of the system:

$$P = \sum_{1 \le i \le N} p_i \ , \ M = \sum_{1 \le i \le N} [r_i p_i] \ .$$

(2.1.8)

Of special importance in the statistical physics are those functions of dynamic variables which satisfy conditions

$$f_N(X,t) \ge 0 \ , \ \int f_N(X,t)dX = 1 \ ,$$

(2.1.9)

and are, therefore, distribution functions. Subscript "N" indicates that we are dealing with distribution functions of variables of N particles of the system.

Further on we will also use local functions of dynamic variables. As follows from their name, they depend not only on the particles' variables, but also on the coordinates r in three-dimensional space, or, in the more general case, on the coordinates $x = (r, p)$ in six-dimensional phase space of positions and momenta.

2.2. Liouville Theorem. Liouville Equation

Using the distribution function introduced above, we write an obvious condition

$$f_N(X,t)dX = f_N(X_0,t_0)dX_0 ,$$ (2.2.1)

to express the fact that the system, which at t_0 occurred in the neighborhood dX_0 of point X_0, will at time t occur in the neighborhood dX of some other point X. The volumes dX and dX_0 are linked by equation

$$dX = DdX_0 ,$$ (2.2.2)

where D is the Jacobian of the transformation. Using the Hamilton equations, we can prove that

$$D = 1 , \text{ and therefore } dX = dX_0 .$$ (2.2.3)

This equation expresses the Liouville theorem. On the strength of this result, we rewrite (2.2.1) as

$$f_N(X,t) = f_N(X_0,t_0) ,$$ (2.2.4)

which implies that the distribution function is the same along the path defined by (2.1.2). In other words, the distribution function $f_N(X,t)$ is the integral of motion, and its total derivative is zero. As a result, we come to the following equation for the distribution function:

$$\frac{\partial f_N}{\partial t} + \sum_{1 \le i \le N} \left(\frac{\partial H}{\partial p_i} \frac{\partial f_N}{\partial r_i} - \frac{\partial H}{\partial r_i} \frac{\partial f_N}{\partial p_i} \right) = 0 ,$$ (2.2.5)

which is the well-known Liouville equation of analytical mechanics.

The Liouville equation for the distribution function is equivalent to the set of equations of motion of the particles (2.1.5). Because of this, the Liouville equation (like the Hamilton equation) is invariant with respect to transformation

$$t \Rightarrow -t \; , \;\; r_i \Rightarrow r_i \; , \;\; p_i \Rightarrow -p_i \; , \;\; i = 1, \; 2, \; ..., \; N \; . \tag{2.2.6}$$

This is usually regarded as a sign of reversibility of the equations. However, the following must be borne in mind.

The directionality of time is a concept more primal than the concepts of reversibility and irreversibility. The replacement $t \Rightarrow -t$ in Nature is impossible, because the time always flows in one direction. So we can only speak of reversibility in the sense of "coming back home". In other words, if after the time lapse $t - t_0$ we reverse the signs of momenta of all particles, then at $2(t - t_0)$ the particles will return to their initial positions with $-p_{i0}$. The reversibility of motion means that, in principle, such "homecoming" is possible. This is a property of Hamilton equations; it is in this sense that we shall speak of the reversibility of Liouville equations. Observe that all equations of the statistical theory are irreversible, and thus differ fundamentally from the equations of mechanics.

And now back to the Liouville equation.

Liouville equation is an equation in partial derivatives of the first order. The solution of the inverse problem (Cauchy problem) for such equation reduces to the solution of equations for characteristics which coincide with Hamilton equations. For this purpose one must use the solution of equations of motion (2.1.5) to express the initial vector X_0 in terms of X — that is, to find the dependence

$$X_0 = X_0\big(X, -(t - t_0)\big) \; , \tag{2.2.7}$$

and substitute this function into the distribution of initial values $f_N(X_0, t_0)$. As a result, we get the solution of inverse problem for Liouville equation:

$$f_N(X, t) = f_N\big(X_0\big(X, -(t - t_0), t_0\big)\big) \; . \tag{2.2.8}$$

So, the solution of Liouville equation is equivalent to solution of the set of equations of motion. As said in the Introduction, this task for a macroscopic system cannot be accomplished, and therefore the transition to statistical description is inevitable.

2.3. Local Dynamic Distribution of States in 6N-Dimensional Phase Space

We use notation $X(t)$ (2.1.1) for the set of values of momenta and coordinates of particles of the system at time t, and X for the set of momentum and position variables in $6N$-dimensional phase space. Then the distribution function of dynamic variables (superscript "d") in the phase space can be defined as

$$f_N^{(d)}(X,t) = \delta(X - X(t)) \; , \; \int f_N^{(d)}(X,t)dX = 1 \; . \tag{2.3.1}$$

This function may be interpreted as microscopic (dynamic) density in $6N$-dimensional phase space. From definition (2.3.1) it follows that the equation for this function is defined by the equations of motion (2.1.5) (Hamilton equations) and can be presented in the form

$$\frac{\partial f_N^{(d)}}{\partial t} + \sum_{1 \leq i \leq N} \left(\frac{\partial H}{\partial p_i} \frac{\partial f_N^{(d)}}{\partial r_i} - \frac{\partial H}{\partial r_i} \frac{\partial f_N^{(d)}}{\partial p_i} \right) = 0 \; . \tag{2.3.2}$$

Using the definition of Hamilton function (2.1.3), this equation can be rewritten in unabridged form. This equation will be often used further as the starting point for the transition from the reversible equations of mechanics to the irreversible equations of statistical theory.

Observe that expression (2.3.1) plays a dual role. On the one hand, it represents the dynamic local distribution in 6N-dimensional phase space. Being a microscopic distribution, it depends not only on the position in space, but also on the complete set of variables $X(t)$ of all particles of the system.

On the other hand, this function can be used for transforming the functions of dynamic variables $A(X,t)$ into the relevant functions in the phase space. In the language of hydrodynamics we may say that this function can be used for performing the transition on microscopic level from Lagrange's to Euler's description.

2.4. Equation for Microscopic Phase Density.
Microscopic Transfer Equations

Along with the dynamic distribution (2.3.1) in 6N-dimensional phase space, one can use a simpler description of dynamic evolution based on microscopic distribution: microscopic density in 6N-dimensional phase space.

This description is exhaustive when the functions of dynamic variables $A(X)$ are symmetrical with respect to the particles' numbers. Such is the Hamilton function (2.1.3), as well as the density of the number of particles, densities of momentum and kinetic energy, densities of charge and current, and other characteristics symmetrical with respect to the particles.

Microscopic phase density occupies a special place among the local functions of dynamic variables. It is defined by the following expression:

$$N(r,p,t) = \sum_{1 \le i \le N} \delta(r - r_i(t)) \delta(p - p_i(t)) \ . \tag{2.4.1}$$

From this definition it follows that

$$\int N(r,p,t)drdp = N \ , \tag{2.4.2}$$

where N is the total number of particles. Using notation $x = (r,p)$ for six-dimensional vector in the phase space, we may introduce the following shorthand:

$$N(x,t) = \sum_{1 \le i \le N} \delta(x - x_i(t)) \ , \quad \int N(x,t)dx = N \ . \tag{2.4.3}$$

Observe that all the main functions of dynamic variables can be expressed in terms of $N(x,t)$. For instance, the Hamilton function (2.1.3) can be written as

$$H(X) = \int \left(\frac{p^2}{2m} + U(r) \right) N(r,p,t)drdp + \frac{1}{2} \int \Phi(|r - r'|) N(x,t)N(x',t)dxdx' \ . \tag{2.4.4}$$

This expression only differs from (2.1.3) in that it includes self-interaction of particles, because the terms with $i = j$ are retained in the summation. If necessary, the contribution of self-interaction can be eliminated in the final results.

Microscopic phase density $N(x,t)$ can also be used for expressing the more simple local dynamic functions — for example, the densities of particles, momentum, kinetic energy:

$$n^m(r,t) = \int N(r,p,t)dp \ ,$$

$$\rho^m u^m = \int p N(r,p,t)dp \ , \tag{2.4.5}$$

$$K^m(r,t) = \int \frac{p^2}{2m} N(r,p,t)dp \ .$$

Let us also quote the relevant expressions for microscopic densities of charge and current:

$$q^m(r,t) = e\int N(r,p,t)dp \ ,$$

(2.4.6)

$$j^m(r,t) = e\int vN(r,p,t)dp \ .$$

Superscript "m" marks off the microscopic local characteristics which depend on the positions and momenta of all particles of the system.

Using the relationship between the Hamilton function and distribution $N(x,t)$ (see (2.4.4)), we may write Hamilton equations for points of phase space where the microscopic phase density (2.4.1) is nonzero in place of Hamilton equations of motions of particles (2.1.5):

$$\frac{dr}{dt} = \frac{\partial}{\partial p}\frac{\delta H(x)}{\delta N(x,t)} = \frac{p}{m} \ ,$$

(2.4.7)

$$\frac{dp}{dt} = -\frac{\partial}{\partial r}\frac{\delta H(x)}{\delta N(x,t)} = F^m(r,t) \ .$$

(2.4.8)

Here we have used functional derivatives and the following definition of the microscopic force $F^m(r,t)$:

$$F^m(r,t) = -\frac{\partial U(r)}{\partial r} - \frac{\partial}{\partial r}\int \Phi(|r - r'|)N(r',p',t)dp' \ .$$

(2.4.9)

Now let us find equation for function $N(x,t)$.

Since the total number of particles is given (that is, given is the integral (2.4.3) of the microscopic distribution $N(x,t)$), the following equation holds, which is similar to (2.2.1):

$$N(x,t)dx = N(x_0,t_0)dx_0 \ .$$

(2.4.10)

The volumes in six-dimensional space are linked by equation (cf. (2.2.2))

$$dx = Ddx_0 \ ,$$

(2.4.11)

where D now is the Jacobian of transformation in six-dimensional space. Using the Hamilton equations, one can prove (see Wiener 1958) that $D = 1$, and therefore $dx = dx_0$. As a result, we find that $N(x,t) = N(x_0,t_0)$, which is equivalent to equation

$$\frac{\partial N(x,t)}{\partial t} + v \frac{\partial N(x,t)}{\partial r} + F^m(r,t) \frac{\partial N(x,t)}{\partial p} = 0 \ . \tag{2.4.12}$$

Together with (2.4.9), this is a closed equation in $N(x,t)$.

Observe that, entirely in the spirit of field theory, equation (2.4.12) for the microscopic phase density (like expression (2.4.4) for the Hamilton function) includes self-interaction. As a matter of fact, it is in the field theory that the method of microscopic phase density is most efficient. This will become clear when we deal with the kinetic theory of fluctuations, and with the kinetic theory of charged particles — the kinetic theory of plasma which, besides the motion of charged particles, requires describing the electromagnetic field. The microscopic force (2.4.9) can then be expressed in terms of microscopic field strengths $E^m(r,t)$, $B^m(r,t)$. The microscopic force (microscopic Lorentz force) depends not only on the coordinate r, but also on the velocity v.

Using (2.4.12), we can obtain the set of equations of transfer in microscopic functions (2.4.5). The simplest of these is the equation in $n^m(r,t)$. It follows from (2.4.12) by integration with respect to momenta, which yields the microscopic equation of continuity — equation of balance for the microscopic density of the number of particles:

$$\frac{\partial n^m}{\partial t} + \mathrm{div}\left(n^m u^m\right) = 0 \ . \tag{2.4.13}$$

The equation for the density $\rho^m = m n^m$ is obtained by multiplying this equation by the mass m.

In the statistical theory, where the description is not complete from the standpoint of mechanics, equation (2.4.12) should be regarded as an equation in random function $N(x,t)$. Accordingly, the set of transfer equations is a set of equations in random functions n^m, u^m, T^m. One of the principal tasks of the statistical theory consists in obtaining approximate dissipative closed equations in the first moments of these random functions.

In Ch. 6, for example, we consider equation in the first moment of random function $N(x,t) \equiv nf(x,t)$ (where $f(x,t)$ is the relevant distribution function), the kinetic Boltzmann equation. An example of a closed set of dissipative equations in the first moments of microscopic functions n^m, u^m, T^m is provided by equations of hydrodynamics (equations of gas dynamics).

Here ends the section devoted to equations for dynamic distributions which are equivalent to equations of mechanics (Hamilton equations). Next we try to make the first step in the transition from these reversible equations to the irreversible equations of statistical theory.

2.5. Ensemble of Identical Macroscopic Systems — the Gibbs Ensemble

The concept of Gibbs ensemble has been discussed in the Introduction. Let us review the material and elaborate on the details.

In the beginning we quoted from Gibbs' definition of the ensemble: "We may imagine a great number of systems of the same nature, but differing in the configurations and velocities..." This implies that from the outset one allows for a certain incompleteness (deficiency) of information about the microstates of individual systems of the ensemble. What is the extent of this incompleteness? What are the factors which determine the deficiency of the description?

We start with an example typical of the thermodynamic description, when the processes are regarded as a sequence of equilibrium states (quasistatic processes).

Let the system in question be a gas contained in a cylinder. The volume of gas V is confined by the walls of cylinder and the piston. As a whole, the system is at rest. Consider an ensemble of such systems.

Now what do we know about the system — in other words, what are the factors which define the deficiency of description of the microstates of individual systems of the ensemble?

First of all, we know the amount of the gas, and therefore the total number of particles. When the piston is fixed, we also know the volume. The potential energy in the field of gravity is $U = mgz$. Finally, we assume that the system is in a thermostat at constant temperature T.

Obviously, this set of values leaves enormous freedom for the values of vector $X(t)$ which characterizes the microstate of individual system. Because of this, we may treat $X(t)$ as a random function of time t, and go over to statistical description.

In order to define this distribution function, we introduce the Gibbs ensemble of identical systems at certain initial conditions. In our current example we specify the following parameters:

$$N, T, V, mg \equiv N, T, a , \tag{2.5.1}$$

where by a we denote all the so-called external parameters, which currently include the volume V and gravity mg.

Let N_{en} be the total number of systems in the ensemble. We are going to pass to the limit $N_{en} \to \infty$. The states of individual systems of the ensemble correspond to certain points in the phase space. We select a volume dX around point X in the phase space. Assume that at a given time this volume contains the number dN_{en} of systems of the ensemble. Then the limit

$$\lim_{N_{en} \to \infty} \frac{dN_{en}}{N_{en}} = f_N(X,t)dX \qquad (2.5.2)$$

defines the desired distribution function (density of distribution) $f_N(X,t)$ of microstates of systems of the ensemble at time t. From this definition it follows that the normalization condition for the distribution function has the form

$$\int f_N(X,t)dx = 1 . \qquad (2.5.3)$$

Recall once again that subscript "N" indicates that the system under consideration consists of N particles (N small elements).

Naturally, this definition of the distribution function is entirely formal. As a matter of fact, the linkage between this statistical distribution and the dynamic distribution (2.3.1) for the same system remains unclear. After all, the Liouville equation for the dynamic distribution is equivalent to the set of the Hamilton equations of motion.

What is the equation for the statistical distribution $f_N(X,t)$?

If we use the conventional definition of the Gibbs ensemble, when the uncertainty of the statistical description is determined by the distribution of the initial values (1.1.6), and the distribution function $f_N(X,t)$ itself is linked with the dynamic distribution $f_N^{(d)}(X,t)$ by (1.1.7), we run into difficulties.

To wit, averaging with respect to initial values does not change the dynamic Liouville equation (2.3.2), since the initial values can be regarded as independent parameters. As a result, the equation for the statistical distribution (1.1.7) coincides in form with the Liouville equation (2.2.5) and is therefore a reversible equation.

Then we are faced with the problem of reconciling this definition of ensemble for nonequilibrium processes and the definition of Gibbs ensemble for the equilibrium state (as already indicated in Sect. 1.1.2).

It would be more natural to define Gibbs ensemble for nonequilibrium processes as done in Sect. 1.1.2: this definition is based on the physical concept of continuous medium, which links the "point" with physically infinitesimal volume. Then equation for $f_N(X,t)$ is definitely irreversible. Is there a general form of equation for this distribution?

We shall see that there is no general recipe; such equations, however, can be constructed for many systems: rarefied gas, plasma, a system of atoms interacting with electromagnetic field. In all these cases the equations for statistical distributions $f_N(X,t)$ are irreversible and describe evolution of the system towards the equilibrium state.

It is worthwhile to begin with the question of the structure of equilibrium statistical distributions. This problem is much simpler — if only because the equilibrium distribution function is time-independent. This implies that information about the distribution of initial values disappears as the system evolves towards the state of equilibrium, which points to the fact that the form of equilibrium statistical distribution depends on but a small number of parameters (2.5.1).

Our immediate task will consist in studying the basics of the statistical theory of equilibrium state. We shall see that, depending on the external conditions, there are two particular structures of equilibrium statistical distributions, which are the famous Gibbs distributions. They offer the possibility of giving statistical foundation to the laws of thermodynamics for quasistatic processes viewed as a sequence of equilibrium states.

It seems that we have worked out a clear and comprehensive plan. Our progress, however, will not be entirely smooth. One of the fundamental questions is whether the state of complete equilibrium is actually feasible in Nature. If the answer is positive, then our plan is quite sound. If, however, complete equilibrium is not achievable, then questionable is the very existence of irreversible equations for statistical distributions which describe the process of evolution towards the equilibrium state.

In this intriguing environment we will eventually prove that our plan is justified. It does not mean, of course, that we are building something which will not have to be revamped. Whatever is in store, however, let us go ahead!

CHAPTER 3

STATISTICAL THEORY OF EQUILIBRIUM STATE

The present chapter is devoted to the basics of the statistical theory of equilibrium state of macroscopic systems. The equilibrium state is established spontaneously as a result of time evolution provided that the external conditions remain unchanged.

As a rule, macroscopic systems approach the state of equilibrium in several stages, each of which is characterized by its own relaxation time τ_{rel}. The completion of this process is only possible if the external conditions remain unchanged for the time which exceeds the maximum relaxation time $(\tau_{rel})_{max}$.

In the state of equilibrium the distribution function $f_N(X,t)$ is time-independent. Consequently, time-independent are all thermodynamic functions, since they are moments of this most general distribution. This circumstance simplifies considerably the statistical description of the system, and allows one to find the concrete form of distribution functions $f_N(X)$ at the given external conditions. For practical applications, of special interest are the following two cases:

(1) Isolated system, when the total energy of the system has a certain fixed value.

(2) System in a thermostat, when the ambient temperature is fixed.

The relevant statistical distributions were established by Gibbs and are known, respectively, as the *microcanonical* and the *canonical* Gibbs distributions.

In principle, these distributions can be used for calculating the thermodynamic functions and giving statistical background for the laws of thermodynamics in case of quasistatic (reversible) processes when each intermediate state is a state of equilibrium.

As we shall see, the concept of entropy occupies a very special place in the statistical theory. It is entropy that can serve as the measure of uncertainty of the statistical description. Because of this fundamental property, entropy can be widely used also in the statistical theory of nonequilibrium processes in open systems, in particular, as the measure of the relative degree of order in processes of self-organization.

Without going into the finer details of the statistical theory of equilibrium state, we will only discuss those points which are relevant for our further discussion of the statistical theory of open systems. Whatever remains beyond our scope can be found in literature listed at the end of this volume.

3.1. Microcanonical Gibbs Distribution

Assume that the system under consideration is energy-insulated. This means that the Hamilton function (the energy of the system) is fixed at a certain value E. The main presumption of the Gibbs theory is that the dependence of the distribution function on the set of microscopic variables X in the state of equilibrium is completely determined by the Hamilton function. Then the distribution function for our isolated system has the following form:

$$f_N(X,a,E) = \frac{1}{\Omega(a,E)} \delta(H(X,a) - E) ; \quad \int f_N(X,a,E)dX = 1 . \tag{3.1.1}$$

As before, subscript "N" indicates that the system under consideration consists of N particles; a is the set of external parameters (for instance, the volume or the pressure); δ–function singles out the microstates X at the given energy E of the system. The expression for $\Omega(a,E)$ follows from the normalization condition:

$$\Omega(a,E) = \int \delta(H(X,a) - E)dX . \tag{3.1.2}$$

We see that this function defines the area of hypersurface in the phase space X corresponding to the given constant energy E.

Observe once again that the dependence on the microscopic variables in the microcanonical distribution (3.1.1) is wholly defined by one and only function of dynamic variables, the Hamilton function. This special role of the Hamilton function is due to the fact that it determines the form of equations of motion of individual particles of the system, the reversible dynamic Hamilton equations (2.1.5). This assumption, however, remains a hypothesis which, as we shall see, is justified in case of reversible quasistatic processes, when all intermediate states are equilibrium. For nonequilibrium states it is not possible to establish the general form of the distribution.

Microcanonical Gibbs distribution (3.1.1) is widely used for calculating the thermodynamic functions in cases when the systems under consideration are so extensive that the boundary conditions are relatively unimportant, and it suffices to use the idealized condition of closedness of the system. Now we are going to investigate a more realistic situation.

3.2. System in Thermostat. Canonical Gibbs Distribution

Assume now that the macroscopic system in question is capable of exchanging energy with surrounding bodies, under provision that the temperature of the environment is constant. In other words, the system under consideration is placed in a thermostat.

In this way, now the distribution function of the system depends on thermal motion of the surrounding bodies. Let us investigate the conditions under which this dependence can be reduced to a function of the temperature of the thermostat.

We use the following notation.

The system of N particles constitutes a small but macroscopic ($N \gg 1$) part of the large system. Let N_1 be the number of particles in the surrounding bodies ($N_1 \gg N$). The entire system of $N_1 + N$ particles is closed, and its energy is E. Then for the entire system we can use the microcanonical distribution

$$f_{N_1+N}(X,X_1,a,E) = \frac{1}{\Omega(a,E)} \delta\big(H(X,X_1)-E\big) ,$$

(3.2.1)

$$\int f_{N_1+N}(X,X_1,a,E)dXdX_1 = 1 .$$

Here f_{N_1+N} is the distribution function of the entire closed system; X_1 is the set of microscopic variables of the surrounding bodies.

Generally, the Hamilton function of the entire system can be presented as a sum of three parts:

$$H(X,X_1) = H(X) + H_1(X_1) + H_{\text{int}}(X,X_1) .$$

(3.2.2)

The first term on the right-hand side is the Hamilton function of the system under consideration, the second is the Hamilton function of the surrounding bodies, the third takes care of the interaction of the system under consideration with the surrounding bodies.

Our task is to find a closed expression for the distribution function $f_N(X)$ for the microstates of the system under consideration. This expression follows from (3.2.1) after integration with respect to X_1:

$$f_N(X) = \int f_{N+N_1}(X,X_1)dX_1 .$$

(3.2.3)

It is only possible to carry out the integration and thus obtain a closed expression for the desired distribution function under an additional condition that the energy of interaction in (3.2.2) should be negligibly small. For such an assumption there are good enough reasons.

Indeed, the interaction between macroscopic atomic bodies occurs via the atoms of surface layer. The thickness of this layer is of the order of the effective size of the atoms r_0, or something like 10^{-8} cm. Since the volume of the surface layer is $\sim r_0 V^{2/3}$, the contribution of the interactive term is proportional to small parameter $r_0/V^{1/3}$.

If the interaction occurs, for instance, via the electromagnetic field, then the interactive term in (3.2.2) is small because of the weak coupling of atoms with the field.

With this provision we carry out integration of (3.2.3) with respect to X_1 in zero interaction approximation (see Ch. 4 in Klimontovich IV). As a result, we obtain the sought-for canonical Gibbs distribution:

$$f_N(X,a,T) = \frac{1}{Z(a,T)} \exp\left(-\frac{H(X,a)}{\kappa T}\right) . \tag{3.2.4}$$

Function $Z(a,T)$ is called the statistical integral, and follows from the normalization condition for the distribution function:

$$Z(a,T) = \int \exp\left(-\frac{H(X,a)}{\kappa T}\right) dX . \tag{3.2.5}$$

Further on we shall often use another form of the canonical Gibbs distribution,

$$f_N(X,a,T) = \exp\frac{F(a,T) - H(X,a)}{\kappa T} . \tag{3.2.6}$$

Function $F(a,T)$ and the statistical integral are linked by the following relation:

$$F(a,T) = -\kappa T \ln Z(a,T) . \tag{3.2.7}$$

We shall see that $F(a,T)$ coincides with the free energy, one of the basic thermodynamic potentials.

Finally, we must define the parameter T which enters the canonical Gibbs distribution. This parameter is determined by the mean kinetic energy of particles which surround the system under consideration,

$$\frac{3}{2}\kappa T = \lim_{N_1 \to \infty} \frac{1}{N_1} \sum_{1 \le i \le N_1} \frac{p_i^2}{2m} , \tag{3.2.8}$$

and thus characterizes the temperature of the thermostat.

Thus defined, the temperature T in the canonical distribution is constant, and therefore the temperature fluctuations are zero. This rigorous constraint on the temperature is due to the physical assumptions under which the canonical Gibbs distribution was derived.

There are alternative ways leading to the canonical Gibbs distribution and based on less restrictive provisions.

Let the macroscopic system in question be comprised of n macroscopic subsystems, the interaction between which is negligibly small. Then the Hamilton function $H(X,a)$ of the system can be represented as the sum of Hamilton functions of subsystems,

$$H(X,a) = \sum_{1 \leq i \leq n} H_i(X_i,a) .$$

(3.2.9)

Under this condition, the distribution function $f_N(X,a,T) = f_N(H(X,a),T)$ can be represented as the product of distribution functions for individual subsystems:

$$f_N(H(X,a)) = \prod_{1 \leq i \leq n} f_{N_i}(H_i(X_i,a)) , \quad \sum_{1 \leq i \leq n} N_i = N .$$

(3.2.10)

The solution of this functional equation brings us to the exponential distribution similar to (3.6.1). This time, however, the result is only based on the assumption that the structure of the distribution function is wholly determined by the Hamilton function. Because of this we can give a different interpretation to the parameter T in the Gibbs distribution.

In the above reasoning, T was the temperature of thermostat, thus being a constant external parameter characterizing the state of the surrounding bodies. Now we may regard it as an internal parameter of the system, the absolute thermodynamic temperature of the body under consideration. Like any other thermodynamic characteristic of a system, this temperature may exhibit fluctuations. How small are the relative fluctuations of the temperature depends on the condition of macroscopicity of the system in question (condition $N \gg 1$).

Further on we shall discuss yet another way of deriving the canonical Gibbs distribution, based on the principle of maximum of entropy. Prior to that, however, we must speak about the statistical approach to the laws of thermodynamics.

3.3. First Law of Thermodynamics

The first law of thermodynamics states that the internal energy $U(a,T)$ as the function of state variables a, T may change because of (1) heat transfer dQ, and (2) work of thermodynamic forces

$$dA = \sum_{1 \leq i \leq n} F_i(a,T)da_i \ , \quad a = (a_1,...,a_n) \ . \tag{3.3.1}$$

The external parameters here act as generalized coordinates. In thermodynamics the work done does not entirely account for the change in the internal energy, because the forces depend not only on the generalized coordinates but also on the temperature. In other words, the internal energy also changes because of the transfer of heat, and so

$$dU = dQ - dA \ . \tag{3.3.2}$$

Since thermodynamic forces depend on the temperature, the amount of work dA depends on the path of transition from one state to another. For this reason dA and dQ, taken separately, are not total differentials.

Let us give a statistical definition of thermodynamic forces in the mathematical expression of the first law of thermodynamics. It would be natural to define the internal energy as the mean value of Hamilton function,

$$U(a,T) = \int H(X,a)f_N(X,a,T)dX \ . \tag{3.3.3}$$

Hence follows the expression for the total differential:

$$dU(a,T) = \int \sum_{1 \leq i \leq n} \left(\frac{\partial H(X,a)}{\partial a_i} da_i \right) f_N(X,a,T)dX$$
$$+ \int H(X,a)d_{a,T}f_N(X,a,T)dX. \tag{3.3.4}$$

In the first term on the right-hand side we can use the microscopic definition of thermodynamic force,

$$F_i^m(X,a) = -\frac{\partial H(X,a)}{\partial a_i} \ . \tag{3.3.5}$$

By thermodynamic force we understand the mean value of the relevant thermodynamic force:

$$F_i(a,T) = \int F_i^m(X,a)f_N(X,a,T)dX \ . \tag{3.3.6}$$

Analyzing these expressions we see that the first term on the right-hand side of (3.3.4) answers for the work of thermodynamic forces (3.3.6), whereas the amount of heat transfer is given by

$$dQ = \int H(X,a) d_{a,T} f_N(X,a,T) dX \ . \tag{3.3.7}$$

We see that the heat transfer term accounts for the change in the internal energy due not to the work of thermodynamic forces, but rather to the change in the distribution function caused by the thermodynamic variables a, T.

So, it is possible to give statistical interpretation to all thermodynamic functions in the mathematical expression of the first law of thermodynamics. Now let us turn our attention to the second law.

3.4. Second Law of Thermodynamics for Quasistatic Processes

Recall that the second law of thermodynamics for quasistatic processes states that

$$dQ = TdS(a,T) \ . \tag{3.4.1}$$

This implies that there exists a function of state $S(a,T)$ called entropy which remains constant in adiabatic processes. The absolute temperature acts as integration factor.

Let us demonstrate that (3.4.1) follows from the statistical definition of dQ (3.3.7).

For the distribution function in (3.3.7) we take the canonical Gibbs distribution, and show that (3.3.7) can then be reduced to (3.4.1).

We rewrite (3.3.7) in the equivalent form

$$dQ = -\kappa T \int \frac{F(a,T) - H(X,a)}{\kappa T} d_{a,T} f_N(X,a,T) dX \ . \tag{3.4.2}$$

The new term added into this expression is equal to zero because of the normalization condition of distribution function $f_N(X,a,T)$. Now, using (3.2.6), we rewrite (3.4.2) in the form

$$dQ = -\kappa T \int \ln f_N(X,a,T) d_{a,T} f_N(X,a,T) dX \ . \tag{3.4.3}$$

By virtue of the same normalization condition, we can factor out the differential and get

$$dQ = -\kappa T d_{a,T} \int \ln f_N(X,a,T) f_N(X,a,T) dX \ . \tag{3.4.4}$$

We see that the expression for dQ is integrable. If we take $1/T$ for the integration factor, thus identifying T with the absolute thermodynamic temperature, then, using (3.4.1) and (3.4.4), we can give the statistical definition of entropy:

$$S(a,T) = -\kappa \int \ln f_N(X,a,T) f_N(X,a,T) dx + S_0 \ . \tag{3.4.5}$$

Here S_0 is the contribution to the entropy which does not depend on the variables a, T, but may depend on the number of particles in the system N.

Observe that the expression for entropy is equivalent to the mean value of microscopic function

$$S^m(X,a,T) = -\ln f_N(X,a,T) \ . \tag{3.4.6}$$

This is not a conventional function of dynamic variables X, since it depends not only on the generalized coordinates a, but also on a statistical characteristic, the absolute temperature T.

Finally, let us prove that function $F(a,T)$ in the canonical Gibbs distribution (3.2.6) coincides with the thermodynamic free energy $F = U - TS$.

Using the Gibbs distribution, we find that

$$F_{\text{therm}} = U - TS = \int (H + \kappa T \ln f_N) f_N dX = F \int f_N dX = F(a,T) \ . \tag{3.4.7}$$

So, we have given statistical definitions to all the main thermodynamic functions including the entropy. Shortly we shall demonstrate that entropy plays a special role in the statistical theory, serving as measure of uncertainty of the statistical description. Of course, this is by far not the only useful role of the entropy.

3.5. Entropy as Measure of Uncertainty of Statistical Description of States of the System

Recall that expression (3.4.5) for entropy was obtained for the equilibrium state, when $f_N(X,a,T)$ is the canonical Gibbs distribution. We shall see that this expression holds also for an arbitrary distribution $f(X)$. Therefore, in place of (3.4.5) we may use

$$S[X] = -\kappa \int \ln f(X) f(X) dX \ , \quad \int f(X) dX = 1 \ . \tag{3.5.1}$$

Here we have dropped the constant S_0 and introduced new notation for $S[X]$ to emphasize that the entropy S is defined for the set of random variables X.

Equation (3.5.1) defines the entropy in terms of a distribution of continuous variable X. In many cases (for instance, in quantum mechanics), the set of random variables takes on discrete values. If we denote the relevant distribution function by f_n, the expression for the entropy takes the form

$$S[n] = -\kappa \sum_n \ln f_n f_n \, , \quad \sum_n f_n = 1 \, . \tag{3.5.2}$$

Let us now demonstrate that the entropy thus defined possesses a number of properties which allow one to use it as the measure of uncertainty of statistical description of the states of the system.

Properties of functions $S[X]$, $S[n]$.

(1) If the distribution function f_n is unity for one of the possible values of $n = n_1$, and zero for all other n, which corresponds to zero uncertainty of statistical description, then the entropy is zero:

$$S[n] = 0 \, . \tag{3.5.3}$$

(2) Let the number of possible values of variable n be M, and let them all be equally probable, which corresponds to the topmost uncertainty of statistical description. Let us show that the entropy in this case is at maximum.

Because the distribution is uniform, the distribution function and the entropy are given by

$$f_N = \frac{1}{M} \, , \quad S_0 = \ln M \, . \tag{3.5.4}$$

To compare this entropy with the entropy for an arbitrary distribution, we take advantage of the fact that

$$\ln a \geq 1 - \frac{1}{a} \tag{3.5.5}$$

and set $a = Mf_n$. Then, with due account for the normalization conditions, we find that

$$S_0 - S[n] = \sum_n \ln(Mf_n) f_n \geq \sum_n \left(1 - \frac{1}{Mf_n}\right) f_n = 0 \, . \tag{3.5.6}$$

(3) Let the state of the system be determined by two sets of variables: X, Y for continuous variables, or n, m for discrete variables. The corresponding entropies we denote by $S[X,Y]$ and $S[n,m]$. Then it is straightforward that, as long as X and Y (or n and m) are statistically independent, and

$$f(X,Y) = f(X)f(Y) , \quad f_{n,m} = f_n f_m ,$$ (3.5.7)

we have

$$S[X,Y] = S[X] + S[Y] , \quad S[n,m] = S[n] + S[m] .$$ (3.5.8)

In other words, the entropies for statistically independent sets of variables X and Y (or n and m) are additive. This result is in accord with the obvious fact that the uncertainties for statistically independent states are also additive.

(4) We shall also introduce the concept of conditional entropy, which characterizes the uncertainty of X on condition that Y is given. For this purpose we use the definition of conditional distribution function

$$f(X,Y) = f(X|Y)f(Y) , \quad f(Y,X) = f(Y|X)f(X) .$$ (3.5.9)

If X, Y are statistically independent, then the conditional distribution coincides with the non-conditional distribution.

Conditional entropy is defined as

$$S[X|Y] = -\kappa \ln f(X|Y)f(X,Y)dXdY ,$$ (3.5.10)

with a similar expression for $S[Y|X]$. Using the definitions of $S[X|Y]$, $S[X]$, $S[Y]$, we find that

$$S[X,Y] = S[X|Y] + S[Y] = S[Y|X] + S[X] .$$ (3.5.11)

The implication is that the total uncertainty of description of state X, Y equals to the uncertainty of state X at given Y plus the uncertainty of Y, and vice versa.

Now the fourth property can be formulated as follows. Additional conditions may either reduce the uncertainty of statistical description, or leave it unchanged. This property is expressed by inequalities which are easily proved using (3.5.5):

$$S[X|Y] \le S[X] , \quad S[Y|X] \le S[Y] .$$ (3.5.12)

(5) Information about the state X at given Y is defined by

$$I[X] = S[X] - S[X|Y] .$$ (3.5.13)

Information, therefore, reduces the uncertainty of definition of state X when the value of Y is known.

The fifth property states that information is always positive,

$$I[X] \geq 0 .$$ (3.5.14)

Information is zero when X and Y are statistically independent.

We see that the entropy as defined by (3.5.1) possesses properties (1) – (5). It is these properties of the entropy that allow one to use it as the measure of uncertainty of the statistical description. Observe that none of the basic thermodynamic functions, including the free energy, displays all of these properties.

We assumed that $f(X)$ is an arbitrary distribution function. For equilibrium states this can be, in particular, the canonical Gibbs distribution. For a macroscopic system of N particles in general, this can be the distribution function $f_N(X,t)$ which is the main statistical characteristic of an arbitrary nonequilibrium process. Because of this, we can give the following definition of entropy for nonequilibrium processes:

$$S(t) = -\kappa \int \ln f_N(X,t) f_N(X,t) dX .$$ (3.5.15)

Now what is the equation satisfied by this most general distribution function? We will start answering this question in Sect. 3.7.

Modern views about entropy and information emerged from the synthesis of concepts introduced by Boltzmann and Gibbs in the statistical theory, and, much later, in the general theory of communication in the nineteen fifties. The new field of science came to be known as the theory of information, which owes much to the work of Claude Shannon.

The linkage between the entropy and the distribution function, and the fact that entropy may serve as the measure of uncertainty of statistical description of states of the system, allow one to draw comparison between the degrees of chaoticity of equilibrium and nonequilibrium states. The possibility of such comparison was first indicated by Gibbs. The statement formulated and proved by this scholar we shall call the Gibbs theorem.

3.6. Gibbs Theorem

Consider a system characterized by an arbitrary Hamilton function $H(X)$ (as before, X is the complete set of variables). The equilibrium state is described by the canonical Gibbs distribution (3.2.6). By $f(X,t)$ we denote the "arbitrary" nonequilibrium distribution, which may be time-dependent and is normalized in the same way as the Gibbs distribution,

$$\int f_N(X,a,T)dX = 1 \;, \quad \int f(X,t)dX = 1 \;. \tag{3.6.1}$$

We have put the word "arbitrary" in quotes, because actually we impose a certain constraint on this function:

$$\int H(X)f_N(X,a,T)dX = \int H(X)f(X,t)dX \;, \tag{3.6.2}$$

which means that the states of the system will be compared at the same value of mean energy.

Now we introduce appropriate notation for the entropies of equilibrium and nonequilibrium states:

$$S_0 = -\kappa \int \ln f_N(X,a,T)f_N(X,a,T)dX \;,$$

$$S(t) = -\kappa \int \ln f(X,t)f(X,t)dX \;. \tag{3.6.3}$$

The Gibbs theorem states that the highest value of entropy corresponds to the equilibrium state:

$$S_0 - S(t) \geq 0 \;. \tag{3.6.4}$$

The "equals" sign corresponds to the case when distribution f coincides with the canonical Gibbs distribution.

To prove this statement, we use conditions (3.6.1), (3.6.2) and inequality (3.5.5), and get

$$S_0 - S(t) = \kappa \int \ln \frac{f(X,t)}{f_N(X,a,T)} f(X,t)dX \geq 0 \;. \tag{3.6.5}$$

The following comment is important for our future discussion. We have

$$\int \ln \frac{f}{f_0} f dX \geq 0 \ , \quad \text{when} \quad \int f_0 dX = \int f dX = 1 \tag{3.6.6}$$

— that is, when the arbitrary distribution functions are normalized in the same way. Without the additional condition (3.6.2), however, this integral is not equal to the difference in the entropies, and therefore inequality (3.6.6) does not by itself express the statement of the Gibbs theorem.

To repeat, the entropy is maximum for the equilibrium state provided that the mean energy remains constant. Since entropy is the measure of chaoticity (uncertainty of statistical description), the state of equilibrium is more chaotic than any nonequilibrium state whose mean energy is the same.

3.7. Change in Entropy in the Course of Time Evolution

Recall that (3.5.15) defines the entropy for an arbitrary distribution $f_N(X,t)$. According to Gibbs theorem, the value of entropy for the state of equilibrium is greater than that for any other state. Now, what is the form of the equation for $f_N(X,t)$ which describes evolution of the system towards equilibrium, the most chaotic state.

Using the Liouville equation (2.2.5) we come to a contradiction with the Gibbs theorem. Indeed, let us supplement the Liouville equation with boundary conditions for the distribution function $f_N(X,t)$. The first of these expresses the condition of closedness, condition that the particles do not leave the confines of the system. This condition can be formulated by setting to zero the distribution function of any particle which goes to infinity:

$$f_N(X,t)\big|_{r_{i,\alpha}=\pm\infty} = 0 \ , \quad \alpha = x,y,z \ , \quad i = 1,2,...,N \ . \tag{3.7.1}$$

The other boundary condition must be used in integration with respect to momenta:

$$f_N(X,t)\big|_{p_{i,\alpha}=\pm\infty} = 0 \ , \quad \alpha = x,y,z \ , \quad i = 1,2,...,N \ , \tag{3.7.2}$$

which implies that the probability of infinite values of momenta (and kinetic energies) of particles is zero.

Given these boundary conditions, from Liouville equation we find the following equation for the entropy as defined by (3.5.15):

$$\frac{dS}{dt} = 0 \ , \tag{3.7.2}$$

which has already been quoted in the Introduction (see (1.1.10)).

We see that the entropy remains constant in the course of time evolution according to Liouville equation. This means that the uncertainty of statistical description, as defined by the initial distribution $f_N(X_0, t_0)$, does not change with the time. Obviously, in general it will not coincide with the entropy of the equilibrium state.

This result is a consequence of the reversibility of the Liouville equation. Reversible equations do not describe the transition towards the state of equilibrium, which, according to Gibbs theorem, corresponds to the maximum value of the entropy, and therefore to the highest degree of chaos.

So we face the task of constructing irreversible equations which describe, in particular, the transition towards the state of equilibrium. If the mean energy is conserved in the course of the evolution (as is the case, for instance, with the kinetic Boltzmann equation), the prerequisites of the Gibbs theorem are satisfied. Then the time evolution leads towards the state which corresponds to the highest entropy value — that is, the most chaotic state.

Inequality (3.6.4) may be regarded as the statistical formulation of the second law of thermodynamics. Open, however, remains the question whether the increase in the entropy is monotonic in the course of time evolution towards the state of equilibrium — in other words, whether it is possible to obtain a result similar to Boltzmann's H-theorem on the basis of equations for many-particle distributions $f_N(X, t)$.

As we shall see in due course, such irreversible equations can be constructed for a number of cases, given that the mean energy of the system remains constant in the process of evolution. We come to a result more general than (3.6.4):

$$\frac{dS(t)}{dt} \geq 0 , \quad S(t) = -\kappa \int \ln f_N(X, t) f_N(X, t) dX .$$
(3.7.4)

As above, the "equals" sign corresponds to the state of equilibrium.

A dissipative equation for $f_N(X, t)$ was first proposed by M.A. Leontovich in 1935. His equation includes the famous Boltzmann equation; moreover, it can be used as a basis for the theory of kinetic fluctuations.

Now we use the properties of the entropy to give an alternative independent derivation of the canonical Gibbs distribution.

3.8. Principle of Maximum of Entropy. Derivation of Canonical Gibbs Distribution

On the basis of Gibbs theorem, let us solve the inverse problem of reconstructing the canonical Gibbs distribution from the principle of maximum of entropy.

Let the entropy of the system in the state of equilibrium be given by (3.4.5). Now we assume, however, that the form of the distribution function $f_N(X,a,T)$ is not known, but known are the normalization condition and the mean value of energy (mean value of the Hamilton function $H(X,a)$. In other words, we have

$$\int f_N(X,a,T)dX = 1 , \quad \int H(X,a)f_N(X,a,T)dX = U(a,T) .$$ (3.8.1)

Now we demonstrate that these conditions together with the principle of maximum of entropy are sufficient for finding the form of the distribution function of the equilibrium state.

In order to find the maximum of entropy (3.4.5) under the two conditions of (3.8.1) we use the method of Lagrangian multipliers, which, as will become clear later, are conveniently written as $\lambda + \kappa$ and $-\lambda_1$.

Now the problem reduces to finding the extremum of the functional

$$\int [-\kappa \ln f_N(X,a,T)f_N(X,a,T)+(\lambda + \kappa)f_N(X,a,T)-\lambda_1 H(X,a)f_N(X,a,T)]dX .$$
(3.8.2a)

We vary this functional with respect to the distribution function and equate the result to zero, thus getting the expression for the distribution function which contains undetermined Lagrangian multipliers. These depend on the external parameters a and on the temperature T, since the internal energy in (3.8.1) depends on a, T:

$$\lambda = \lambda(a,T) , \quad \lambda_1 = \lambda_1(a,T) .$$ (3.8.2b)

For this reason the sought-for distribution function also depends on a, T via the so far undetermined Lagrangian multipliers:

$$f_N(X,a,T) = \exp\left(\frac{\lambda}{\kappa} - \frac{\lambda_1 H(X,a)}{\kappa}\right) .$$ (3.8.3)

So, we have once again come to a distribution which exhibits exponential dependence on the Hamilton function. The comparison between (3.8.3) and the canonical distribution (3.2.4) (or (3.2.6)) leads to the following natural interpretation of Lagrangian multipliers:

$$\lambda = \frac{F(a,T)}{T} = -\kappa \ln Z(a,T) , \quad \lambda_1 = \frac{1}{T} .$$ (3.8.4)

Obviously, the two conditions of (3.8.1) are then satisfied.

As we shall see further, the principle of maximum of entropy plays an important role in the statistical theory of open systems, since it allows one to find the form of the distribution functions for most diverse characteristics of systems under consideration directly from experimental data.

3.9. Principle of Indistinguishability of Particles in Statistical Theory

In classical mechanics the particles are assumed to be, in principle, distinguishable even in a one-component system, when all the particles are similar to each other. This means that one can assign a number to each particle and trace its path.

In quantum mechanics, by contrast, all particles are identical and indistinguishable from one another, since the principle of uncertainty makes it impossible to follow the path of any selected particle.

Now what is the situation like in the statistical theory?

Observe that Gibbs distribution functions are symmetrical functions of variables of individual particles. Indeed, from the structure of Hamilton function for a system of identical particles (2.1.3) we see that the function is invariant to permutations of particles. Because of this, the distribution functions of the equilibrium state also remain unchanged, since they depend on the microscopic variables X only via the Hamilton function $H(X,a)$. For this reason, in the statistical theory the states which only differ because of permutation of particles are physically indistinguishable.

In the statistical integral (3.2.5), however, the integration is carried out with respect to the variables of all particles. In this way care is taken of the contribution of those states which differ because of permutation of particles and are therefore the same physically.

In order to eliminate the contribution from nonphysical states, we divide the statistical integral into the number of nonphysical states $N!$, getting thus in place of (3.2.5) the following definition of the statistical integral in the canonical Gibbs distribution:

$$Z(a,T) = \frac{1}{N!}\exp\left(-\frac{H(X,a)}{\kappa T}\right)dX .$$
(3.9.1)

The free energy is, as before, related to the statistical integral by (3.2.7).

We end this section with one final comment.

The above-defined Gibbs distribution functions have the dimensions determined by the normalization condition for f_N. In many cases, especially in quantum theory, it would be more natural to use the dimensionless distribution functions. This can be done by scaling the phase space of the relevant variables. In quantum theory the gauge is given by Planck's constant $\hbar$. Then the transition to dimensionless distribution function is accomplished by making the following substitution in the statistical integral (3.9.1):

$$dX \Rightarrow \frac{dX}{(\hbar)^{3N}} , \qquad (3.9.2)$$

and for the transition from statistical integral to statistical sum we substitute

$$\frac{1}{N!} \int \frac{dX}{(\hbar)^{3N}} \cdots \Rightarrow \sum_n \cdots . \qquad (3.9.3)$$

We shall presently see that neglect of indistinguishability of particles in the statistical theory leads to nonphysical results, the so-called Gibbs paradox being one example. A general remark must be made, however, before we continue.

3.10. Dependence of Thermodynamic Functions on the Number of Particles

In thermodynamics and in statistical theory one can distinguish a class of the so-called additive functions which are proportional to the number of particles N in the system. These include the energy characteristics (internal energy, free energy, ...), as well as the entropy S, since the temperature in the energy-like product TS does not depend on the number of particles.

In dealing with *local processes* (such as do not depend explicitly on the size of the system), one may pass to the thermodynamic limit

$$N \to \infty , \quad V \to \infty , \quad \text{but } n = \frac{N}{V} = \text{const} , \qquad (3.10.1)$$

thus assuming that the number of particles and the volume may tend to infinity, provided that the mean density of particles remains constant.

We shall see that in passing to the thermodynamic limit we lose the possibility to take care of the fluctuations of thermodynamic variables, thus making it impossible to describe many physical phenomena which depend on the fluctuations — the Brownian motion, for example.

From arguments developed above it follows that internal energy, free energy and entropy as functions of the relevant independent variables in the thermodynamic limit have the following structure:

$$U(S,T) = Nu\left(\frac{S}{N},T\right), \quad F(V,T) = Nf\left(\frac{V}{N},T\right), \quad S(U,V) = Ns\left(\frac{U}{V},\frac{V}{N}\right), \tag{3.10.2}$$

where the lower-case characters denote thermodynamic quantities per one particle.

3.11. Thermodynamic Functions of Ideal Gas

Let N particles of ideal gas be contained within the volume V. Then, in the absence of external field, the Hamilton function is the sum of kinetic energies. In this approximation all thermodynamic functions can be found using the canonical Gibbs distribution.

We use (3.2.4) for the distribution function, and (3.2.5) for the statistical integral. Then

$$Z(V,T) = V^N (2\pi m\kappa T)^{3N/2} . \tag{3.11.1}$$

Hence follows the expression for the free energy:

$$F = -\kappa T \ln Z = -\kappa T N \ln V - \frac{3}{2} N\kappa T \ln(2\pi m\kappa T) . \tag{3.11.2}$$

We use this to find the pressure

$$p = \frac{\partial F}{\partial V}\bigg|_T = \frac{N\kappa T}{V} = n\kappa T . \tag{3.11.3}$$

We see that equation of state found with the aid of the canonical Gibbs distribution coincides with the equation established experimentally for a rarefied gas. Accordingly, our model of ideal gas coincides to the phenomenological ideal gas approximation.

The expression for the free energy, however, is not quite satisfactory. As a matter of fact, the expression (3.11.2) does not satisfy condition (3.10.2), because in the thermodynamic limit the first term on the right-hand side of (3.11.2) is proportional to $N \ln N$ rather than to N. Because of this, the condition of additivity is violated.

This snag can be overcome if we take advantage of the indistinguishability of particles and multiply (3.11.1) by $1/N!$. Then for a macroscopic system ($N \gg 1$) we may use the Stirling formula

$$\ln N! = N \ln N - N = N \ln \frac{N}{e} \tag{3.11.4}$$

to bring the expression for the free energy to the form

$$F(V,T) = -\kappa TN \ln\left(\frac{V}{N}e\right) - \frac{3}{2}\kappa TN \ln(2\pi m\kappa T) .$$ (3.11.5)

Now we have V/N in place of N under the logarithm, and the controversy is solved.

Next we take the example of entropy calculation to show that neglect of indistinguishability of particles in the statistical description may lead to nonphysical results.

3.12. Entropy of Ideal Gas. Gibbs Paradox

We use (3.11.2) to find the entropy:

$$S = -\left(\frac{\partial F}{\partial T}\right)_V = \kappa N\left[\ln V + \frac{3}{2}\ln(2\pi m\kappa T) + \frac{3}{2}\right] .$$ (3.12.1)

Again we come to a violation of additivity, and, as a consequence, to the so-called *Gibbs paradox* which can be described as follows.

Let a vessel be divided with a wall into two equal parts of the volume $V/2$ each. The number of particles in each half of the vessel is $N/2$; both parts of the vessel are filled with the same gas. If we remove the wall, the physical state of the system from the point of view of thermodynamics will not change: after all, one and the same gas occupies the vessel with or without the wall.

It would be natural to expect that the entropy of the system will not change after the wall is removed. This will not be the case, however, if we calculate the entropy according to the above formula.

Let S_1 be the entropy of the system with the wall in place. By (3.12.1),

$$S_1 = 2\kappa\frac{N}{2}\left[\ln\frac{V}{2} + \frac{3}{2}\ln(2\pi m\kappa T) + \frac{3}{2}\right] .$$ (3.12.2)

If S_2 is the entropy of the system with the wall removed, then

$$S_2 = \kappa N\left[\ln V + \frac{3}{2}\ln(2\pi m\kappa T) + \frac{3}{2}\right] ,$$ (3.12.3)

whence it follows that

$$S_2 - S_1 = \kappa N \ln 2 \; , \tag{3.12.4}$$

which means that the entropy of the system increases when the wall is taken away.

This result is known as the *Gibbs paradox*. The paradox is that in case of thermodynamic description, when the molecular structure is not taken into account and thus the model of continuous medium is employed, the thermodynamic state remains the same when the partition is removed. One should expect, therefore, that the entropies at states "1" and "2" are the same, and so $S_1 - S_2 = 0$.

From the molecular standpoint, however, there is nothing paradoxical, since the states "1" and "2" are different. Different are, for instance, the times of self-diffusion of atoms of the gas: at state "1" the diffusion takes place in the volume $V/2$, and at state "2" in the volume V. Because of this, the degree of chaoticity increases when the wall is removed, since each atom is free to wander over the space twice as large as before.

This time we calculate the entropy using (3.11.5) for the free energy:

$$S = \kappa N \left[\ln\left(\frac{V}{N}e\right) + \frac{3}{2}\ln(2\pi m\kappa T) + \frac{3}{2} \right]. \tag{3.12.5}$$

We see that this expression satisfies the additivity condition.

Consider three possible situations.

(1) Assume once again that the gas of N particles is distributed uniformly in a vessel divided by a wall. Let its entropy be S_1, as given by (3.12.5). Remove the wall. Since the concentration of particles N/V and their total number N remain the same, the entropy S_2 is the same as S_1, and

$$S_2 - S_1 = 0 \; . \tag{3.12.6}$$

In other words, the entropy does not change, which is quite natural since the physical state of the system in case of thermodynamic description, when self-diffusion is not taken into account, does not change when the wall is removed.

(2) Let the two halves of the vessel be filled with different gases. Note that we do not specify in what respect the gases differ: the difference must just be associated with some information. For instance, the particles may be exactly the same in all their characteristics except for the color. Since the concentrations N/V of the particles on the right and on the left are the same, and their total number N is the same, the entropy S_1 is again given by (3.12.5).

When we remove the wall, the concentration of particles of each gas, now occupying the entire vessel, will be one-half of the initial value, whereas the total number of particles is N, as before. Then

$$S_2 = \kappa N\left[\ln\left(\frac{V}{2N}e\right) + \frac{3}{2}\ln(2\pi m\kappa T) + \frac{3}{2}\right].$$

(3.12.7)

The difference in the entropies is given by (3.12.4). The increase in the entropy is a consequence of *mutual diffusion*, the *mixing* of particles of different color.

(3) Let us now consider an example of self-diffusion, and show that the entropy in this case also increases.

Assume that the vessel is divided in two halves with the wall, while all particles are identical and occupy, for instance, the left-hand half of the vessel. Thus in the initial state we have a *gradient of density*. The concentration of particles is $2N/V$, and the total number of particles is N, so the entropy S_1 is given by

$$S_1 = \kappa N\left[\ln\left(\frac{V}{2N}e\right) + \frac{3}{2}\ln(2\pi m\kappa T) + \frac{3}{2}\right].$$

(3.12.8)

When we remove the wall, the concentration of particles will become uniform throughout the vessel because of self-diffusion. As a result, spatially homogeneous distribution of identical particles will be established, with the entropy given by (3.12.5). The difference in the entropies of the initial and final states is

$$S_2 - S_1 = \kappa N \ln 2 .$$

(3.12.9)

So, we have considered two irreversible processes: self-diffusion of particles of two different sorts, and self-diffusion of identical particles whose initial distribution is not homogeneous. In both cases the entropy increased by one and the same amount. There is, however, an important dissimilarity between the two irreversible processes. The process of mutual diffusion occurs on molecular level. Unlike that, self-diffusion only occurs in the presence of density gradient — that is, in the presence of a "thermodynamic force".

In chapters to follow we shall repeatedly raise these questions in connection with the analysis of nonequilibrium processes in open systems. In the next chapter, however, we continue our discussion of the statistical theory of equilibrium state.

CHAPTER 4

DISTRIBUTIONS OF FUNCTIONS OF DYNAMIC VARIABLES.
FLUCTUATIONS OF INTERNAL PARAMETERS

This chapter ends that part of the book which deals with the statistical theory of the state of equilibrium. From the diverse material belonging to this domain of statistical physics we carefully select whatever is necessary for the construction of the theory of nonequilibrium processes in open systems. From time to time we will need some additional knowledge from the statistical theory of equilibrium state — for instance, when dealing with the quantum theory and the theory of phase transitions. Currently, however, we are going to treat two fundamental topics from the classical equilibrium theory.

We know by now that all thermodynamic functions can be regarded either as the mean values $\langle A \rangle$ of the relevant functions of dynamic variables, or as the most probable value $A_{\text{m.p.}}$ of variable A. In either case we have to deal with fluctuations relative to these selected values. This is one of the tasks of the present chapter.

Towards the end of this chapter we will generalize the Gibbs theorem (see Sect. 3.6) for the case of arbitrary distributions of values of dynamic variables. In place of the Hamilton function of the system $H(X,a)$ we will use the effective Hamilton function $H_{\text{eff}}(A,a,T)$ which does not have any mechanical analogy, being a function of not only a and A, but also of the temperature T. This result will serve as starting point in introducing the criterion of self-organization, based on juxtaposition of entropies of nonequilibrium states of open systems at a fixed mean value of H_{eff}.

4.1. Distribution Functions of Values of Dynamic Variables

In Sect. 2.1 we introduced notation $A(X)$ for an arbitrary function of a set of dynamic variables X which characterize the microscopic state of the system under consideration. Examples of such functions were given in Sect. 2.1.

We also discussed a few examples of local functions of dynamic variables. In particular, we spoke of the dynamic distribution (2.3.1) in $6N$–dimensional phase space, and of the microscopic density (2.4.1) in 6–dimensional phase space $x = (r,p)$.

It is convenient to use a unified notation $A(r,p,X)$ for a local function of dynamic

68

variables X of the system. The distribution function of its values $A(r,p)$ is expressed in terms of the most general distribution function $f_N(X,t)$:

$$f(A(r,p),t) = \int \delta(A(r,p) - A(r,p,X)) f_N(X,t) dX . \tag{4.1.1}$$

Function f introduced in this way is positive and normalized to unity, because

$$\int f(A(r,p),t) dA = \int f_N(X,t) dX = 1 , \tag{4.1.2}$$

which gives us good reason to call it the distribution function of the value of A at a given point of phase space $x = (r,p)$.

This definition is general since it refers to an arbitrary nonequilibrium distribution $f_N(X,t)$. In case of equilibrium state the concrete form of this function can be specified; further we shall use, as a rule, the canonical Gibbs distribution.

Let us consider some examples of equilibrium distributions.

4.2. Distribution Function of Values of Internal Energy

For the function of dynamic variables we take the Hamilton function $H(X,a)$ of the system under consideration. For f_N we choose the canonical Gibbs distribution in the form of (3.2.4) or (3.2.6). The possible value of internal energy is E. Then the sought-for distribution can be written as

$$f(E,a,T) = \int \delta(E - H(X,a)) f_N(X,a,T) dX , \tag{4.2.1}$$

or, explicitly,

$$f(E,a,T) = \exp\frac{F(a,T) - E}{\kappa T} \int \delta(E - H(X,a)) dX . \tag{4.2.2}$$

This distribution can be used for finding different moments of the values of internal energy. For instance, the mean internal energy is expressed via the statistical integral:

$$U(a,T) = \int E f(E,a,T) dE = \kappa T^2 \frac{\partial \ln Z(a,T)}{\partial T} . \tag{4.2.3}$$

We shall also need the expression for the dispersion (variance) of values of internal energy:

$$\left\langle (\delta E)^2 \right\rangle = \int (E - \langle E \rangle)^2 f(E,a,T) dE = \kappa T^2 \left(\frac{\partial \langle E \rangle}{\partial T} \right)_a . \tag{4.2.4}$$

We see that the dispersion of energy values depends on the temperature and on the derivative of internal energy. This allows obtaining a general estimate for the relative fluctuations of the internal energy. Indeed, in the thermodynamic limit (see Sect. 3.10) the internal energy is proportional to the number of particles N. Accordingly,

$$\frac{\left\langle (\delta E) \right\rangle^2}{\langle E \rangle^2} \sim \frac{1}{N} . \tag{4.2.5}$$

Hence it follows that energy fluctuations of the system on the whole are very small and, as a rule, negligible. The situation, however, is quite different when it comes to calculating the local functions. Then the number of particles N in the above expressions must be replaced by the number of particles N_{ph} within physically infinitesimal volume (see Sect. 1.1.3, 1.1.4).

Now let us give concrete formulas for the case of ideal gas.

For the mean energy and dispersion about the mean we may use (3.11.1), since the $1/N!$ factor is then not significant. This brings us to the well-known results:

$$U(a,T) = \frac{3}{2} \kappa TN , \quad \left\langle (\delta E) \right\rangle^2 = \frac{3}{2} \kappa T^2 N \equiv c_v T^2 N , \tag{4.2.6}$$

where c_v is heat capacity at constant volume per particle.

4.3. Mean and Most Probable Energy Values

Giving statistical foundation to the laws of thermodynamics (see Sect. 3.3, 3.4), we assumed that the thermodynamic functions are represented by the mean values of the relevant functions of dynamic variables. Internal energy, for instance, was defined as the mean value of the Hamilton function $H(X,a)$.

There is, however, an alternative way of defining the thermodynamic functions. As the internal energy, for example, we can take the most probable value of the distribution (4.2.2), which we denote by $E_{m.p.}$.

Now let us evaluate the difference between U and $E_{m.p.}$ for an ideal gas, and show that this difference in the thermodynamic limit is zero. In this way we may justify (if only to some extent) such definition of thermodynamic functions.

Recall that the mean energy of ideal gas is given by (4.2.6). In order to find $E_{m.p.}$ we turn to the distribution (4.2.2), where the Hamilton function is now the sum of kinetic energies. The V^N factor takes care of the integration with respect to coordinates; integration with respect to momenta is carried out over the surface of hypersphere in $3N$–dimensional space. As a result, we come to the following dependence of the distribution function on energy:

$$f(E) = \exp\left(-\frac{E}{\kappa T}\right) E^{\frac{3N}{2}-1} \ , \ \text{whence} \ \ E_{\text{m.p.}} = \left(\frac{3N}{2}-1\right)\kappa T \ .$$

(4.3.1)

Then

$$U - E_{\text{m.p.}} = \kappa T \ , \ \text{and therefore} \ \lim_{N\to\infty} \frac{1}{N}\left(U - E_{\text{m.p.}}\right) = 0 \ .$$

(4.3.2)

This example demonstrates that the mean and the most probable values of the internal energy are the same for the equilibrium state of ideal gas, thus justifying the definition of thermodynamic internal energy as the mean value of the Hamilton function. There are, however, two questions which have to be answered.

Does the result of (4.3.2) hold for nonideal systems — that is, for arbitrary interaction between the atoms, like in dense fluids.

Does the equality of the mean and the most probable values apply to arbitrary thermodynamic functions in the thermodynamic limit?

We shall see that the answer to both these questions is negative. To wit, the equality

$$\lim_{N\to\infty} \frac{1}{N}\left(\langle A\rangle - A_{\text{m.p.}}\right) = 0$$

(4.3.3)

only holds for those distribution functions $f(A)$ of values of internal parameters A which exhibit one sharp maximum.

In chapters devoted to the theory of phase transitions we show that the distribution functions of values of macroscopic internal parameters may have several maxima. In such cases the statistical definition of thermodynamic quantities must be given in a different way.

4.4. Distribution Function of Values of Entropy

Let us go back to equation (4.1.1), which allows us to find the distribution function of any internal parameter from the many-particle distribution $f_N(X,t)$ for both equilibrium and nonequilibrium states. This requires knowing the form of the relevant function of dynamic variables X. For the internal energy, for instance, this is the Hamilton function. In order to find the function of dynamic variables corresponding to the entropy, we turn to equation (3.4.5) obtained by Gibbs.

This formula defines the mean entropy as the mean logarithm of the distribution function, thereby defining also the sought-for function of dynamic variables:

$$S(X,a,T) = -\kappa \ln f_N(X,a,T) .$$ (4.4.1)

We see that this function of dynamic variables for the equilibrium state depends not only on X and a, but also on the temperature T.

Thus, the function of distribution of entropy values for the equilibrium state is given by

$$f(S,a,T) = \int \delta(S + \kappa \ln f_N(X,a,T)) f_N(X,a,T) dX .$$ (4.4.2)

This distribution can be used for finding the expression for the mean entropy $S(a,T)$, which obviously coincides with the Gibbs formula (3.4.5). The same can be written in an equivalent form:

$$S(a,T) = \int Sf(S,a,T) dS = \frac{F(a,T) - U(a,T)}{T} .$$ (4.4.3)

Now we use (4.4.2) to find the dispersion of entropy. First we obtain the expression for the mean square of entropy:

$$\left\langle S^2 \right\rangle = \int \left(\frac{F(a,T) - U(a,T)}{T} \right)^2 f_N(X,a,T) dX .$$ (4.4.4)

Using the expression for the mean entropy value, we can express the dispersion of entropy in terms of the dispersion of internal energy:

$$\left\langle (\delta S)^2 \right\rangle = \frac{\langle (\delta E) \rangle^2}{T^2} = \kappa \frac{\partial U(a,T)}{\partial T} = \kappa c_a N .$$ (4.4.5)

Here we have used (4.2.4) for the dispersion of internal energy; c_a is the heat capacity per particle at fixed a. In particular, it can be heat capacity at constant volume if $a = V$ (cf. (4.2.6)). We see that for calculating the dispersion of entropy it suffices to know the caloric equation of state in $U(a,T)$.

4.5. Distributions of Local Functions of Dynamic Variables

Now we return to (4.1.1), which defines the distribution function of the values of local function of dynamic variables. Since the distribution depends on the position of only one point, it may be called the one-point distribution.

Let us give some examples of one-point distributions.

For the local function of dynamic variables we take the microscopic phase density (2.4.3) in six-dimensional phase space $x = (r,p)$. Then the relevant one-point distribution is given by

$$f(N(x,t)) = \int \delta\left(N(x) - \sum_{1 \leq i \leq N} \delta(x - x_i) \right) f_N(X,t) dX .$$ (4.5.1)

The one-point distribution function is the simplest in a broad class of complicated and informative distributions of local functions of dynamic variables. One step higher are the two-point distributions, which are constructed in a similar way. The difference is that delta-function will be applied to the product $N(x)N(x')$, and the single sum will be replaced by the double sum over i, j.

Observe also that the time enters the distribution (4.5.1) along with parameter x. If the functions of dynamic variables in a two-point distribution correspond to one and the same time, we speak of the one-time distribution function. Two-time distributions are more complicated and informative.

Let us write the normalization condition for the distribution function $f(N(x))$:

$$\int f(N(x),t) dN(x) = \int f_N(X,t) dX = 1 ,$$ (4.5.2)

and consider also the expression for the first moment

$$\langle N(x,t) \rangle = \int N(x) f(N(x),t) dN(x) = \int \sum_{1 \leq i \leq N} \delta(x - x_i) f_N(X,t) dX .$$ (4.5.3)

This mean value can be expressed in terms of a simpler distribution function. As a matter of fact, all terms in the summand are equivalent with respect to the integration, and so the calculation can be based on the more simple one-particle distribution function:

$$f_1(x_1,t) = V \int f_N(x_1,...,x_N,t) dx_2...dx_N , \quad \frac{1}{V} \int f_1 dx_1 = 1 .$$ (4.5.4)

This function is defined by the integration of N-particle distribution f_N with respect to all variables except, for instance, the variables of particle number one. The integral is multiplied by V with a view to a more convenient representation of our future results. The expression for $\langle N(x,t) \rangle$ can now be written in the form

$$\langle N(x,t) \rangle = n \int \delta(x - x_1) f_1(x_1,t) dx_1 = n f_1(x,t) , \quad n = \frac{N}{V} . \tag{4.5.5}$$

We see that the mean phase density in 6-dimensional phase space is proportional to the one-particle distribution function.

Now let us consider examples of expressions for the mean values of the more simple local functions.

4.6. Mean Density in Space of Momenta: Maxwell Distribution

Recall that the simpler local functions (2.4.5), (2.4.6) are expressed via the local function of dynamic variables $N(r,p,t)$ (2.4.1). They define the densities of various physical quantities in the conventional space r. Let us now define the density in the space of momenta:

$$n^m(p,t) = \int N(r,p,t) dr = \sum_{1 \le i \le N} \delta(p - p_i(t)) . \tag{4.6.1}$$

The mean value of this microscopic density is expressed in terms of the one-particle distribution function of the values of momenta (cf. definition (4.5.4) of function $f_1(x_1,t)$):

$$f_1(p_1,t) = \int f_N(r_1,...,r_N,p_1,...,p_N) dr_1 ... dr_N dp_2 ... dp_N . \tag{4.6.2}$$

The sought-for relationship follows from the last two equations:

$$n(p,t) = N f_1(p,t) . \tag{4.6.3}$$

This equation is exact because the one-particle distribution function $f_1(p,t)$ is defined via the most general distribution function of variables of all N particles of the system. Now let us particularize this result for the state of equilibrium.

With this purpose we substitute the canonical Gibbs distribution (3.2.4) into (4.6.2). If the Hamilton function is given by (2.1.3) and is therefore the sum of kinetic and potential energies, then in (4.6.2) the integrals with respect to coordinates in the numerator and denominator cancel out. Furthermore, the integrals with respect to momenta of all particles but the first also cancel out, because the integral with respect to momenta reduces to the product of integrals for individual particles. As a result, we find the expression for the mean distribution of particles with respect to momenta in the state of equilibrium (subscript "1" is dropped):

$$n(p) = Nf(p) , \quad f(p) = \frac{1}{(2\pi m\kappa T)^{3/2}} \exp\left(-\frac{p^2}{2m\kappa T}\right) ,$$

(4.6.4)

with the normalization condition

$$\int f(p)dp = 1 .$$

(4.6.5)

This is the famous Maxwell distribution. It can be used for finding the moments of distribution with respect to momenta — for example,

$$\langle p \rangle = 0 , \quad \left\langle \frac{p^2}{2m} \right\rangle = \frac{3}{2}\kappa T .$$

(4.6.6)

From arguments developed above it follows that Maxwell distribution holds for arbitrarily strong interaction between particles. Because of this, it holds for gases and liquids provided that the motion of particles obeys the laws of classical mechanics. In quantum systems the distribution can be of quite different nature, and it is not possible to obtain general results which would hold for arbitrary interaction. In case of quantum ideal degenerate gases the Maxwell distribution is replaced by Bose and Fermi statistics.

4.7. Boltzmann Distribution

Let us now find the distribution of the mean density of number of particles in conventional space. The relevant microscopic density is defined by (2.4.5):

$$n^m(r,t) = N(r,p,t)dp = \sum_{1 \le i \le N} \delta(r - r_i(t)) .$$

(4.7.1)

Using the canonical Gibbs distribution (3.2.4) we average this expression, and introduce the appropriate one-particle distribution function (cf. (4.5.4), (4.6.2)):

$$f_1(r_1,t) = V\int f_N(r_1,...,r_N,p_1,...,p_N,t)dr_2...dr_N dp_1...dp_N .$$

(4.7.2)

Using this definition for the equilibrium state, we get the following expression for the mean density:

$$n(r) = \left\langle \int \sum_{1 \le i \le N} \delta(r - r_i) \right\rangle = n f_1(r) , \quad \frac{1}{V} \int f_1(r) dr = 1 . \tag{4.7.3}$$

In order to find function f_1, we substitute the canonical Gibbs distribution into (4.7.2). The integrals with respect to momenta in the numerator and denominator cancel out. Integration over coordinates cannot be carried out in the general case. Let us therefore consider two special cases.

(1) External field is absent. Then the potential energy in the Hamilton function only depends on the interaction between particles. Since this interaction is a function of the distance between particles, the integration with respect to coordinates is facilitated by the change of variables $R_2 \Rightarrow r - r_2,..., R_N \Rightarrow r - r_N$. The integrals with respect to $R_2,...,R_N$ cancel out, and the expression for $n(r)$ takes on the form

$$n(r) = \frac{N}{\int dr_1} = \frac{N}{V} , \quad f_1(r) = 1 . \tag{4.7.4}$$

In the presence of external field the spatial distribution is no longer homogeneous. Integration with respect to coordinates, however, cannot be carried out in the general case.

(2) Ideal gas in external field. In this case the potential energy of interaction is zero, and so the integrals with respect to $r_2,..., r_N$ in the numerator and denominator cancel out. As a result, we come to the distribution

$$n(r) = n f(r) , \quad f(r) = \frac{\exp\left(-\dfrac{U(r)}{\kappa T}\right)}{\dfrac{1}{V} \int \exp\left(-\dfrac{U(r)}{\kappa T}\right) dr} , \tag{4.7.5}$$

called the Boltzmann distribution and defining the spatial distribution of ideal gas particles in external field.

An example of Boltzmann distribution is the well known barometric formula which defines the distribution of classical ideal gas particles in the field of gravity, when $u = mgz$.

The combined Maxwell – Boltzmann distribution also holds for classical ideal gas for the distribution function $f(r,p) = f(r)f(p)$.

So, we have used the canonical Gibbs distribution to obtain the expressions for the mean densities $n(p)$ and $n(r)$. In the same manner one can derive expressions for the higher moments — for instance, for the dispersion of the local functions of dynamic variables. We shall do this in Ch. for nonequilibrium states. Now we are going to give some useful formulas for calculating the fluctuations.

4.8. Gibbs Formulas

(1) Let $A(X,a)$ be a function of dynamic variables and external parameters. We express the mean of this function via the canonical Gibbs distribution and differentiate with respect to T. After some straightforward algebra we get one of the Gibbs formulas:

$$\frac{\partial \langle A(X,a) \rangle}{\partial T} = \frac{1}{\kappa T^2} [\langle AH \rangle - \langle A \rangle \langle H \rangle] \ . \tag{4.8.1}$$

To calculate the fluctuations of, for instance, internal energy, we set $A(X,a) = H(X,a)$ and arrive at the earlier obtained expression (4.2.4).

(2) Fluctuations of microscopic force. Recall that the microscopic force which corresponds to an external parameter a is given by (3.3.5). The thermodynamic force is found by averaging over the canonical Gibbs distribution. In the same way we obtain the dispersion of the microscopic force:

$$\left(\langle \delta F \rangle^2 \right) = \kappa T \left(\frac{\partial \langle F \rangle}{\partial a} - \left\langle \frac{\partial F^m}{\partial a} \right\rangle \right) \ . \tag{4.8.2}$$

This formula is more complicated than (4.2.4) since it includes, besides the derivative of the mean force, the mean derivative of the microscopic force. Because of this, the calculation of fluctuations of microscopic force is not possible for the general case.

The situation is much simplified when the Hamilton function is a linear function of the external parameter,

$$H(X,a) = H(X) - F^m(X)a \ . \tag{4.8.3}$$

Then

$$\left\langle (\delta F)^2 \right\rangle = \kappa T \frac{\partial \langle F \rangle}{\partial a} \ , \tag{4.8.4}$$

which is entirely similar to (4.2.4) for the dispersion of internal energy.

Expressions (4.2.3), (4.2.4) are useful for many practical purposes. The external parameter a can be represented, for instance, by the constant electric field or constant magnetic field. Then (4.8.4) will define the dispersion of polarization vector, or the dispersion of magnetization vector.

4.9. Fluctuations of Volume and Pressure

In order to use (4.8.4) for calculating the dispersion of the volume, we must use the pressure as the external parameter. This can be accomplished as follows.

Fluctuations of volume. Imagine a cylinder filled with gas. For the external parameter we take the pressure on the piston from the side of surrounding medium, $a = P$. The second external parameter is the temperature T. Fluctuations of the volume correspond to thermal motion of the piston; the momentum of the piston is included into the set of variables X. Then the Hamilton function can be represented as

$$H(X,P) = H(X) + PV ,\tag{4.9.1}$$

where we assume that the role of the microscopic force is played by $-V$, since the amount of work of the system is $-VdP$. Accordingly,

$$F^m = -V , \text{ and therefore } F(P,T) = -\langle V \rangle .\tag{4.9.2}$$

As a result, we obtain the following expression for the dispersion of volume:

$$\left\langle (\delta V)^2 \right\rangle = -\kappa T \left(\frac{\partial \langle V \rangle}{\partial P} \right)_T .\tag{4.9.3}$$

Fluctuations of pressure. In connection with the calculation of local hydrodynamic functions in Ch. we shall see that fluctuations of pressure are important for fast processes, when the adiabatic approximation can be used. It is natural therefore to choose, along with $a = V$, the entropy as the second independent parameter in calculations of pressure fluctuations. Then in (4.8.4) we have

$$F^m = -P , \quad F = -\langle P \rangle ,\tag{4.9.4}$$

and therefore the dispersion of pressure is

$$\left\langle (\delta P)^2 \right\rangle = -\kappa T \left(\frac{\partial \langle P \rangle}{\partial V} \right)_S .$$

(4.9.5)

We see that the dispersion of pressure is proportional to the derivative of the mean pressure with respect to the volume at constant entropy.

From the above discussion it follows that the formulas for fluctuations fall into two classes. First, it is the exact formulas, which include expressions for the dispersion of internal energy and the dispersion of entropy.

The second class includes, for instance, expressions for fluctuations of volume and pressure. These are approximate because they are derived under assumption that the Hamilton function is a linear function of the external parameter (pressure or volume, respectively). This constraint corresponds to the Gaussian approximation for fluctuations, which certainly does not hold in the neighborhood of the critical point, as can be seen from the expression for fluctuations of volume.

Indeed, at the critical point the isothermal compressibility is infinite and so the formula predicts infinite fluctuations of volume, which is obviously nonsensical. Correct calculation of fluctuations in the critical region must take into account deviations from the Gaussian distribution. This point will be discussed in Ch. 21 (see also Klimontovich V).

4.10. Boltzmann Principle

Now we introduce, along with the thermodynamic free energy, the conditional free energy at fixed value of some function of dynamic variables.

We go back to the general equation (4.4.1), which defines the distribution function of a local function of dynamic variables X for an arbitrary nonequilibrium state, and rewrite it for the equilibrium distribution of a simpler function of dynamic variables $A(X)$:

$$f(A,a,T) = \exp\frac{F(a,T)}{\kappa T} \int \delta(A - A(X)) \exp\left(-\frac{H(X,a)}{\kappa T} \right) dX .$$

(4.10.1)

Recall that the free energy is related to the statistical integral by (3.2.7), which we now rewrite in the explicit form:

$$F(a,T) = -\kappa T \ln \int \exp\left(-\frac{H(X,a)}{\kappa T} \right) dX .$$

(4.10.2)

The conditional free energy

$$F(a,T|A) = -\kappa T \ln \int \delta(A - A(X)) \exp\left(-\frac{H(X,a)}{\kappa T}\right) dX \qquad (4.10.3)$$

differs from the thermodynamic free energy in that the integration with respect to X is constrained by condition $A = A(X)$. The initial distribution (4.10.1) for the values of function of dynamic variables can be rewritten in an equivalent form

$$f(A,a,T) = \exp\frac{F(a,T) - F(a,T|A)}{\kappa T}. \qquad (4.10.4)$$

Hence follows a practically important relationship between the distribution functions of two arbitrary values A_1, A_2 of function of dynamic variables $A(X)$:

$$f(A_1,a,T) = f(A_2,a,T)\exp\frac{F(a,T|A_2) - F(a,T|A_1)}{\kappa T}. \qquad (4.10.5)_1$$

This statement is known as the Boltzmann principle. Interestingly enough, it had been formulated before the Gibbs' classic, *Elementary Principles in Statistical Mechanics*.

The Boltzmann principle implies that the more probable state corresponds to a lesser value of conditional free energy. As before, by $A_{m.p.}$ we denote the most probable state. Then the probability of an arbitrary state A is related to the distribution function for the most probable state by

$$f(A,a,T) = f\left(A_{m.p.},a,T\right)\exp\frac{F\left(a,T|A_{m.p.}\right) - F(a,T|A)}{\kappa T}. \qquad (4.10.5)_2$$

Hence it follows that the most probable value of A corresponds to the minimum value of conditional free energy. In the next section we shall use this result for calculating the fluctuations of thermodynamic quantities in Gaussian approximation. Here we just mark the following.

The distribution function (4.10.1) (or (4.10.4)) can be represented as the canonical Gibbs distribution,

$$f(A,a,T) = \exp\frac{F(a,T) - H_{eff}(A,a,T)}{\kappa T}, \qquad (4.10.6)$$

where

$$H_{\text{eff}}(A,a,T) = F(a,T|A) \qquad (4.10.7)$$

is the effective Hamilton function, which has no exact counterpart in mechanics since it depends not only on A and a, but also on T. By virtue of (4.10.7), the effective Hamilton function is related to the conventional Hamilton function by (4.10.3).

Using the effective Hamilton function, we can rewrite mathematical formulation of the Boltzmann principle as

$$f(A,a,T) = f\left(A_{\text{m.p.}},a,T\right)\exp\frac{H_{\text{eff}}\left(A_{\text{m.p.}},a,T\right) - H_{\text{eff}}(A,a,T)}{\kappa T}. \qquad (4.10.8)$$

Thus, the most probable value of A corresponds to the least value of the effective Hamilton function.

The Boltzmann principle and the effective Hamilton function will be much used in describing the equilibrium and nonequilibrium phase transitions which, as emphasized in the Introduction, are inseparable from processes of self-organization. The concept of effective Hamilton function has been use in the Introduction for formulating the entropy-based criterion of self-organization (the S-theorem).

4.11. Gaussian Distribution for Fluctuations of Values of Internal Parameters

We saw that there are two equivalent formulations of the Boltzmann principle based on the conditional free energy and the effective Hamilton function. Exact calculation of either of these is, as a rule, not possible because of the enormous complexity of multidimensional integrals with respect to X. There are, however, two practical ways of approximating these functions based on mathematical models or on direct experimental data. Physically, the method based on the effective Hamilton function is more adequate.

Let us return to (4.10.8) and expand the effective Hamilton function in small deviation of A from its most probable value. Since the minimum of the effective Hamilton function corresponds to the most probable state, we come to the Gaussian distribution about the most probable value:

$$f(A,a,T) = f\left(A_{\text{m.p.}},a,T\right)\exp\left(-\frac{A - A_{\text{m.p.}}^2}{2\left\langle(\delta A)^2\right\rangle}\right), \qquad (4.11.1)$$

where

$$\left\langle (\delta A)^2 \right\rangle = \frac{\kappa T}{\left. \dfrac{\partial^2 H_{\text{eff}}}{\partial A^2} \right|_{A=A_{\text{m.p.}}}} \tag{4.11.2}$$

is the dispersion.

Now we face the problem of normalization of the distribution function. Here we may distinguish two cases.

(1) The distribution function of A has one sharp maximum. The normalization condition has the usual form

$$\int_{-\infty}^{+\infty} f(A,a,T)dA = 1 \; ; \quad \text{then} \quad f\left(A_{\text{m.p.}},a,T\right) = \left(\frac{1}{2\pi\left\langle (\delta A)^2 \right\rangle} \right)^{\!\!1/2}, \tag{4.11.3}$$

and the expression for the distribution function is

$$f(A,a,T) = \frac{1}{\left(2\pi\left\langle (\delta A)^2 \right\rangle\right)^{1/2}} \exp\left(-\frac{\left(A - A_{\text{m.p.}}\right)^2}{2\left\langle (\delta A)^2 \right\rangle} \right). \tag{4.11.4}$$

(2) The distribution function of A has a number ($i = 1,2,...,n$) of prominent maxima. Then the expansion must be done for each maximum, and (4.11.1) will be replaced by n Gaussian distributions corresponding to each maximum,

$$f_i(A,a,T) = f_i\left(\left(A_{\text{m.p.}}\right)_i,a,T\right) \exp\left(-\frac{\left(A - \left(A_{\text{m.p.}}\right)_i\right)^2}{2\left\langle (\delta A)^2 \right\rangle_i} \right). \tag{4.11.5}$$

The normalization condition now includes contributions from all n distributions,

$$\sum_i \int_{-\infty}^{+\infty} f_i(A,a,T)dA = 1 . \tag{4.11.6}$$

By way of example, let us consider a case when there are two symmetrical maxima, and the dispersions are the same:

$$\left(A_{\text{m.p.}}\right)_i = \pm A_0 \ , \ \left\langle (\delta A)^2 \right\rangle_i = \sigma \ , \ i = 1,2 \ . \tag{4.11.7}$$

Distribution (4.11.5), with due account for the normalization condition, takes on the form

$$f(A,a,T) = \frac{1}{2} \frac{1}{(2\pi\sigma)^{\frac{1}{2}}} \left[\exp\left(-\frac{(A - A_0)^2}{2\sigma} \right) + \exp\left(-\frac{(A + A_0)^2}{2\sigma} \right) \right] . \tag{4.11.8}$$

We see that, because of the symmetry of the distribution, the mean value $\langle A \rangle$ is zero, while the most probable values are nonzero.

In this example, therefore, equation (4.3.3) does not hold, and the definition of the thermodynamic quantity as the mean value is inadequate. We shall encounter more of such cases in the theory of equilibrium and nonequilibrium phase transitions.

So far we have confined our examples to one internal parameter. Many-dimensional Gaussian distributions are introduced in a similar way. We shall use them as necessary in connection with the theory of Brownian motion and theory of nonequilibrium phase transitions.

4.12. Fluctuations of Number of Particles. Poisson Distribution

In our system we select a volume V, and consider the problem of fluctuations of the number of particles N_V within this volume. Since $N_V = nV$, we use (4.9.3) and obtain the following expression:

$$\left\langle (\delta N_V)^2 \right\rangle = \left\langle N_V \right\rangle \frac{\kappa T}{\left(\dfrac{\partial P}{\partial n} \right)_T} \ , \tag{4.12.1}$$

which, like (4.9.3) and (4.9.5), only holds in the approximation of Gaussian distribution.

We see that the dispersion of the number of particles depends considerably on the isothermal compressibility of the medium. For ideal gas, the second multiplier on the right-hand side is unity, and the expression for the dispersion of the number of particles takes on the form

$$\left\langle (\delta N_V)^2 \right\rangle = \left\langle N_V \right\rangle \ . \tag{4.12.2}$$

Let us now write the Gaussian distribution for the distribution function of the number of particles:

$$f(N_V) = \frac{1}{\left(2\pi\left\langle(\delta N_V)^2\right\rangle\right)^{1/2}} \exp\left(-\frac{\left(N_V - \langle N_V \rangle\right)^2}{\left\langle(\delta N_V)^2\right\rangle}\right).$$

(4.12.3)

The Gaussian distribution only holds provided that the number of particles N_V is much greater than one. This condition ensures that the relative fluctuations are small. For ideal gas it is possible to establish a more general distribution which holds also when N_V is small. It is the Poisson distribution

$$f(N_V) = \frac{\langle N_V \rangle^{N_V}}{N_V!} \exp\left(-\langle N_V \rangle\right),$$

(4.12.4)

where $\langle N_V \rangle$ is the mean number of particles for the Poisson distribution. Dispersion for the Poisson distribution is linked to the mean number of particles by the same expression (4.12.2) as obtained for the Gaussian distribution for ideal gas.

It is important that the Gaussian distribution holds also for nonideal gas. The only correction is that the dispersion of the number of particles is then given by the more complicated formula (4.12.1) (see Ch. 8). Naturally, the Poisson distribution can as well be extended to nonideal systems, although the character of this distribution will be considerably modified (Stratonovich 1961, 1967).

4.13. Generalization of Gibbs Theorem for Distributions of Values of Functions of Dynamic Variables

Let us go back to (4.10.6) for the distribution function of the values A of the function $A(X)$ in the state of equilibrium. Using the effective Hamilton function $H_{\text{eff}}(A,a,T)$ it can be represented in the form of canonical Gibbs distribution. Being a function of the temperature, the effective Hamilton function has no mechanical analogy.

By analogy with the canonical Gibbs distribution, we can generalize the Gibbs theorem for distributions of values of arbitrary functions of dynamic variables. Recall that the statement of the Gibbs theorem is expressed by inequality (3.6.5) under the two conditions of (3.6.1) and (3.6.2).

Let us formulate similar conditions for the equilibrium distribution function (4.10.6) and the distribution function of nonequilibrium state $f(A,t)$ which describes, for instance, the time evolution towards the state of equilibrium:

$$\int f(A,a,T)dA = \int f_1(A,t)dA \ , \tag{4.13.1}$$

$$\int H_{\text{eff}}(A,a,T)f(A,a,T)dA = \int H_{\text{eff}}(A,a,T)f_1(A,t)dA \ . \tag{4.13.2}$$

It is assumed therefore that the mean value of the effective Hamilton function does not change in the course of time evolution towards the equilibrium.

Next we introduce the entropies of equilibrium and nonequilibrium states:

$$S(a,T) = -\kappa \int \ln f(A,a,T)f(A,a,T)dA \ , \tag{4.13.3}$$

$$S_1(t) = -\kappa \int \ln f_1(A,t)f_1(A,t)dA \ . \tag{4.13.4}$$

Under these conditions, the difference in the entropies of equilibrium and nonequilibrium states is

$$S(a,T) - S_1(t) = \kappa \int \ln \frac{f_1(A,t)}{f(A,a,T)} f_1(A,t)dA \geq 0 \ . \tag{4.13.5}$$

So, we have obtained the result similar to (3.6.5) which expresses the statement of the Gibbs theorem. This time, however, the entropies are being compared not at constant Hamilton function of the system, bur rather at constant *effective* Hamilton function which is a function of the temperature. This generalization is useful for the formulation of entropy-based criterion of self-organization.

The effective Hamiltonian (the more apt term in the classical theory is the effective Hamilton function) was first introduced by L.D. Landau in 1958 in his theory of phase transitions (see Sect. 147 in Landau – Lifshitz 1976). The function of dynamic variables $A(r,p,t)$ in Landau's theory was represented by the class of local functions $\eta(r)$ regarded as "parameters of order" (see Ch. 18).

Observe that the effective Hamilton function as introduced here is entirely general, since no restrictions are imposed on the choice of functions of dynamic variables. The effective Hamilton function will hereinafter play an important role, especially in the statistical theory of open systems. It will be worthwhile therefore to demonstrate the use of this concept in the theory of equilibrium state.

4.14. Effective Hamilton Function in Statistical Theory of Equilibrium State

The effective Hamilton function is defined by (4.10.6) in terms of conditional free energy. The latter is in turn defined by (4.10.2). Hence follows the relationship between the effective and the conventional Hamilton functions:

$$H_{eff}(A,a,T) = -\kappa T \ln \int \delta(A - A(X)) \exp\left(-\frac{H(X,a)}{\kappa T}\right) dX .$$
(4.14.1)

We see that the effective Hamilton function carries information about the microstates of particles of the system only in that part of the phase space which satisfies condition $A = A(X)$. This information is complete only when

$$A(X) = X , \text{ and therefore } H_{eff} = H(X,a) .$$
(4.14.2)

Because of this, the distribution function $f(A,a,T)$, defined by (4.10.6) in terms of the effective Hamilton function, carries information about the statistical state only for a restricted region of the phase space. This function will coincide with the canonical Gibbs distribution only under condition (4.14.2).

Recall that in Ch. 3 we used the canonical Gibbs distribution to give a statistical definition of all thermodynamic functions in the first law of thermodynamics, and offered statistical validation of the second law of thermodynamics for quasistatic processes. In this connection we defined, in particular, the most important thermodynamic functions — the entropy and the free energy.

Now, how will these definitions be modified if we use the distribution function $f(A,a,T)$ in place of the canonical Gibbs distribution?

To answer this question, we go back to distribution (4.10.6). The free energy $F(a,T)$ will be given by the same expression as before, whereas the entropy will be defined by (4.13.3), in a much different way. In order to emphasize the difference, we shall use (by analogy with (3.5.1)) the new notation $S(a,T,[A])$. Then the free energy can be represented as

$$F(a,T) = \langle H_{eff}(A,a,T) \rangle - TS(a,T,[A]) .$$
(4.14.3)

Comparing this equation with (3.4.7), obtained using the canonical Gibbs distribution, we see that the role of internal energy is played by the mean value of the effective Hamilton function, and the role of Gibbs entropy $S(a,T,[X])$ is played by the entropy $S(a,T,[A])$.

Recall that for the canonical Gibbs distribution the differential of internal energy (3.3.4) is represented as a sum of two parts. The first of these accounts for the work of mean thermodynamic forces F_i corresponding to external parameters a_i, and the second represents the contribution of heat transfer. Now the situation is different, because the effective Hamilton function depends not only on the external parameters, but also on the temperature. Equation (3.3.4) therefore takes on the form

$$d_{a,T}\langle H_{eff}(A,a,T)\rangle = \int\left[\sum_{1\leq i\leq n}\left(\frac{\partial H_{eff}}{\partial a_i}da_i\right) + \frac{\partial H_{eff}}{\partial T}dT\right]f(A,a,T)dA$$

$$+\int H_{eff}(A,a,T)d_{a,T}f(A,a,T)dA\ . \tag{4.14.4}$$

Now the equation includes an additional contribution of "thermal force". Because of this, the contributions of thermodynamic forces and heat transfer into the change of the mean effective energy are modified.

Further we shall often use the effective Hamilton function in the theory of phase transitions, and in connection with various aspects of dissipative structures in open systems (nonequilibrium phase transitions).

Now we are through with the statistical theory of equilibrium state, and ready to begin a systematic presentation of the statistical theory of nonequilibrium processes.

CHAPTER 5

METHODS OF DISTRIBUTION FUNCTIONS AND MICROSCOPIC PHASE DENSITY

In the present we start a systematic presentation of the statistical theory of nonequilibrium processes in open systems. Our main task will consist in the transition from the reversible equations of mechanics for the particles of macroscopic systems to the irreversible equations of the statistical theory. Two methods are available for this purpose.

The first is based on the Liouville equation (2.2.5) for the distribution function $f_N(X,t)$ of the possible states of all particles of the system. This method has been developed in the works of N.N. Bogolyubov, M. Born and H. Green, J. Kirkwood, I. Yvon. A most comprehensive presentation of the mathematical background of this method can be found in Bogolyubov's classic entitled *The Dynamic Methods in Statistical Physics*.

The other method relies on the equation (2.4.14) for microscopic phase density (2.4.1), in which the microscopic force is given by (2.4.9).

In the statistical theory, where the complete information about the motion of particles is not available, functions $N(r,p,t)$, $F^m(r,t)$ are treated as random functions. The transition to statistical description requires averaging the equations in these functions over the Gibbs ensemble.

The two approaches are equivalent, and the choice between them is based on considerations of greater convenience for a particular application.

After this brief preamble we go straight to the basics of the statistical theory of nonequilibrium processes, and begin with the method named first.

5.1. Sequence of Distribution Functions

For our starting point we take the Liouville equation (2.2.5) for the distribution function of the states of the system. The solution of this equation is as complicated as the solution of the set of equations of motion (the set of Hamilton equations for all particles of the system). As indicated before, such a solution is not feasible.

Moreover, there is really no need to get this solution. Indeed, the amount of information carried by the many-particle distribution $f_N(X,t)$ is excessive. Much simpler (one-particle and two-particle) distribution functions are entirely suitable for all practical purposes of the statistical theory of nonequilibrium processes.

In this connection we face the problem of obtaining closed equations for these simple distribution functions in the basis of Liouville equation in $f_N(X,t)$.

We denote one-particle and two-particle distribution functions by f_1 and f_2, respectively. They are linked with the many-particle distribution f_N by integral relations

$$f_1(x_1,t) = V \int f_N(x_1,...,x_N,t)dx_2...dx_N \ , \tag{5.1.1}$$

$$f_2(x_1,x_2) = V^2 \int f_N(x_1,...,x_N,t)dx_3...dx_N \ . \tag{5.1.2}$$

Similar relations can be written for the three-and-more-particle distribution functions. Factors V, V^2 are introduced here for a more convenient presentation of our subsequent results.

Normalization conditions for the new functions follow from their definition and the normalization condition for function f:

$$\frac{1}{V} \int f_1(x_1,t)dx_1 = 1 \ , \quad \frac{1}{V^2} \int f_2(x_1,x_2,t)dx_1dx_2 = 1 \ . \tag{5.1.3}$$

Recall that dimensionless distribution functions are used along with the dimensional distribution functions in both classical and quantum statistics. For instance, the many-particle distribution function (assuming that the particles are indistinguishable) is normalized as follows:

$$\frac{1}{(2\pi\hbar)^{3N}} \int f_N(X,t)dX = N! \ . \tag{5.1.4}$$

Dimensionless one, two-and-more-particle distribution functions are defined in a similar way.

Now we demonstrate that the two simplest distribution functions f_1 and f_2 are sufficient for the definition of the main hydrodynamic and thermodynamic functions.

5.2. Connection of Hydrodynamic and Thermodynamic Functions with One-Particle and Two-Particle Distribution Functions

5.2.1. Gasdynamic Functions

Among the local functions of dynamic variables we select the so-called additive functions. The most important of these is the microscopic phase density (2.4.3) in 6-dimensional space $x = (r,p)$. In terms of this density it is possible to express the simpler densities in conventional space r (see (2.4.5), (2.4.6)), as well as, for instance, the microscopic density in the space of momenta (see (4.6.1)).

Recall that the mean value of microscopic phase density (that is, the first moment of random function $N(x,t)$) is defined by (4.5.5) in terms of one-particle distribution function f_1 at the given point of phase space x. Hence it follows that the mean values of local dynamic functions (2.4.5) are expressible in terms of one-particle distribution function f_1: the density of number of particles

$$\rho(r,t) = mn(r,t) = mn\int f_1(r,p,t)dp \; , \tag{5.2.1}$$

the density of momentum

$$\rho(r,t)u(r,t) = n\int pf_1(r,p,t)dp \; , \tag{5.2.2}$$

the density of kinetic energy

$$\frac{K(r,t)}{V} = n\int \frac{p^2}{2m} f_1(r,p,t)dp \; . \tag{5.2.3}$$

In (5.2.2) we introduced notation for the mean local velocity. The local temperature is determined by the mean kinetic energy of relative motion:

$$\frac{3}{2}n(r,t)\kappa T(r,t) = n\int \frac{(p-mu(r,t))^2}{2m} f_1(r,p,t)dp \; . \tag{5.2.4}$$

Local density, velocity and temperature comprise a complete set of variables for the description of nonequilibrium processes in ideal gas, and are therefore referred to as gasdynamic functions. Other local thermodynamic functions of ideal gas can be expressed in terms of these — for instance, the density of internal energy and the pressure:

$$\frac{U(r,t)}{V} = \frac{3}{2}n(r,t)\kappa T(r,t) , \quad P(r,t) = n(r,t)\kappa T(r,t) , \tag{5.2.5}$$

as well as the density of entropy

$$\frac{S(r,t)}{V} = n\left[\ln\frac{e}{n(r,t)} + \frac{3}{2}\ln(2\pi m\kappa T(r,t)) + \frac{3}{2}\right] . \tag{5.2.6}$$

In writing this expression we made use of (3.12.5) assuming its validity for the state of local equilibrium. This assumption will be verified in Sect. 6.3, in connection with definition of the local Maxwell distribution.

The above expressions for thermodynamic functions U, P, S only hold for the ideal gas, when the potential energy of interaction is zero. As shown, the gasdynamic and thermodynamic functions are then expressible via the one-particle distribution function. Let us now consider a more general case of a nonideal system, when the potential energy of interaction is nonzero.

5.2.2. Internal Energy

Assume that the Hamilton function is given by (2.1.3), whereas the external field is zero, $U(r_i) = 0$. If we average this expression using function $f_N(X,t)$, the resulting formula for the internal energy is the sum of the kinetic and potential energies. The local kinetic energy is linked with the one-particle distribution f_1 by (5.2.3). The integral which defines the mean potential energy contains $N(N-1)/2$ terms, each of which can only be expressed via the two-particle distribution function (5.1.2), since each term in the integrand depends on the variables of two particles. Accordingly, the potential energy of interaction is expressed via the two-particle distribution function f_2, and the density of total internal energy of the system (after integration over volume in (5.2.3)) is

$$\frac{U}{V} = n\int\frac{p^2}{2m}f_1(x_1,t)\frac{dx_1}{V} + \frac{1}{2}n^2\int\Phi(|r_1 - r_2|)f_2(x_1,x_2,t)\frac{dx_1dx_2}{V} . \tag{5.2.7}$$

In the second term on the right-hand side we replaced $N(N-1)/V^2$ by n^2.

The expression for the internal energy is considerably simplified in the equilibrium state, when (in the absence of external field) the two-particle distribution function can be represented as

$$f_2(x_1,x_2) = f_2(|r_1 - r_2|)f(p_1)f(p_2) . \tag{5.2.8}$$

The first multiplier on the right-hand side is the so-called radial distribution function; the other two are Maxwell distributions. Substituting this into the general expression for internal energy we obtain the desired result:

$$\frac{U}{V} = \frac{3}{2}n\kappa T + \frac{n^2}{2}\int \Phi(|r|)f_2(|r|)dr \ .$$

(5.2.9)

Observe that the integral with respect to r is finite, and therefore the right-hand side in the thermodynamic limit ($N \gg 1$) the right-hand side is proportional to N. This is in accord with the general requirement of (3.10.2).

5.2.3. Pressure

The pressure for nonideal system is also expressed in terms of f_1, f_2, and is given by

$$P = \frac{n}{3}\int mv^2 f_1(x,t)\frac{dx}{V} - \frac{n^2}{6}\int r\frac{\partial\Phi}{\partial r}f_2(x_1,x_2,t)\frac{dx_1 dx_2}{V} \ , \quad r = |r_1 - r_2| \ .$$

(5.2.10)

This result will be proved at a later point. At equilibrium, (5.2.10) becomes

$$P = n\kappa T\left(1 - \frac{2\pi n}{3\kappa T}\int_0^\infty r\frac{d\Phi(|r|)}{dr}f_2(|r|)r^2 dr\right) \ .$$

(5.2.11)

We see that at the state of equilibrium the pressure is also expressible via the radial distribution function.

5.2.4. Free Energy

Free energy, unlike such characteristics as internal energy, pressure, or entropy, can be consistently defined only for the state of equilibrium. Proof can be found in (Klimontovich IV); in the present book, however, we are not going to use it.

We shall see that in some cases free energy can be introduced for nonequilibrium processes. Examples will be given in Ch. 12, devoted to the theory of Brownian motion, and in Ch. 15.

So, we have shown that the knowledge of one-particle and two-particle distribution functions f_1, f_2 is sufficient for calculation of the main hydrodynamic and thermodynamic functions, which explains our aspiration to obtain equations in these distribution functions.

This will be done in two stages, first for the state of equilibrium, and then for nonequilibrium processes.

5.3. Chain of Equations for Sequence of Equilibrium Distribution Functions

We start with the canonical Gibbs distribution for the equilibrium distribution function $f_N(X,a,T)$. Since the Hamilton function is represented as a sum of kinetic and potential energies, the canonical Gibbs distribution can be written as

$$f_N(X) = f_N(r_1, r_2, ..., r_N) \prod_{1 \le i \le N} f(p_i) , \qquad (5.3.1)$$

where $f(p_i)$ is the Maxwell distribution, f_N is the N-particle distribution function

$$f_N(r_1, ..., r_N) = \frac{1}{Q_N} \exp\left(-\frac{U(r_1, ..., r_N)}{\kappa T} \right) , \qquad (5.3.2)$$

and Q_N is the so-called configuration integral which follows from the normalization condition,

$$Q_N = \int \exp\left(-\frac{U(r_1, ..., r_N)}{\kappa T} \right) dr_1 ... dr_N . \qquad (5.3.3)$$

Basing on this distribution, we introduce a sequence of distribution functions of coordinates for a system of particles $1, ..., s$:

$$f_s(r_1, ..., r_s) = V^s \int f_N dr_1 ... dr_N , \quad V^s \int f_s dr_1 ... dr_s = 1 . \qquad (5.3.4)$$

In order to obtain equations in these functions we proceed as follows. Differentiating with respect to r_1, we convert distribution (5.3.2) into equation

$$\frac{\partial f_N}{\partial r_1} = -\frac{1}{\kappa T} \frac{\partial U(r_1, ..., r_N)}{\partial r_1} f_N . \qquad (5.3.5)$$

From equation in f_N, using definitions of one-particle distribution function and total potential energy of the system, we obtain the desired equation in f_1:

$$\frac{\partial f_1}{\partial r_1} + \frac{1}{\kappa T}\frac{\partial U_0(r_1)}{\partial r_1}f_1 = -\frac{n}{\kappa T}\int\frac{\partial \Phi_{12}}{\partial r_1}f_2(r_1, r_2)dr_2 \ . \tag{5.3.6}$$

The term on the right-hand side accounts for the action exerted on the selected particle from the side of surrounding particles. This term is proportional to $N - 1$, which we replace here by N because the number of particles in the system is very large.

This equation in f_1 is not closed, since it involves the two-particle distribution function. Equation in f_2 will include, in turn, the three-particle distribution function f_3, etc. As a result, the initial equation in f_N is replaced by a chain of equations in f_1, f_2, ... The advantage of the chain of equations over the initial distribution (5.3.2) becomes clear when approximate methods are used, based on the available small parameters.

Consider, for example, a rarefied gas. Then the small parameter is the parameter of density (1.1.12). The use of perturbation theory is justified by the following considerations.

The range of integration with respect to r_2 on the right-hand side of equation in f_1 is determined by the potential of interaction, and is therefore proportional to the volume r_0^3. Since the right-hand side is multiplied by n, it is proportional to the small parameter of density.

In zero approximation with respect to the small parameter of density (approximation of ideal gas) the right-hand side of (5.3.6) is zero, and the equation in f_1 is closed. It is easily proved that in this approximation the equation is satisfied by the Boltzmann distribution.

In order to find the solution in the first approximation with respect to parameter of density we must use, along with equation in f_1, the equation in the two-particle distribution f_2. The right-hand side of the latter contains the three-particle distribution function f_3, and is also proportional to the small parameter of density. In the first approximation with respect to the parameter of density we use the equation in f_1 as is, and set the right-hand side of equation in f_2 equal to zero. As a result, we get a closed set of equations in the distribution functions f_1, f_2.

The solution of this set of equations is especially elegant when the external field is zero. Then, according to (4.7.4), the one-particle distribution function is equal to one, and the solution for f_2 is

$$f_2(r_1, r_2) = C\exp\left(-\frac{\Phi(|r_1 - r_2|)}{\kappa T}\right) . \tag{5.3.7}$$

Constant C is found from normalization condition for f_2:

$$C = \left[\frac{1}{V^2} \int \exp\left(-\frac{\Phi(|r_1 - r_2|)}{\kappa T} \right) dr_1 dr_2 \right]^{-1} . \tag{5.3.8}$$

Since the exponential function in the integrand is zero within the volume of an atom, and unity outside, constant C equals to one in the zero approximation with respect to small ratio of the atom's volume to the volume of the system. So,

$$f_1 = 1 , \quad f_2(r_1, r_2) = \exp\left(-\frac{\Phi(|r_1 - r_2|)}{\kappa T} \right) . \tag{5.3.9}$$

In Ch. 8 we shall use this solution for calculating the thermodynamic functions of nonideal gas. Next we consider the chain of equations for a sequence of nonequilibrium distribution functions.

5.4. Equations for Sequence of Nonequilibrium Distribution Functions

In place of (5.3.5) we now use the Liouville equation (2.2.5) for nonequilibrium distribution function $f_N(X, t)$. Using the Hamilton function (2.1.3) and the relevant equations of motion, we rewrite the Liouville equation in expanded form:

$$\frac{\partial f_N}{\partial t} + \sum_{1 \leq i \leq N} \left[v_i \frac{\partial f_N}{\partial r_i} + F_0 \frac{\partial f_N}{\partial p_i} - \sum_{\substack{1 \leq j \leq N \\ i \neq j}} \frac{\partial \Phi_{ij}}{\partial r_i} \frac{\partial f_N}{\partial p_i} \right] = 0 . \tag{5.4.1}$$

In order to find the equation in the one-particle distribution function, we multiply the Liouville equation by V, and carry out integration in all terms with respect to $x_2, \ldots, x_N$ — that is, with respect to the variables of all particles but the first.

For integration we use boundary conditions (3.7.1), (3.7.2). Then all terms with $i \neq 1$ in the first sum disappear. The interaction term involves summation over j. Since all j–terms in the integral are equivalent, we may retain just one of them (say, with $j = 2$), and multiply the result times the number of surrounding particles $N - 1$. Finally, we use definition (5.1.2) for the two-particle distribution function, and replace $(N-1)/V$ with n (this is justified because $N \gg 1$). As a result, we get the following equation for the one-particle distribution:

$$\left(\frac{\partial}{\partial t} + v_1\frac{\partial}{\partial r_1} + F_0(r_1)\frac{\partial}{\partial p_1}\right)f_1 = n\int\frac{\partial \Phi_{12}}{\partial r_1}\frac{\partial f_2}{\partial p_1}dx_2 \ , \tag{5.4.2}$$

where F_0 is the external force.

This equation in f_1 is not closed, since its right-hand side contains the two-particle distribution function f_2. Equation in f_2 can also be derived from the Liouville equation and has the following form:

$$\left(\frac{\partial}{\partial t} + v_1\frac{\partial}{\partial r_1} + v_2\frac{\partial}{\partial r_2} + F_0\frac{\partial}{\partial p_1} + F_0\frac{\partial}{\partial p_2} - \frac{\partial \Phi_{12}}{\partial r_1}\frac{\partial}{\partial p_1} - \frac{\partial \Phi_{12}}{\partial r_2}\frac{\partial}{\partial p_2}\right)f_2$$

$$= n\int\left(\frac{\partial \Phi_{13}}{\partial r_1}\frac{\partial}{\partial p_1} + \frac{\partial \Phi_{23}}{\partial r_2}\frac{\partial}{\partial p_2}\right)f_3(x_1,x_2,x_3,t)dx_3 \ . \tag{5.4.3}$$

The right-hand side of this equation describes interaction of the selected pair of particles "1" and "2" with the surrounding particles. It is proportional to the number of surrounding particles $N-2$, and includes the three-particle distribution function. Because the system is macroscopic ($N \gg 1$), we may replace factor $(N-2)/V$ with n.

It can be shown that the equation in f_3 contains the four-particle distribution function f_4, etc. This chain of equations is called the Bogolyubov – Born – Green – Kirkwood – Yvon chain, or BBGKY for short.

Exact solution of this set of equations is as complicated as the solution of the initial Liouville equation. Its advantage is realized when small parameters are available which allow using the methods of perturbation theory. Then the infinite chain of equations can be reduced to a set of several equations in the most simple distribution functions.

It should be noted that the BBGKY chain of equations, like the Liouville equation itself, is reversible. This is natural, since our manipulations did not violate the time symmetry of the system under consideration.

5.5. Correlation Functions

In many cases, in place of the sequence of distribution functions it is convenient to use the relevant correlation functions

$$g_1 \equiv f_1 \ , \ \ g_2 \ , \ \ g_3 \ ,\dots \ . \tag{5.5.1}$$

Correlation function g_2 characterizes the statistical coupling of the selected pair of particles and is defined by equation

$$f_2(x_1,x_2,t) = f_1(x_1,t)f_1(x_2,t) + g_2(x_1,x_2,t) . \qquad (5.5.2)$$

If the particles of the selected pair are statistically independent, the correlation function is zero, and

$$f_2(x_1,x_2,t) = f_1(x_1,t)f_1(x_2,t) .$$

To simplify the notation, we shall further replace the set of particle's variables by just the number of particle: $f_2(x_1,x_2,t) \equiv f_2(1,2)$, etc. Then the three-particle correlation function can be written as

$$f_3(1,2,3)$$
$$= f_1(1)f_1(2)f_1(3) + f_1(1)g_2(2,3) + f_1(2)g_2(1,3) + f_1(3)g_2(1,2) + g_3(1,2,3) . \qquad (5.5.3)$$

Correlation function g_3 is nonzero only when all the three particles are statistically dependent. If, for instance, particle "3" is separated from particles "1" and "2" by distance much larger than the relevant characteristic "radii" of correlation, then $g_3 = 0$, $g_2(1,3) = 0$, and

$$f_3(1,2,3) = f_1(1)f_1(2)f_1(3) + f_1(3)g_2(1,2) \equiv f_2(1,2)f_1(3) . \qquad (5.5.4)$$

Now back to equation (5.4.2) for the one-particle distribution function. In the right-hand side we use (5.5.2), move the term with the one-particle function to the left-hand side, and introduce notation for the "mean force"

$$F(r,t) = F_0(r,t) - n\frac{\partial}{\partial r}\int \Phi(|r - r'|)f_1(r',p',t)dr'dp' , \qquad (5.5.5)$$

which, in addition to the external force, accounts for the collective action of surrounding particles not associated with their correlation. The force of this kind is called the self-correlated or Vlasov force, in commemoration of contribution made by A.A. Vlasov to the theory of plasma where such forces are of special importance. "Self-correlated" implies that the force is completely defined by the one-particle distribution function.

Finally we come to the following equation:

$$\left(\frac{\partial}{\partial t} + v_1\frac{\partial}{\partial r_1} + F(r_1,t)\frac{\partial}{\partial p_1}\right)f_1(r_1,t) = I(x_1,t) . \qquad (5.5.6)$$

Here $I(x_1,t)$ is the so-called "collision integral", which depends on the contribution from the correlation function:

$$I(x_1,t) = n \int \frac{\partial \Phi_{12}}{\partial r_1} \frac{\partial g_2}{\partial p_1} dx_2 \ .$$ (5.5.7)

The name "collision integral" comes from the kinetic theory of rarefied gas, where the two-particle correlations may be attributed to collisions in pairs of particles. The same term is also used in plasma theory, is spite of the fact that the model of pairwise collisions is no longer adequate since the interaction between charged particles of plasma is of a collective nature.

We see that the equation for the one-particle distribution function is not closed, because the collision integral involves the higher two-particle correlation. It will be shown, however, that the collision integral for a gas or plasma can be approximated using the one-particle distribution function. In this way a closed kinetic equation can be obtained. In the next chapter we shall discuss the first example of such equation, the famous Boltzmann kinetic equation. Before that, however, we shall consider an alternative method for constructing the theory of nonequilibrium processes, as mentioned in the beginning of the present chapter.

5.6. Method of Microscopic Phase Density

Let us look at equation (2.4.14) for the microscopic phase density. Recall that the microscopic force therein is linked with the microscopic phase density by (2.4.9).

At the beginning of this chapter we indicated that statistical theory regards functions in these equations as random functions. To obtain kinetic equations, these random functions must be averaged over the Gibbs ensemble.

The first moment of microscopic phase density is linked with the one-particle distribution function by (4.5.5). Using this relationship, we average all terms in the equation for microscopic phase density $N(x,t)$ over the Gibbs ensemble. However, this operation will not yield a closed equation in the first moment — in the one-particle distribution function.

Indeed, the last term in (2.4.14) contains the product of two random functions, $N(x,t)$ and $F^m(r,t)$. Because of this, averaging over the Gibbs ensemble involves the second moment

$$\langle F^m(r,t)N(x,t) \rangle = F(r,t)nf(r,t) + \langle \delta F(r,t)\delta N(r,p,t) \rangle \ .$$ (5.6.1)

Here we made use of the identity which relates the second moment to the correlator (second central moment) of fluctuations about the mean.

Given this, the equation for microscopic phase density after averaging over the Gibbs ensemble can be written as

$$\left(\frac{\partial}{\partial t} + v\frac{\partial}{\partial r} + F(r,t)\frac{\partial}{\partial p}\right)f_1(x,t) = I(x,t) ,$$ (5.6.2)

where

$$I(x,t) = -\frac{1}{n}\frac{\partial}{\partial p}\langle \delta F(r,t)\delta N(r,p,t)\rangle$$ (5.6.3)

is the relevant "collision integral".

Let us compare this equation with the first equation in the BBGKY chain (5.5.6). The left-hand sides of equations are the same. The only difference is that in the method of microscopic phase density, in the spirit of field theory, it would be natural not to associate the points in space with the numbered particles. Accordingly, in place of x_1, x_2,... we use x', x'',....

Next we compare the "collision integrals" and show that the apparently different definitions (5.5.7) and (5.6.3) are actually equivalent. For this purpose we transform the expression for the correlator of fluctuations of force and phase density in (5.6.3).

First of all we demonstrate that this correlator is linked with the more general correlator of fluctuations of phase density. The relationship between the relevant fluctuations, which follows from (2.4.9), is

$$\delta F(r,t) = -\frac{\partial}{\partial r}\int \Phi(|r - r'|)\delta N(r',p',t)dr'dp' .$$ (5.6.4)

By virtue of this relation, the correlator in the collision integral can be represented as

$$\langle \delta F(r,t)\delta N(r,p,t)\rangle = -\frac{\partial}{\partial r}\int \Phi(|r - r'|)\langle \delta N(r',p',t)\delta N(r,p,t)\rangle dr'dp' .$$ (5.6.5)

Now we seek the relationship between the correlator of fluctuations of phase density and the two-particle correlation function, and start with the expression for the second two-point moment of phase density:

$$\langle N(x,t)N(x',t)\rangle = \int \sum_{1\leq i\leq N}\sum_{1\leq j\leq N} \delta(x-x_i)\delta(x'-x_j)f_N(X,t)dX \ . \tag{5.6.6}$$

Two contributions may be distinguished on the right-hand side. The first comes from the diagonal terms of the double sum. The number of these terms is N, and they are proportional to $\delta(x-x')$. This contribution is, of course, expressible via the one-particle distribution function. The second contribution comes from the nondiagonal elements of the double sum; these elements are $N(N-1)$ in number and are expressed in terms of the two-particle distribution function. On this basis we establish the linkage between the second moment of phase density and the distribution functions:

$$\langle N(x,t)N(x',t)\rangle = \frac{N}{V}\delta(x-x')f_1(x,t)+\frac{N(N-1)}{V^2}f_2(x,x',t) \ . \tag{5.6.7}$$

The third moment can be expressed in terms of functions f_1, f_2, f_3.

To obtain the expression for the second central moment which enters (5.6.50, we use the definition

$$\langle \delta N(x,t)\delta N(x',t)\rangle = \langle N(x,t)N(x',t)\rangle - \langle N(x,t)\rangle\langle N(x',t)\rangle \ . \tag{5.6.8}$$

Into the right-hand side we substitute the moments as expressed in terms of the distribution functions, and make use of (5.5.2). As a result, the second central moment is expressed in terms of the one-particle distribution function and the two-particle correlation function:

$$\langle \delta N(x,t)\delta N(x',t)\rangle = \frac{N(N-1)}{V^2}g_2(x,x',t)$$

$$+\frac{N}{V}\left(\delta(x-x')f_1(x,t)-\frac{1}{V}f_1(x,t)f_1(x',t)\right) \ . \tag{5.6.9}$$

In the thermodynamic limit (3.10.1) this expression is much simplified and becomes

$$\langle \delta N(x,t)\delta N(x',t)\rangle = n^2 g_2(x,x',t)+n\delta(x-x')f_1(x,t) \ . \tag{5.6.10}$$

So, we have obtained the desired expression for the second central moment, which we now substitute into the right-hand side of (5.6.5). Note that the contribution of the second term in (5.6.10) is zero: integration with respect to coordinates yields

$$\int \frac{\partial \Phi(|r-r'|)}{\partial r}\delta(r-r')dr' = 0 \ , \tag{5.6.11}$$

since the integrand is an odd function of vector $r - r'$.

This gives us the relationship between the correlator in the "collision integral" of (5.6.3) and the two-particle correlation function:

$$\langle \delta F(r,t)\delta M(r,p,t)\rangle = -n\int \frac{\partial \Phi(|r-r'|)}{\partial r} g_2(r,p,r',p',t)dr'dp' \; . \tag{5.6.12}$$

Substituting this into (5.6.3) we obtain an expression for the collision integral which coincides with (5.5.7).

Now we see that the equation for the one-particle distribution function, derived by averaging the equation in the microscopic phase density, coincides with the first equation in the BBGKY chain. It is also possible to prove the equivalence of equations for the higher distribution functions obtained with the two different methods (see Klimontovich II).

Equation (5.6.2) with the collision integral (5.6.3) is the first in the sequence of equations in moments derived from the equation for microscopic phase density. To obtain equations in higher moments, we write equations for the fluctuations whose correlator defines the collision integral. These latter result from subtracting the equations for the mean values from the relevant equations for the microscopic functions:

$$\left(\frac{\partial}{\partial t} + v\frac{\partial}{\partial r} + F\frac{\partial}{\partial p}\right)\delta N(x,t) =$$

$$- \delta F(r,t)\frac{\partial n f_1(x,t)}{\partial p} - \frac{\partial}{\partial p}\left(\delta F(r,t)\delta N(x,t) - \langle \delta F(r,t)\delta N(x,t)\rangle\right)$$

$$\tag{5.6.13}$$

$$\delta F(r,t) = \frac{\partial}{\partial r}\int \Phi(|r-r'|)\delta N(r',p',t)dr'dp' \; .$$

This set of equations is nonlinear, since the first equation includes terms quadratic in fluctuations. Because of this, the equations in moments are "meshing" — that is, equations for the second moments involve third moments, etc.

Naturally, the comparative efficiency of the two methods for description of nonequilibrium processes depends on the structure of the particular system. The method of moments is more efficient when collective interactions play an important role.

Finally, let us reiterate that the equations discussed so far are reversible. Now we are going over to equations which can be used for describing dissipative processes in macroscopic open systems. It would be natural to start with the simplest model, Boltzmann's rarefied gas of structureless particles.

CHAPTER 6

BOLTZMANN KINETIC EQUATION

The present chapter is devoted to one of the central topics in the statistical theory of nonequilibrium processes, the Boltzmann kinetic theory.

Ludwig Boltzmann is one the founding fathers of contemporary statistical physics. He proposed the first kinetic equation which describes, in particular, the irreversible process of transition towards the state of equilibrium. He was also the first to give the statistical definition of entropy, and he used his kinetic equation to show that the entropy increases as the system evolves towards the equilibrium state. This result came to be known under the name of H-theorem.

Boltzmann's results were based more on physical intuition than on consistent mathematical treatment. Mathematically, it was not clear how the reversible equations of motion of particles of the system could be used for deriving the irreversible equations. This was the reason why an unprecedentedly fierce debate started around Boltzmann's papers.

Among his opponents was Henri Poincaré, the famous mathematician. For illustration, let us quote some of his statements.

"In one of his papers Poincaré went so far as to write that he could not recommend the study of Boltzmann's paper because the premises in Boltzmann's considerations clashed with his conclusions" (Poincaré 1893 as quoted in Prigogine 1980).

On the basis of the reversible equations of mechanics Poincaré maintained that the theory of irreversible processes and mechanics are incompatible. One of his arguments derived from the fact that a Lyapunov function, which would act as H-function (entropy) in Boltzmann's theory, cannot be constructed within the framework of Hamiltonian mechanics.

Another version of Poincaré's statement is quoted in (Misra – Prigogine 1984):

"In this connection it would be amusing to recall Poincaré's words that he would not advice anyone to read Boltzmann's paper, since he would never recommend studying a proof in which the conclusions contradict the assumptions". Unfortunately, the authors do not specify the source of their quotation.

The drastic change in the attitude towards Boltzmann's results is reflected in the words of Erwin Schrödinger, an outstanding representative of the next generation of physicists:

"His [Boltzmann's] line of thought may be called my first love in science. No other has ever thus enraptured me or will ever do so again" (see Prigogine 1980).

Many years have passed since 1873 when Boltzmann proposed his kinetic equation. Nevertheless, the problem of validation of the kinetic equation retains its importance even today. At the same time, our admiration with Boltzmann's physical intuition only grows with time.

Our presentation of Boltzmann's kinetic theory will be focused not on the mathematical details which can be found in numerous textbooks, but rather on the important physical implications.

6.1. Rarefied Gas. Boltzmann Kinetic Equation

The kinetic Boltzmann equation is a closed irreversible equation in the one-point distribution function $f(x,t)$ in six-dimensional phase space $x = (r,p)$:

$$\left(\frac{\partial}{\partial t} + v\frac{\partial}{\partial r} + F_0\frac{\partial}{\partial p}\right)f(x,t) = I_B(x,t) .$$

(6.1.1)

Here $I_B(x,t)$ is the Boltzmann collision integral which accounts for the dissipation due to redistribution of particles' velocities because of interactions (collisions) between the particles. We shall see that in Boltzmann's theory the collision integral is a functional of the one-point distribution function $f(x,t)$; this equation is therefore a closed equation, unlike those for the one-particle distributions in the preceding chapter.

Observe that the kinetic equation (6.1.1) coincides with (5.4.2), (5.6.2) for the one-particle distributions f_1 only in form. Because of the fundamental importance of this dissimilarity, it should be also noted that we distinguish between the one-particle distributions in equations (5.4.2) and (5.6.2), and the one-point distribution in the Boltzmann equation. This distinction ought to emphasize the fact that the BBGKY equations are just one of the possible representations of the reversible equations of motion of particles of the system. Hence the terms one-particle, two-particle, etc. distribution functions, numbered f_1, f_2,....

In the Boltzmann equation the distribution function has an essentially different meaning. This function characterizes the mean distribution of particles in a continuous medium — not, however, a conventional hydrodynamic continuous medium, but rather a six-dimensional phase space $x = (r,p)$. As said in the Introduction, equations of continuous medium are inevitably irreversible, since the information about the motion of particles within the "points" is not available, and the "homecoming" cannot be accomplished by reverting the velocities of all particles of the system.

We shall see that Boltzmann's theory does not actually emphasize this distinction, and

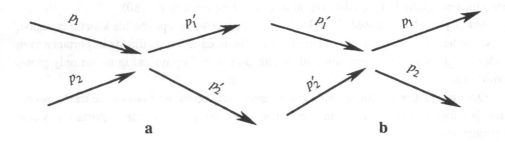

Figure 1: Two types of collisions.

the crucial step from "mechanical" to "statistical interpretation" of the distribution function is hidden behind Boltzmann's postulates. As early as 1935, long before the advent of the BBGKY theory, M.A. Leontovich wrote in this connection:

"The kinetic theory considers processes in gases. It is a statistical theory, since the Boltzmann equation is based on a statistical assumption, *der Stoßzahlansatz*. The structure of this theory, however, is certainly far from perfection. The quantity $f(r,p,t)drdp$ has to be assigned the meaning of a certain statistical mean (mathematical expectation) of the number of particles contained within volume $drdp$ of phase μ-space — only then it is possible to understand the irreversible nature of the Boltzmann equation and the implications of this fact. Within the framework of the theory, however, it is not clear what is meant by this "mathematical expectation", because nothing is said about the probabilities on which such "mathematical expectations" are based. For this reason, the theory is also unable to give any information about fluctuations in the gas and their trends."

In the chapters to follow we shall return both to Leontovich's paper, and to the fundamental problems of validation of kinetic equations. And now we go directly to exposition of Boltzmann's theory.

Boltzmann's theory considers a rarefied gas of structureless particles, the most simple example of a macroscopic system.

Let us review some results discussed in the Introduction.

A gas is considered rarefied if the parameter of density (1.1.12) is small, and therefore inequalities (1.1.13) hold.

After Boltzmann, we shall only take into account the collisions in pairs of particles, assuming that triple and multiple collisions in a rarefied gas are unlikely.

A collision may have two outcomes (Fig. 1). Collisions of type "a" reduce the number of particles with momentum p_1; collisions of type "b" increase the number of particles with momentum p_1. Because of this, the collision integral can be represented as

$$I_B(r,p_1,t) = I_b(r,p_1,t) - I_a(r,p_1,t) \ . \tag{6.1.2}$$

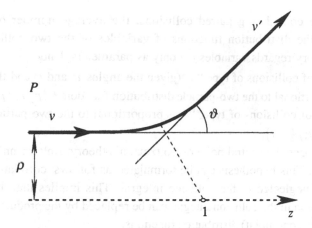

Figure 2: Definition of Boltzmann collision integral.

Subscript "1" is only used for identification of the colliding particle.

For collisions of type "a" let p_1, p_2 be the momenta before the collision, and p_1', p_2' the momenta after the collision. Collisions obey the laws of conservation of momentum and kinetic energy:

$$p_1 + p_2 = p_1' + p_2' , \quad \frac{p_1^2}{2m} + \frac{p_2^2}{2m} = \frac{p_1'^2}{2m} + \frac{p_2'^2}{2m} . \tag{6.1.3}$$

We shall describe collisions in cylindrical coordinates (Fig. 2). Let the origin coincide with the position of particle "1". The path of the second particle lies in the plane P which passes through the z-axis. Orientation of plane P is given by azimuthal angle φ. Let v and v' be the relative velocities before and after the collision, and ϑ the angle between these two vectors. Finally, ρ is the target distance. Target distance as function of angle ϑ is found by solving the equations of motion of the two particles with interaction $\Phi(r)$.

The conservation laws (6.1.3) comprise four scalar equations which relate the momenta after collision to the momenta before collision provided that two additional parameters are specified (for example, angles ϑ and φ):

$$p_{1,2}' = p_{1,2}(\vartheta, \varphi) . \tag{6.1.4}$$

In calculating the average number of local collisions (at point r and time t) the following must be taken into account.

This number is proportional to the magnitude of relative velocity $|v_1 - v_2|$ which is conserved during collisions, and to the cross section of collision (cross section of elastic scattering). Finally, there is another important factor.

Since we are considering paired collisions, the average number of collisions is proportional to the distribution functions of variables of the two colliding particles. Boltzmann's theory regards variables r, t only as parameters. Under this assumption the average number of collisions of type "a" (given the angles ϑ and φ and the momenta of particles) is proportional to the two-particle distribution function $f_2(p_1,r,t;p_2,r,t)$, and the average number of collisions of type "b" is proportional to the two-particle distribution function $f_2(p_1',r,t;p_2',r,t)$.

Now we approach the central point of Boltzmann's theory: Boltzmann's *hypothesis of molecular chaos*. This hypothesis can be formulated as follows: correlation of colliding particles can be neglected in the collision integral. This implies that the two-particle distribution function in the collision integral can be replaced by the product of one-particle (or, more precisely, one-point) distribution functions:

$$f_2(p_1,r,t;p_2,r,t) = f(p_1,r,t)f(p_2,r,t) \; ,$$

$$(6.1.5)$$

$$f_2(p_1',r,t;p_2',r,t) = f(p_1',r,t)f(p_2',r,t) \; .$$

As a result, we come to the following expression for Boltzmann's collision integral:

$$I(r,p_1,t) = n\int_0^\infty \rho d\rho \int_0^{2\pi} d\varphi \int dp_2 |v_1 - v_2| \{ f(r,p_2',t)f(r,p_1',t) - f(r,p_2,t)f(r,p_1,t) \} \; . (6.1.6)$$

The arguments $p_{1,2}'$ are expressible in terms of the angle variables by virtue of (6.1.4), and so the number of arguments (excluding p_1) in the integrand is equal to the number of variables over which the integration is carried out.

So, with certain assumptions which will be analyzed further, we have obtained a closed equation in the one-point distribution $f(r,p,t)$, the Boltzmann kinetic equation.

6.2. Properties of Boltzmann Collision Integral

The Boltzmann equation is an extremely complicated nonlinear integrodifferential equation, and so the knowledge of its general properties is very useful. This will be particularly helpful in the transition from the kinetic description to the more coarse hydrodynamic description.

Multiply the collision integral by an arbitrary function $\Phi(p_1)$ and integrate with respect to p_1, denoting the resulting integral by $I(r,t)$. Then carry out double symmetrization in the integrands:

(1) exchange variables $p_1, p_1' \Leftrightarrow p_2, p_2'$;

(2) exchange variables $p_1, p_2 \Leftrightarrow p_1', p_2'$.

Observe that the Jacobian of transformation of variables "before" and "after" collision is unity.

After the double symmetrization the integral $I(r,t)$ becomes

$$I(r,t) = \frac{n^2}{4} \int dp_1 dp_2 \int_0^\infty \rho d\rho \int_0^{2\pi} d\varphi |v_1 - v_2| \{\Phi(p_1) + \Phi(p_2) - \Phi(p_1') - \Phi(p_2')\}$$

$$\times \{f(r,p_2',t)f(r,p_1',t) - f(r,p_2,t)f(r,p_1,t)\} \qquad (6.2.1)$$

We see that this integral is zero for such functions $\Phi(p)$ which are conserved at paired collisions:

$$I(r,t) = n \int \Phi(p_1) I(r_1, p_1, t) dp_1 = 0 \quad \text{for} \quad \Phi = 1, \ p, \ \frac{p^2}{2m} . \qquad (6.2.2)$$

These properties will be used in transition from the kinetic equation to equations of gas dynamics.

Let us establish another property of Boltzmann collision integral which will be used in proving the H-theorem. We set $\Phi(p)$ as the logarithm of the distribution function,

$$\Phi(p) = -\kappa \ln f(r,p,t) , \qquad (6.2.3)$$

where the multiplier $-\kappa$ is introduced for convenience in the algebra of H-theorem.

Given this definition of $\Phi(p)$, the embraced terms in (6.2.1) will always have opposite signs, and hence the product $\{...\} \times \{...\}$ is negative for all values of the distribution functions. As a result, we come to inequality

$$I(r,t) \geq 0 \quad \text{for} \quad \Phi(p) = -\kappa \ln f(r,p,t) . \qquad (6.2.4)$$

Let us show that the "equals" sign corresponds to the case when the distribution function is the Maxwell distribution. This will enable us to find the equilibrium solution of Boltzmann equation.

6.3. Equilibrium Solution of Boltzmann Equation

The "equals" sign in (6.2.4) occurs for all distribution functions for which the term in braces $\{...\}$ is zero. This condition leads to two equivalent functional equations. Let us write one of these, dropping r and t:

$$\ln f(p_1) + \ln f(p_2) = \ln f(p_1') + \ln f(p_2') \ . \tag{6.3.1}$$

We can show that the solution of this functional equation is the Maxwell distribution.

From the structure of this functional equation it follows that the logarithm of the distribution function can be represented as a linear combination of three functions, 1, p, $p^2/2m$, which are conserved at paired collisions of particles. Hence

$$\ln f(p) = \alpha + \beta p + \gamma p^2 \ . \tag{6.3.2}$$

Here α, β, γ are five arbitrary quantities (two scalars and one three-vector). If we replace these variables according to the following rules:

$$\alpha = \ln a - bc^2 \ , \quad \beta = 2bc \ , \quad \gamma = -b \ , \tag{6.3.3}$$

the expression for the distribution function becomes

$$f(p) = a \exp\left[-b(p - c)^2\right] \ . \tag{6.3.4}$$

Now if we relate variables a, b, c to certain physical quantities, and make use of equations

$$\int f dp = 1 \ , \quad \int \frac{p}{m} f dp = u \ , \quad \int \frac{m(v - u)^2}{2} f dp = \frac{3}{2} \kappa T \tag{6.3.5}$$

(of which the first is the normalization condition, the second is the definition of mean velocity, and the third is the definition of temperature), we come to the Maxwell distribution:

$$f(p) = \frac{1}{(2\pi m \kappa T)^{3/2}} \exp\left[-\frac{(p - mu)^2}{2\kappa T}\right] \ . \tag{6.3.6}$$

At equilibrium, u is the mean velocity of the center of mass of the system of particles. If the system as a whole is at rest, then $u = 0$, and (6.3.6) coincides with the Maxwell distribution of (4.6.4).

In the presence of external field $U(r)$, Boltzmann equation at the state of equilibrium (when the distribution function is time-independent) takes on the form

$$v\frac{\partial f(r,p)}{\partial r} - \frac{\partial U}{\partial r}\frac{\partial f(r,p)}{\partial p} = I(r,p) \ . \tag{6.3.7}$$

The right-hand side is brought to zero by substituting the Maxwell distribution, and the left-hand side is zero for the combined Maxwell-Boltzmann distribution. Hence the equilibrium solution of the Boltzmann equation in the presence of external field is the combined Maxwell-Boltzmann distribution:

$$f(r,p) = \frac{1}{(2\pi m\kappa T)^{3/2}} \frac{\exp\left[-\dfrac{\dfrac{p^2}{2m}+U(r)}{\kappa T}\right]}{\dfrac{1}{V}\int \exp\left[-\dfrac{U(r)}{\kappa T}\right]dr} \ . \tag{6.3.8}$$

Finally, let us make use of the property (6.2.4) of Boltzmann collision integral. We shall show that this property enables one to obtain the equation of entropy balance, which expresses the second law of thermodynamics for nonequilibrium processes.

6.4. Increase in Entropy During Time Evolution to the State of Equilibrium. Boltzmann's H–Theorem. Lyapunov Functional

Recall that the entropy for an arbitrary nonequilibrium state of a system with distribution $f_N(x,t)$ is defined by (3.5.15). It is noteworthy that such definition of entropy was first given by Boltzmann for the model of rarefied gas, when the state of the system is characterized by the one-point distribution $f(x,t)$. Accepting Boltzmann's definition, we use the following expression for the entropy of rarefied gas:

$$S(t) = -\kappa n \int \ln[nf(x,t)]f(x,t)drdp \ . \tag{6.4.1}$$

Multiplication by n is used here to reconcile this expression with (3.12.5) for the entropy of equilibrium ideal gas. Indeed, substituting here the Maxwell distribution we get

$$S = \kappa N\left\{\ln\frac{1}{n} + \frac{3}{2}(2\pi m\kappa T) + \frac{3}{2}\right\} \ . \tag{6.4.2}$$

This only differs from (3.12.5) by the κN constant, which does not affect the calculation of thermodynamic functions.

We shall also need the expression for the local entropy

$$S(r,t) = -\kappa n \int \ln\big(nf(r,p,t)\big)f(r,p,t)dp \ . \tag{6.4.3}$$

Now we employ the kinetic Boltzmann equation to find the equation for local entropy. Making use of inequality (6.2.4) and the continuity equation for the local density $n(r,t) = n \int f(r,p,t)dp$, we get the following equation for entropy balance in ideal gas:

$$\frac{\partial S(r,t)}{\partial t} + \frac{\partial}{\partial r}\int \frac{p}{m}(-\kappa n)\ln\big(nf(r,p,t)\big)f(r,p,t)dp = \sigma(r,t) \ . \tag{6.4.4}$$

Here $\sigma(r,t)$ denotes the local entropy production

$$\sigma(r,t) = -\kappa n \int \ln\big[nf(r,p,t)\big]I(r,p,t)dp \geq 0 \ . \tag{6.4.5}$$

It is at this point that we take heed of the property (6.2.4).

From the equation of balance (6.4.4) it follows that the time dependence of total entropy (6.4.1) for a closed system (when the flow of matter across the boundary of the system is zero) is characterized by inequality

$$\frac{dS(t)}{dt} \geq 0 \ . \tag{6.4.6}$$

This means that the entropy of a closed system cannot decrease in the course of time evolution towards the state of equilibrium. The entropy either increases, or, after the state of equilibrium has been attained, remains constant.

In place of S Boltzmann used the function $H = -S$; for this reason the result expressed by inequality (6.4.6) is called the H–theorem.

The name of H–theorem was suggested by the British physicist called Burbury in 1884, and a year later was accepted by Boltzmann. The letter "H" here stands for "heat', as a reminder that the theorem deals with the transition to the state of thermal equilibrium.

Some remarks are due in connection with Boltzmann's H–theorem.

In Sect. 1.1, 1.5 we have already spoken of the inevitability of transition to irreversible equations when Gibbs ensemble was defined for nonequilibrium states. Irreversibility of motion arises as a result of the loss of information because of smoothing over physically infinitesimal volumes.

The introduction of physically infinitesimal volumes (1.1.17) was in its turn justified by the estimated characteristic times of development of dynamic instability of motion of ball atoms of the Boltzmann gas.

So it is the development of dynamic instability which results in exponential divergence of trajectories, combined with the existence of uncontrollable external factors, that necessitates the transition to irreversible equations.

Recall also that in Sect. 1.5 we spoke of the constructive role of dynamic instability of motion of atoms. The loss of information because of transition to irreversible equations is generously paid off, since the resulting equations are much better coordinated with the real experimental methods used for studying the processes in macroscopic systems.

Another remark points to the possibility of an alternative formulation of Boltzmann's H–theorem, based on the introduction of Lyapunov functional defined as the difference in the entropies of equilibrium and nonequilibrium states. This has been discussed in Sect. 1.6, and now we shall just recall the main points.

Taking advantage of the properties (6.2.2) of Boltzmann collision integral, we may express Boltzmann's H–theorem with the two inequalities of (1.6.3), (1.6.4) for the Lyapunov functional Λ_S. This is possible because the mean energy is conserved in the course of evolution towards the state of equilibrium (see (1.6.2)).

Since it is the mean energy $\langle E \rangle$ that is conserved in the course of evolution as described by the Boltzmann equation, fluctuations of energy are possible. Therefore, the definition of the closed system which assumes complete energy isolation of the system is not applicable.

Fluctuations of energy in Boltzmann gas may be associated not only with the external factors, but also with the deficiency of information about the internal motion of the particles. We may therefore say the the system is not closed internally. This internal non-closedness may be due both to the loss of information because of transition from the description of a system of particles to the description of a continuous medium, and to the incompleteness of description of processes in the approximation of continuous medium. Later on we shall return to the discussion of this important matter.

6.5. Relaxation Time and Length in Boltzmann Gas

Further we shall require an estimate of the time taken by the Boltzmann gas to relax to the state of equilibrium. We shall distinguish between the relaxation time of the kinetic stage which leads to incomplete (local) equilibrium, and the relaxation time governed by the gasdynamic processes. In the former case the characteristic time is expressed in terms of internal parameters of the system, whereas in the latter case the relaxation time depends also on the external scales — for instance, on the dimensions of the system, or on the size of the obstacle.

We denote the characteristic relaxation time and length scales by τ_{rel} and $l_{rel} = v_T \tau_{rel}$ respectively. To evaluate the relaxation time we represent the Boltzmann integral in the form

$$I_B(r,p,t) \sim \frac{1}{\tau_{\text{rel}}} f(r,p,t) \ . \tag{6.5.1}$$

In (6.1.6) we set $|v_1 - v_2| \sim v_T$, $\rho \sim r_0$, $\int f dp \sim 1$. This yields the following estimates for the time and length of relaxation:

$$\tau_{\text{rel}} \sim \frac{1}{n r_0^2 v_T} = \frac{l}{v_T} = \tau \ , \quad l_{\text{rel}} = \tau_{\text{rel}} v_T \sim \frac{1}{n r_0^2} = l \ . \tag{6.5.2}$$

In this way, the relaxation time is equal by the order of magnitude to the mean path time τ, and the relaxation length to the mean free path l. Recall that it is these values that we have used for the relaxation scales when defining physically infinitesimal scales in Sect. 1.1 on the kinetic level of description.

We should remind once again that these scales of relaxation characterize only the transition to local equilibrium. Similar scales for the hydrodynamic range will be obtained in Ch. 11. Here we just note that the expression for the time of relaxation has the structure of (1.1.18), and the relaxation time, therefore, is proportional to the square of the characteristic length of the system.

As indicated before, the solution of the kinetic Boltzmann equation is an extremely complicated task. For this reason it is necessary to investigate the possibility of reducing the kinetic equation to simpler equations. Let us show that this can be done in the two extreme cases.

6.6. Approximations of Free Molecular Flow and Gas Dynamics

As indicated in Sect. 1.1, a rarefied gas is characterized by three internal parameters of length which satisfy inequalities (1.1.11). We also introduced the external parameter of length L. By T we denoted the respective parameter of time. Finally we introduced the appropriate dimensionless parameter, the Knudsen number (1.2.1). Depending on the magnitude of the Knudsen number, two extreme cases may be distinguished. Let us discuss them in due course.

The range of large Knudsen numbers (which means that the free path length is much greater than L) corresponds to free molecular flow. At the same time, it is important that

$$l_{\text{ph}} \ll L \ll l \ . \tag{6.6.1}$$

The right-hand inequality states that the Knudsen number is large, whereas the left-hand inequality permits us to consider the free molecular flow in the approximation of continuous medium. Is the left-hand inequality is not satisfied, the kinetic approach cannot be used, and one has to return to the initial model of the gas as a system of particles.

We rewrite the kinetic Boltzmann equation in dimensionless variables $t' = t/T$, $r' = r/L$, $v' = v/v_T$, and assume that $L = v_T T$. Then the Boltzmann equation can be represented as

$$\text{Kn}\left(\frac{\partial}{\partial t'} + v'\frac{\partial}{\partial r'} + F'\frac{\partial}{\partial p'}\right)f = I' . \tag{6.6.2}$$

This equation shows explicit dependence on the dimensionless Knudsen number; this enables us to analyze the kinetic equation in the two extreme cases.

(1) Zero approximation in inverse Knudsen number. This approximation corresponds to free molecular flow. Returning to dimensional variables, we get the kinetic equation

$$\frac{\partial f}{\partial t} + v\frac{\partial f}{\partial r} + F\frac{\partial f}{\partial p} = 0 . \tag{6.6.3}$$

(2) Zero approximation in Knudsen number. In this approximation the kinetic Boltzmann equation reduces to a homogeneous integral equation

$$I_B(r,p,t) = 0 . \tag{6.6.4}$$

As we shall see in Ch. 11, this equation serves as the starting point for the construction of the set of equations of gas dynamics. Its solution is represented in the form of local Maxwell distribution, which differs from the equilibrium Maxwell distribution in that all parameters are functions of coordinates and time.

Let us return to equation (6.6.3), which we consider for the stationary case and assume that $F = 0$. We supplement this equation with the boundary condition which crucially depends on the way the particles are reflected by the wall. Consider two extreme cases.

(1) Mirror (elastic) reflection. We direct the unit vector n_0 inwards and normal to the wall (Fig. 3a). Let p be the momentum of particle incident on the wall; then the momentum of particle bounced off the wall is $p' = p - 2n_0(n_0 p)$. The boundary condition for the distribution function on the wall can then be written as

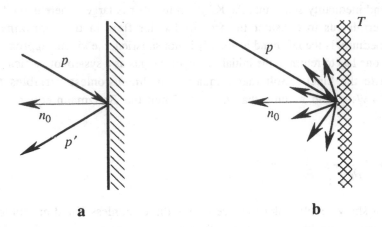

Figure 3: Mirror reflection (a), and diffuse reflection (b).

$$f(r_s, p) = f(r_s, p - 2n_0(n_0 p))$$
$$n_0 p < 0 \qquad n_0 p > 0 \qquad\qquad\qquad (6.6.5)$$

Of course, this is an idealized boundary condition. It is non-dissipative, and for this reason the flow described by the kinetic equation (6.6.3) occurs without friction. This means that the flow in the tube may continue with zero pressure head between the ends of the tube.

Now let us give an example of a more realistic boundary condition.

(2) Diffuse reflection. With a rough wall surface, a situation is possible when the distribution of momenta of reflected particles is not related to the distribution function of the incident particles, but depends solely on the conditions of the wall — in particular, the wall temperature T (Fig. 3b). This situation can be expressed by the following boundary condition:

$$f(r_s, p) = \frac{1}{(2\pi m \kappa T)^{3/2}} \exp\left(-\frac{p^2}{2m\kappa T}\right).$$
$$n_0 p > 0 \qquad\qquad\qquad\qquad\qquad (6.6.6)$$

This is the condition of diffuse reflection. Of course, intermediate situations are possible, when the boundary condition is a linear combination of conditions (1) and (2).

The condition of diffuse reflection is dissipative, and therefore the corresponding free molecular flow is a flow with friction. Let us illustrate this point with a simple example.

Let a steady flow occur in a tube of radius R and length L, and let the pressure difference at the ends of the tube be $P_1 - P_2$. Then the mass of gas passing through the tube in unit time is (see Sect. 22 in Silin 1971):

$$Q_m = \frac{8R^3}{3} \left(\frac{\pi m}{2\kappa T} \right)^{1/2} \frac{P_1 - P_2}{L} . \tag{6.6.7}$$

Let us compare this result with the known Poiseuille formula. The latter is the solution of a similar problem in hydrodynamics, and corresponds therefore to the limit of small Knudsen number. The Poiseuille formula has the form

$$Q_m = \frac{\pi}{8} \frac{R^4}{v} \frac{P_1 - P_2}{L} , \tag{6.6.8}$$

where v is the kinematic viscosity. By order of magnitude, $v \sim v_T l$.

Comparing the two expressions for Q_m, we notice that

$$v \Leftrightarrow v_T R , \text{ and therefore } l \Leftrightarrow R . \tag{6.6.9}$$

We see that the free path length in hydrodynamic description corresponds to the radius of the tube in the approximation of free molecular flow.

Free molecular flow may be referred to as collisionless flow, since the collisions between particles are of no importance. This term is widely used in plasma theory. It must be remembered, however, that it is the collisions between particles that are neglected, whereas collisions of particles with the wall determine the dissipation. This fact is reflected by the condition of correspondence (6.6.9).

In Ch. 11 we shall consider the opposite extreme, when the Knudsen number is small, and the collisions between particles are important.

6.7. Summary. Unsolved Problems

(1) Recall that the Boltzmann kinetic equation is derived under the assumption of paired collisions. This implies that collisions of three or more particles are too unlikely to be taken into consideration. This simplification is actually based on the assumption that the parameter of density $\varepsilon = n r_0^3$ is small.

(2) The Boltzmann equation is derived under the tacit assumption that the system of particles can be replaced with the continuous medium. In particular, this applies to the replacement of the one-particle distribution used in the BBGKY equations by the one-point distribution of a continuous medium. The question concerning the structure of the

continuous medium remains, therefore, open. We started discussing this matter in the Introduction, and will continue this discussion in the chapters to follow.

(3) Open remains the question of the definition of Gibbs ensemble for nonequilibrium processes, and, as a consequence, the question of the physical background of irreversibility in the kinetic theory, and of the meaning of hypothesis of molecular chaos.

(4) Boltzmann's collision integral involves the coordinates and the time as parameters. Because of this, the changes in the distribution function on the scales of r_0, $\tau_0 = r_0/v_T$, characteristic of the interaction between particles, are not taken into account. We shall see that this is equivalent to neglecting the contribution of interaction to the thermodynamic functions.

It follows that in Boltzmann's approximation the interaction only determines the dissipative process as described by the collision integral. The contribution of interaction to the thermodynamic functions is neglected. This implies that the Boltzmann equation describes nonequilibrium processes which occur only in an ideal gas (in thermodynamic sense). The problem of constructing the kinetic theory of nonideal gas thus remains open.

(5) In Boltzmann's approximation, the function $f(r,p,t)$ is a one-point deterministic (nonrandom) distribution. Not clarified therefore remains the role of fluctuations of the distribution function — that is, the problem of the kinetic theory of fluctuations. Fluctuations turn out to be important in the description of highly nonequilibrium processes, as well as phenomena which depend on the structure of the continuous medium, such as Brownian motion.

Notwithstanding these limitations, the Boltzmann equation remains one of the most elegant and most important equations in theoretical physics. At a conference in Vienna in 1973, dedicated to the centennial of Boltzmann's kinetic equation, the prominent physicist Professor G.E. Uhlenbeck expressed his praise in the following words:

"The Boltzmann equation has become such a generally accepted and central part of statistical mechanics that it almost seems seem blasphemy to question its validity and to seek out its limitations."

CHAPTER 7

FROM BBGKY EQUATIONS TO KINETIC EQUATIONS FOR BOLTZMANN GAS

In the preceding chapter we considered the derivation of the kinetic equation for rarefied gas as proposed by Boltzmann himself. This derivation is largely based on Boltzmann's powerful physical intuition, rather than on a consistent mathematical treatment. Of course, at that time this was the only available option. Such approach, however, raised a lot of questions, many of which were highlighted even by Boltzmann's contemporaries. As the theory matured, the questions grew in number. The most important have been enumerated at the end of the preceding chapter.

Now we shall try to answer these questions, which is not an easy task at all. Many things have only become more or less clear quite recently.

The first important step forward (not appreciated for decades) was made in an excellent paper by Leontovich (1935). Instead of the reversible Liouville equation, which is equivalent to the set of Hamilton equations for rarefied gas, he started with the irreversible Smoluchowski equation. The importance of Leontovich's work was recognized half a century later (Klimontovich 1973). Its main results will be discussed in Ch. 10.

. The irreversible kinetic Boltzmann equation was first derived from the reversible Liouville equation independently by Bogolyubov (1946), Born and Green (1946), Kirkwood (1946). These seminal papers set the guidelines for the development of kinetic theory of irreversible processes for years to come. The basics of this theory constitute the main theme of the present chapter. We shall see that only some of the questions formulated in Sect. 6.7 can be answered on this level of the statistical theory. The remaining problems will be dealt with later on in this book.

7.1. BBGKY Equations in Approximation of Paired Interactions

Boltzmann's kinetic theory is based on the approximation of paired interactions, when it is considered that three or more particles are unlikely to take part in one collision. This approximation may be justified for a rarefied gas — that is, on condition that the dimensionless parameter of density $\varepsilon = nr_0^3$ is small (see (1.1.13)).

Let us show that the BBGKY chain of equations in the approximation of paired interactions is truncated and reduces to a closed set of equations in the simplest distribution functions f_1, f_2.

Recall that the first equation in the BBGKY chain can be reduced to equation (5.5.6) with the "collision integral" (5.5.7). This representation is convenient, since all contributions determined by the one-particle distribution function are assembled in the left-hand side of the equation. The right-hand side is determined by the two-particle distribution function. In the approximation of paired interactions the latter is naturally nonzero, and it defines the statistical coupling in pairs of particles. Simplifications are therefore possible only in the next equation of the BBGKY chain — that is, in (5.4.3).

In this equation, the information about triple (and more complicated) collisions enters via the three-particle distribution function. Using (5.5.3), we express the three-particle distribution function in terms of the correlation functions. Observe that in the approximation of paired interactions the three-particle correlation function g_3 can be neglected, since it is nonzero only for interactions between three particles. This simplification is good enough by itself, since the chain of BBGKY equations is then reduced to the closed set of equations in functions f_1, f_2 (or functions f_1, g_2). However, further simplifications are possible.

Indeed, when we substitute (5.5.3) into the right-hand side of (5.4.3), we obtain such combinations as

$$\Phi\left(|r_1 - r_2|\right)g_2\left(r_2,p_2,r_3,p_3,t\right) , \quad \Phi\left(|r_2 - r_3|\right)g_2\left(r_1,p_1,r_3,p_3,t\right) ,$$

which are nonzero only when three particles come together close enough to interact. For this reason, the terms which contain these functions can also be dropped.

Four of the remaining terms on the right-hand side we unite into the two-particle distribution functions $f_2(1,2)$, move them to the right-hand side, and use definition of the Vlasov force. As a result, only the two terms with functions $g_2(1,3)$, $g_2(2,3)$ remain on the left-hand side, which can be expressed in terms of the "collision integral" (5.5.7). These combinations are eliminated with the aid of (5.5.6); as a result, we come to the following equation for the two-particle distribution function in the approximation of paired interactions:

$$\left(\frac{\partial}{\partial t} + v_1\frac{\partial}{\partial r_1} + v_2\frac{\partial}{\partial r_2} + F\frac{\partial}{\partial p_1} + F\frac{\partial}{\partial p_2} - \frac{\partial\Phi_{12}}{\partial r_1}\frac{\partial}{\partial p_1} - \frac{\partial\Phi_{12}}{\partial r_2}\frac{\partial}{\partial p_2}\right) f_2(1,2)$$

$$= \left(\frac{\partial}{\partial t} + v_1\frac{\partial}{\partial r_1} + v_2\frac{\partial}{\partial r_2} + F\frac{\partial}{\partial p_1} + F\frac{\partial}{\partial p_2}\right) f_1(1)f_1(2) . \qquad (7.1.1)$$

So, the two equations (5.5.6) and (7.1.1) (with relation (5.5.2)) comprise a closed set of equations in functions f_1, f_2 (or functions f_1, g_2). Relation (5.5.2) is useful for rewriting (7.1.1) in terms of the two-particle correlation function g_2.

7.2. Separation of Dissipative Contribution of Interaction of Particles

From the closed set of equations in the distribution functions, which, like the initial Liouville equation, is reversible, it is possible to move to the irreversible Boltzmann kinetic equation. At the same time, further simplifications of the set of equations in f_1, f_2 are feasible. This possibility is based on the following considerations.

Let us return to the formulas of (6.5.2) which define the characteristic time and length of relaxation of a rarefied gas towards the state of equilibrium. Using the definition of the density parameter, these can be rewritten as

$$\tau_{rel} \sim \frac{r_0}{v_T \varepsilon} \, , \ l_{rel} \sim \frac{r_0}{\varepsilon} \, . \tag{7.2.1}$$

We see that in zero approximation in the parameter of density ε the relaxation length is infinite, since the size of atoms r_0 is fixed. Infinite is also the relaxation time (at any fixed value of temperature). This implies that it is not possible to describe the irreversible transition towards the state of equilibrium in zero approximation in ε.

By contrast, zero approximation in ε is physically reasonable for calculating the nondissipative characteristics — for instance, the thermodynamic functions. This corresponds to the approximation of ideal gas with respect to the nondissipative thermodynamic functions.

So we see that in transition from the reversible equations in distribution functions f_1, f_2 to the kinetic Boltzmann equation it is sufficient to take only partial account of the interaction — only as far as the interaction determines the dissipative processes. This allows us to make further simplifications in the set of equations in distribution functions.

Observe that the situation is somewhat paradoxical. Indeed, we are going to separate the dissipative and the nondissipative contributions in equations in functions f_1, f_2 which are reversible and therefore do not contain dissipative terms. Is there a way out of this situation?

The solution to this problem will be given in the next section. First of all, however, we shall eliminate the "unnecessary" terms in (7.1.1).

At the end of the preceding section we indicated (see paragraph (4)) that the Boltzmann equation is derived without considering the variation of the distribution function $f(r,p,t)$ on the characteristic scales of interactions (collisions) of particles r_0, τ_0. Let us estimate the relative magnitude of such variations. Since the characteristic scales of variation of the distribution function $f(r,p,t)$ are τ and l (see (6.5.2)), the relative magnitude of variation of the distribution function $f(r,p,t)$ on the characteristic scales of interactions (collisions) of particles r_0, τ_0 is

$$\frac{\tau_0 \frac{\partial f}{\partial t}}{f} \sim \frac{\tau_0}{\tau} \sim \varepsilon \ , \qquad \frac{r_0 \frac{\partial f}{\partial r}}{f} \sim \frac{r_0}{r} \sim \varepsilon \ . \tag{7.2.2}$$

It follows that these terms are of the same order of magnitude as the dissipative terms which define the collision integral (see (7.2.1)), but they determine the contribution of the interaction to the thermodynamic (nondissipative) functions. This will be verified by calculations in the next chapter, which is devoted to the kinetic theory of nonideal gas.

Observe that for derivation of the Boltzmann kinetic equation it suffices to know the solutions of equation (7.1.1) for $g_2 = f_2 - f_1 f_1$ only on the small scales r_0, τ_0. Then by virtue of estimate of (7.2.2) we may discard the derivatives of the one-particle distribution functions in (7.1.1).

In the same equation on the small scales r_0, τ_0 we may also neglect the terms with the mean force. Indeed, we may assume that these forces are important in the kinetic equation when the characteristic scale is l. To make an estimate, let us assume that the work of force F over distance l is of the order of κT. Then

$$\frac{Fl}{\kappa T} \sim 1 \ , \quad \text{and therefore in (7.1.1)} \quad \frac{Fr_0}{\kappa T} \sim \varepsilon \ . \tag{7.2.3}$$

This allows us to discard also the terms with the mean force, since they are of the same order of magnitude as those of (7.2.2), and are nondissipative.

On the basis of arguments developed above, for deriving the kinetic equation (the closed equation in the one-particle distribution function) we may use in place of (7.1.1) a much simpler equation:

$$\left(\frac{\partial}{\partial t} + v_1 \frac{\partial}{\partial r_1} + v_2 \frac{\partial}{\partial r_2} \right) (f_2 - f_1 f_1) - \left(\frac{\partial \Phi_{12}}{\partial r_1} \frac{\partial}{\partial p_1} + \frac{\partial \Phi_{12}}{\partial r_2} \frac{\partial}{\partial p_2} \right) f_2 = 0 \ ,$$

$$\tag{7.2.4}$$

$$f_2 = g_2 + f_1 f_1 \ .$$

Equations (5.5.6), (7.2.4) for the one-particle distribution function and the two-particle correlation function comprise a closed set of equations. This set has been obtained under two simplifying assumptions: the assumption of paired collisions, and the neglect of derivatives of one-particle distributions in the equation for the correlation function. As follows from its structure, this set of equations is reversible. The next step is now the transition to the closed equation in the one-particle distribution — the kinetic Boltzmann equation.

7.3. Transition to Boltzmann Equation

In equation (7.2.4) we have eliminated all terms which are negligibly small in the approximation under consideration. To facilitate the solution, however, we add "zero terms" with the derivatives of functions f_1 with respect to r and t, and obtain a homogeneous equation for f_2:

$$\left(\frac{\partial}{\partial t} + v_1 \frac{\partial}{\partial r_1} + v_2 \frac{\partial}{\partial r_2} - \frac{\partial \Phi_{12}}{\partial r_1} \frac{\partial}{\partial p_1} - \frac{\partial \Phi_{12}}{\partial r_2} \frac{\partial}{\partial p_2}\right) f_2 = 0 \ . \tag{7.3.1}$$

This is a linear equation of the first order. To solve it, we must specify the distribution function at the initial time t_0:

$$f_2(X_1, X_2, t_0) = f_1(X_1, t_0) f_1(X_2, t_0) + g_2(X_1, X_2, t_0) \ . \tag{7.3.2}$$

Here and further we use upper-case characters to denote the initial values of variables of the two particles. To find the unknown distribution function $f_2(x_1, x_2, t)$ we substitute into (7.3.2) the values of X expressed from the solution of equations for characteristics in terms of variables at current time:

$$X_{1,2} = X_{1,2}(x_1, x_2, -(t - t_0)) \ . \tag{7.3.3}$$

Equation (7.3.2) is the definition of the two-particle distribution function at the initial time. In transition to the kinetic equation, the relative importance of the first and the second terms on the right-hand side of this equation is quite different.

As a matter of fact, the solution of Cauchy problem for the Boltzmann equation depends on the initial value of only the one-point function $f(x, t)$, whereas the same for the set of equations under consideration depends on the initial values of two functions: the distribution function f_1, and the two-particle correlation function g_2. So, in order to transit to the kinetic equation, one has to eliminate the initial correlation function. This is possible owing to the existence of small parameter ε, but only for a certain class of nonequilibrium processes.

Let us go back to expressions (5.3.9) for the distribution functions of rarefied gas when the external field is zero, and write the relevant expression for the two-particle space correlation function:

$$g_2(r_1,r_2) = f_2 - f_1 f_1 = \exp\left(-\Phi(|r_1 - r_2|)\right) - 1 . \tag{7.3.4}$$

Since the interaction disappears over the distance of r_0, the correlation radius r_{cor} at the state of equilibrium is of the order of r_0. Hence follow the estimates for the ratio of the correlation radius to the relaxation length (free path length), and the ratio of the respective time scales:

$$\frac{r_{cor}}{r_{rel}} \sim \frac{r_0}{l} \sim \varepsilon , \quad \frac{\tau_{cor}}{\tau_{rel}} \sim \frac{r_0}{v_T \tau} \sim \varepsilon . \tag{7.3.5}$$

We see that the correlation scales are small compared to the respective relaxation scales of the kinetic theory, and that the measure of smallness is again the density parameter ε. Now what is the situation in the nonequilibrium theory?

For the transition from the set of equations in f_1, f_2 to the kinetic Boltzmann equation we use the hypothesis of complete damping of the initial correlations — the Bogolyubov hypothesis. This hypothesis amounts to the assumption that the correlation scales for the nonequilibrium states are also defined by the microscopic scales r_0, r_0/v_T, and therefore the estimates of (7.3.5) hold.

Under these conditions we may consider the solution of (7.3.1) on the time interval $t - t_0$ which satisfies the following two-way constraint:

$$\tau_{cor} \sim \frac{r_0}{v_T} \ll t - t_0 \ll \tau . \tag{7.3.6}$$

Within this time the initial correlations in zero approximation in ε will be completely damped, whereas the distribution function $f_1(r,t)$ will remain practically the same. This allows us to drop the term with g_2 in the solution of (7.3.2) with $X_{1,2}$ from (7.3.3), and to use condition (7.2.2) of smooth time and space variation of the one-particle distribution. Then the solution of (7.3.1) is

$$f_2(x_1,x_2,t) = f_1(r_1,P_1(-\infty),t) f_1(r_1,P_2(-\infty),t) , \tag{7.3.7}$$

where

$$P_{1,2}(-\infty) = \lim_{t-t_0 \to \infty} P_{1,2}\left(-(t - t_0)\right) \tag{7.3.8}$$

denotes the initial momenta of particles which collide at the time t. Transition to the limit is possible by virtue of the left-hand inequality in (7.3.6).

Equation (7.3.8) implies that the two particles are too far from one another to interact. This allows writing the conservation law in the form

$$\frac{P_1^2}{2m} + \frac{P_2^2}{2m} = \frac{p_1^2}{2m} + \frac{p_2^2}{2m} + \Phi(|r_1 - r_2|) . \tag{7.3.9}$$

Knowing the solution of equations of motion for the two selected particles, we can use (7.3.3) to express the initial momenta via the values of variables at current time:

$$P_{1,2} = P_{1,2}(r_1 - r_2, p_1, p_2, -\infty) . \tag{7.3.10}$$

Substituting these expressions into the right-hand side of (7.3.7), we obtain the final expression for the unknown two-particle distribution function. Observe that the two-particle distribution function is expressed as the product of one-particle distribution functions with the arguments defined by (7.3.10).

Making use of the relation between functions f_2 and g_2, we find the expression for the correlation function:

$$g_2(x_1, x_2, t) = f_1(r_1, P_1(-\infty), t) f_1(r_1, P_2(-\infty), t) - f_1(r_1, p_1, t) f_1(r_1, p_2, t) . \tag{7.3.11}$$

Finally, we substitute this into (5.5.7) and obtain a closed kinetic equation in the one-particle distribution. However, the expression for the collision integral obtained in this way differs from that obtained by Boltzmann. We shall demonstrate that the two expressions are in fact equivalent. But first let us make some comments.

We assumed that the one-particle distribution functions with respect to variables r, t are smooth enough (see (7.2.2)), so that their variation on the correlation scales is negligibly small in zero approximation in ε. This does not apply to the variation of these functions with respect to momenta: it is these variations that are definitive for the nonzero correlation function (7.3.11).

Let us demonstrate that the correlation in (7.3.11) is determined by the change in the momenta of particles due to collision, whereas the purely spatial correlation (which is not associated with the distribution with respect to momenta, like in (7.3.4)) is absent.

For this purpose we carry out integration in (7.3.11) with respect to p_1, p_2, and take advantage of the fact that the Jacobian of transition from these momenta to the initial momenta P_1, P_2 is unity. After integration the two terms on the right-hand side of (7.3.11) cancel out, and therefore the space correlation function is null for the distributions integrated over the momenta.

Observe also that in the approximation under consideration, when the variation of the one-particle distribution functions over the distance r_0 is negligibly small, we may discard the second term in the expression for the Vlasov force (5.5.5) which takes care of the contribution not related to the interaction of particles. For the same reason we may drop the second term in (7.3.11) which is substituted into (5.5.7). As a result we come to the following expression for the collision integral in the form obtained by Bogolyubov (1946):

$$I(r_1, p_1, t) = n \int \frac{\partial \Phi_{12}}{\partial r_1} \frac{\partial}{\partial p_1} f(r_1, P_1(-\infty), t) f(r_1, P_2(-\infty), t) dr_2 dp_2 \ . \tag{7.3.12}$$

As said, this expression for the collision integral can be reduced to Boltzmann's form, and this will be done presently. First of all we observe that all properties of the Boltzmann integral as enumerated in Sect. 6.2 – 6.4 remain valid for the collision integral in Bogolyubov's form. Let us begin with (6.2.1) and show that this integral goes to zero for the same functions $\Phi(p)$ as in (6.2.2). For $\Phi = 1$ this is proved directly by integrating with respect to p_1. For $\Phi = p$ we first carry out integration over p_1, p_2. The remaining integral over r_2 is null, since we neglect the variation of the distribution function over the length r_0. For $\Phi = p^2/2m$ we can also prove that the integral (6.2.2) is null, although the proof is a little lengthier (see Sect. 1.11 in Klimontovich II).

To prove that substitution of the Maxwell distribution nullifies the collision integral in Bogolyubov's form, we note that for the Maxwell distribution, given the conservation law (7.3.9), the two-particle distribution function in (7.3.12) is

$$f_2 = f_1(P_1) f_1(P_2) = f(p) f(p) f_2(|r_1 - r_2|) \ . \tag{7.3.13}$$

Here $f(p)$ is the Maxwell distribution, and $f_2(|r_1 - r_2|)$ is the equilibrium spatial two-particle distribution (5.3.9) for a rarefied gas. Now we substitute this distribution into (7.3.12). Since the integrand is an odd function, integration yields a null result.

We see that the equilibrium distribution with respect to momenta nullifies the collision integral in Bogolyubov's form as well. In the presence of external field the equilibrium solution is, of course, the Maxwell – Boltzmann distribution.

In the next chapter we shall see that Bogolyubov's representation of the collision integral is very useful in the kinetic theory of nonideal gas. At this point, however, let us prove the equivalence of the two forms of collision integral.

Using (7.3.1) for the two-particle distribution function and integrating with respect to r_2, p_2, we obtain a relation which allows us to rewrite (5.5.7) (replacing g_2 with f_2) in the form

$$I = n \int \frac{\partial \Phi_{12}}{\partial r_1} \frac{\partial f_2}{\partial p_1} dr_2 dp_2 = n \int (v_1 - v_2) \frac{\partial f_2}{\partial (r_1 - r_2)} dr_2 dp_2 \ . \tag{7.3.14}$$

Then Bogolyubov's collision integral can be represented as

$$I(r_1, p_1) = n \int (v_1 - v_2) \frac{\partial}{\partial (r_1 - r_2)} \left[f_1(P_1(-\infty)) f_1(P_2(-\infty)) \right]_{r,t} dr_2 dp_2 \ . \tag{7.3.15}$$

Now in place of $r_1 - r_2$ we introduce cylindrical coordinates z, ρ, φ, direct the z–axis along the vector of relative velocity $v_1 - v_2$ and integrate over z. As a result, we come to the following expression:

$$I(r_1, p_1, t) = n \int\limits_0^\infty \rho d\rho \int\limits_0^{2\pi} d\varphi \int dp_2 |v_1 - v_2| \left[f_1(P_1(-\infty)) f_1(P_2(-\infty)) \right]_{r,t} \Big|_{z = \pm\infty} \ . \tag{7.3.16}$$

Observe that substitution of $z = \pm\infty$ splits this integral into two parts, one of which enters with the plus, and the other with the minus, which corresponds to the representation of Boltzmann collision integral in the form of (6.1.2).

At $z = -\infty$ we obtain the contribution defined by the collisions of type "a". In this case the initial momenta $P_{1,2}$ (the momenta before collision of type "a", see Fig. 1) are given by

$$P_{1,2}\big|_{z=-\infty} = p_{1,2} \ . \tag{7.3.17}$$

At $z = \infty$ we obtain the contribution defined by the collisions of type "b". Then the initial momenta $P_{1,2}$ (the momenta before collision of type "b", see Fig. 1) are given by a different equation,

$$P_{1,2}\big|_{z=\infty} = p'_{1,2} \ . \tag{7.3.18}$$

After all these manipulations, the expression for the Bogolyubov collision integral becomes

$$I(r_1, p_1, t) = n \int\limits_0^\infty \rho d\rho \int\limits_0^{2\pi} d\varphi \int dp_2 |v_1 - v_2|$$

$$\times \left[f_1(r_1, p'_2, t) f_1(r_1, p'_1, t) - f_1(r_1, p_2, t) f_1(r_1, p_1, t) \right] \ . \tag{7.3.19}$$

To make the coincidence with the Boltzmann collision integral (6.1.6) complete, we have to drop subscript "1" in f_1. This will emphasize the fact that the Boltzmann equation is written for the distribution function in a continuous 6–dimensional medium, whereas subscript "1" in Bogolyubov's form reminds us that we started with a model of gas as a population of particles. For the same reason we may drop subscripts at variables r. We shall continue this discussion in the section to follow.

7.4. Results. Unsolved Problems. Boltzmann's Hypothesis of Molecular Chaos

So, by using the perturbation theory in small parameter ε it is possible to go from the chain of the reversible BBGKY equations to the irreversible Boltzmann equation. This is accomplished in two steps.

First we use the approximation of paired collisions to obtain a closed set of reversible equations in the two distribution functions, f_1 and f_2. At the second step, approximations in the same small parameter ε are used for the transition to the irreversible Boltzmann equation.

Recall that the reversibility was lost when we applied Bogolyubov's assumption of complete damping of initial correlations g_2. This assumption is justified by the two-way inequality (7.3.6). Two important remarks are due in this connection.

First, the separation of fast (τ_{cor}) and slow (τ) time scales is performed in terms of the elapsed time $t - t_0$ rather than in terms of the characteristic time scale of the system. Because of this, the definition of the nonequilibrium Gibbs ensemble corresponding to the kinetic level of description of nonequilibrium processes remains vague (if not at all absent).

The situation becomes much clearer when we introduce information about the concrete structure of the continuous medium — information about physically infinitesimal elements. The concept of physically infinitesimal elements for a rarefied gas was introduced in Sect. 1.1.3. This concept allows us to draw a natural watershed between fast and slow processes, and thus to improve the structure of irreversible kinetic equations, as briefly discussed in the Introduction. In the chapters to follow we shall discuss thus matter in greater detail.

Secondly, the condition of complete damping of initial correlations amounts to a tacit assumption that the characteristic time of all correlations which determine the dissipation in the kinetic theory are small, $\tau_{cor} \sim \tau_0 \sim r_0/v_T$. It would be natural to call these the small-scale correlations. There also exist, however, the large-scale correlations, or the relevant large-scale fluctuations. Such are, for instance, the kinetic fluctuations with $\tau_{cor} \sim \tau$, or the fluctuations related to hydrodynamic and diffusion processes.

Thus, the condition of complete damping of initial correlations actually implies that the large-scale fluctuations do not play any significant role in the kinetic theory. In other words, it is assumed that the fluctuations of the distribution function and the fluctuations of the hydrodynamic functions are negligibly small.

Because of this, such fundamental physical phenomena as, for instance, Brownian motion and turbulence, fall outside the scope of consideration. Brownian motion can only be explained by considering the structure of the continuous medium, and, as a direct consequence, the fluctuations of the medium. In case of turbulence, the large-scale fluctuations are the essential part of motion itself.

Subsequent chapters will deal with the theory of large-scale (kinetic, hydrodynamic and diffusion) fluctuations. To round up our present discussion, let us return to the hypothesis of molecular chaos, which is the main assumption made by Boltzmann in the derivation of the kinetic equation.

As indicated earlier, in deriving the Boltzmann kinetic equation from the chain of BBGKY equations, we tacitly replace the one-particle distribution, which reflects the structure of the gas as a system of particles, with the one-point distribution $f(r,p,t)$ which has been introduced by Boltzmann. The one-point distribution in fact portrays the gas as a continuous medium in 6–dimensional space of coordinates and momenta.

The important point now is that the use of the continuous medium model is based on the introduction of physically infinitesimal volume. Let us show that proper implementation of physically infinitesimal volume will explain the essence of Boltzmann's hypothesis, and will give it some justification.

For this purpose we use information about the structure of rarefied gas (see Sect. 1.1). Two-way inequality (1.1.13) holds for a rarefied gas, which defines the relationship between three internal length scales. A fourth scale, l_{ph}, comes up when we take into account the structure of continuous medium. This scale is defined by the second expression in (1.1.17). The hierarchy of scales (1.1.13) can then be extended to include l_{ph}:

$$r_0 \ll r_{av} \ll l_{ph} \ll l . \tag{7.4.1}$$

Similar inequalities hold for the time scales:

$$\frac{r_0}{v_T} \ll \tau_{av} \ll \tau_{ph} \ll \tau . \tag{7.4.2}$$

Comparing these with (7.3.6), we find that when the structure of continuous medium is taken into account, the fast and the slow processes are being set apart on the basis of the internal parameters of the system rather than in terms of the current time $t - t_0$.

When physically infinitesimal scales are used, the one-point distribution function in the kinetic Boltzmann equation should be regarded as a dynamic distribution smoothed over physically infinitesimal volume. Using the unit weight smoothing function, we get

$$\tilde{N}(r,p,t) = \int N(r+\rho,p,t)\frac{d\rho}{V_{\text{ph}}} \ . \tag{7.4.3}$$

Thus, the smoothed one-point distribution function is defined by the microscopic phase density smoothed over physically infinitesimal volume. Obviously, a function defined in this way is, in general, a random function. To obtain a deterministic distribution, we additionally must carry out averaging over the Gibbs ensemble. As a result, we obtain the following definition of the one-point distribution in terms of the one-particle distribution used by the BBGKY theory:

$$\int \langle N(r+\rho,p,t)\rangle d\rho = \langle \tilde{N}(r,p,t)\rangle = n\tilde{f}_1(r,p,t) \ , \tag{7.4.4}$$

where we have used the relation (4.5.5) between the first moment of the microscopic phase density and the one-particle distribution function.

Ergo, the one-point distribution function in the kinetic Boltzmann equation must be regarded as the one-particle distribution function of the BBGKY theory, smoothed over physically infinitesimal volume.

To test the validity of Boltzmann's hypothesis of molecular chaos, we introduce the appropriate smoothed two-particle correlation function,

$$\tilde{g}_2(r_1,p_1,r_2,p_2,t) = \int g_2(r_1+\rho_1,r_2+\rho_2,p_1,p_2,t)\frac{d\rho_1}{V_{\text{ph}}}\frac{d\rho_2}{V_{\text{ph}}} \ , \tag{7.4.5}$$

where we have the two-particle distribution function of the BBGKY theory in the integrand.

Recall that Boltzmann's hypothesis of molecular chaos (see (6.1.5)) allows us to assume that the two-point correlation functions can be neglected. Observe that we are speaking of the two-point rather than two-particle correlation functions of the BBGKY theory. In other words, we must check whether the contribution of the smoothed correlation function $\tilde{g}_2$ can be neglected.

To this end, we represent the correlation function as the product of space correlation function and the relevant one-particle distribution functions with respect to coordinates. Then in (7.4.5) we only have to evaluate the integral of space correlation function with respect to coordinates, taking due account of the fact that Boltzmann's theory leaves out the fluctuations of the distribution function. This implies that the large-scale (kinetic) fluctuations are negligibly small. Under this condition we have $r_0 \ll l_{\text{ph}}$, and therefore

$$\tilde{g}_2(r_1, r_2)\big|_{r_1 = r_2} = \int g_2(r_1 + \rho_1, r_2 + \rho_2) \frac{d\rho_1}{V_{ph}} \frac{d\rho_2}{V_{ph}} \sim \frac{r_0^3}{V_{ph}} \sim \varepsilon^{3/2} . \tag{7.4.6}$$

Here we have used the definition of physically infinitesimal scales (1.1.17).

We see that the contribution from the correlation function into the collision integral for rarefied gas is small, which validates Boltzmann's hypothesis of molecular chaos.

So, the irreversibility of Boltzmann kinetic equation is associated with the implicit transition to the model of continuous medium. The equations for nonequilibrium processes in the approximation of continuous medium are inevitably irreversible, since information about motion within "particles" of continuous medium is irretrievably lost in transition from the system of particles to continuous medium.

In BBGKY theory, the irreversibility results from complete damping of initial correlations. This is, of course, a very important factor, which may also be discerned in Boltzmann's theory, since the time evolution only depends on the one-point initial distribution. It should be noted that the one-particle distribution in the BBGKY theory is tacitly replaced by the one-point distribution, and so the transition from the system of particles to continuous medium also takes place.

It follows that the transition to continuous medium is made both in Boltzmann's theory and in the BBGKY theory. Neither of these theories, however, studies the very feasibility of the approximation of continuous medium. Undefined, therefore, remain the concept of the "point" of continuous medium and the physically infinitesimal scales for a rarefied gas.

Here we once again face the question of the role of dynamic instability of motion of atoms of rarefied gas, which leads to mixing in the phase space and, as a consequence, necessitates the use of irreversible equations for nonequilibrium processes in macroscopic systems.

In Sect. 1.3, 1.5 we estimated the minimum time of development of instability of motion of atoms of Boltzmann gas, which turned out to be of the order of physically infinitesimal time scale. Because of this, the existence of mixing due to dynamic instability makes it necessary and possible to carry out smoothing in transition to the approximation of continuous medium.

These questions will be discussed in greater detail in subsequent chapters. We shall see that recognition of the structure of continuous medium brings about an important change in the kinetic equations: they must include additional dissipative terms which take care of the redistribution of particles in the conventional rather than in the phase space.

So, we have set forth the guidelines for further development of the kinetic theory. Let us begin with generalizing the kinetic Boltzmann equation for the nonideal gas.

CHAPTER 8

KINETIC THEORY OF NONIDEAL GAS

We know now that the density parameter ε in rarefied gas plays a variety of roles. It characterizes dissipative processes, since the time and length of relaxation are expressed in terms of ε. As follows from (7.2.1), relaxation towards the state of equilibrium cannot be described in zero approximation in ε, since the time of relaxation in this approximation is infinite.

By contrast, calculation of nondissipative characteristics (for instance, thermodynamic functions) can be carried out in zero approximation in ε. As a matter of fact, internal energy, pressure, entropy and other thermodynamic characteristics in this approximation are nonzero and describe the thermodynamic state of ideal gas.

The kinetic Boltzmann equation corresponds exactly to this approximation. It describes nonequilibrium processes in ideal gas. In this connection we have to deal with two questions.

The closed set of equations (5.5.6), (7.1.1), obtained from the BBGKY chain of equations in the approximation of paired collisions, fully accounts for the contribution due to interaction. We assumed that the contribution of interaction into the thermodynamic functions dropped out because we neglected the variation of one-particle distribution function $f_1(r,p,t)$ on the microscopic scales $\tau_0 = r_0/v_T$, r_0. According to (7.2.2), this corresponds to the zero approximation in ε, or to the approximation of ideal gas. This assumption must, of course, be verified.

The second problem consists in the construction of the more general kinetic equation for nonideal gas, which would take into account the contribution of interaction to the thermodynamic functions. This problem is dealt with in the present chapter.

8.1. Equilibrium State. Perturbation Theory in Density Parameter

At equilibrium, there are two possibilities for calculating the thermodynamic functions. First, one can use the canonical Gibbs distribution. For approximate calculations, in many cases it is more convenient to use the chain of BBGKY equations for the sequence of equilibrium distribution functions.

In case of rarefied gas, this chain of equations is solved by perturbation theory in small parameter ε. Recall that in the absence of external field, when the equilibrium distribution of particles is spatially homogeneous, the distribution functions f_1, f_2 are defined by (5.3.9).

These results correspond to the zero approximation in the density parameter ε. In principle, for the state of equilibrium it is possible to write expressions for these functions in any arbitrary approximation in small parameter ε. For illustration, let us quote here the result corresponding to the first approximation:

$$f_2(r_1,r_2) = \exp\left(-\frac{\Phi_{12}}{\kappa T}\right)\left[1+n\int\left(\exp\left(-\frac{\Phi_{13}}{\kappa T}\right)-1\right)\left(\exp\left(-\frac{\Phi_{23}}{\kappa T}\right)-1\right)dr_3\right],$$

$$(8.1.1)$$

$$g_2(r_1,r_2) = f_2(r_1,r_2)-1 .$$

We see that, as compared with (5.3.9), this expression contains an additional term which is proportional to the density parameter.

As indicated, the solution can be formally obtained in any approximation in ε. Higher-order approximations, however, pose formidable computational difficulties. Because of this, the two-particle distributions for dense gases are obtained from model integral equations in f_2, g_2. Examples of such equations can be found in (Klimontovich V, Ch. 20, and Martynov 1992). Now we are going to perform exemplary calculations of thermodynamic functions of nonideal rarefied gas.

8.2. Thermodynamic Functions of Nonideal Rarefied Gas

Let us give the results of calculations of the main thermodynamic functions for the state of equilibrium the first approximation in ε. We assume that the potential of interaction has the following form:

$$\Phi(|r|) = \begin{cases} \infty & \text{at } r \leq r_0 \\ \Phi(|r|) < 0, \ |\Phi(|r|)| << \kappa T & \text{at } r > r_0 \end{cases} .$$

$$(8.2.1)$$

As opposed to the model of elastic balls (2.1.6), here at $r > r_0$ we have weak attraction.

Internal energy. We substitute (5.3.9) for the distribution function f_2 into expression (5.2.9) for the internal energy. The result can be represented as

$$\frac{U}{V} = \frac{3}{2}n\kappa T - an^2 \,, \tag{8.2.2}$$

where

$$a = 2\pi \int_{r_0}^{\infty} |\Phi(r)| r^2 dr \tag{8.2.3}$$

is one of the Van der Waals' constants. We see that because of interaction the internal energy decreases.

Pressure. To find the equation of state of nonideal rarefied gas we turn to (5.2.11). Using once again the expression (5.3.9) for the two-particle distribution and the potential of interaction (8.2.1), after simple manipulations we find that

$$P = n\kappa T\left[1 + n\left(b - \frac{a}{\kappa T}\right)\right] \,, \tag{8.2.4}$$

where

$$b = \frac{2\pi}{3}r_0^3 \tag{8.2.5}$$

is the other Van der Waals' constant.

The inclusion of interaction adds two terms to the equation of state. One term is positive and depends on the repulsion of particles, since it is proportional to b. The other is negative and depends on the attraction of particles.

Fluctuations of the number of particles within given volume. We return to expression (5.6.9) which defines the correlator of fluctuations of microscopic phase density in 6–dimensional space (r, p). Now we use it for calculating the fluctuations of the number of particles within the volume ΔV in rarefied nonideal gas.

Since the distribution with respect to momenta at equilibrium is the Maxwell distribution, we may carry out integration over momenta in (5.6.9) to obtain the expression for the correlator of fluctuations of local density of the number of particles:

$$\langle \delta n \delta n \rangle_{r,r'} = n^2 g_2(r,r') + \left[\delta(r-r')n(r) - \frac{1}{N}n(r)n(r')\right] \,. \tag{8.2.6}$$

To obtain the sought-for equation, we introduce appropriate notation for the number of particles within the selected volume ΔV:

$$N_{\Delta V} = \int\limits_{\Delta V} n(r)dr \ , \quad \delta N_{\Delta V} = \int\limits_{\Delta V} \delta n(r)dr \ . \tag{8.2.7}$$

Under the condition that the number of particles within the selected volume $N_{\Delta V}$ is much less than the total number of particles in the system, after some straightforward algebra we obtain the expression for the dispersion of the number of particles within the selected volume:

$$\left\langle (\delta N_{\Delta V})^2 \right\rangle = N_{\Delta V}\left[1 + n\int g_2(r)dr\right] \ . \tag{8.2.8}$$

In case of ideal gas we must set here $g_2 = 0$.

In the first approximation in density parameter, we must use (7.3.4) for the correlation function. If the potential of interaction is again given by (8.2.1), the expression for the dispersion of the number of particles becomes

$$\left\langle (\delta N_{\Delta V})^2 \right\rangle = N_{\Delta V}\left[1 + 2n\left(\frac{a}{\kappa T} - b\right)\right] \ . \tag{8.2.9}$$

Here, like in (8.2.4) for the pressure, the interaction is accounted for by two terms having opposite signs, one of which depends on the repulsion and the other on the attraction of particles.

Observe finally that the dispersion of the number of particles can be expressed in terms of isothermal compressibility β_T. To do this we use the equation

$$\beta_T = \frac{1}{n}\left(\frac{\partial n}{\partial P}\right)_T = \frac{1}{n\kappa T}\left(1 + n\int g(|r|)dr\right) \ , \tag{8.2.10}$$

which allows us to rewrite (8.2.8) in the form of (4.12.1). In any case, (8.2.9) is justified when Gaussian distribution holds for the fluctuations of thermodynamic functions (see Sect. 4.11). The question of its validity near the critical point, when Gaussian distribution is no longer valid, we shall leave open for the time. This will be discussed in the chapter devoted to the theory of phase transitions.

8.3. Entropy and Free Energy of Nonideal Gas

Boltzmann defined the entropy of nonequilibrium rarefied gas by (6.4.1). It is for this definition that the H–theorem was proved. As we already know, however, the Boltzmann equation only holds for the ideal gas. This implies that the interaction defines only the dissipative processes. The latter cannot be described in the ideal gas approximation, since the time of transition to the state of equilibrium is infinite.

The fact that Boltzmann's approximation corresponds to the thermodynamic functions of ideal gas also follows from Boltzmann's definition of entropy. To wit, for the state of equilibrium (that is, for the Maxwell distribution), Boltzmann's formula (6.4.1) goes over into (6.4.2) for the entropy of ideal gas.

How then do we define the entropy of nonideal gas?

For this purpose we turn to the general definition of entropy (3.5.15) which holds for an arbitrary nonequilibrium state of the system of N interacting particles with the distribution function $f_N(X,t)$. We express the N–particle distribution in terms of the one-particle distribution and the correlation functions g_2, g_3, ... For a rarefied gas we only take paired collisions into account, and thus $g_3 = g_4 = ... = 0$. Then the expression for the entropy takes on the following form (Green 1953; Klimontovich 1975):

$$S(t) = S_B(t) - \frac{\kappa n^2}{2} \int \ln \frac{f_2(x_1, x_2, t)}{f_1(x_1, t) f_1(x_2, t)} f_2(x_1, x_2, t) dx_1 dx_2 , \qquad (8.3.1)$$

where S_B is the Boltzmann entropy (6.4.1). As in case of ideal gas, the entropy (8.3.1) is proportional to the total number of particles in the system N. For ideal gas, when $g_2 = 0$, expression (8.3.1) coincides with Boltzmann's formula.

Since

$$\int \ln \frac{f_2}{f_1 f_1} f_2 dx_1 dx_2 \geq 0 ,$$

we have

$$S(t) \leq S_B .$$

Thus, the entropy of nonideal gas for arbitrary potential of interaction is less than the entropy of ideal gas. This does not imply, however, that the state of nonideal gas is more ordered. To assess the relative degree of order, we must compare the values of entropy related to the same value of the mean energy. We shall return to this question in the chapter devoted to criteria of self-organization.

Let us reduce the general expression (8.3.1) to the state of equilibrium. With this purpose we first carry out integration over the momenta, and recall that the correlation function at equilibrium only depends on the difference in the coordinates. Then

$$S = S_{id} - V \frac{\kappa n^2}{2} \int_0^\infty \ln f_2(r) f_2(r) 4\pi r^2 dr .$$

(8.3.2)

Finally, we use (5.3.9) for the two-particle distribution to get

$$S = S_{id} - V \frac{\kappa n^2}{2} \int_0^\infty \left[\frac{\Phi}{\kappa T} \exp\left(-\frac{\Phi}{\kappa T}\right) + \left(1 - \exp\left(-\frac{\Phi}{\kappa T}\right)\right) \right] 4\pi r^2 dr \leq S_{id} .$$

(8.3.3)

In particular, for the potential of interaction as defined by (8.2.1) we have

$$S = S_{id} - V \kappa n^2 b - V \kappa n^2 2\pi \int_{r_0}^\infty \frac{(|\Phi(r)|)^2}{(\kappa T)^2} r^2 dr \leq S_{id} ,$$

(8.3.4)

where b is the Van der Waals' constant.

Let us now consider the pertinent expressions for the free energy. For the state of equilibrium the free energy can be derived from the above expressions for internal energy and entropy:

$$F = U - TS = F_{id} - V \frac{n^2}{2} \kappa T \int_0^\infty \left(1 - \exp\left(-\frac{\Phi}{\kappa T}\right)\right) 4\pi r^2 dr .$$

(8.3.5)

For the potential (8.3.1) the expression for the free energy becomes

$$F = F_{id} - V n^2 \kappa T \left(b - \frac{a}{\kappa T}\right) .$$

(8.3.6)

We see that the contribution of interaction to the free energy is, like it was with the pressure, expressed in terms of the two Van der Waals' constants.

Let us now analyze the changes in the structure of the kinetic Boltzmann equation brought about by taking into account the contribution of interaction to the nondissipative (thermodynamic) characteristics.

8.4. Higher Approximations in Density Parameter in Kinetic Theory

Recall that the distribution functions f_1, f_2 in the zero approximation in density parameter for equilibrium rarefied gas are defined by (5.3.9). The more general expression (8.1.1) for the two-particle distribution function accounts also for the contribution in the first order in ε. For the state of equilibrium this procedure can be continued. As a result, we come to the so-called virial (in density parameter) expansion for, say, the two-particle distribution function. This expansion generates the corresponding virial expansion for internal energy and pressure, since the interaction enters the expressions for these quantities via the distribution function f_2 (see (5.2.9), (5.2.11)).

So, for the state of equilibrium it is possible to find the distribution functions (and hence the thermodynamic functions) in any approximation in density parameter, and thus to find virial expansions. Can the same be done for nonequilibrium states?

At first it seemed that there are no insurmountable obstacles in solving this problem. As a matter of fact, we already know the solution for the two-particle distribution. This solution is given by (7.3.11). This approximation is sufficient for obtaining the kinetic equation which describes nonequilibrium processes in the ideal gas.

The next step was also a success.

To wit, G.E. Uhlenbeck, G.F. Choh, and, independently, M.S. Green found the expression for the two-particle distribution function in the first approximation in density parameter ε. This approximation includes not only the dissipation due to triple collisions, but also the first-order corrections in ε to the nondissipative characteristics (thermodynamic functions). The resulting more general kinetic equation takes into account not only paired, but also triple collisions.

This stirred up hope that the same way could be pursued to obtain kinetic equations in any approximation in ε. Further progress, however, ran into formidable difficulties. Investigations showed that the collision integrals diverge for collisions between four and more particles (Cohen 1971; see references to Section 17 in Klimontovich II, and Lifshitz – Pitayevsky 1979).

Divergence of the collision integral for four-particle interaction is logarithmic:

$$I_{(4)} \propto \ln \frac{t}{\tau_{(0)}} , \tag{8.4.1}$$

where the subscript refers to the number of particles participating in collisions. The collision integrals for a greater number of particles diverge even faster, as a power function.

The appearance of diverging (at $t \to \infty$) term in the kinetic equation indicates that collisions of as many as four particles cannot be treated separately from collisions involving five, six, and more particles. In other words, collective interactions of particles must be taken into account. Such calculations have been carried out using the so-called method of summation of the fastest-diverging integrals in all terms of the expansion in density parameter (Kawasaki – Oppenheim; see also Section 18 in Lifshitz – Pitayevsky 1979). Summation allows one to obtain a finite expression for the collision integral, because (8.4.1) is replaced by the time-independent function

$$I_{(4)} \propto \ln \frac{\tau}{\tau_{(0)}} \sim \ln \frac{1}{\varepsilon} . \tag{8.4.2}$$

We see that the dependence of thus obtained collision integral on small parameter ε is non-analytical. This is the mathematical reason why it is not possible to obtain a kinetic equation for dense gas by using the perturbation theory in density parameter.

So, we cannot use expansions in density parameter for constructing kinetic equations for nonideal gases. What is the physical background of this setback?

To answer this question we go back to the problem of structure of continuous medium in the context of kinetic description. In Sect. 1.1 we introduced physically infinitesimal scales in the kinetic description of ideal gas. The relevant time interval is given by (1.1.16), where τ is the free path time for paired collisions. Hence it follows that the definition of physically infinitesimal scales, which determine the structure of continuous medium, depends on the nature of the relaxation process — in the given context, on the number of colliding particles.

Now let us in the same way define physically infinitesimal scales for collisions of three and four particles. We find that (Klimontovich II, Sect. 16; Klimontovich IV, Ch. 20):

$$\tau_{ph}^{(3)} \sim (\varepsilon)^{1/4} \tau , \ \ l_{ph}^{(3)} \sim (\varepsilon)^{1/4} l , \ \ \tau_{ph}^{(4)} \sim \tau , \ \ l_{ph}^{(4)} \sim l . \tag{8.4.3}$$

These relations indicate that in case of triple collisions it is possible to define physically infinitesimal scales and, consequently, the structure of the continuous medium, because the relevant scales are "much less" (by a factor of $\varepsilon^{1/4}$) than the scales of paired collisions τ, l.

In case of four-particle collisions, however, this is no longer possible, since the pertinent physically infinitesimal scales are not small compared to the minimum relaxation scales of rarefied gas. This means that, to incorporate four-and-more-particle collisions into the kinetic equations, we must revise drastically our description of nonequilibrium processes.

Such modification requires taking into account the effect of damping (dissipation) due to paired collisions on the collisions of a larger number of particles. This procedure removes the difficulties which arise when we try to include the collisions between four and more particles into consideration. For details, the interested reader may refer to (Klimontovich 1973; Klimontovich II).

8.5. Kinetic Equations for Nonideal Rarefied Gas

In the preceding section we learned that the perturbation theory in density parameter is only capable of yielding finite results when the collision between two or three particles are taken into consideration. Even then, however, the description of kinetic processes is not quite consistent.

Indeed, to derive the Boltzmann equation in the approximation of paired collisions, we describe the dissipative processes in the first order in density parameter, whereas the nondissipative processes are described in zero order. Because of this, the kinetic equation only holds for ideal gas. In the approximation of triple collisions, the dissipative processes are described in the second order in density parameter, and the nondissipative processes in the first.

We may say that there is a kind of "lagging" in the description of nondissipative processes. Is it possible to be more coherent, and describe nonequilibrium processes in rarefied gases using the same order in the density parameter (or, in other words, the same approximation in terms of the number of colliding particles) for both dissipative and nondissipative terms in the kinetic equations? This would result in more consistent kinetic equations for rarefied nonideal gases.

The relevant results are presented in detail in (Klimontovich II). They are based on the above-discussed methods of BBGKY equations and microscopic phase density. Kadanov and Beim (1964) have based their theory of nonequilibrium processes in nonideal systems on the method of Green's functions, widely used in the statistical theory (see Zubarev 1971; Lifshitz – Pitayevsky 1978). Here we shall briefly discuss the meaning of this generalization and quote the main results. The particulars can be found in (Klimontovich II) and references cited therein. Let us begin with the most simple case.

Spatially homogeneous distribution of the gas. In place of (7.3.1) we use a more general equation (7.1.1) for the two-particle distribution function. As before, we assume that condition (7.2.3) is satisfied. Then, given that the distribution of the gas is spatially uniform, we only have to add the time lag to our former solution (7.3.7). As a result, we get the following expression for the two-particle distribution function:

$$f_2(x_1, x_2, t) = f_1(P_1(-\infty), t) f_1(P_2(-\infty), t)$$

$$-\frac{\partial}{\partial t} \int_0^\infty \tau \frac{d}{d\tau} f_1(P_1(-\tau), t) f_1(P_2(-\tau), t) d\tau \ . \tag{8.5.1}$$

As compared with the Bogolyubov's solution, here we have an additional term which takes care of the time change of the distribution function in the process of collisions. The corresponding additional term appears also in the collision integral,

$$I(p_1, t) = I_B(p_1, t) + I_{(1)}(p_1, t) \ , \tag{8.5.2}$$

where

$$I_{(1)} = -n \frac{\partial}{\partial t} \int_0^\infty \frac{\partial \Phi_{12}}{\partial r_1} \frac{\partial}{\partial p_1} \tau \frac{d}{d\tau} f_1(P_1(-\tau), t) f_1(P_2(-\tau), t) d\tau dr_2 dp_2 \ . \tag{8.5.3}$$

Subscript (1) at the additional collision integral indicates that another additional term will appear if we take spatial inhomogeneity into account.

Let us consider the properties of the generalized collision integral. The first term in (8.5.2) coincides with Boltzmann's collision integral and has therefore the properties defined in (6.2.2). The additional term $I_{(1)}$ has the same properties only for $\varphi(p) = 1, p$. At $\varphi = p^2/2m$ the relevant expression is nonzero and can be represented as

$$n \int \frac{p^2}{2m} I_{(1)}(p, t) dp = -\frac{\partial}{\partial t} \frac{n^2}{2} \int \Phi_{12} f_2(x_1, x_2, t) \frac{dx_1 dx_2}{V} \ . \tag{8.5.4}$$

We see that the integral in question is defined by the time derivative of the mean potential energy of interaction of particles. Recall that the total internal energy is given by (5.2.7). Since we are concerned with the first approximation in density parameter, we may retain only the first term in expression (8.5.1) for the two-particle distribution.

In Ch. 11, which is devoted to equations of gas dynamics, we shall see that the last property ensures conservation of internal energy of nonideal gas. Now let us consider a more general case.

Spatially inhomogeneous distribution of the gas. In this case the expression for the two-particle distribution and the collision integral include additional terms, which take care of the change in the one-particle distribution functions over distances of the order of r_0. These terms are proportional to spatial derivatives, and the collision integral can be represented as

$$I(r,p,t) = I_B(r,p,t) + I_{(1)}(r,p,t) + I_{(2)}(r,p,t) ,$$ (8.5.5)

Once again, the first term coincides with the Boltzmann collision integral, this time for spatially inhomogeneous gas. The other two are defined by space-time changes of the one-particle distribution functions over the atomic scales τ_0, r_0.

Consider the properties of this generalized collision integral. The integral is zero for $\varphi = 1$. For $\varphi = p$ we get

$$n\int \varphi(p)I(r,p,t)dp = -\frac{\partial}{\partial r}\frac{n^2}{6}\int r\frac{d\Phi(r)}{dr}f_2(x_1,x_2,t)\frac{dx_1dx_2}{V} .$$ (8.5.6)

We see that the integral in question is defined by the spatial derivative of that part of the total pressure (5.2.10) which is due to the interaction of particles. This property will enable us to obtain the Navier – Stokes equation for nonideal gas (see Ch. 11).

In the first approximation in ε, it is sufficient to retain only the first term on the right-hand side of expression (8.5.1) for f_2.

Finally, at $\varphi = p^2/2m$ we come to the following equation:

$$n\int \frac{p^2}{2m}I(p)dp = -\frac{n^2}{2}\int \Phi_{12}\left(\frac{\partial}{\partial t} + \frac{v_1+v_2}{2}\frac{\partial}{\partial r}\right)f_2(x_1,x_2,t)\frac{dx_1dx_2}{V}$$

$$-\frac{n^2}{6}\frac{\partial}{\partial r}\int \frac{v_1+v_2}{2}r\frac{d\Phi}{dr}f_2(x_1,x_2,t)\frac{dx_1dx_2}{V} .$$ (8.5.7)

Here also we may use Bogolyubov's formula for f_2 — that is, the zero approximation in density parameter. This property of the generalized collision integral will be important for the construction of equations of gas dynamics for nonideal gas.

We see that it is possible to construct the kinetic theory of nonideal gas in the first approximation in density parameter (that is, in the approximation of paired interactions), in such a way that the interaction between particles defines both the dissipative processes and the pertinent contributions to the thermodynamic functions. In Ch. 11 this kinetic equation will be employed for obtaining the set of equations of gas dynamics for nonideal gas.

A further step is also possible in the construction of kinetic equations for nonideal gas. This step corresponds to the second approximation in density parameter, which means that paired and triple collisions are taken into account. These results are presented in Sect. 17 in (Klimontovich II). The inclusion of triple interactions will also give rise to additional terms in equations of gas dynamics (Ch. 11).

As indicated earlier, collective processes become important when multiple collisions are taken into consideration. Because of this, the structure of the corresponding kinetic equations is considerably modified. We shall touch upon these questions in the forthcoming chapters. A comprehensive theory of complicated nonequilibrium processes has not yet been developed.

CHAPTER 9

KINETIC THEORY OF FLUCTUATIONS

In the preceding chapters we considered two approaches to the derivation of kinetic equation for rarefied gas. The first is based on physical intuition and goes back to Boltzmann himself. The second approach, called the Bogolyubov – Born – Green – Kirkwood – Yvon method, relies on the reversible Liouville equation.

Distribution function in the Boltzmann equation is deterministic (nonrandom). Because of this, fluctuations of the distribution function fall out of consideration. Since the initial dynamic equations contain, in principle, complete information about the system, we may ask where on our way towards the kinetic equation we lost information about fluctuations of the distribution function.

We have already discussed this question. Let us repeat the main points.

Boltzmann's theory actually treats rarefied gas as a continuous medium in 6–dimensional space of coordinates and momenta $x = (r, p)$. The state of the medium is characterized by the one-point distribution $f(r, p, t)$. The problem of the structure of "continuous medium" — that is, the definition of a "point" — is left out of consideration. This circumstance gave rise to many questions. Let us recall the words of Leontovich (1935) quoted in the beginning of Ch. 6:

"The structure of this theory, however, is certainly far from perfection. The quantity $f(r, p, t) dr dp$ has to be assigned the meaning of a certain statistical mean (mathematical expectation) [...] Within the framework of the theory, however, it is not clear what is meant by this "mathematical expectation" [...] For this reason, the theory is also unable to give any information about fluctuations in the gas and their trends."

So, where did we lose the fluctuations in the method of BBGKY? We have already pointed to two pitfalls:

(1) The kinetic equation is derived under a very severe Bogolyubov's condition of complete damping of initial correlations. To construct the kinetic theory of fluctuations, we must replace this condition with a less restrictive assumption of partial damping of initial correlations (Klimontovich II; 1971; 1973). This implies that in transition to the kinetic description we may use the condition of damping of only the small-scale fluctuations, whose correlation scales are much smaller than the relaxation scales for the one-particle distribution function.

142

(2) Derivation of the kinetic equation is associated with the implicit replacement of the Gibbs ensemble related to the initial Liouville equation by another Gibbs ensemble which reflects the transition from the system of particles to the continuous medium.

Interestingly, in his seminal paper M.A. Leontovich proposed a dissipative equation for the N–particle distribution of all particles of the gas $f_N(X,t)$ which could be used for formulating the kinetic theory with due account for the fluctuations. These results were published long before the works of Bogolyubov, Born and Green, and Kirkwood, but remained practically unnoticed. Only much later was the Leontovich equation rediscovered by other students (see Klimontovich 1983; Spohn 1991).

We shall consider two approaches to the construction of the kinetic theory of fluctuations. The first is based on the Leontovich equation for the N–particle distribution function. The other relies on the Boltzmann equation for a random (non-deterministic) distribution function, which can be presented as the Langevin equation (Boltzmann equation for a random function with the Langevin source).

We shall see that the two approaches are equivalent. The choice between them will depend on the particular problem in question. The second is more efficient in case of complex systems.

9.1. Leontovich Kinetic Equation

Let us go back to the Liouville equation. Recall that the Liouville equation is reversible, and its solution is determined by the solution of the Hamilton equations. For a rarefied gas it is also possible to write an irreversible equation in $f_N(X,t)$.

To this end, we use the principle of partial damping of initial correlations, which means neglecting only the small-scale fluctuations whose characteristic times τ_{cor} are much smaller than the relaxation time (the free path time τ).

The division of fluctuations into small-scale and large-scale is based on the definition of physically infinitesimal scales, which at the kinetic stage of relaxation for a rarefied gas are given by (1.1.17). The concept of physically infinitesimal scales allows us to define the Gibbs ensemble for the statistical description of nonequilibrium processes.

In case of rarefied gas, the condition of partial damping of initial correlations allows us to go over from the reversible Liouville equation to the irreversible equation for the N–particle distribution function (Klimontovich II; 1983; Spohn 1991):

$$\hat{L}_{x_1,\dots,x_N} f_N(X,t) = I_N(x_1,\dots,x_N,t) \;. \tag{9.1.1}$$

The left-hand side of this equation coincides in form with the Liouville equation, whereas the right-hand side contains the collision integral as defined by

$$I_N(x_1,\ldots,x_N,t) = \sum_{1\leq i<j\leq N} \int_0^\infty \rho_{ij}d\rho_{ij} \int_0^{2\pi} d\varphi_{ij}\delta(r_i - r_j)|v_i - v_j|$$

$$\times \left[f_N(x_1,\ldots,r_i,p_i',\ldots,r_j,p_j',\ldots,x_N,t) - f_N(x_1,\ldots,x_i,\ldots,x_j,\ldots,x_N,t) \right] . \tag{9.1.2}$$

It was an equation of this type that was proposed by Leontovich. He started, however, not with the Liouville equation, but rather with the integral equation in the probabilities of transition from one state of the system to another. Such equations had been first introduced by Smoluchowski in the theory of Brownian motion; they are also called the Chapman – Kolmogorov equations. We shall discuss them in the chapter devoted to the theory of Brownian motion.

Let us consider some of the properties of the Leontovich equation.

Observe that, unlike the Boltzmann equation, it is a *linear* integro-differential equation. Instead of the product of one-particle distributions, the integrand now contains the difference in the N–particle distribution functions. The arguments of these functions include the momenta of pairs of colliding particles which obey the conservation laws of (6.1.3). Summation in the collision integral (9.1.2) is carried out for all pairs of colliding particles.

Like the Boltzmann equation, this equation does not take into account the space-time change of the distribution functions. Indeed, the collision integral neglects the time lag. Delta-function $\delta(r_i - r_j)$ indicates that the colliding particles belong to one and the same point. Thus, the Leontovich equation, like the Boltzmann equation, only holds for a thermodynamically ideal gas.

At the same time, the Leontovich equation is much more general than the Boltzmann equation, since it takes care of the statistical correlation of positions and momenta of different particles. If we choose to neglect these correlations, and define the many-particle distribution as the product of the one-particle distributions,

$$f_N(x_1,\ldots,x_N,t) = \prod_{1\leq i\leq N} \left(\frac{1}{V} f_1(x_i,t)\right) , \quad \frac{1}{V}\int f_1(x_i,t)dx_i = 1 , \tag{9.1.3}$$

then, substituting this into the Leontovich equation for the one-particle distributions, we obtain an equation which formally coincides with the Boltzmann equation. To make the coincidence complete, we must identify the one-particle distribution, which relates to the states of the particles, with the one-point distribution, which characterizes the state of the continuous medium.

Let us now evaluate the minimum relaxation time.

In a system of many particles, the process of relaxation towards the state of equilibrium falls into several stages. The minimum relaxation time is determined by the interval within which one of the N particles of the system completes another act of collision,

$$(\tau_{rel})_{min} \sim \frac{\tau}{N} \, . \tag{9.1.4}$$

Obviously, such short time intervals are inconsequential in the kinetic context. This estimate just serves to illustrate how short is the time lapse on which the irreversibility sets in. In the kinetic theory, the pertinent estimate is obtained by using the number of particles within the physically infinitesimal volume, N_{ph}, instead of the total number of particles N. This brings us to the definition of the physically infinitesimal time for the Boltzmann gas (see (1.1.16), (1.1.17)).

9.2. Equilibrium Solution of Leontovich Equation

Assume that the external forces are absent. Then the collision integral is zero at the state of equilibrium. This is possible when the difference in the distribution functions in (9.1.2) is zero. Let us write the relevant equation in logarithms:

$$\ln f_N\left(p_1,...,p_i',...,p_j',...,p_N\right) = \ln f_N\left(p_1,...,p_i,...,p_j,...,p_N\right) \, . \tag{9.2.1}$$

Given the laws of conservation of momentum and energy of pairs of colliding particles (6.1.3), the solution of this functional equation can be represented in the form

$$\ln f_N(p_1,...,p_N) = 1 + \beta \sum_{1 \leq i \leq N} p_i + \gamma \sum_{1 \leq i \leq N} \frac{p_i^2}{2m} \, . \tag{9.2.2}$$

Assume that the gas as a whole is at rest. Then the total momentum is zero, and the sought-for distribution function is represented as the N–dimensional Maxwell distribution:

$$f_N(p_1,...,p_N) = \left(\frac{1}{2\pi m \kappa T}\right)^{3N/2} \exp\left(-\frac{\displaystyle\sum_{1 \leq i \leq N} \frac{p_i^2}{2m}}{\kappa T}\right) \, . \tag{9.2.3}$$

This distribution allows finding not only the one-particle, but also many-particle moments of momenta of particles.

9.3. H-Theorem. Lyapunov Functional

Using the general definition of entropy (3.5.15) and the Leontovich equation, we find the equation of entropy balance. Subscript "L" will be used to remind that these results are based on the Leontovich equation. Then

$$\frac{dS_L}{dt} = -\kappa \int \ln f_N(X,t) I_N(X,t) dX \equiv \sigma_L(t) \ . \tag{9.3.1}$$

Substituting the collision integral (9.1.2) into the expression for entropy production, we carry out symmetrization with respect to momenta before and after collisions and use the known inequality $\ln(b/a)(b-a) \geq 0$. The resulting inequality expresses the H–theorem:

$$\frac{dS_L}{dt} = \sigma_L(t) \geq 0 \ . \tag{9.3.2}$$

The "equals" sign corresponds to the state of equilibrium, when the state of the system is described by the N–dimensional Maxwell distribution.

Now let us construct the Lyapunov functional for the Leontovich equation. Recall that Boltzmann's H-theorem can be reduced to two inequalities (1.6.3), (1.6.4) for the Lyapunov functional Λ_S, which is defined by the difference in the entropies of the equilibrium and nonequilibrium states of the Boltzmann gas. This is possible because the mean kinetic energy of the particle (1.6.2) is conserved in the course of time evolution of the Boltzmann gas.

Along with equation (1.6.2), which follows from energy conservation at paired interactions, a more general condition holds for the Leontovich equation:

$$\langle E \rangle = \left\langle \sum_{1 \leq i \leq N} \frac{p_i^2}{2m} \right\rangle = \text{const} \ . \tag{9.3.3}$$

It is this condition that allows us to define the Lyapunov functional,

$$\Lambda_S = S_0 - S_L(t) = \kappa \int \ln \frac{f_N(X,t)}{f_N(p_1,...,p_N)} f_N(X,t) dX \geq 0 \ , \tag{9.3.4}$$

which is defined as the difference in the total entropies of the entire gas for the equilibrium and nonequilibrium states. Using the H-theorem, we get the second inequality,

$$\frac{d\Lambda_S}{dt} = \frac{d}{dt}\left(S_0 - S_L(t)\right) \le 0 \ . \tag{9.3.5}$$

Thus, we a have complete analogy with Boltzmann's H–theorem. However, owing to the fact that the correlations of the one-particle distributions are taken into consideration, the Leontovich equation contains more information about the system. Let us prove this statement.

9.4. Comparison Between Entropies of Boltzmann and Leontovich. Correlations and Order of States of Gas

As before, by S_B and S_L we denote the entropies of Boltzmann and Leontovich. To find the Boltzmann entropy for the entire gas, we substitute expression (9.1.3) for the N–particle distribution (neglecting fluctuations) into the general definition of non-equilibrium entropy. As a result, we get:

$$S_B(t) = -\kappa \int \ln\left[\prod_{1 \le i \le N} \frac{1}{V} f_1(x_i,t)\right] \prod_{1 \le i \le N}\left[\frac{1}{V} f_1(x_i,t)\right] dx_1 ... dx_N \ .$$

Let us transform this expression.

Note that the logarithm reduces to the sum over individual particles. In each term, averaging reduces to averaging with respect to one-particle distribution. Hence,

$$S_B(t) = -\kappa \sum_{1 \le i \le N} \int \ln\left[\frac{1}{V} f_1(x_i,t)\right]\frac{1}{V} f_1(x_i,t) dx_i \ . \tag{9.4.1}$$

Next we use the relationship between the one-particle distribution and the N–particle distribution,

$$\frac{1}{V} f_1(x_1,t) = \int f_N(x_1,...,x_N,t) dx_2 ... dx_N \ . \tag{9.4.2}$$

This relationship holds for a particle with any number. Because of this, the expression for the Boltzmann entropy can be written in the form

$$S_B(t) = -\kappa \int \ln\left[\prod_{1 \le i \le N} \frac{1}{V} f_1(x_i,t)\right] f_N(x_1,...,x_N,t) dx_1 ... dx_N \tag{9.4.3}$$

Now, using the general definition of the Leontovich entropy S_L and the last expression for the Boltzmann entropy S_B, we can compare the values of the two:

$$S_B(t) - S_L(t) = \kappa \int \ln \frac{f_N(x_1,...,x_N,t)}{\prod_{1 \le i \le N} \frac{1}{V} f_1(x_i,t)} f_N(x_1,...,x_N,t) dX \ge 0 . \tag{9.4.4}$$

Here we have taken advantage of the known fact that $\ln a \ge 1 - 1/a$, where a is the ratio of the distribution functions in the approximations of Leontovich and Boltzmann.

We see that the entropy of the gas in Boltzmann's approximation is greater than that in Leontovich's approximation. This implies that the entropy is less when the kinetic correlations are taken into account. This is natural, since the Leontovich equation carries additional information about the system.

The difference between the two entropies is zero for the state of equilibrium. Then the N–dimensional distribution reduces to the Maxwellian N–dimensional distribution, which does not carry any additional information, being representable as the product of the one-particle Maxwell distributions. This fact indicates that the correlations described by the Leontovich equations are nonequilibrium characteristics of the gas.

In the presence of external forces the steady distribution may differ considerably from the equilibrium distribution. In this case, however, the Boltzmann entropy is also greater than the Leontovich entropy.

Now we go over to the description of nonequilibrium fluctuations on the basis of the Leontovich kinetic equation. For this we shall need some facts from the BBGKY theory.

9.5. Chain of Dissipative Equations for a Sequence of Distribution Functions

Now we go over from the Leontovich equation to the chain of equations for a sequence of distribution functions for one, two, and more particles. As before, these functions are related to the N–particle distribution function by (5.1.1), (5.1.2). This time, however, f_N satisfies the dissipative Leontovich equation rather than the Liouville equation.

Using (5.1.1) and the Leontovich equation, we obtain the following equation for function $f_1(r,p,t)$:

$$\frac{\partial f_1}{\partial t} + v_1 \frac{\partial f_1}{\partial r_1} + F_0 \frac{\partial f_1}{\partial p_1} = n \int \frac{\partial \Phi_{12}}{\partial r_1} \frac{\partial}{\partial p_1} f_2(r_1, p_1, r_2, p_2, t) dr_2 dp_2$$

$$+ n \int_0^\infty \rho d\rho \int_0^{2\pi} d\varphi \int dr_2 dp_2 \delta(r_1 - r_2)|v_1 - v_2| \qquad (9.5.1)$$

$$\times \left[f_2(r_1, p_1', r_2, p_2', t) - f_2(r_1, p_1, r_2, p_2, t) \right]$$

Let us compare this with the first equation in the BBGKY chain, equation (5.4.2).

Equation (9.5.1) is also not closed; what is more, it is irreversible because on the right-hand side there is a dissipative term. This term is determined by paired interactions, and differs from the Boltzmann collision integral in that the integrand contains the two-particle correlation g_2.

In Ch. 7 we assessed the contribution of the two-particle correlation to the collision integral (see (7.4.6)), and showed it to be negligibly small for a rarefied gas (proportional to $\varepsilon^{3/2}$). This conclusion bears out Boltzmann's hypothesis of molecular chaos.

It follows that the dissipative term on the right-hand side of (9.5.1) to a good accuracy coincides with Boltzmann's collision integral. Let us now transform the first term on the right-hand side. The following must be taken into account.

The Leontovich equation was derived under the assumption of partial damping of initial correlations. This means that the small-scale correlations are excluded from consideration — that is, the correlations whose characteristic scales are smaller than the physically infinitesimal scales. The division of correlations (or the corresponding fluctuations) into small-scale and large-scale was carried out in Sect. 1.1.3, where the physically infinitesimal scales were used for this purpose.

As a consequence, the distribution functions are already smoothed over the physically infinitesimal volume when we are using the chain of dissipative equations obtained on the basis of the Leontovich equation. This allows us to treat f_1 in (9.5.1) as the one-point distribution function for continuous medium. The linkage between this function and the one-point distribution of the BBGKY theory is given by (7.4.3), (7.4.4).

Accordingly, the two-particle distribution must be regarded as a smoothed function, and the corresponding correlation function (see (7.4.5)) will be defined as

$$\tilde{f}_2(x_1, x_2, t) = f(x_1, t) f(x_2, t) + \tilde{g}_2(x_1, x_2, t) . \qquad (9.5.2)$$

Then the equation for the one-particle distribution can be written in the form

$$\frac{\partial f}{\partial t} + v \frac{\partial f}{\partial r} + F(r, t) \frac{\partial f}{\partial p} = I_B(r, p, t) + \tilde{I}(r, p, t) . \qquad (9.5.3)$$

The first term on the right-hand side is the Boltzmann collision integral. The additional "collision integral" is given by

$$\tilde{I}(x_1,t) = n\int \frac{\partial \Phi |r_1 - r_2|}{\partial r_1} \frac{\partial}{\partial p_1} \tilde{g}_2(r_1,p_1,r_2,p_2,t)dr_2 dp_2 \ . \tag{9.5.4}$$

The "~" sign on top of I indicates that the additional "collision integral" is determined by the large-scale fluctuations; F is the Vlasov force.

So, the first equation in the dissipative chain, based on the Leontovich equation, coincides with the corresponding Boltzmann equation only when $\tilde{g}_2 = 0$ — that is, when the large-scale correlations are completely neglected.

Now we may continue the chain and write the equation for $\tilde{g}_2$. This equation will contain the three-particle correlation function $\tilde{g}_3$, and so forth. As a result, we come to a chain of dissipative equations, and once again face the problem of truncation.

For truncating the chain of the BBGKY equations we used the approximation of paired interactions, which brought us to a closed set of reversible equations in f_1, f_2. In order to obtain a closed equation in the one-particle distribution we additionally postulated complete damping of initial correlations. In this way, the Boltzmann equation was derived.

Now the situation is much different. As a matter of fact, the dissipation due to paired collisions has already been taken into account in the Leontovich equation itself. In order to include the additional dissipation due to the large-scale fluctuations, we must consider many-particle interactions. To understand how the chain of dissipative equations can be cut short, let us first discuss some examples of closing the chain of the BBGKY equations.

9.6. Approximation of Perturbation Theory in Interaction, and Approximation of Second Correlation Functions in BBGKY Theory

We go back to the chain of the BBGKY equations obtained in Ch. 5 from the reversible Liouville equation. The first equation in (5.5.6), (5.5.7) remains intact. The second equation, given the definitions of the correlation functions g_2, g_3, is transformed into

$$\left(\hat{L}_{x_1 x_2} - \hat{\Theta}_{12}\right) g_2(x_1,x_2,t) = \hat{\Theta}_{12} f_1(x_1,t) f_1(x_2,t)$$

$$+ n\int \frac{\partial \Phi_{13}}{\partial r_1} \frac{\partial}{\partial p_1} \left[f_1(x_1,t) g_2(x_2,x_3,t) + g_3(x_1,x_2,x_3,t) \right] dx_3$$

$$+ n\int \frac{\partial \Phi_{23}}{\partial r_2} \frac{\partial}{\partial p_2} \left[f_1(x_2,t) g_2(x_1,x_3,t) + g_3(x_1,x_2,x_3,t) \right] dx_3 \tag{9.6.1}$$

where we have introduced the operator

$$\hat{\Theta}_{ij} = \frac{\partial \Phi_{ij}}{\partial r_i}\frac{\partial}{\partial p_i} + \frac{\partial \Phi_{ij}}{\partial r_j}\frac{\partial}{\partial p_j} , \tag{9.6.2}$$

and the operator of evolution of a pair of particles under the action of the mean force (5.5.5) (Vlasov force)

$$\hat{L}_{x_1 x_2} = \frac{\partial}{\partial t} + v_1\frac{\partial}{\partial r_1} + v_2\frac{\partial}{\partial r_2} + F(r_1,t)\frac{\partial}{\partial p_1} + F(r_2,t)\frac{\partial}{\partial p_2} . \tag{9.6.3}$$

Note that the first term on the right-hand side of equation for g_2 is defined by the one-particle distributions. Such terms will be referred to as *sources*. The remaining terms include the correlation functions.

Now we can discuss different methods used for truncating the chain of equations in f_1, g_2, g_3,....

Perturbation theory in interactions. Since the correlation functions are zero in the absence of interaction, they are proportional to the interaction when the latter is weak. Because of this, in the linear approximation with respect to interaction (first approximation of the perturbation theory) only the first term in retained on the right-hand side of (9.6.1), and this equation takes on the form

$$\hat{L}_{x_1 x_2} g_2(x_1, x_2, t) = \hat{\Theta}_{12} f_1(x_1, t) f_1(x_2, t) . \tag{9.6.4}$$

In spite of its outward simplicity, the straightforward perturbation theory in interactions is inefficient. This can be illustrated for a system of charged particles with Coulombian interaction. The pertinent kinetic equation was obtained by Landau in 1937. It was found that the "collision integral" in the first approximation of the perturbation theory is divergent for both short and long distances. This question will be discussed in greater detail in chapters dealing with the theory of plasma. At this point we just mark the following.

Divergence on small distances is easily eliminated. For large distances, however, the problem is more fundamental, because the divergence stems from collective interactions between the Coulombian particles. This circumstance cannot be taken into account in the first approximation of the perturbation theory.

At the same time, the perturbation theory cannot be used for short distances because the interaction is strong (of the order of the mean kinetic energy of the particle).

From arguments developed above it follows that one has to take care simultaneously of the short-range strong interactions, and the long-range collective interactions. This is an extremely complicated venture. Let us first consider two extreme cases.

Approximation of paired interactions. This approximation has already been used for deriving the kinetic Boltzmann equation (7.2.4), which can be rewritten in a form better suited for our current purpose:

$$\left(\hat{L}_{x_1 x_2} - \hat{\Theta}_{12}\right) g_2(x_1, x_2, t) = \hat{\Theta}_{12} f_1(x_1, t) f_1(x_2, t) . \tag{9.6.5}$$

In the first approximation of the perturbation theory the second term on the right-hand side is negligibly small, and this equation coincides with (9.6.4).

Approximation of second correlation functions. Let us consider another extreme case, when the interaction is weak but collective. This means that many particles are involved into the sphere of interaction. A vivid example of such system is a completely ionized plasma.

Let us turn to the exact equation (9.6.1). Since the interaction is weak, the second term on the left-hand side can be dropped. In the right-hand side we use the approximation of small correlations,

$$g_2(x_1, x_2, t) << f(x_1, x_2, t) f(x_2, t) , \quad g_3 = 0 , \quad g_4 = 0 . \tag{9.6.6}$$

As a result, (9.6.1) is much simplified, and becomes

$$\hat{L}_{x_1 x_2} g_2(x_1, x_2, t) = \hat{\Theta}_{12} f_1(x_1, t) f_1(x_2, t)$$

$$+ n \int \left[\frac{\partial \Phi_{13}}{\partial r_1} \frac{\partial f_1(x_1, t)}{\partial p_1} g_2(x_2, x_3, t) + \frac{\partial \Phi_{23}}{\partial r_2} \frac{\partial f_1(x_2, t)}{\partial p_2} g_2(x_1, x_3, t) \right] dx_3 . \tag{9.6.7}$$

In the left-hand side, this equation coincides with (9.6.4) which has been obtained in the approximation of weak interaction. It describes the behavior of the correlation function at small distances. The term on the right is proportional to the concentration of particles, and describes the collective effect of correlation of the selected particle with all surrounding particles. In Boltzmann's approximation such correlations were neglected. We shall see that they may be important or even definitive in the description of many phenomena (Brownian motion is one example).

For describing the large-scale fluctuations it is convenient to use another form of (9.6.7). The fact is that the two-particle correlation function is a characteristic of the statistical linkage between particles. More adequate for describing the processes in the continuous medium is the method of moments of microscopic phase density.

In Sect. 5.6 we established the linkage between the moments and the correlation functions. In particular, expression (5.6.9) defines the relationship between the correlator of fluctuations of microscopic phase density and the two-particle correlation function. Importantly, this correlator is nonzero also when $g_2 = 0$. This implies that the approximation of the second correlation functions and the approximation of the second moments are not equivalent. The difference between the two reflects the existence of the atomic structure of the "continuous medium", and may be expressed as

$$\left\langle \delta N(x,t)\delta N(x',t)\right\rangle_{\text{source}} = n\left[\delta(x-x')f(x,t) - \frac{1}{V}f(x,t)f(x',t)\right]. \tag{9.6.8}$$

Subscript "source" indicates that this correlator characterizes the atomic structure of the continuous medium in equations for fluctuations of phase density, and thus serves as a source of fluctuations of phase density. In the thermodynamic limit the second term in this expression is negligibly small (cf. (5.6.10)).

Equation (9.6.7) will be transformed in two steps. First, we use (5.6.4) and (5.6.6) to relate the correlator of fluctuations of force and phase density to the two-particle distribution function:

$$\left\langle \delta F(r,t)\delta N(x',t)\right\rangle = -\int \frac{\partial \Phi(|r-r''|)}{\partial r}\left\langle \delta N(x',t)\delta N(x'',t)\right\rangle dx''$$

$$= -n\frac{\partial \Phi(|r-r''|)}{\partial r}f(x,t) - n^2\int \frac{\partial \Phi(|r-r''|)}{\partial r}g_2(x',x'',t)dx'' . \tag{9.6.9}$$

We also use a similar expression for the correlator $\left\langle \delta N\delta F\right\rangle_{x,r',t}$. Then (9.6.7) can be rewritten in a more concise form:

$$\hat{L}_{xx'}g_2(x,x') + \left\langle \delta F\,\delta N\right\rangle_{r,x',t}\frac{\partial f(x,t)}{\partial p} + \left\langle \delta N\delta F\right\rangle_{x,r',t}\frac{\partial f(x',t)}{\partial p'} = 0 . \tag{9.6.10}$$

At the second step, we use (5.6.9) and the notation for correlator (9.6.8) which characterizes the role of the atomic structure of "continuous medium". Then (9.6.10) can be represented as

$$\hat{L}_{xx'}\left\langle \delta N\delta N\right\rangle_{x,x',t} + \left\langle \delta F\,\delta N\right\rangle_{r,x',t}\frac{\partial f(x,t)}{\partial p} + \left\langle \delta N\delta F\right\rangle_{x,r',t}\frac{\partial f(x',t)}{\partial p'}$$

$$= \hat{L}_{xx'}\left\langle \delta N\delta N\right\rangle_{x,x',t}^{\text{source}} . \tag{9.6.11}$$

We see that all terms on the left-hand side of this equation contain moments of fluctuations of phase density and force. The term on the right is the source which characterizes the structure of "continuous medium". In the approximation of second moments this term is zero. The source plays a crucial role in the description of dissipative processes with due account for the large-scale fluctuations.

Now we can apply this knowledge to the formulation of the closed set of equations based on the Leontovich kinetic equation. This will pave the way to the kinetic theory of fluctuations.

9.7. Leontovich Equation. Approximation of Second Correlation Functions

Now we return to the chain of dissipative equations for the sequence of distribution functions, obtained on the basis of the Leontovich equation.

The first equation (9.5.3) of the chain contains two "collision integrals" on the right-hand side. The first coincides with Boltzmann's collision integral. As we know, this term describes the dissipation due to the small-scale correlations. The second "collision integral", whose structure is yet unknown, is defined by the large-scale fluctuations.

Since the pertinent correlation scales are much greater than the physically infinitesimal scales, the correlation volume contains many particles. For this reason the large-scale correlations are small. Since the correlator of these fluctuations is linked with the correlation function by equation (9.5.2), there are good reasons for choosing the approximation of second correlation functions, when condition (9.6.6) is satisfied, as the tool for truncating the dissipative chain of equations.

Recall that in the approximation of second correlation functions in the BBGKY theory, either (9.6.10) for the two-particle correlation function, or the equivalent equation (9.6.11) for the two-point correlator of fluctuations of phase density can be chosen as the second equation which closes the set. The second option is more consistent with the model of continuous medium, and will be used in our analysis of the chain of dissipative equations based on the Leontovich equation.

Derivation of equations for the two-particle correlation function and the two-point correlator of fluctuations of phase density is described in detail in Ch. 3 and 4 in (Klimontovich II). Here we shall give the final equation for the two-point correlator of fluctuations of phase density and compare it with the corresponding equation (9.6.11) from the BBGKY theory:

$$\left(\hat{L}_{xx'} + \delta\hat{i}_p + \delta\hat{i}_{p'}\right)\langle\delta N\delta N\rangle_{x,x',t} + \langle\delta F\delta N\rangle_{r,x',t}\frac{\partial nf(x,t)}{\partial p}$$

$$+ \langle\delta N\delta F\rangle_{x,r',t}\frac{\partial nf(x',t)}{\partial p'} = A(x,x',t) .$$

(9.7.1)

This equation differs from (9.6.11) of the BBGKY theory in two respects.

First, dissipative terms (linearized collision operators) have appeared on the left-hand side. The action of this operator upon an arbitrary function $F(p)$ is defined by

$$-\delta\hat{i}_{p_1}F(p_1) = n\int_0^\infty \rho d\rho \int_0^{2\pi} d\varphi \int dx_2 \delta(r_1 - r_2)|v_1 - v_2|$$

$$\times\left[f(p_1')F(p_2') + F(p_1')f(p_2') - f(p_1)F(p_2) - F(p_1)f(p_2)\right]_{r_1,t} .$$

(9.7.2)

Its structure is defined by the Boltzmann collision integral. This circumstance allows us to maintain that the additional terms in this equation define the dissipation of large-scale fluctuations due to the small-scale correlations, the elimination of which results in the Boltzmann collision integral.

The second difference is associated with the modification of the structure of the source, the right-hand side of equation. The source is now represented as the sum of two terms:

$$A(x,x',t) = \left(\hat{l}_{xx'} + \delta\hat{i}_p + \delta\hat{i}_{p'}\right)\langle\delta N\delta N\rangle_{x,x',t}^{\text{source}}$$

$$+ n\left\{I_B(x,x',t) - \frac{1}{V}\left[I_B(x,t)f(x',t) + f(x,t)I_B(x',t)\right]\right\} .$$

(9.7.3)

The first term here differs from the source in (9.6.11) by inclusion of the dissipative terms similar to those on the left-hand side of (9.7.1). The second term is new; it reflects the atomic structure of "continuous medium". Along with the Boltzmann collision integrals, it contains the so-called non-integrated collision integral

$$I_B(x_1,x_2,t) = n\int_0^\infty \rho d\rho \int_0^{2\pi} d\varphi \; \delta(r_1 - r_2)|v_1 - v_2|\left[f(p_1')f(p_2') - f(p_1)f(p_2)\right]_{r_1,t} .$$

(9.7.4)

which, by contrast to the Boltzmann collision integral, does not include integration with respect to x_2.

This expression for the source is complicated, and defies a clear-cut physical interpretation. It can, however, be reduced to a form which may be called canonical (Klimontovich II – IV). In the canonical form the source is represented as the sum of two terms,

$$A(x,x',t) = A_B(x,x',t) + \tilde{A}(x,x',t) \ . \tag{9.7.5}$$

The first term with subscript "B" (for Boltzmann) is defined by the operator of paired collisions and can be written as

$$A_B(x,x',t) = \left[\left(\delta\hat{I}_p + \delta\hat{I}_{p'}\right) - \left(\delta\hat{I}_p + \delta\hat{I}_{p'}\right)_0\right]\langle\delta N \delta N\rangle^{\text{source}}_{x,x',t} \ . \tag{9.7.6}$$

Subscript "0" indicates that the collision operators in brackets only act upon the regular functions, and do not apply to the delta-function. Recall that delta-function enters the expression (9.6.8) for the correlator of fluctuations of phase density, which acts as the source.

Since the second term on the right does not contain the delta-function, the action of operator $\left[(...) - (...)_0\right]$ on this term gives a null contribution. As a result, we come to a simpler expression,

$$A_B(x,x',t) = \left[\left(\delta\hat{I}_p + \delta\hat{I}_{p'}\right) - \left(\delta\hat{I}_p + \delta\hat{I}_{p'}\right)_0\right]n\delta(x - x')f(x,t) \ . \tag{9.7.7}$$

Further simplification is possible for the equilibrium state, when f is the Maxwell distribution. Then the term with operator $(...)_0$ is null, and

$$A_B(x,x',t) = \left(\delta\hat{I}_p + \delta\hat{I}_{p'}\right)n\delta(x - x')f(p) \ . \tag{9.7.8}$$

We see that this source of fluctuations is nonzero even for the state of equilibrium. This ought to be expected, since this source reflects the atomic structure of "continuous medium".

Let us now analyze the second term on the right-hand side of (9.7.5). This term takes care of the contribution of the large-scale fluctuations, and is given by

$$\tilde{A}(x,x',t) = n\delta(x - x')\tilde{I}(x,t) - \frac{n}{V}\left[\tilde{I}(x,t)f(x',t) + f(x,t)\tilde{I}(x',t)\right] \ . \tag{9.7.9}$$

Recall that the "collision integral" $\tilde{I}(x,t)$ is determined by the large-scale correlation function $\tilde{g}_2(x,x',t)$ according to (9.5.4). At the state of equilibrium this integral vanishes (Klimontovich II; 1971), and so does the source $\tilde{A}(x,x',t)$.

Assume that the state of the system is so close to the equilibrium that the large-scale fluctuations are negligibly small. Then the corresponding "collision integral" is also negligibly small, and the kinetic equation (9.5.3) coincides with the Boltzmann kinetic equation, but includes also the Vlasov force:

$$\frac{\partial f(x,t)}{\partial t} + v\frac{\partial f}{\partial r} + F(r,t)\frac{\partial f}{\partial p} = I_B(x,t) . \tag{9.7.10}$$

In this approximation, the correlator of fluctuations of force and microscopic phase density is also zero, since the "collision integral" $\tilde{I}(x,t)$ is expressed in terms of this correlator (we shall prove this in the next chapter). This does not mean, however, that the more general correlator $\langle \delta N \delta N \rangle_{x,x',t}$ is zero: the other source of fluctuations $A_B(x,x',t)$ does not vanish, but the equation for this source assumes a much simpler form:

$$\left(\hat{L}_{x,x'} + \delta\hat{I}_p + \delta\hat{I}_{p'}\right)\langle \delta N \delta N \rangle_{x,x',t} = A_B(x,x',t) . \tag{9.7.11}$$

As follows from (9.7.6), in this approximation, when the large-scale fluctuations are assumed to be negligibly small, the source is completely determined by the solution of the kinetic Boltzmann equation.

So, for calculating the small-scale nonequilibrium fluctuations we have a closed set of equations (9.7.10), (9.7.11), which must be used together with (9.7.6) for the source of fluctuations. Note that these equations are based on the Leontovich equation in the approximation of second correlation functions.

It must be emphasized that this approximation differs considerably from the approximation of second moments, which corresponds to the approximation of continuous medium. In the latter the source $A_B(x,x',t)$, which reflects the structure of "continuous medium", is zero, and equation (9.7.11) for the correlator of fluctuations of phase density becomes homogeneous. Because of this, it is not possible to take fluctuations into account in the approximation of second moments.

Now let us look at the general set of equations (9.5.3), (9.7.1) with the source (9.7.5), which corresponds to the approximation of second correlation functions. We see that this approximation is a generalization of Boltzmann's theory in two respects.

First, the atomic structure of "continuous medium" is taken into account. This allows one to calculate the fluctuations which are definitive for many physical phenomena, like the molecular scattering of light or the Brownian motion. By Brownian motion we understand not just the classical Brownian motion of small macroscopic particles in a fluid, but also the "Brownian motion" of most diverse characteristics — for example, of phase and amplitude in self-oscillatory systems. The corresponding source of fluctuations $A_B(x,x',t)$ may be referred to as the *molecular source*.

The second generalization consists in the inclusion of the large-scale fluctuations, which become important at states far from equilibrium. Such states may arise, in particular, when instability develops — for example, when the instability of laminar motion give rise to a much more complex turbulent motion.

For this reason we call $\tilde{A}(x,x',t)$ the *turbulent source*. There are many physical varieties of turbulent motion. In all cases, however, the common feature is the high intensity of large-scale (coarse-grain) fluctuations. We discussed some aspects of turbulent motion in Sect. 1.7, and will have more to say in the forthcoming chapters.

So, we have gone through the basics of the kinetic theory of fluctuations. For the starting point we used the Leontovich equation — the dissipative kinetic equation for the N–particle distribution $f_N(X,t)$ of rarefied (dilute) gas.

There is another approach to the kinetic theory of fluctuations, based on the kinetic equation for $\tilde{N}(x,t)$, the microscopic phase density smoothed over physically infinitesimal volume. Like the microscopic phase density $N(x,t)$, function $\tilde{N}(x,t)$ is a random function, but it it much smoother. This will be the subject of the next chapter.

In the next chapter we shall show that the problem of calculation of kinetic fluctuations may be regarded as a problem of Brownian motion. The main tool is the kinetic equation with an additional random source, the so-called Langevin force, and the distribution function is regarded as Brownian particle.

CHAPTER 10

LANGEVIN METHOD IN KINETIC THEORY OF FLUCTUATIONS

In Ch. 5 and 7, devoted to the basics of the statistical theory of nonequilibrium processes and validation of the kinetic Boltzmann equation, we used two alternative methods.

The first is the BBGKY method, based on the Liouville equation — the dynamic equation for the N–particle distribution function $f_N(X,t)$. The second method is based on the dynamic equations for the microscopic phase density in 6–dimensional phase space. The two methods are equivalent, although each of them has certain advantages for particular applications.

Similarly, there are two alternative approaches to the construction of the kinetic theory of fluctuations, based on dissipative equations. One is based on the Leontovich equation for the distribution of states of rarefied gas. The other relies on the equation for the microscopic phase density smoothed over physically infinitesimal volume,

$$\tilde{N}(x,t) \equiv n\tilde{f}(x,t) = \int N(r-\rho,\rho,t)F(\rho)d\rho \ . \tag{10.0.1}$$

This equation differs from (7.4.3) in that here we have introduced the smoothing function. This function is normalized to unity, and its width is determined by physically infinitesimal length scale. The dissipative equation in this function will now form the basis of the kinetic theory of fluctuations.

10.1. Dissipative Equation for Smoothed Microscopic Phase Density

Let us go back to equation (2.4.14) for the microscopic phase density. The microscopic force in this equation is related to the microscopic phase density by (2.4.9). In each term of these equations we are going to carry out smoothing over physically infinitesimal volume.

Smoothing destroys information about the microstates of the particles within the points of "continuous medium". As a result, the equation becomes dissipative. Like with the Boltzmann equation in the BBGKY theory, we assume that the dissipation is due to the damping of initial correlations. The undertone of this assumption is that the large-scale correlations do not play any significant role. This circumstance considerably narrows the class of phenomena under consideration.

As we know, the construction of the kinetic theory of fluctuations requires replacing the Bogolyubov condition by a less severe condition of partial damping of initial correlations. Since damping only concerns the small-scale fluctuations, the distribution function in the kinetic equation must be regarded as a random function. Consequently, for the smoothed distribution function we use the familiar kinetic equation

$$\left(\frac{\partial}{\partial t} + v\frac{\partial}{\partial r} + \tilde{F}(r,t)\frac{\partial}{\partial p}\right)\tilde{f}(r,p,t) = \tilde{I}_{B}(r,p,t) ,$$

$$\frac{1}{V}\int \tilde{f}(r,p,t)drdp = 1 .$$

(10.1.1)

Here we have used notation for the smoothed microscopic force which is similar to the Vlasov force:

$$\tilde{F}(r,t) = F_0 - n\frac{\partial}{\partial r}\int \Phi\big(|r - r'|\big)\tilde{f}(r',p',t)dr'dp' .$$

(10.1.2)

The dissipative term $\tilde{I}_{B}(r,p,t)$ coincides in form with the Boltzmann collision integral:

$$\tilde{I}_{B}(r,p_1,t) = n\int_0^\infty \rho d\rho \int_0^{2\pi} d\varphi \int dp_2 |v_1 - v_2| \big[\tilde{f}(p_1')\tilde{f}(p_2') - \tilde{f}(p_1)\tilde{f}(p_2)\big]_{r,t} .$$

(10.1.3)

So, the use of the less rigorous condition of partial damping of initial correlations allows treating these equations as equations in random functions. Because of this, it becomes possible to construct a sequence of equations in the moments of random functions $\tilde{f}(r,p,t)$, $\tilde{F}(r,p,t)$, which will form the basis for the kinetic theory of fluctuations.

Recall that a similar situation is encountered in thermodynamics. According to Reynolds' hypothesis, the equations of hydrodynamics in the description of turbulence are regarded as equations in random functions. Averaging of these equations results in a sequence of meshing equations in the moments of random thermodynamic functions.

The set of equations obtained in this manner is, however, so complicated that one is faced with the problem of closure — the problem of cutting short the infinite chain of equations.

10.2. Sequence of Equations in Moments of Smoothed Microscopic Phase Density

In (10.1.1) – (10.1.3) we carry out averaging with respect to Gibbs ensemble, and use the following notation for the large-scale fluctuations:

$$\delta \tilde{f} = \tilde{f} - f \ , \quad \delta \tilde{F} = \tilde{F} - F \ . \tag{10.2.1}$$

Swung dash "~" above the fluctuations indicates that here we are dealing with the large-scale fluctuations. Small-scale fluctuations have already been eliminated in the definition of collision integral I_B; hereinafter we drop the swung dash.

Now we use the definition of one-point distribution (7.4.4) and the identity

$$\left\langle \tilde{F}(r,t)\tilde{f}(r,p,t) \right\rangle = F(r,t)f(r,p,t) + \left\langle \delta F(r,t)\delta f(r,p,t) \right\rangle \ . \tag{10.2.2}$$

Then we get the following equations for the distribution function $f(r,p,t)$ and the mean force:

$$\frac{\partial f}{\partial t} + v\frac{\partial f}{\partial r} + F(r,t)\frac{\partial f}{\partial p} = I_B(r,p,t) + \tilde{f}(r,p,t) \ , \tag{10.2.3}$$

$$F(r,t) = F_0 - n\frac{\partial}{\partial r}\int \Phi(|r - r'|)f(r',p',t)dr'dp' \ . \tag{10.2.4}$$

Two comments are due at this point.

First, in Boltzmann collision integral we neglected the contribution from correlator of large-scale fluctuations, which is equivalent to neglecting the two-particle correlation function $\tilde{g}_2(x,x',t)$ in the integrand. This contribution to the collision integral has been evaluated in (7.4.6); it is of the order of $\varepsilon^{3/2}$ and is consequently very small for a rarefied gas.

The additional "collision integral" is now defined by the correlator of fluctuations of force and phase density:

$$\tilde{I}(r,p,t) = -\frac{1}{n}\frac{\partial\langle \delta F \delta N \rangle}{\partial p} \ . \tag{10.2.5}$$

Since fluctuations of force and phase density are linked by

$$\delta F(r,t) = -\int \frac{\partial \Phi(|r-r'|)}{\partial r} \delta N(r',p',t) dr' dp' \; , \tag{10.2.6}$$

the additional collision integral can be expressed in terms of the correlator of fluctuations of phase density,

$$\bar{I}(r,p,t) = \frac{1}{n} \int \frac{\partial \Phi(|r-r'|)}{\partial r} \frac{\partial}{\partial p} \langle \delta N \delta N \rangle_{r,p,r',p',t} dr' dp' \; . \tag{10.2.7}$$

Recall that earlier the additional collision integral was given by (9.5.4). The equivalence of these two expressions can be proved using the relation (5.6.10) between the correlator of fluctuations of phase density and the two-particle correlation function for large-scale correlations.

So we arrived at the same equation for the one-particle distribution as that obtained from the Leontovich equation. The equivalence of the two approaches can be proved in general.

10.3. Equation for Fluctuations of Distribution Function

As we have seen, equation (10.2.3) for the one-particle distribution $f(x,t)$, like the corresponding equation (9.5.3) obtained from the Leontovich equation, is not closed. Indeed, the additional collision integral (10.2.5) is determined by the one-time correlator of fluctuations of smoothed phase density and smoothed microscopic force.

To calculate this correlator using (10.1.1) – (10.1.3) for smoothed random functions and equations in deterministic functions $f(x,t)$, $F(r,t)$, we write the appropriate equations for fluctuations. They have the following form:

$$\left(\frac{\partial}{\partial t} + v \frac{\partial}{\partial r} + F(r,t) \frac{\partial}{\partial p} \delta \hat{I}_p \right) \delta N(r,p,t) + \delta F(r,t) \frac{\partial n f(r,p,t)}{\partial p} =$$

$$- \frac{\partial}{\partial p} \left[\delta N(r,p,t) \delta F(r,t) - \langle \delta N(r,p,t) \delta F(r,t) \rangle \right] \; , \tag{10.3.1}$$

$$\delta F(r,t) = -\int \frac{\partial \Phi(|r-r'|)}{\partial r} \delta N(r',p',t) dr' dp' \; . \tag{10.3.2}$$

We see that equations for fluctuations are nonlinear, since the right-hand side of the first of these contains terms quadratic in fluctuations. Because of this, equations in the second moments, derived from equations from fluctuations, include third moments. As a result we once again come to a chain of meshing dissipative equations in the moments of smoothed random functions. They can be shown (see, for example, Klimontovich II) to be equivalent to the chain of dissipative equations based on the Leontovich equation.

In the approximation of second correlation functions (under conditions (9.6.6)) we arrive at the closed set of equations (9.5.3), (9.7.1) for the one-point distribution $f(x,t)$ and the second moment $\langle \delta N(x,t)\delta N(x',t)\rangle$. The source in equation for second moments is defined by (9.7.5) – (9.7.9) and characterizes the two levels of superstructure over the classical Boltzmann theory which correspond to the inclusion of, respectively, molecular and turbulent large-scale fluctuations.

Now let us consider another approach to the description of large-scale fluctuations. This approach is similar to the Langevin method in the theory of Brownian motion; this time, however, it is applied to the description of much more complicated fluctuative processes.

10.4. Langevin Method in Kinetic Theory of Fluctuations

We go back to equation (10.3.1) for fluctuations of smoothed phase density. In the approximation of second moments the right-hand side of this equation is zero. As we have more than once pointed out earlier, the approximation of second moments does not take into account the atomic structure of "continuous medium". So we chose to consider a more adequate approximation of second correlation functions, which gave rise to a source $A(x,x',t)$ on the right-hand side of (9.7.1).

Now we describe the atomic structure of "continuous medium" in a different way. To wit, we retain only the linear terms in (10.3.1), which will correspond to the approximation of second moments, but include a so far unknown random (Langevin) source into the right-hand side to take care of the atomic structure. This brings us to the following equation:

$$\left(\frac{\partial}{\partial t} + v\frac{\partial}{\partial r} + F(r,t)\frac{\partial}{\partial p} + \delta\hat{I}_p\right)\delta N + \delta F\frac{\partial nf}{\partial p} = y(x,t) \ . \tag{10.4.1}$$

Fluctuation of the force δF is linked with the fluctuation of phase density by our former relation (10.3.2). The Langevin source is represented as a random Gaussian process. This is justified because the approximation in question leads to a closed set of equations only for the two first moments. Then the statistical properties of the Langevin source are completely defined by the two first moments:

$$\langle y(x,t)\rangle = 0 \ , \ \langle y(x,t)y(x',t')\rangle = A_L(x,x',t)\delta(t-t') \ . \tag{10.4.2}$$

When the structure of "continuous medium" is taken into account, the width of δ-function in this expression is defined by physically infinitesimal time interval τ_{ph}, so the definition of the Langevin correlator corresponds to the zero approximation in small parameter of time τ_{ph}/τ of the continuous medium in kinetic theory.

The intensity of Langevin source $A_L(x,x')$ in (10.4.2) is as yet undefined; actually, it is the same as that of the source $A(x,x',t)$ in equation (9.7.1) for the second moment.

To show this, we pay attention to the fact that juxtaposition of the Langevin equation (10.4.1) and equation (9.7.1) for the second moment results in the following equation:

$$\langle y(x,t)\delta N(x',t)\rangle + \langle \delta N(x,t)y(x',t)\rangle = A(x,x',t) \ . \tag{10.4.3}$$

In order to obtain the equation for the sources, we transform the expression for the correlator on the left-hand side of (10.4.3).

Since the source is delta-correlated (in the above sense), it suffices to know the solution of (10.4.1) on the small time scale of the order of τ_{ph}. Since all terms of (10.4.1), except the Langevin source, operate on the time scale characteristic of the kinetic stage of evolution, this equation in the zero approximation in small parameter τ_{ph}/τ can be written in "abridged" form

$$\frac{\partial \delta N(r,p,t)}{\partial t} = y(r,p,t) \ . \tag{10.4.4}$$

The sought-for correlator does not depend on the initial condition, and so the solution of (10.4.4) is

$$\delta N(x,t) = \int_{-\infty}^{t} y(x,t')dt' = \int_{0}^{\infty} y(x,t-\tau)d\tau \ . \tag{10.4.5}$$

Using (10.4.2) for the correlator of Langevin source, we obtain the desired expression for correlators on the left-hand side of (10.4.3):

$$\langle y(x,t)\delta N(x',t)\rangle = \langle \delta N(x,t)y(x',t)\rangle = \frac{1}{2}A_L(x,x',t) \ . \tag{10.4.6}$$

Juxtaposition of (10.4.3) and (10.4.6) shows that

$$A(x,x',t) = A_L(x,x',t) \ , \tag{10.4.7}$$

which is an important result: the source in (9.7.1) for the second moment and the Langevin source are the same.

By virtue of this, the intensity of the Langevin source can be represented, like in (9.7.5), as a sum of two contributions. The first of these is given by (9.7.8) and reflects the existence of atomic structure of "continuous medium". This is the *molecular source*.

The second contribution to the intensity of the Langevin source is given by (9.7.9) and is important for states far from equilibrium, when the large-scale fluctuations play an important, often decisive role because of the development of various instabilities. For this reason this source of fluctuations is referred to as *turbulent*.

Large-scale fluctuations are, as a rule, of minor importance in the states close to equilibrium. This circumstance allows us to neglect the additional "collision integral" $\tilde{I}(x,t)$ in the kinetic equation (10.2.3), and the term with $\delta\tilde{F}(r,t)$ in the Langevin equation (10.4.1). Then the kinetic equation (10.2.3) coincides with the Boltzmann equation, and the Langevin equation (10.4.1) takes on the form

$$\left(\frac{\partial}{\partial t} + v\frac{\partial}{\partial r} + F(r,t)\frac{\partial}{\partial p} + \delta\hat{I}_p\right)\delta\tilde{N}(x,t) = y(x,t) . \tag{10.4.8}$$

The left-hand side coincides with the linearized Boltzmann equation, and the intensity of the Langevin source is given by (9.7.7). The latter is, of course, nonzero even in the state of equilibrium, when it is defined by a simpler expression (9.7.8) (Kadomtsev 1957).

We see that the calculation of fluctuations in the state close to equilibrium reduces to the following:
(1) solution of the kinetic Boltzmann equation;
(2) solution of the Langevin equation with the specified intensity of the source.
For the state of equilibrium the task is much simplified by the fact that the solution of the Boltzmann equation is known: it is the Maxwell distribution.

10.5. Entropy and Entropy Production for Nonequilibrium States Including Fluctuations

10.5.1. *Generalized Equation of Entropy Balance*

Instead of the kinetic Boltzmann equation, we start with equation (10.1.1) for smoothed distribution function $\tilde{f}(x,t)$ and introduce the appropriate definition of the entropy:

$$\tilde{S}_B(r,t) = -\kappa n \int \ln\left(n\tilde{f}(x,t)\right)\tilde{f}(x,t)dp . \tag{10.5.1}$$

Then, in place of (6.4.4), we get the following equation of balance for the mean value of entropy of the entire system:

$$\frac{d\langle \tilde{S}_B \rangle}{dt} = -\kappa n \int \langle \ln\left(n\tilde{f}\right) \tilde{I}_B \rangle dr dp \equiv \int \sigma_B(r,t) dr \geq 0 \ . \tag{10.5.2}$$

We see that now the mean entropy production $\langle \sigma_B \rangle$ depends on the higher moments of smoothed distribution $\tilde{f}(x,t)$. Obviously, the mean value of entropy production can only be calculated under certain simplifying assumptions. Let us discuss some special cases which are useful for many applications.

10.5.2. *Entropy and Entropy Production for Nonequilibrium States Excluding Kinetic Fluctuations*

For the "initial point" we take the state of local equilibrium with the distribution function $f^{(0)}(r,p,t)$. Then the corresponding entropy is given by

$$S_B^{(0)} = -\kappa n \int \ln\left(n f^{(0)}(x,t)\right) f^{(0)}(x,t) dx \ . \tag{10.5.3}$$

Assume that the nonequilibrium state differs little from the distribution at local equilibrium:

$$f(x,t) = f^{(0)}(x,t) + f^{(1)}(x,t) \ , \quad f^{(1)} << f^{(0)} \ . \tag{10.5.4}$$

Recall that the local distribution $f^{(0)}$ is defined by

$$n f^{(0)} = \frac{n(r,t)}{\left(2\pi m \kappa T(r,t)\right)^{3/2}} \exp\left(-\frac{(p - mu(r,t))^2}{2m\kappa T(r,t)}\right) \ . \tag{10.5.5}$$

From the normalization condition of functions f, $f^{(0)}$ it follows that

$$\int f^{(1)} dx = 0 \ . \tag{10.5.6}$$

Taking advantage of the properties (6.2.2) of Boltzmann collision integral, from Boltzmann equation we find that

$$\int \varphi(p)f dx = \int \varphi(p)\left(f^{(0)} + f^{(1)}\right)dx = \text{const} \quad \text{at} \quad \varphi(p) = 1, \ p, \ \frac{p^2}{2m} \ . \tag{10.5.7}$$

It is natural to assume that the gasdynamic functions $\rho(r,t)$, $u(r,t)$, $T(r,t)$ in the local Maxwell distribution are moments of the complete distribution (10.5.4). Then (10.5.4) imposes constraints on the function $f^{(1)}$:

$$\int \varphi(p)f^{(1)}(x,t)dx = 0 \quad \text{at} \quad \varphi(p) = 1, \ p, \ \frac{p^2}{2m} \ . \tag{10.5.8}$$

As a result, in the first nonvanishing approximation in $f^{(1)}$ we come to the following expression for the entropy:

$$S_B = S_B^{(0)} - \frac{\kappa n}{2} \int \frac{1}{f(x,t)}\left(f^{(1)^2}(x,t)\right)dx \le S_B^{(0)} \ , \tag{10.5.9}$$

where $S_B^{(0)}$ is Boltzmann entropy for the state of local equilibrium.

We see that Boltzmann entropy decreases as $f^{(1)}$ increases — that is, as the system recedes from the state of local equilibrium. This result is due to conditions (10.5.7), of which one establishes constancy of the mean energy in the course of evolution towards local equilibrium according to the Boltzmann equation. This condition, as we have seen in the proof of Gibbs theorem, is necessary for comparing the relative degree of order on the basis of entropy of the states, and for a system with arbitrary interaction between particles.

Let us consider the expression for total (within the entire volume of the system) entropy production $\sigma_B(t) = \int \sigma_B(r,t)dr$ in the same approximation. In the first nonvanishing approximation we come to the following expression:

$$\sigma_B(t) = \frac{\kappa n}{2} \int \frac{\delta(x - x')}{f^{(0)}}\left(\delta \hat{l}_p + \delta \hat{l}_{p'}\right)f^{(1)}(x,t)f^{(1)}(x',t)dxdx' \ . \tag{10.5.10}$$

In the state of local equilibrium, when $f^{(1)} = 0$, the value of entropy production is also zero.

In the last expression we included a unit factor $\int \delta(x - x')dx' = 1$ into the integrand, and carried out symmetrization with respect to variables x, x'. We shall appreciate the expedience of this form when fluctuations are taken into account.

10.5.3. Entropy and Entropy Production Including Kinetic Fluctuations

Now we return to equation of entropy balance (10.5.2), and quote at once the final expressions for entropy and entropy production with the inclusion of kinetic fluctuations (the details can be found in Klimontovich V, Sect. 1.9):

$$S_B^{(0)} - \langle \tilde{S}(t) \rangle = \frac{\kappa n}{2} \int \frac{\delta(x - x')}{f^{(0)}} \Big[f^{(1)}(x,t) f^{(1)}(x',t) + \langle \delta f \delta f \rangle_{x,x',t} \Big] \geq 0 \ , \qquad (10.5.11)$$

$$\langle \tilde{\sigma}_B \rangle = \frac{\kappa n}{2} \int \frac{\delta(x - x')}{f^{(0)}} \Big\{ \big(\delta \hat{I}_p + \delta \hat{I}_{p'} \big)$$

$$\times \Big[f^{(1)}(x,t) f^{(1)}(x',t) + \langle \delta \tilde{f} \delta \tilde{f} \rangle_{x,x',t} \Big] - \frac{1}{n^2} A(x,x',t) \Big\} dx dx' \ . \qquad (10.5.12)$$

Let us compare these expressions with those we have obtained earlier neglecting the kinetic fluctuations.

As compared with expression (10.5.9), inclusion of fluctuations resulted in a certain decrease in the entropy. This ought to be expected, since additional information about the system has been taken into account.

Two new terms appear in expression for entropy production (10.5.12) as against (10.5.10). The first of these is proportional to the correlator of fluctuations of phase density and acts to increase the entropy production. The second is defined by the intensity of the Langevin source and partly compensates the rise in the production of entropy. Obviously, compensation must be complete in the state of equilibrium, since the value of entropy production must then be zero. This circumstance can be used for defining the intensity of the Langevin source at equilibrium. Let us explore this possibility.

10.5.4. Entropy Production and Intensity of Langevin Source

Let the system occur in the state of equilibrium. Then the value of entropy production, as defined by (10.5.12), must be equal to zero. How can this become possible?

Only the term with $f^{(1)}$ is zero in the state of equilibrium, whereas the other two are not. Let us consider the first of these. Correlator of fluctuations of phase density $\langle \delta N \delta N \rangle_{x,x',t} \equiv n^2 \langle \delta f \delta f \rangle_{x,x',t}$ in the general case is defined by (5.6.9). Thermodynamically, however, Boltzmann gas is an ideal gas, and therefore the correlation function g_2 in the state of equilibrium is zero. Because of this, for the correlator of fluctuations of phase density we may use expression (9.6.8) which is simpler than (5.6.9). Maxwell distribution should be used here as the one-particle distribution.

So, the second term on the right-hand side of expression for entropy production is nonzero in the state of equilibrium as well. This brings us to the following definition for the intensity of the Langevin source:

$$A(x,x',t) = \left(\delta\hat{I}_p + \delta\hat{I}_{p'}\right)\langle\delta N \delta N\rangle^{\text{source}}_{x,x',t}$$
$$= \left(\delta\hat{I}_p + \delta\hat{I}_{p'}\right)\left[n\delta(x-x')f(x,t) - \frac{1}{V}f(x,t)f(x',t)\right]. \qquad (10.5.13)$$

Observe that the second term in brackets drops out when f is the Maxwell distribution. As a result, the expression for the intensity of the Langevin source, obtained in this way for the state of equilibrium, coincides with our earlier expression (9.7.8).

Now let us summarize the results.

First, the expressions for intensities of sources of fluctuations are the same, whether obtained by the method of moments or by the Langevin method. The intensities of sources reflect both the existence of atomic structure of "continuous medium", and the effects of large-scale nonequilibrium fluctuations.

Secondly, the source of fluctuations in the state of equilibrium can be determined from condition that the value of entropy production at equilibrium should be zero. A similar approach can be used for introducing the effective source of nonequilibrium fluctuations, based on condition that the mean value of entropy production should be zero. This will help us to define the structure of nonequilibrium source of fluctuations, such that the mean value of entropy $\langle S \rangle$ remains the same in the course of time evolution.

10.5.5. Effective Source of Fluctuations

Let us return to formula (10.5.12) for the mean value of entropy production in the nonequilibrium system under consideration, and use it to find the expression for the effective source of fluctuations which nullifies the mean value of entropy production. This expression has the form

$$A_{\text{eff}}(x,x',t) = \left(\delta\hat{I}_p + \delta\hat{I}_{p'}\right)\left[f^{(1)}(x,t)f^{(1)}(x',t) + \langle\delta N \delta N\rangle_{x,x',t}\right]. \qquad (10.5.14)$$

In chapters to follow we shall see that the effective source of fluctuations simplifies the calculation of steady-state fluctuations. This approach is useful, for example, in hydrodynamics for the description of well-developed turbulence.

To end this section, let us mark the following.

The integrand in (10.5.10), (10.5.12) contains $\delta(x-x')$, and therefore the expressions for sources of fluctuations obtained by assuming that the mean entropy production is zero are only valid when x and x' are close to each other (within physically infinitesimal volume). In this respect the method of moments and the Langevin method are more general, since they do not impose constraints on the separation between x and x'.

At the same time, the method based on the equation of entropy balance has its own advantages, because it allows finding those conditions under which the mean value of entropy production is zero, and so the mean value of entropy remains constant. Of course, entropy fluctuations are possible, which define the higher moments of the entropy.

Further on we shall again return to these issues. Presently we shall just particularize the condition of smallness of deviation from equilibrium, on which the entire discussion of this Section 10.5 is based.

10.5.6. *Criterion of "Small Deviation from Equilibrium"*

Let us return to equation (10.5.4), where we have introduced function $f^{(1)}$, which characterizes the deviation of distribution $f(x,t)$ from the equilibrium distribution $f^{(0)}$. Along with the averaged distribution $f^{(1)}$ we also took into account the contribution of fluctuations δf, which inevitably occur even at the state of equilibrium.

Let us find the condition at which the deviation $f^{(1)}$ becomes about equal to the fluctuation δf, but still remains much less than $f^{(0)}$. To this end we make use of the above definition $N_{ph} = nV_{ph}$ of the number of particles in physically infinitesimal volume.

For functions smoothed over the physically infinitesimal volume V_{ph} we have $\delta f / f^{(0)} \sim \left(N_{ph}\right)^{1/2}$. Hence it follows that $f^{(1)}$ and δf become equal by order of magnitude when

$$\frac{f^{(1)}}{f^{(0)}} \sim \frac{1}{\left(N_{ph}\right)^{1/2}} \ll 1 . \tag{10.5.15}$$

We see that even when the deviations from equilibrium are very small, and the inequality in (10.5.4) holds with an ample margin, the contributions from $f^{(1)}$ to entropy and entropy production may compete with the contributions from large-scale fluctuations.

Now we go over to one of the main problems of statistical system of open systems — the connection between kinetic and hydrodynamic descriptions of nonequilibrium processes. We shall start with the traditional derivation of gasdynamic equations from the kinetic Boltzmann equation. This derivation is based on various perturbation theories in Knudsen number. Recall that we have touched upon this matter in the Introduction.

As we shall see, the traditional approach to validation of equations of gas dynamics is very elegant and efficient. Open, however, remain important problems which call for a unified description of kinetic and hydrodynamic processes, without resorting to the perturbation theory in Knudsen number. This method will be discussed in Ch. 13.

CHAPTER 11

FROM KINETIC BOLTZMANN EQUATION TO
EQUATIONS OF GAS DYNAMICS

Let us briefly return to Sect. 1.2 of the Introduction. Recall first of all that the kinetic and the gasdynamic equations are employed for describing nonequilibrium processes in continuous medium. We saw, however, that the definition of a "point" of continuous medium is not universal, but depends on the adopted level of description. In other words, the kinetic and the gasdynamic descriptions are actually based on different models of continuous medium, but this difference is, as a rule, overlooked.

To wit, Boltzmann equation serves as the starting point, and some kind of perturbation theory in small Knudsen number (1.2.1) is used to accomplish the transition to equations of gas dynamics.

In this way, the kinetic equation enables one to establish the general structure of gasdynamic equations, and, what is more, to calculate the main characteristics, such as the dissipative coefficients of viscosity and thermal conductivity, and the speed of sound.

This traditional approach to validation of equations of gas dynamics is, however, not without a flaw, since some important questions remain open. Some of these issues have been enumerated in the Introduction.

In view of this, we divide the problem of validation of gasdynamic equations into two parts. In this chapter we consider the conventional procedure of derivation of gasdynamic equations, whereas in Ch. 13 we shall demonstrate the necessity and feasibility of unified description of kinetic and gasdynamic processes based on the definition of the structure of "continuous medium". This will enable us to get answers to some of the problems which up to now have remained unsolved.

11.1. Gasdynamic Functions. Transfer Equations

We begin with the Boltzmann equation (6.1.11), with the "collision integral" (6.1.6). Our task consists in using this equation for deriving a closed set of equations for the more simple gasdynamic functions (1.2.2).

Recall that these gasdynamic functions are related to the one-point distribution function, for which the Boltzmann equation has been written, by equations (5.2.1) – (5.2.4). Recall also that in the derivation of the kinetic equation we replaced the one-particle distribution $f_1(x,t)$ by the one-point distribution $f(x,t)$ which is more pertinent to the approximation of continuous medium (see Ch. 7).

From definitions (5.2.1) – (5.2.4) we see that the gasdynamic functions ρ, u, T are moments of velocity v or of deviation of velocity from its mean value,

$$\delta v = v - u(r,t) . \tag{11.1.1}$$

As we shall presently see, equations for functions ρ, u, T include also the more complicated moments of velocity. One of these is the so-called *tensor of internal stress*

$$P_{ij}(r,t) = mn\int \delta v_i \delta v_j f(r,p,t)dp , \tag{11.1.2}$$

which can be represented as a sum of two parts,

$$P_{ij}(r,t) = \delta_{ij}p(r,t) + \pi_{ij}(r,t) , \quad \pi_{ii}(r,t) = 0 , \tag{11.1.3}$$

where π_{ij} is the tensor of viscous stress. The reason for this name will soon become clear. From (11.1.3), (5.2.4) it follows that p is defined by

$$p(r,t) = \frac{\rho(r,t)}{m}\kappa T(r,t) , \tag{11.1.4}$$

which expresses the equation of state of ideal gas in local thermodynamic characteristics.

Finally, we require the expression for the so-called *vector of thermal flux*

$$q(r,t) = n\int \delta v \frac{m(\delta v)^2}{2} f(r,p,t)dp . \tag{11.1.5}$$

As pointed out in Sect. 1.2 of the Introduction, the name is not quite consistent with the structure of this expression; we shall discuss this matter in the next chapter.

From the above definitions it follows that the equations in gasdynamic functions ρ, u, T can be derived from the kinetic Boltzmann equation. To obtain the equation in ρ, we multiply Boltzmann equation by nm, integrate with respect to p, and use the first of the properties (6.2.2) of the Boltzmann collision integral. As a result, we arrive at the well-known continuity equation

$$\frac{\partial \rho}{\partial t} + \text{div}(\rho u) = 0 \ , \tag{11.1.6}$$

which may be regarded as the equation of balance of matter (transfer equation).

In the derivation of continuity equation from the kinetic Boltzmann equation we have used boundary conditions which are similar to (3.7.1), (3.7.2) for the distribution function $f_N(X,t)$.

In addition to ρ, the continuity equation includes the density of momentum ρu. To obtain the equation for the density of momentum, we multiply Boltzmann equation by nmv_i, and integrate with respect to p. Taking advantage of the second of the properties (6.2.2) of Boltzmann collision integral and the boundary condition for $f(r,p,t)$, we get the following equation for the components of the vector of density of momentum:

$$\frac{\partial \rho u_i}{\partial t} + \frac{\partial \rho u_i u_j}{\partial r_i} = -\frac{\partial p}{\partial r_i} - \frac{\partial \pi_{ij}}{\partial r_j} + \frac{\rho}{m} F_i \ . \tag{11.1.7}$$

Finally, to write down the equation of balance of density of kinetic energy we multiply Boltzmann equation by $nmv^2/2$, and integrate with respect to p. Taking advantage of the last of the properties (6.2.2) of Boltzmann collision integral and equations

$$n \int \frac{mv^2}{2} f dp = \frac{\rho u^2}{2} + \frac{\rho}{m} \varepsilon \ , \quad \varepsilon = \frac{3}{2} \kappa T \ , \tag{11.1.8}$$

$$n \int v_i \frac{mv^2}{2} f dp = u_i \left[\frac{\rho u^2}{2} + \frac{\rho}{m} \varepsilon + p \right] + \pi_{ij} u_j + q_i \tag{11.1.9}$$

(where we have used the definitions of tensor of viscous stress and vector of thermal flux, and symbol ε for the internal energy per one particle), we come to the equation of balance of density of kinetic energy of particles of Boltzmann gas:

$$\frac{\partial}{\partial t}\left(\frac{\rho u^2}{2} + \rho \varepsilon\right) + \frac{\partial}{\partial r_i}\left[u_i\left(\frac{\rho u^2}{2} + \rho \varepsilon + p\right) + \pi_{ij} u_i + q_i \right] = \frac{\rho}{m} Fu \ . \tag{11.1.10}$$

The equations of balance of density of particles, density of momentum and density of kinetic energy derived from the kinetic Boltzmann equation make up a set of five equations, which, however, is not closed. Indeed, these equations, apart from the gasdynamic functions ρ, u, T, include also the tensor of viscous stress and the vector of thermal flux.

In order to express the "excessive" functions $\pi_{ij}(r,t)$, $q_i(r,t)$ via the "main" gasdynamic functions ρ, u, T, we use the approximate solution of the Boltzmann equation based on the perturbation theory in Knudsen number.

11.2. Zero Approximation. Local Maxwell Distribution

In Boltzmann equation (6.1.1) we switch to dimensionless variables $t' = t/T$, $r' = r/L$. Assuming that the characteristic scales of the problem T, L are linked via the thermal velocity, $L = v_T T$, the Knudsen number is

$$\text{Kn} \sim \frac{l}{L} \sim \frac{\tau}{T} \ . \tag{11.2.1}$$

We also use the estimate (6.5.1) for the collision integral. Then the Boltzmann equation (dropping the prime) takes on the form

$$\text{Kn}\left(\frac{\partial f}{\partial t} + v\frac{\partial f}{\partial r} + F\frac{\partial f}{\partial p}\right) = I_B(r,p,t) \ . \tag{11.2.2}$$

Hence it follows that in the zero approximation in Knudsen number the Boltzmann equation reduces to a nonlinear inhomogeneous integral equation

$$I_B(r,p,t) = 0 \ . \tag{11.2.3}$$

As we know, the solution of this equation at the state of equilibrium, when there is no dependence on r, t, is the Maxwell distribution (6.3.6).

In a nonequilibrium state the equation (11.2.3) retains its structure, since it includes the variables r, t only as parameters. Because of this, the solution of this equation is the same function of p — that is, the Maxwell distribution. This time, however, all the parameters of distribution $nf(x,t)$ are functions of coordinates and time. As a result, we come to the local Maxwell distribution (10.5.5). Using the definition of relative velocity (11.1.1), we can write this distribution in the form

$$nf^{(0)}(r,p,t) = \frac{\frac{\rho}{m}}{(2\pi m\kappa T)^{3/2}} \exp\left[-\frac{m(\delta v)^2}{2\kappa T}\right], \quad \delta v = v - u(r,t) \ , \tag{11.2.4}$$

where superscript "0" stands to indicate that this distribution corresponds to the zero approximation in Knudsen number.

We see that the solution of Boltzmann equation in zero approximation in Knudsen number contains three unknown functions $\rho(r,t)$, $u(r,t)$, $T(r,t)$. Let us show that a closed set of equations is available for determining these functions.

We return to expressions (11.1.4), (11.1.5) for tensor of viscous stress and vector of thermal flux, and substitute there the local Maxwell distribution. It is straightforward that in our approximation both these expressions are zero,

$$\pi_{ij}^{(0)} = 0 \ , \ q^{(0)} = 0 \ . \tag{11.2.5}$$

It follows that the "excessive" functions in transfer equations are null in the zero approximation in Knudsen number. As a result, we obtain a closed set of equations in functions ρ, u, T. Let us bring them to the "canonical" form.

The continuity equation remains intact. Using the continuity equation, we reduce the equation of balance of density of momentum to the form

$$\frac{\partial u}{\partial t} + (u \ \text{grad}) \ u = -\frac{1}{\rho} \ \text{grad} \ p + \frac{F}{m} \ . \tag{11.2.6}$$

Next we use this and the continuity equation to convert the equation of balance of density of energy into the equation for local pressure:

$$\frac{\partial p}{\partial t} + (u \ \text{grad}) \ p + \frac{5}{3} p \ \text{div} \ u = 0 \ , \ p = \frac{\rho}{m} \kappa T \ . \tag{11.2.7}$$

In this way, we have built a closed set of equations (11.1.6), (11.2.6), (11.2.7) for density, velocity, and pressure. Using the equation of state, we can eliminate the pressure and write a closed set of equations in ρ, u, T.

These equations do not contain dissipative terms, and therefore are reversible. They have been obtained for the Boltzmann gas, which thermodynamically is an ideal gas. This fact is revealed in the form in which the equation of state is written. If we use a more general equation of state $p = p(\rho,T)$, the closed set of equations can be applied to nonideal gases and simple (isotropic) liquids, but only if the dissipative terms are insignificant. Sometimes such liquids are referred to as "ideal", meaning not the absence of interaction between the particles, but rather the absence of dissipative terms.

The equations for ideal liquid have been first formulated by Euler, and are commonly known as the Euler equations.

11.3. First Approximation in Knudsen Number. Equations of Gas Dynamics Including Viscosity and Thermal Conductivity

Two variants of perturbation theory are widely used for deriving the equations of gas dynamics with dissipative terms, the well-known Chapman – Enskog and Grad methods (in addition to the original papers, see Silin 1971; Zhdanov 1982; Ferziger – Kaper 1972, 1976; Lifshitz – Pitayevsky 1989; Klimontovich I, IV). Here we shall use Grad's method, which is more comprehensible and, unlike the Chapman – Enskog method, does not require solving integral equations. Instead, the task consists merely in computing definite integrals.

In Grad's method the distribution function is represented as a series in a complete set of orthonormal polynomials. The choice of polynomials is predetermined by the form of the solution in the zero approximation, which is the local Maxwell distribution. As a result, we deal with Hermitian polynomials in vector variable

$$\xi = \frac{\delta v}{(\kappa T/m)^{1/2}} \equiv \frac{p - mu}{(m\kappa T)^{1/2}} , \tag{11.3.1}$$

which are defined by the formula

$$H^{(n)}_{i_1,\ldots,i_n}(\xi) = (-1)^n \exp\left(\frac{\xi^2}{2}\right) \frac{\partial^n}{\partial \xi_{i_1} \ldots \partial \xi_{i_n}} \exp\left(-\frac{\xi^2}{2}\right) . \tag{11.3.2}$$

We see that the polynomial of power n is a tensor of rank n. Such polynomials comprise a complete orthonormal set. Owing to this fact, an arbitrary distribution function $f(r,t)$ can be represented as an expansion in Hermitian polynomials. In general, the coefficients of expansion will be functions of coordinates and time r, t.

Substituting the expansion into the kinetic Boltzmann equation one can reduce the Boltzmann equation to a generally infinite chain of meshing equations for coefficients of the expansion. The solution of this set is obviously no less complicated than the solution of the initial kinetic equation. The task therefore consists in approximating the distribution function with a finite number of terms in the expansion in Hermitian polynomials.

The efficiency of this technique depends on the judicious choice of the finite number of terms from the infinite series. Grad's method is based on the so-called thirteen-moment approximation.

The most coarse approximation, based on just the first term of the expansion, brings us back to the local Maxwell distribution which corresponds, as we know, to the zero approximation in Knudsen number. What will be the next step?

Grad assumes that there are thirteen unknown functions in the equations of transfer for the density of particles, density of momentum and density of kinetic energy: five functions ρ, u, T; five components of the symmetrical tensor of viscous stress π_{ij} (because the sum of the diagonal elements is $\pi_{ij} = 0$); three components of the vector of thermal flux q_i. Accordingly, only those terms are retained in the expansion of the distribution function which contain the above thirteen functions, and the distribution function is represented as

$$nf(r,p,t) = \frac{\frac{\rho}{m}}{(2\pi mкT)^{3/2}} \exp\left[-\frac{(p - mu)^2}{2mкT}\right]$$

$$\times \left[1 + \frac{\pi_{ij}}{2p}\frac{m\delta v_i \delta v_j}{3кT} + \frac{m(\delta vq)}{pкT}\left(\frac{m(\delta v)^2}{5кT} - 1\right)\right]. \tag{11.3.3}$$

The first term in this expression coincides with the local Maxwell distribution and corresponds to the zero approximation in Knudsen number.

The next two terms, which are proportional, respectively, to the tensor of viscous stress and the vector of thermal flux, are regarded as small perturbations (proportional to Knudsen number). Calculations support this assumption.

So, we have to demonstrate that the tensor of viscous stress and the vector of thermal flux are expressible via the gasdynamic functions ρ, u, T, and give small (in terms of Knudsen number) contributions to the equations of gas dynamics. What is more, such additional terms in equations of gas dynamics will take care of the dissipation due to viscous friction and heat conduction. Let us start with the tensor of viscous stress.

Tensor of viscous stress. We go back to the kinetic Boltzmann equation (11.2.2) in dimensionless variables. As prescribed by the perturbation theory in small Knudsen number, we substitute the zero-approximation function (the local Maxwell distribution) into the left-hand side of this equation, and the distribution (11.3.3) into the right-hand side.

This done, we must carry out linearization in the collision integral. As a result, we get two terms, one of which is a linear function of tensor π_{ij}, and the other is a linear function of vector q_i. In order to separate these contributions we proceed as follows.

We obtain the equation for the tensor of internal stress (11.1.2) by multiplying the Boltzmann equation by mnv_iv_j and integrating with respect to momenta. Then, making use of the transfer equations, we eliminate time derivatives of density and mean velocity, and get the following equation (see Eq. (4.26) in Ch. 10 of Klimontovich IV):

$$\left(\frac{\partial p}{\partial t}+u_k\frac{\partial p}{\partial r_k}+\frac{5}{3}\frac{\partial u_k}{\partial r_k}\right)\delta_{ij}+p\left(\frac{\partial u_i}{\partial r_j}+\frac{\partial u_j}{\partial r_i}-\frac{2}{3}\delta_{ij}\frac{\partial u_k}{\partial r_k}\right)=mn\int \delta v_i\delta v_j I_B(r,p,t)dp \ .$$

$$(11.3.4)$$

The right-hand side of this equation is a symmetrical tensor of second rank, and in the linear approximation is thus proportional to the tensor π_{ij}. The coefficient of proportionality we denote by p/η, where p is the pressure, and η, as we shall presently see, plays the role of dynamic viscosity. Then the right-hand side of (11.3.4) is

$$mn \ \int \delta v_i\delta v_j I_B(r,p,t)dp = -\frac{p}{\eta}\pi_{ij}(r,t) \ . \tag{11.3.5}$$

We also note that, by virtue of (11.2.7), the first term on the left-hand side of (11.3.4) is zero. Then we get the following expression for the tensor of viscous stress:

$$\pi_{ij}=-\eta\left(\frac{\partial u_i}{\partial r_j}+\frac{\partial u_j}{\partial r_i}-\frac{2}{3}\delta_{ij}\frac{\partial u_k}{\partial r_k}\right), \ \ \pi_{ii}=0 \ . \tag{11.3.6}$$

In accordance with Newton's law, the tensor of viscous stress is proportional to the derivatives of velocity. The coefficient of dynamic viscosity η depends on the form of the potential of interaction between the particles of Boltzmann gas. For the model of rigid spheres, when the potential of interaction is given by (2.1.4), the coefficient of dynamic viscosity η is

$$\eta \equiv \rho v = \frac{5m}{16\pi^{1/2}r_0^2}\left(\frac{\kappa T}{m}\right)^{1/2}, \tag{11.3.7}$$

where v is the kinematic viscosity.

Vector of thermal flux. The calculation of thermal flux is similar, so here we just quote the final result:

$$q=-\lambda \ \mathrm{grad} \ T \ , \ \ \lambda \equiv c_p\rho\chi=\frac{75}{64}\frac{\kappa}{\pi^{1/2}r_0^2}\left(\frac{\kappa T}{m}\right)^{1/2}, \tag{11.3.8}$$

where, along with the coefficient of thermal conductivity λ, we also use the coefficient of temperature conductivity χ. The two coefficients are linked via the value of heat capacity at constant pressure. This expression for thermal flux is consistent with the experimental Fourier law.

Substituting the above expressions into the equations of balance of density of particles, density of momentum, and density of kinetic energy, we obtain a closed set of equations of gas dynamics with dissipation due to irreversible effects of viscosity and heat conduction. Let us write out this set in a conventional form.

The continuity equation (11.1.6) remains as it is. Using this equation, and the identity

$$\frac{\partial}{\partial r_j}\left(\frac{\partial u_i}{\partial r_j}+\frac{\partial u_j}{\partial r_i}-\frac{2}{3}\delta_{ij}\frac{\partial u_k}{\partial r_k}\right)=\left[\Delta u_i+\frac{1}{3}\text{ grad div } u_i\right],\tag{11.3.9}$$

we get the following equation for the hydrodynamic velocity:

$$\frac{\partial u}{\partial t}+(u \text{ grad}) u=-\frac{1}{\rho}\text{ grad } p+v\left[\Delta u+\frac{1}{3}\text{ grad div } u\right]+\frac{F}{m}.\tag{11.3.10}$$

This is the so-called Navier – Stokes equation, which differs from the Euler equation in that it takes the viscous dissipation into account.

Instead of equation of balance of density of kinetic energy it is more convenient to use the appropriate equation for pressure or temperature which can be written in the form (see Ch. 10 in Klimontovich IV)

$$\frac{\partial p}{\partial t}+(u \text{ grad}) \rho+\frac{5}{3}p \text{ div } u=\frac{\eta}{3}\left(\frac{\partial u_i}{\partial r_j}+\frac{\partial u_j}{\partial r_i}-\frac{2}{3}\delta_{ij}\frac{\partial u_k}{\partial r_k}\right)^2+\frac{2}{3}\lambda\Delta T.\tag{11.3.11}$$

Now, equations (11.1.6), (11.3.10), and (11.3.11) represent the canonical set of equations of gas dynamics with dissipation due to viscous friction and heat conduction.

Recall that these equations have been derived on the basis of Boltzmann kinetic equation using perturbation theory in Knudsen number. Therewith we assumed that the dissipative terms are small (of the first order in Knudsen number). Let us now check whether our results are consistent with this assumption.

We look at the Navier – Stokes equation and compare the nondissipative term $(u \text{ grad})u$ with the dissipative term $v\Delta u$. Along with the Knudsen number Kn we introduce Mach number $M = u/v_T$, and also Reynolds number Re, which can be represented as a combination of Mach and Knudsen numbers,

$$\text{Re}=\frac{uL}{v}\sim\frac{M}{\text{Kn}}\text{ , since } v\sim v_T l.\tag{11.3.12}$$

As a result, we come to the following estimate:

$$\frac{\nu|\Delta u|}{|(u\,\mathrm{grad})u|} \sim \frac{\mathrm{Kn}}{M} \sim \frac{1}{\mathrm{Re}} \; . \tag{11.3.13}$$

We see that in general the ratio of dissipative and nondissipative contributions is characterized by the inverse Reynolds number. In case of transonic flows, however, when Mach number is about unity, the smallness of the ratio of dissipative and nondissipative contributions actually depends on how small the Knudsen number is.

So, by way of solving approximately the kinetic Boltzmann equation we have obtained a closed set of equations in gasdynamic functions which are simpler than the equation for the distribution function $f(x,t)$. The dissipation is due to two irreversible processes, internal friction (viscosity) and heat conduction, which are characterized by two appropriate kinetic coefficients η, λ (or ν, χ). This result holds for a gas of structureless particles. If a more sophisticated model is used, which takes into account the existence of internal degrees of freedom, one has to deal with the so-called second viscosity, characterized by the coefficient of internal viscosity ξ.

This method of validation of equations of gas dynamics is certainly very elegant and efficient. There are, however, some associated problems which will be discussed in Ch. 13.

Recall that the kinetic Boltzmann equation can be used for obtaining the equation of balance of entropy, which describes the entropy increase as a closed system evolves in time towards the state of equilibrium. This result constitutes the statement of Boltzmann's famous H–theorem. Let us consider the appropriate balance equation in case of gasdynamic description of irreversible processes.

11.4. Equation of Entropy Balance in Gas Dynamics

In connection with the derivation of equations of gas dynamics we noticed that the equations of state for local gasdynamic functions have the same form as those for the state of equilibrium. Namely, the expressions for pressure and internal energy per one particle are (see (11.1.4), (11.1.8)):

$$p(r,t) = \frac{\rho(r,t)}{m}\kappa T(r,t) \; , \quad U(r,t) = \frac{\rho}{m}\varepsilon \; , \quad \varepsilon(r,t) = \frac{3}{2}\kappa T(r,t) \; . \tag{11.4.1}$$

For this reason one may say that gasdynamic description corresponds to local thermodynamic equilibrium. Then we may also use the equilibrium form of the second law of thermodynamics, expressed in local functions.

We denote the density of entropy and entropy per one particle by, respectively, $S(r,t)$ and $s(r,t)$. Now the second law of thermodynamics for local equilibrium has the form

$$T ds = d\varepsilon + pd\left(\frac{m}{\rho}\right), \quad S(r,t) = \frac{\rho(r,t)}{m} s(r,t) .$$

(11.4.2)

The corresponding differential equation is

$$T\frac{ds}{dt} = \frac{d\varepsilon}{dt} + p\frac{d}{dt}\left(\frac{m}{\rho}\right),$$

(11.4.3)

where we use the notation for a total derivative — for example,

$$\frac{ds}{dt} = \frac{\partial s}{\partial t} + (u \ \mathrm{grad}) \ s .$$

(11.4.4)

Using equations of gas dynamics, we reduce (11.4.2) to the form of equation of balance of entropy (Landau – Lifshitz 1986; Klimontovich IV):

$$\frac{\partial}{\partial t}\left(\frac{\rho}{m} s\right) + \mathrm{div}\left[\frac{\rho}{m} us - \frac{\lambda \ \mathrm{grad} T}{T}\right] = \sigma(r,t) \geq 0 ,$$

(11.4.4)

$$\sigma(r,t) = \frac{\eta}{2}\left(\frac{\partial u_i}{\partial r_j} + \frac{\partial u_j}{\partial r_i} - \frac{2}{3}\delta_{ij}\frac{\partial u_k}{\partial r_k}\right)^2 + \lambda\left(\frac{\mathrm{grad} T}{T}\right)^2 \geq 0 .$$

We see that entropy production is represented as a sum of two positive terms, defined, respectively, by the processes of viscous friction and heat conduction. Equality sign corresponds to the case of complete equilibrium.

From the above equation of entropy balance it follows that total entropy of closed system (that is, when the flux of entropy across the surface is zero) can either increase or, in the state of equilibrium, remain constant.

Observe that the flux of entropy in the equation of balance consists of two parts. The first term may be referred to as convective, since it is proportional to the flux of density of entropy. The structure of the other term, however, is notably different: it is proportional to thermal flux as defined by the Fourier law (11.3.8), and is therefore dissipative. In the next chapter we shall see that this structure of thermal flux is retained in the unified description of kinetic and hydrodynamic processes. The dissipative part of the flux of entropy is then proportional to entropy gradient, and only coincides with the above expression under certain important constraints. In the Ch. 13 we shall also obtain a more general expression for entropy production.

Since the kinetic Boltzmann equation describes nonequilibrium processes only in ideal gas, when the contribution of the interaction between particles to thermodynamic functions is negligibly small, the corresponding equations of gas dynamics only hold for ideal gas. This is reflected, in particular, in the structure of local equations of state (11.4.1).

Recall that in Ch. 8 we also considered the more general kinetic equations which hold for nonideal gases. The difference from ideal gas is accounted for by introducing additional "collision integrals" into the Boltzmann equation. These "collision integrals" take care of the nonlocal nature of collisions between the gas particles, and the time lag due to the finite duration of a "collision". These equations may serve as basis for constructing the equations of gas dynamics for nonideal gases.

11.5. Equations of Gas Dynamics for Nonideal Gas

Recall that the collision integral in the kinetic Boltzmann equation exhibits properties (6.2.2). Because of this, such collision integral does not contribute to the equations of transfer of density of particles, density of momentum, and density of kinetic energy.

In case of nonideal gas the situation is different, since the structure of collision integral is not the same. Changed are also the properties (6.2.2). To take this into account, we return to Sect. 8.5 dealing with the kinetic equation for nonideal gas, and look into the changes in the structure of equations of gas dynamics.

The continuity equation remains unchanged, because the property (6.2.2) remains unchanged when $\varphi(p) = 1$. At $\varphi(p) = p$ the integral (6.2.2) is not zero, but is instead defined by expression (8.6.2) which includes the two-particle distribution function of nonequilibrium state. For Boltzmann gas this function may be taken in Bogolyubov's form — in other words, only the first term is retained on the right-hand side of (8.5.1). Then (8.5.6) is expressed via the one-particle distribution function which satisfies the kinetic equation for nonideal gas.

In the approximation of gas dynamics this expression can be further simplified. Indeed, for the distribution function it suffices to use the solution of Boltzmann equation in zero approximation in Knudsen number. This brings us to the zero approximation in Knudsen number for the distribution function $f(x,t)$ — that is, the local Maxwell distribution (10.5.5) or (11.2.4). As a result, with due account for conservation of momentum and energy in pairs of interacting particles, we come to the following expression for the two-particle distribution function in the approximation of local equilibrium:

$$f_2(r_1,p_1,r_2,p_2,t) = \exp\left[-\frac{\Phi|r_1 - r_2|}{\kappa T}\right] \times f^{(0)}(r_1,p_1,t)f^{(0)}(r_2,p_2,t) \ . \tag{11.5.1}$$

This distribution must be substituted into (8.5.6) to obtain the contribution from the "collision integral" of nonideal gas to the equation of balance of density of momentum. Then we carry out integration with respect to momenta of particles, and introduce variable $r = r_1 - r_2$. The resulting expression can be represented as the gradient of that part of the pressure which is due to the interaction of particles:

$$n\int pI(r,p,t)dp = -\frac{\partial p_{\text{int}}}{\partial r} \ , \tag{11.5.2}$$

where p_{int} is the relevant part of the pressure,

$$p_{\text{int}} = -\frac{2\pi n^2}{3}\int_0^\infty r\frac{d\Phi(r)}{dr}\ \exp\left[-\frac{\Phi(r)}{\kappa T(r,t)}\right]r^2 dr \ , \tag{11.5.3}$$

or, after integration by parts,

$$p_{\text{int}} = -2\pi n^2 \kappa T(r,t)\int_0^\infty\left[\exp\left(-\frac{\Phi(r)}{\kappa T(r,t)}\right) - 1\right]r^2 dr \ . \tag{11.5.4}$$

Now we may return to equation of transfer of momentum (11.1.7) or to the corresponding Navier – Stokes equation (11.2.6). The pressure p for Boltzmann gas in these equations was determined by the local equation of state of ideal gas (11.1.4) or (11.4.1). In case of nonideal gas the form of these equations remains the same, whereas the pressure is determined by the equation of state of nonideal gas,

$$p(r,t) = \frac{\rho(r,t)}{m}\kappa T(r,t) + p_{\text{int}} \ . \tag{11.5.5}$$

The contribution of interaction is defined by (11.5.4).

To avoid misunderstanding, let us once again indicate that the contribution of interaction into thermodynamic functions is taken here in the first approximation in density parameter. This implies that the above expression for the pressure corresponds to inclusion of the first term of virial expansion, this time for the local equilibrium.

If the potential of interaction is defined by (8.2.1), which regards atoms as elastic spheres with weak interaction, the pressure can be represented as

$$p(r,t) = \frac{\rho(r,t)}{m} \kappa T(r,t) \left[1 + n \left(b - \frac{a}{\kappa T(r,t)} \right) \right],$$ (11.5.6)

where we once again use the Van der Waals constants (8.2.3), (8.2.5). At equilibrium this expression coincides with (8.2.4).

So, in the first virial approximation we have obtained the Navier – Stokes equation for nonideal gas. Let us consider the relevant equation of balance of energy. Let us first see how the property (6.2.2) is changed when $\varphi(p) = p^2/2m$. In the same approximation of local equilibrium, the expression (8.5.7) becomes

$$n \int \frac{p^2}{2m} I(r,p,t)dp = -\frac{\partial u_{\text{int}}}{\partial t} - \frac{\partial}{\partial r} \left[u(u_{\text{int}} + p_{\text{int}}) \right].$$ (11.5.7)

The local pressure, due to the interaction between particles, is defined by (11.5.4). The local contribution of interaction to the density of internal energy u_{int} in the approximation of local equilibrium is given by

$$u_{\text{int}}(r,t) = 2\pi \left(\frac{\rho(r,t)}{m} \right)^2 \int_0^\infty \Phi(r) \exp\left[-\frac{\Phi(r)}{\kappa T(r,t)} \right] r^2 dr.$$ (11.5.8)

If the potential of interaction is given by (8.2.1), this result is much simplified and becomes

$$u_{\text{int}}(r,t) = -a \left(\frac{\rho(r,t)}{m} \right)^2,$$ (11.5.9)

where a is one of the Van der Waals constants. The complete expression for the local density of internal energy in the first approximation of virial expansion is

$$\frac{U(r,t)}{V} = u(r,t) = \frac{\rho(r,t)}{m} \frac{3}{2} \kappa T(r,t) + u_{\text{int}}(r,t),$$ (11.5.10)

which at equilibrium coincides with the expression given in Ch. 8.

If we use property (11.5.7) for the "collision integral" of nonideal gas, the equation of balance of density of internal energy (11.1.10) will include additional contributions to the density of internal energy and the pressure which are due to the interaction of particles.

So, in the first virial approximation based on the generalized Boltzmann equation it is possible to obtain a set of equations of gas dynamics with due account for the contribution of interaction between particles to the local thermodynamic functions in the appropriate approximation.

Recall that the generalized Boltzmann equation was derived in Ch. 8 under assumption of paired interactions as formulated by Boltzmann himself. By contrast to the conventional treatment, however, which leads to the kinetic equation for ideal gas, we took into account the interaction between particles, and used the nondissipative terms to define all local thermodynamic functions.

Obviously, this approximation corresponds to a slightly nonideal gas. Now the question is whether this approach to the description of nonequilibrium processes is suitable for more dense nonideal gases.

11.6. Equations of Gas Dynamics in Approximation of Triple Collisions

Let us go back to Sect. 8.4 where we discussed the construction of kinetic equations in higher approximation in density parameter. We saw that perturbation theory in density parameter allows one to move just one step ahead of the Boltzmann approximation, by taking triple interactions into account in the construction of kinetic equations. The inclusion of collisions between four and more particles leads to divergence of collision integrals.

To overcome these complications, one has to give a more comprehensive treatment to the structure of "continuous medium", and, in particular, to include collective processes into consideration.

It follows that the extension of kinetic equations and the corresponding gasdynamic equations to denser gases along the above guidelines is only possible for nonideal gases of moderate density, when triple interactions are taken into account together with paired interactions. As pointed out in Ch. 8, the pioneering steps towards the solution of this problem have been made by G. Uhlenbeck and S. Chow, and in the paper by M. Green (see Uhlenbeck – Ford 1965).

This theory, however, is not quite consistent, because triple collisions are only included into those dissipative terms which are defined by the relevant "collision integrals". At the same time, the contributions to nondissipative terms are taken in the first approximation in density parameter. In the state of local equilibrium this corresponds to the inclusion of only the first terms of virial expansions into thermodynamic functions. These are precisely those contributions that we have discussed in the preceding section.

This is of course an important accomplishment, since, upon transition from the kinetic equation with triple interactions to the corresponding equations of gas dynamics, the general structure of such equations remains the same, and refined are only the expressions for kinetic coefficients. To wit, the coefficients of viscosity and thermal conductivity now have the following form:

$$\eta = \eta_0 + \varepsilon\eta_1 , \quad \lambda = \lambda_0 + \varepsilon\lambda_1 , \quad \varepsilon = nr_0^3 \ll 1 , \tag{11.6.1}$$

where η_0 and λ_0 are the coefficients of viscosity and thermal conductivity obtained with the aid of the kinetic Boltzmann equation (in the approximation of paired interactions). These coefficients for the model of rigid spheres are given by (11.3.7), (11.3.8).

When triple interactions are included into consideration, we have to deal with additional contributions which are proportional to the density parameter. The additional coefficients for the model of rigid spheres are:

$$\eta_1 = 9.6\eta_0 , \quad \lambda_1 = 2.9\lambda_0 . \tag{11.6.2}$$

In order to obtain a complete description of nonequilibrium processes in the approximation of triple collisions one has to apply the technique described above for the case of paired collisions. The resulting kinetic equation will then serve as basis for constructing the equations of gas dynamics in which the local thermodynamic functions include second-order terms of virial expansions (Klimontovich – Slin'ko 1974; Klimontovich II).

As indicated, collective processes become all-important when more complicated interactions are taken into consideration. This applies to dense gases, and especially to liquids and glass-like condensed media. So far the problem of construction of the theory of nonequilibrium processes in such media for the most part remains open. As a rule, it is not possible to separate explicitly the kinetic level of description of nonequilibrium processes. This is the reason why semi-empirical methods are widely used for constructing the equations of mechanics of condensed media. This takes us to that domain of the theory of nonequilibrium processes which is commonly referred to as the thermodynamics of nonequilibrium irreversible processes. The next chapter is devoted to certain issues of this relatively young branch of thermodynamics and statistical theory.

CHAPTER 12

THERMODYNAMICS OF NONEQUILIBRIUM
IRREVERSIBLE PROCESSES

In this chapter we are going to get an overview of the feasibility of thermodynamic and statistical theory for describing dynamic and fluctuative nonequilibrium processes in arbitrary dissipative open systems.

The efficiency of this approach will be illustrated with a number of concrete examples. We start with introducing one of the main concepts of the theory: the effective Hamilton function, which can be defined either from phenomenological equations or directly from experimental data.

12.1. Thermodynamic Method in Theory of Irreversible Processes

In the preceding chapters we set forth the basics of the statistical theory of nonequilibrium processes in a rarefied (Boltzmann) gas. The degree of rarefaction was characterized by the density parameter $\varepsilon = nr_0^3$, the ratio of atom's volume to average space per one atom, $V/N = 1/n$. Thermodynamically, Boltzmann gas is an ideal gas. We also considered generalization of Boltzmann theory for the description of processes taking place in slightly nonideal gases. The problem of nonequilibrium processes in systems of higher density (condensed media) remained open.

Another important example of a macroscopic system, for which theoretical construction of kinetic and gasdynamic equations on the basis of microscopic equations is possible, is rarefied electron-ion plasma.

This system differs from Boltzmann gas in two important respects.

First, an important feature of a plasma is the interaction between charged particles (electrons e and ions i) according to Coulomb's law

$$\Phi_{ab}(|r|) = \frac{e_a e_b}{|r|} , \quad a = e, i .$$ (12.1.1)

Interaction of this kind is referred to as long-range interaction, since it decreases slowly as the distance between the particles increases. Because of this, the interaction of charged particles in a plasma is collective: many particles, both electrons and ions, fall within the range of action of any charged particle. Because of the interference between charged particles, the Coulombian interaction is screened off at a certain characteristic distance called the Debye radius r_D. In case of rarefied plasma this distance is much greater than the mean separation between the charged particles. Accordingly, within the sphere of interaction there are many particles, and the interaction of charged particles in a rarefied plasma is collective, by contrast to the Boltzmann gas where the interactions between particles are reduced to paired collisions.

It follows that the models of rarefied gas and rarefied plasma are based on diametrically opposite assumptions. Indeed, the main assumption for the gas is the assumption of paired interactions, which is based on the smallness of density parameter $\varepsilon = nr_0^3$.

By contrast, in a rarefied plasma the small parameter is the so-called plasma parameter $\mu = 1/nr_D^3$, which is the ratio of space per one particle to the volume defined by the Debye radius r_D. For a rarefied plasma the number of particles in the sphere of action is large, and so the parameter μ is small.

Secondly, in many cases the interaction through electromagnetic field in a plasma is no less important than the Coulombian interaction. In this situation the field is regarded as one of the components of plasma.

A considerable part of the second volume will be devoted to the statistical theory of plasma. This will give us an opportunity to expand the scope of phenomena under consideration.

The problem of kinetic description of nonequilibrium processes on the basis of microscopic models in systems with strong interaction of particles remains to a large extent open. More efficient are the methods of nonequilibrium thermodynamics, which include also the kinetic level of description.

Recall that thermodynamics is characterized by the universality of its laws and relations, which are independent of the particular structure of medium in question.

Important developments have been made recently in a new domain of statistical theory called the nonlinear nonequilibrium fluctuation thermodynamics (Zubarev 1971; Nicolis – Prigogine 1977, 1979; De Groot – Mazur 1956; Grechanny 1989; Haken 1978, 1980, 1983, 1985; Stratonovich 1985, 1992; Keiser 1987, 1990; Grandy I 1987, II 1988; Klimontovich IV, V; Zubarev – Morozov 1994).

The key words in the name of this new branch of thermodynamics are *nonlinear* and *fluctuation*. The first implies that the main equations of the theory are, as a rule, nonlinear. The second emphasizes the important, often definitive (in the theory of well-developed turbulence, for example) role of fluctuations in the description of nonequilibrium processes in open systems.

The sections to follow will be devoted mainly to the fundamentals of fluctuation thermodynamics of nonequilibrium nonlinear processes. We shall consider a number of general issues, as well as examples of equations of thermodynamics of irreversible processes. Like in the theory of gases, we shall use different levels of description: kinetic, thermodynamic and diffusion. We shall also consider the feasibility of unified description of all these processes.

The generality of classical thermodynamics is to a large degree preserved in nonequilibrium thermodynamics. Our immediate task will consist in proving this statement. This approach will allow us to advance considerably towards the description of nonequilibrium dynamic and fluctuative processes in various open systems, and, in particular, in condensed media.

The description of nonequilibrium processes in condensed media is based on equations of mechanics and electrodynamics of continuous medium — that is, on closed sets of equations in local thermodynamic functions. Examples of such equations are equations of thermodynamics of simple liquids, equations of theory of elasticity of solids, and the more complicated equations of hydrodynamics for liquid crystals, equations of chemical kinetics, and the like. All these can be referred to generically as the equations of thermodynamics of nonequilibrium irreversible processes.

Semi-empirical approach is commonly used for constructing the relevant equations for condensed media, so that the general principles of nonequilibrium thermodynamics are employed together with the information derived from experimental results. It is in such way that the equations are established for describing nonequilibrium processes in diverse condensed media. The so-called simple liquids offer an elementary example of such media.

12.2. Equations of Thermodynamics of Simple Liquids

The main assumption in the construction of equations of hydrodynamics is the hypothesis of local thermodynamic equilibrium, which allows one to formulate the second law of thermodynamics for local thermodynamic functions:

$$Tds = d\varepsilon + pd\left(\frac{m}{\rho}\right), \quad S(r,t) = \frac{\rho(r,t)}{m}s(r,t) \ . \tag{12.2.1}$$

Earlier (see (11.4.2)) we introduced this equation for ideal gas, as an implication of the local Maxwell distribution. Now this equation is assumed to hold for nonideal gases of arbitrary density, and for liquids.

In place of equations of state of ideal gas (11.4.1), we now use empirical equations of state

$$p = p(r,T) \; , \quad \varepsilon = \varepsilon(r,T) \; . \tag{12.2.2}$$

Naturally, various models can also be used in case of condensed media for calculating the equations of state. Examples of such models will be given in Ch. 21; as a rule, they are only suitable for the state of local equilibrium.

Now let us turn to equations of transfer obtained earlier in this chapter on the basis of the kinetic Boltzmann equation. They express the laws of conservation of density of matter, density of momentum, and density of internal energy (density of mean kinetic energy in case of Boltzmann gas). It would be natural to assume that the general form of equations of balance (transfer equations) will be preserved for condensed media as well.

So, for the construction of hydrodynamic equations we use the same equations of balance (11.1.6) – (11.1.8). Of course, in case of condensed media we deal with internal energy, assuming arbitrarily strong interaction of particles.

Let us write out the functions which occur in these general transfer equations. There are three scalar functions, two vectors, and one tensor:

$$\rho(r,t) \; , \quad u(r,t) \; , \quad \varepsilon(r,t) \; , \quad p(r,t) \; , \quad q(r,t) \; , \quad \pi_{ij}(r,t) \; . \tag{12.2.3}$$

The number of these functions is greater than the number of equations of balance, which make up a set of five. Where do we gain additional information?

For one thing, there are two equations of state (12.2.2) which describe the experimentally observed behavior of thermodynamic functions at quasistatic (reversible) processes. Then we are left with two "excessive" functions. This is about the same situation as we encountered in the derivation of equations of gas dynamics. The difference is that the equations of state for ideal gas follows from the solution of Boltzmann equation in zero approximation in Knudsen number — that is, from the local Maxwell distribution.

In the kinetic theory of gases, the expressions for the "excessive" functions $\pi_{ij}(r,t)$, $q(r,t)$ (11.3.6), (11.3.8) also follow from Boltzmann equation (in the first approximation in Knudsen number). In this way, the equations of gas dynamics for ideal gas are based entirely on the kinetic Boltzmann equation. In order to define these "excessive" functions in the theory of condensed media one has to turn to the experiment.

Let us start with the definition of tensor $\pi_{ij}(r,t)$.

According to Newton's law, the force of viscous friction is proportional to the relative velocity of motion of layers of liquid (that is, to the gradient of the relevant component of velocity). We further assume that tensor $\pi_{ij}(r,t)$ is symmetrical, and characterizes the dissipation due to viscous friction (which is why we call it the tensor of viscous stress).

Arguments in favor of symmetry of the tensor of viscous stress ($\pi_{ij} = \pi_{ji}$) can be found in textbooks on hydrodynamics (see, for example, Landau – Lifshitz 1986). The evidence, however, is not quite conclusive, since the boundary conditions for viscous fluid are left out of consideration.

The most general form of the tensor of viscous stress, which satisfies the above requirements, is determined by two constants which can be chosen so that

$$\pi_{ij} = -\eta \left(\frac{\partial u_i}{\partial r_j} + \frac{\partial u_j}{\partial r_i} - \frac{2}{3} \delta_{ij} \frac{\partial u_k}{\partial r_k} \right) - \xi \delta_{ij} \frac{\partial u_k}{\partial r_k} \ . \tag{12.2.4}$$

This expression differs from (11.3.6) for Boltzmann gas in that it contains a second coefficient of viscosity ξ, called the coefficient of internal viscosity because it reflects the existence of internal structure of atoms. In case of structureless particles (like atoms in Boltzmann gas), $\xi = 0$.

The definition of the other "excessive" function, the vector of heat flux q, is based on the empirical Fourier law

$$q(r,t) = -\lambda \frac{\partial T}{\partial r} \ . \tag{12.2.5}$$

This expression coincides with (11.3.8) for Boltzmann gas. This time, however, the value of heat conductivity λ cannot be calculated, but must instead be found experimentally.

Now we have a closed set of equations for functions $\rho(r,t)$, $u(r,t)$, $T(r,t)$ — a set of equations of hydrodynamics. Equations for other condensed media can be obtained in a similar way. Examples are the equations of theory of elasticity for solids, hydrodynamic equations for liquid crystals, etc.

Observe that the equations of hydrodynamics can be used for deriving the appropriate equation of balance of entropy. The procedure is similar to that described in Sect. 11.4 for equations of gas dynamics. The details can be found in Landau – Lifshitz 1986b (Ch. 5); Klimontovich IV (Ch. 13). Here we just make the following remark.

In hydrodynamics, the equation of entropy balance still has the form of (11.4.4), and expresses the second law of thermodynamics for hydrodynamic processes. The second law of thermodynamics requires that entropy production should be positive; hence it follows that all kinetic coefficients must be positive,

$$\eta > 0 \ , \ \ \xi > 0 \ , \ \ \lambda > 0 \ . \tag{12.2.6}$$

In case of Boltzmann gas, these constraints are confirmed by calculations.

With this example we have demonstrated that the equations of hydrodynamics of nonequilibrium processes can be built on the following principles:

1. Hypothesis of local equilibrium.

2. Equations which express the basic laws for the medium in question. In hydrodynamics, these are the equations of balance of density of matter, density of momentum, and density of energy.

Equations of thermodynamics of nonequilibrium processes are extremely diverse. In particular, they include Maxwell equations of macroscopic electrodynamics which express fundamental laws of electromagnetic processes. For the description of processes in conducting liquids, Maxwell equations are used together with equations of hydrodynamics. 3. Equations of thermodynamics of nonequilibrium processes form a closed set only when supplemented by the so-called material equations. The latter are sometimes divided into two classes: (a) relations between local thermodynamic functions, such as the equation of state (12.2.2) in hydrodynamics, and (b) dissipative relations between the fluxes and the main functions, such as the tensor of viscous stress (12.2.4) and the vector of thermal flux (12.2.5) in hydrodynamics, or Ohm's law in electrodynamics.

Observe finally that while the equations of hydrodynamics are nonlinear, the dissipative processes are described by linear relations (tensor of viscous stress (12.2.4) and vector of thermal flux (12.2.5)). We shall see many examples of equations of thermodynamics of nonequilibrium processes in which the dissipative terms are also nonlinear. Such are, for instance, almost all equations for nonequilibrium processes in active media.

12.3. Boltzmann's Principle and Effective Hamilton Function

Boltzmann's principle is expressed by (4.10.4) and (4.10.5) which include, along with the thermodynamic free energy $F(a,T)$, the conditional free energy (4.10.3).

It is convenient to replace the conditional free energy by the effective Hamilton function $H_{eff}(A,a,T)$ using equation (4.10.7). Then the distribution function (4.10.6) assumes the form of the canonical Gibbs distribution.

Recall that the concept of the effective Hamilton function was discussed in Sect. 4.14. Then, however, we were dealing with the state of equilibrium. As this concept will be central for the statistical theory of nonequilibrium processes in open systems, we shall reproduce here some of our earlier findings.

We start with the formula (4.10.6) for the "canonical Gibbs distribution" in variable A:

$$f(A,a,T) = \exp \frac{F(a,T) - H_{eff}(A,a,T)}{\kappa T} . \tag{12.3.1}$$

Free energy is expressed via the effective Hamilton function as

$$F(a,T) = -\kappa T \int \exp \left[-\frac{H_{eff}(A,a,T)}{\kappa T} \right] dA . \tag{12.3.2}$$

Function H_{eff} is related to the Hamilton function $H(a,T)$ by expression (4.14.1),

$$H_{\text{eff}}(A,a,T) = -\kappa T \ln \int \delta(A - A(X)) \exp\left(-\frac{H(X,a)}{\kappa T}\right) dX \ . \tag{12.3.3}$$

Substitution of this expression into (12.3.2) brings us back to the initial definition of free energy:

$$F(a,T) = -\kappa T \int \exp\left(-\frac{H(X,T)}{\kappa T}\right) dX . \tag{12.3.4}$$

What is the reason for introduction of the effective Hamilton function?

For a small number of variables A the "canonical Gibbs distribution" (12.3.1) is much simpler than the initial canonical distribution for function $f_N(X,a,T)$. This is because the effective Hamilton function carries information about the statistical distribution just for a restricted region of phase space, and only coincides with the conventional Hamilton function when we choose $A(X) = X$ (see (4.14.2)).

With rare exceptions, microscopic calculation of the effective Hamilton function is practically impossible. This task is as complicated as the calculation of statistical integral for finding the thermodynamic functions of a given system.

The common practice consists therefore in using approximate expressions for the Hamilton function, such as can be derived from semiempirical kinetic equations (based on mathematical models of processes under consideration), or directly from experimental data — for example, from time realizations of the processes. Of course, in such cases the relation (12.3.3) between the effective Hamilton function and the Hamilton function of the system $H(X,a)$ will not hold in general.

Definition of free energy for nonequilibrium states can be based on the concept of the effective Hamilton function. Then the difference between equilibrium and nonequilibrium free energies is a Lyapunov functional whose least value corresponds to the state of equilibrium.

12.4. Change of Free Energy and Entropy in Processes of Time Evolution. Lyapunov Functionals

The canonical distribution $f_N(X,a,T)$ includes the free energy "according to Gibbs", $F(a,T) \equiv F_G(a,T)$. We know (see (3.4.7)) that it coincides with the thermodynamic definition of free energy,

$$F_{\text{therm}}(a,T) = U(a,T) - TS_G(a,T) , \tag{12.4.1}$$

where the internal energy U and the Gibbs entropy S_G are expressed via the canonical

Gibbs distribution $f_N(X,a,T)$ by (3.3.3), (3.4.5).

Now, how do we define the thermodynamic functions in case of incomplete statistical description, when a simpler distribution (12.3.1) is used in place of the canonical Gibbs distribution?

In this case the internal energy and entropy are given by the following expressions:

$$U(a,T,[A]) = \int H_{\text{eff}}(A,a,T)f(A,a,T)dA \,, \tag{12.4.2}$$

$$S(a,T,[A]) = -\kappa \int \ln f(A,a,T)f(A,a,T)dA \,, \tag{12.4.3}$$

where A is put in brackets to indicate that we are dealing not with the canonical Gibbs distribution, but rather with the distribution $f(A,a,T)$ of values A of a certain function of dynamic variables $A(X)$. Let us compare the entropy (12.4.3) with the Gibbs entropy S_G which corresponds to the complete statistical description.

Into (12.4.3) we substitute expression (4.10.6) for $f(A,a,T)$ and integrate with respect to A. Then

$$S(a,T,[A]) = -\kappa \int \ln f(A(X),a,T)f_N(X,a,T)dX \,, \tag{12.4.4}$$

Using (12.4.4), (3.4.5) for Gibbs entropy S_G, and inequality (3.5.5), we find that

$$S(a,T,[A]) - S_G(a,T) = \int \frac{f_N(X,a,T)}{f(A(X),a,T)} f_N(X,a,T)dX \geq 0 \,. \tag{12.4.5}$$

It follows that the deficiency of statistical description increases when the simpler distribution function $f(A,a,T)$ is used, with fixed values of parameters a, T. Equality takes place in the particular case when the function of dynamic variables is $A(X) = X$.

We see that the transition to a simplified description results in an increase in the relevant entropy value. Recall that in Ch. 9 we have already arrived at a similar conclusion, as expressed by inequality (9.4.4) which pertains to the domain of nonequilibrium processes. Inequality (9.4.4) states that the entropy in Boltzmann's approximation is greater than the entropy in Leontovich's (more general) approximation. This result stems from the fact that Leontovich's equation takes care of nonequilibrium kinetic correlations (nonequilibrium fluctuations of one-particle distributions) which are neglected in Boltzmann's theory.

Making use of (12.4.2), (12.4.3), and the "canonical Gibbs distribution" (12.3.1), we may give an appropriate thermodynamic definition of free energy (4.14.7):

$$F(a,T,[A]) = U(a,T,[A]) - TS(a,T,[A])$$

$$= \int H_{eff}(A,a,T)f(A,a,T)dA + \kappa T \int \ln f(A,a,T)f(A,a,T)dA$$

$$= F_G(a,T) \ .   \quad (12.4.6)$$

We see that the value of free energy is the same both for the original canonical Gibbs distribution and for the simpler distribution $f(A,a,T)$. The change in the entropy (12.4.5) is balanced out by the change in the free energy when the initial Hamilton function $H(X,a)$ is replaced by the effective Hamilton function $H_{eff}(A,a,T)$:

$$U(a,T,[A]) - U(a,T) = TS(a,T,[A]) - S_G(a,T) \geq 0 \ .   \quad (12.4.7)$$

Recall that the equals sign corresponds to the special case when the function of dynamic variables is $A(X) = X$.

Observe that relation (12.4.7) comes up also in the calculation of fluctuations of entropy and internal energy, which has been carried out in Sect. 4.4. More complete is the statistical description which includes fluctuations of entropy and internal energy, whereas the description based on the mean values is "coarse". If we take due account of this fact and use expression (12.4.7) for the fluctuations, the relation between dispersions of entropy and internal energy will be given by (4.4.5). Fluctuations of free energy are then zero, which corresponds to fulfilment of (12.4.6).

At equilibrium, the transition from the complete statistical description based on the canonical Gibbs distribution to the incomplete statistical description based on the simpler distributions $f(A,a,T)$ allows us to redefine the relevant thermodynamic functions. The mean values of entropy and effective energy are greater than their values corresponding to the canonical Gibbs distribution, whereas the value of free energy, defined by (12.4.6), remains the same as for the Gibbs distribution.

Equilibrium distributions $f(A,a,T)$ are established as a result of time evolution which is described by the appropriate kinetic equations. We have considered two examples of such equations: equations of Boltzmann and Leontovich. Later we shall discuss many other kinetic equations of thermodynamics of irreversible processes.

At specified constant external conditions in open systems, equations of this kind will describe evolution towards stationary states which may be essentially different from the state of equilibrium. Is it possible to define the relevant nonequilibrium thermodynamic characteristics, and to establish general relations between the respective equilibrium and nonequilibrium thermodynamic functions.

Let us begin with the case when time evolution is described by the distribution function $f(A,t)$, and leads to the equilibrium state with the "canonical Gibbs distribution" (12.3.1).

The relevant free energy is then given by (12.4.6). We use a similar equation to define the free energy of nonequilibrium state,

$$F(t) = \int H_{eff}(A,T)f(A,t)dA - \kappa T \int \ln f(A,t)f(A,t)dA .$$ (12.4.8)

Using (12.4.6), the normalization conditions for functions $f(A,a,T)$, $f(A,t)$, and inequality (3.5.5), we may express the difference in the free energies of equilibrium and nonequilibrium states as

$$\Lambda_F = F(t) - F(a,T,[A]) = \kappa T \int \ln \frac{f(A,t)}{f(A,a,T)} f(A,t)dA \geq 0 ,$$ (12.4.9)

where the sign of equality corresponds to the state of equilibrium.

We see that the least value of nonequilibrium free energy as defined by (12.4.8) corresponds to the state of equilibrium. Moreover, in the study of Brownian motion (Ch. 15) we shall prove that free energy decreases monotonically in the course of evolution towards equilibrium:

$$\frac{d\Lambda_F}{dt} = \frac{d}{dt}\left(F(t) - F(a,T,[A])\right) \leq 0 .$$ (12.4.10)

We see that the difference in equilibrium and nonequilibrium free energies is a Lyapunov functional. Subscript "F" indicates that this Lyapunov functional is defined by the difference in free energies rather than by the difference in entropies as was the case in the formulation of H-theorem in the theory of Boltzmann gas (see (1.6.3), (1.6.4)).

Now let us consider a case when the stationary state established in the result of time evolution is different from the state of equilibrium. There is a class of systems (like Brownian motion in Van der Pol oscillator; see Ch. 16) for which the stationary solution can be represented as

$$f(A,a,T_{eff}) = \exp\frac{F_{eff} - H_{eff}(A,a)}{T_{eff}} , \quad \int fdA = 1 .$$ (12.4.11)

Here all the characteristics are effective. In case of oscillator, the quantity A may be the energy of oscillations, and T_{eff} the intensity of the source of noise. The free energy of stationary state is linked with the mean effective energy and entropy by a "thermodynamic" equation similar to (12.4.6). The free energy $F(t)$ in the course of evolution is given by equation similar to (12.4.8). Such definition is again justified by the fact that the difference in the free energies so defined is a Lyapunov functional

$$\Lambda_F = F(t) - F_{\text{eff}} = \kappa T_{\text{eff}} \int \ln \frac{f(A,t)}{f(A,a,T_{\text{eff}})} f(A,t) dA \geq 0 \ , \quad \frac{d\Lambda_F}{dt} \leq 0 \ . \qquad (12.4.12)$$

In this way one may use the difference Λ_F in "thermodynamically" defined free energies for assessing the nature of nonequilibrium states. The following, however, must be borne in mind.

Lyapunov functionals Λ_F may serve as measure of how far the nonequilibrium state in question is from the equilibrium or stationary state. They are often used in this capacity, being commonly referred to as *information gain* or *Kullback information* (*Kullback entropy*) (Kullback 1951; Haken 1988; Stratonovich 1985, 1992; Klimontovich IV, V).

By contrast to entropy, however, the free energy does not possess all the features which could have allowed using it as the measure of uncertainty in the statistical description of nonequilibrium processes. In other words, the functional Λ_F cannot be a measure of the relative degree of order of equilibrium and nonequilibrium states. As we know (see Sect. 3.5), only the entropy has all the required properties.

Entropy has another important advantage over other thermodynamic potentials, including the free energy. Recall that the definition of free energy for nonequilibrium states is not unique: here, for instance, we have used two definitions: one "according to Gibbs", and the other "thermodynamical". At the same time, the entropy is unambiguously defined for any nonequilibrium state as the mean value of logarithm of the distribution function. For this reason it seems only natural to base the comparison of different states of open systems on the associated values of entropy.

Assume now that the only available source of information is straightforward experimental data. Let the experimental results be represented as time dependences of the characteristic internal parameters A, and the set of external parameters a:

$$A(t,a) \ , \quad A = A(A_1, ..., A_m) \ , \quad a = a(a_1, ..., a_n) \ . \qquad (12.4.13)$$

The external parameters a are also called *controlling parameters*, since the state of the system can be modified by adjusting their values. Ambient temperature (thermostat setting) can be one of the controlling parameters. Another example is the time taken by the open system to reach a stationary state.

Assume that functions $A(t,a)$ are known for long stretches of time, so that the time realizations of the process can be used for constructing the appropriate distribution functions

$$f(A,a) \ , \quad \int f(A,a) dA = 1 \ . \qquad (12.4.14)$$

Then we can redefine the effective Hamilton function as

$$f(A,a) = \exp(-H_{\text{eff}}(A,a)) \ , \ \int \exp(-H_{\text{eff}}(A,a))dA = 1 \ . \tag{12.4.15}$$

In this way it is possible to find the effective Hamilton function directly from the experimental data.

Recall that we have already used a similar definition of the effective Hamilton function in Sect. 1.6.6 in connection with the formulation of S–theorem as criterion of self-organization. According to this criterion, the relative degree of order of states of open systems is assessed by comparing the values of entropy corresponding to a specified value of the mean effective Hamilton function. The initial distribution for the state taken for physical chaos is renormalized and reduced to the form of canonical distribution (12.3.1) with the effective Hamilton function as defined by (12.4.15).

We shall study criteria of self-organization more closely in the forthcoming chapters. As mentioned in Sect. 1.6.6 of the Introduction, one of the main criteria will be formulated as S–theorem. By way of illustration we shall now consider a concrete example of nonequilibrium distribution of velocities of atoms in Boltzmann gas in the presence of a source of heat.

12.5. Distribution Function of Velocity in System with Heat Source

Let us return to Boltzmann's kinetic equation (6.1.11) and replace it with a simple model equation. Namely, in place of the collision integral (6.1.6) we use a nonlinear differential expression having the same properties as Boltzmann's integral. As a result, for spatially homogeneous gas we come for the following equation for the nonlinear distribution function of velocities:

$$\frac{\partial f(v,t)}{\partial t} = \frac{\partial}{\partial v}\left[D_{(v)}(t)\frac{\partial f}{\partial v}\right] + \frac{\partial}{\partial v}[\gamma(v - u(t))f] \ , \ \int f dv = 1 \ , \tag{12.5.1}$$

where $D_{(v)}$ is the coefficient of diffusion in phase space of velocities which is defined by Einstein's relation

$$D_{(v)}(t) = \gamma\kappa T(t) \ , \ \frac{3}{2}\kappa T(t) = \int \frac{m(v - u(t))^2}{2} f(v,t)dv \ , \tag{12.5.2}$$

and

$$u(t) = \int vf(v,t)dv \tag{12.5.3}$$

is the mean velocity.

The right-hand side of (12.5.1) acts as the "collision integral" $I(t)$, and has the following property (cf. (6.2.2)):

$$\int \varphi(v) I_{(v)}(v,t)dv = 0 \quad \text{for} \quad \varphi(v) = 1, \ v, \ \frac{mv^2}{2} . \tag{12.5.4}$$

By virtue of properties (12.5.4), we use the kinetic equation to obtain the following equations for mean velocity and temperature:

$$\frac{du(t)}{dt} = 0 , \quad \frac{dT(t)}{dt} = 0 . \tag{12.5.5}$$

It follows that the mean velocity and temperature of spatially homogeneous system remain constant in the process of evolution towards the state of equilibrium which is characterized by Maxwell distribution

$$f(v) = \left(\frac{m}{2\pi\kappa T}\right)^{3/2} \exp\left(-\frac{mv^2}{2\kappa T}\right) . \tag{12.5.6}$$

Of course, this result is fully consistent with the equations of transfer (11.1.7), (11.1.8).

And now to our illustration.

Assume that in a thermostat with constant temperature T_0 there is a heat sink which is characterized by the coefficients γ, β of, respectively, linear and nonlinear (proportional to energy E) heat escape. In addition, there is a heat source which is characterized by the coefficient of feedback a_f. Now the coefficient of friction can be represented as

$$\gamma(E) = \gamma - a_f + \beta E , \quad E = \frac{mv^2}{2} , \quad \gamma > 0 , \quad \beta > 0 . \tag{12.5.7}$$

If the coefficient of diffusion is $D_{(v)} \equiv 0$, then the relaxation of energy towards its equilibrium value is described by dynamic equation

$$\frac{dE}{dt} = (a_f - \gamma - \beta E)E \equiv (a - \beta E)E . \tag{12.5.8}$$

Now let us take care of diffusion in phase space of velocities. It would be natural to assume that in the absence of heat source ($a_f = 0$) the equilibrium state is characterized by Maxwell distribution (12.5.6). Then it is straightforward that the coefficient of diffusion is nonlinear and is given by

$$D_{(v)}(E) = \gamma\left(1 + \frac{\beta}{\gamma}E\right)\frac{\kappa T}{m}, \quad E = \frac{mv^2}{2}. \tag{12.5.9}$$

As a result, the kinetic equation becomes

$$\frac{\partial f(v,t)}{\partial t} = \frac{\partial}{\partial v}\left[D_{(v)}(E)\frac{\partial f}{\partial v}\right] + \frac{\partial}{\partial v}\left[(\gamma - a_f + \beta E)vf\right]. \tag{12.5.10}$$

Let us discuss the implications of this equation.

Equation (12.5.10) can be used for constructing equations for mean energy. Since this is an equation with variable coefficients, the resulting equation for the first moment will include the second moment, and so on. Thus, we get a meshing chain of equations, and have to deal with the "problem of closure". The simplest, and at the same time the most efficient, is the approximation of first moments. In this approximation we get the following closed equation:

$$\frac{d\langle E\rangle}{dt} = 2\left[\frac{3}{2}mD_{(v)}(\langle E\rangle) + (a_f - \gamma - \beta\langle E\rangle)\langle E\rangle\right]. \tag{12.5.11}$$

This simple example is useful for demonstrating the importance of diffusion for the processes of time evolution of open nonlinear systems. With this purpose we compare (12.5.11) with the corresponding dynamic equation (12.5.8).

From the dynamic equation it follows that the null of parameter $a = a_f - \gamma$ is a point of bifurcation: at $a < 0$ the stable value is $E = 0$, whereas at $a > 0$ stable is a nonzero value of $E = a/\beta$. What happens at the point of bifurcation itself remains not clear.

The stationary solution of equation for the first moment yields essentially correct information about the mean value of energy (the temperature of the system) at all values of the controlling parameter a_f, including the point of bifurcation. At $a_f = 0$ we now have $\langle E\rangle = \frac{3}{2}\kappa T$ instead of $E = 0$; at the bifurcation point the value of $\langle E\rangle$ is determined by the temperature and the coefficient β, and $\langle E\rangle = a/\beta$ at large overruns (in the regime of stabilization of thermal instability).

Let us now consider the stationary solution of the kinetic equation in the presence of heat source, which can be represented as

$$f(v) = C \exp\left[-\frac{1}{\kappa T}\left(\frac{mv^2}{2} - \frac{a_f}{\beta}\ln\left(1 + \frac{\beta}{\gamma}\frac{mv^2}{2}\right)\right)\right] \equiv C\exp\left(-\frac{U(v)}{\kappa T}\right). \tag{12.5.12}$$

In the absence of heat source ($a_f = 0$) this coincides with Maxwell distribution. The distribution becomes essentially different, however, when the heat source is switched on.

Let us discuss this point in greater detail.

From the three-dimensional distribution we single out a one-dimensional distribution for $v_x \equiv v$ in the plane $v_y = 0$, $v_z = 0$. To see the difference between this distribution and the corresponding one-dimensional Maxwell distribution, we replace distribution (12.5.12) by an approximate Gaussian distribution. From condition of minimum of function $U(v)$ we locate the maxima of one-dimensional distribution function. Then distinguished are two regions separated by the bifurcation point $a_f = \gamma$:

$$v_{max} = 0 \text{ if } a_f < \gamma \,,$$
and $\tag{12.5.13}$
$$v_{max} = \pm\left(\frac{2}{m}\frac{a_f - \gamma}{\beta}\right)^{\frac{1}{2}} \text{ if } a_f > \gamma \,.$$

In order to find the relevant Gaussian distribution, we expand $U(v)$ with respect to the values of v_{max} and retain the terms of up to the second order. As a result, we get the desired distributions for regions "above" and "below" the bifurcation point:

$$f(v) = \left(\frac{m}{2\pi\sigma}\right)^{\frac{1}{2}}\exp\left(-\frac{mv^2}{2\sigma}\right), \quad \sigma = \frac{\gamma}{\gamma - a_f}\kappa T \text{ if } a_f < \gamma \,,$$
and $\tag{12.5.14}$
$$f(v) = \frac{1}{2}\left(\frac{m}{2\pi\sigma}\right)^{\frac{1}{2}}\sum_{\mp}\exp\left(-\frac{m(v \mp v_{max})^2}{2\sigma}\right), \quad \sigma = \frac{1}{2}\frac{a_f}{a_f - \gamma}\kappa T \text{ if } a_f > \gamma \,.$$

Both distributions are normalized to unity.

Here an important comment is due.

The above distributions hold only away from the bifurcation point. Crossing of the point of bifurcation corresponds to a "nonequilibrium phase transition", which is manifested by a change in the structure of the distribution function: at $a_f > \gamma$ the one-dimensional velocity distribution function has two maxima, whereas the relative dispersion of velocity is given by

$$\frac{\sigma}{m(v_{max})^2} = \frac{\beta a_f \kappa T}{4(a_f - \gamma)^2} \quad (\propto \frac{1}{a_f} \text{ when } a_f \gg \gamma) . \tag{12.5.15}$$

This expression remains valid (and even becomes much simpler) far above the bifurcation point.

Now let us return to expressions (12.5.14) for the dispersion of velocity. We see that on approaching the bifurcation point the dispersion increases according to Curie's law for nonequilibrium phase transitions:

$$\sigma \propto \frac{1}{|a_f - \gamma|} \quad (\text{instead of } \frac{1}{|T - T_{cr}|}) . \tag{12.5.16}$$

We see that on approaching the bifurcation point (the critical point of "nonequilibrium phase transition") the dispersion tends to infinity, and the Gaussian approximation is no longer valid.

However, the original distribution (12.5.12) holds for all values of the bifurcation parameter, which allows us to evaluate the dispersion at the critical point, and thus find the limits of applicability of the Gaussian approximation.

Let us give a concrete example for the range of parameters typical of Landau's theory of phase transitions. Namely, we assume that the nonlinear dissipative term is much smaller than the linear, and the "overshoot" above the bifurcation point is small, that is

$$\frac{\beta}{\gamma}\langle mv^2 \rangle \ll 1 , \quad \frac{a_f - \gamma}{\gamma} \ll 1 . \tag{12.5.17}$$

Under these conditions, distribution (12.5.12) can be rewritten as

$$f(v) = C \exp\left(-\frac{\left(\frac{mv^2}{2} - \frac{a_f - \gamma}{\beta}\right)^2}{\frac{2\kappa T \gamma}{\beta}}\right) . \tag{12.5.18}$$

In Landau's theory of phase transitions (Landau – Lifshitz 1976; Patashinskii – Pokrovskii 1982; Klimontovich IV) this is the form to which the distribution of values of parameter of order is reduced. The important difference, however, is that the controlling parameter for equilibrium phase transitions is the temperature. In our case the temperature is a common external parameter, whereas the control is exercised by adjusting the bifurcation parameter a_f ("feedback parameter").

From distribution (12.5.18) and formula (12.5.14) it follows that the dispersion of velocity at and above the bifurcation point (but still in the realm of Landau's theory) is defined by the following expressions:

$$\sigma_{a_f = \gamma} \propto \left(\frac{\kappa T \gamma}{\beta}\right)^{\frac{1}{2}} ; \quad \sigma = \frac{1}{2}\frac{\gamma}{a_f - \gamma}\kappa T \text{ (given that } \frac{a_f - \gamma}{\gamma} \ll 1 \text{)} . \tag{12.5.19}$$

Then the condition of applicability of Landau's approximation is expressed by two inequalities:

$$\frac{\sigma}{\sigma_{a_f = \gamma}} \sim \frac{\kappa T \gamma \beta}{(a_f - \gamma)^2} \ll 1 , \quad \frac{a_f - \gamma}{\gamma} \ll 1 . \tag{12.5.20}$$

The first of these implies that the overshoot above the critical point $a_f = \gamma$ must be considerable if Gaussian approximation is to be used. At the same time it must be not too large (second inequality), so as to regard the nonlinearity as small (first condition in (12.5.17)).

Observe finally that distribution (12.5.12) holds for all values of the bifurcation parameter, and is not restricted to the region of small nonlinearities!

12.6. Distribution of Kinetic Energy (Instant Temperature)

In the preceding section we have studied a nonequilibrium phase transition for a special case of three-dimensional velocity distribution (12.5.12) — a one dimensional distribution in the plane $v_y = 0$, $v_z = 0$. Useful is also another example of one-dimensional distribution: the distribution of values of kinetic energy $E = mv^2/2$ which follows from the three-dimensional distribution (12.5.12) and has the following form:

$$f(E) = C\exp\left[-\frac{E}{\kappa T} + \frac{a_f}{\kappa T \beta}\ln\left(1 + \frac{\beta}{\gamma}E\right) + \frac{1}{2}\ln E\right] , \quad \int_0^\infty f dE = 1 . \tag{12.6.1}$$

At $a_f = 0$ we have the equilibrium distribution of energy which follows from Maxwell distribution.

In "Landau's approximation", when the inequalities of (12.5.18) are satisfied, this can be reduced to a simpler distribution

$$f(E) = C \exp\left(-\frac{\left[E - \frac{a_f - \gamma}{\beta}\right]^2}{\frac{2\gamma\kappa T}{\beta}} + \frac{1}{2}\ln E\right) .$$

(12.6.2)

For both (12.6.1) and (12.6.2) the most probable energy value is

$$E_{\text{m.p.}} = \frac{a_f - \gamma}{\beta} .$$

(12.6.3)

If the relative dispersion is small, the distributions (12.6.1) and (12.6.2) can be approximated by the Gaussian distribution

$$f(E) = \left(\frac{1}{2\pi\sigma}\right)^{\frac{1}{2}} \exp\left[-\frac{\left(E - E_{\text{m.p.}}\right)^2}{2\sigma}\right] , \quad \sigma = \frac{a_f}{\beta}\kappa T .$$

(12.6.4)

The relative dispersions of distributions (12.6.1), (12.6.4) are given, respectively, by

$$\frac{\kappa T \beta}{a_f} \quad \text{if } a_f \gg \gamma ,$$

and

$$\frac{\kappa T \beta a_f}{\left(a_f - \gamma\right)^2} \quad \text{if } a_f - \gamma \ll \gamma .$$

(12.6.5)

We see that the more general distribution (12.6.1) beyond the limits of Landau's approximation is characterized by a slower decrease in the relative dispersion of the energy as the bifurcation parameter increases. This is because the distribution falls off less rapidly with the increasing energy.

Moreover, as the energy increases, the nonequilibrium distribution (12.6.1) falls off slower than the equilibrium distribution. All this implies that there is an "excess" of particles at the tail of the distribution. This excess may be important for many phenomena which depend on the number of particles capable of surmounting a barrier — for instance, in case of chemical reactions.

So we have discussed some examples of nonequilibrium distributions whose departure from equilibrium depends on the values of controlling (bifurcation) parameters. Of course, it would be good to have an instrument for measuring the deviation from equilibrium. Let us show that the above-defined Lyapunov functionals Λ_F, Λ_S can be used in this capacity.

12.7. Lyapunov Functional Λ_S as Measure of Relative Nonequilibrium. Illustration of S–Theorem

Now we represent (12.5.12) in the form of canonical Gibbs distribution (12.3.1). The effective Hamilton function is then given by

$$H_{\text{eff}}(v,a_{\text{f}},T) = \frac{mv^2}{2} - \frac{a_{\text{f}}}{\beta} \ln\left(1 + \frac{\beta}{\gamma}\frac{mv^2}{2}\right), \qquad (12.7.1)$$

and for $a_{\text{f}} = 0$ coincides with the energy of free particle. The effective free energy is defined by (12.3.2).

Recall that this is just one possible definition of nonequilibrium free energy (definition "according to Gibbs"). We also introduced a "thermodynamical" definition of nonequilibrium free energy (12.4.4). This allowed us to define the Lyapunov functional Λ_F (12.4.5).

This functional may serve as a kind of measure of deviation of the given nonequilibrium state from equilibrium. However, as pointed out in the Introduction and in Sect. 12.4, the nonequilibrium free energy does not possess all the properties necessary for using it as a measure of relative degree of order of different nonequilibrium states.

Lyapunov functional Λ_S has certain advantages over Λ_F. First, the entropy can be defined in terms of the distribution function for any nonequilibrium state. Secondly, the "entropy" functional has all the properties necessary for being used as a measure of uncertainty.

There is, however, an important circumstance associated with the formulation of entropy-based criterion of relative order.

Namely, we must not forget that Gibbs theorem (see Sect. 3.6 and 4.13), which states that entropy has its maximum at equilibrium, is based on the assumption that the mean energy of nonequilibrium states under consideration is the same as that at equilibrium. In all other respects the nonequilibrium states in question may be arbitrary.

In practice, the requirement that the mean energies of nonequilibrium and nonequilibrium states should be the same is satisfied as an exception rather than a rule. A rare example is the time evolution of Boltzmann gas towards equilibrium: as we know, the mean energy of this system is conserved in the course of evolution. More typical, however, is a different situation.

Consider the evolution of stationary nonequilibrium states as the controlling parameters are varied. Our task consists in comparing the relative degree of order of different states of this open system. The nonequilibrium distribution (12.5.12) or the corresponding "canonical Gibbs distribution" (12.3.1) with the effective Hamilton function (12.7.1) are two examples. The latter has been obtained as solution of the relevant kinetic equation (12.5.1). Alternatively, the form of nonequilibrium distribution for different values of controlling parameter can also be found from experimental data, by analyzing the sufficiently long time realizations.

In such cases the mean energy in the course of evolution, as the controlling parameter is varied, is usually not conserved, and straightforward application of Gibbs theorem is therefore not possible.

Then the comparison of the relative degree of order of different states may be carried out as follows.

We take the state of equilibrium for the "reference state". The equilibrium distribution follows from (12.5.12) with $a_f = 0$, and coincides with Maxwell distribution (12.4.6). By adjusting the temperature of the equilibrium state $(T \Rightarrow \tilde{T}(a_f))$ we renormalize the equilibrium distribution $(f \Rightarrow \tilde{f})$ in such a way as to ensure conservation of the mean energy in the course of evolution as the controlling parameter a_f is varied. Since the "reference state" is the state of equilibrium, this condition can be expressed as

$$\frac{3}{2}\kappa\tilde{T}(a_f) = \int \frac{mv^2}{2}\tilde{f}_0(v)dv = \int \frac{mv^2}{2}f(v,a_f)dv \ . \tag{12.7.2}$$

Recall that $\tilde{f}_0$ is the renormalized $(T \Rightarrow \tilde{T})$ Maxwell distribution (12.5.6), and the nonequilibrium distribution $f(v,a_f)$ is defined by (12.5.12).

Now by $\tilde{S}_0$ we denote the renormalized entropy value at equilibrium, and by $S(a_f)$ the entropy of nonequilibrium state with the distribution function (12.5.12). Then, under condition (12.7.2), we come to an expression similar to (4.13.5)

$$\Lambda_S = \tilde{S}_0 - S(a_f) = \kappa\int \ln\frac{f(v,a_f)}{\tilde{f}_0(v)}f(v,a_f)dv \geq 0 \ . \tag{12.7.3}$$

The result expressed by the last two formulas constitutes the statement of S–theorem as applied to the example under consideration (see Sect. 1.6.6 of the Introduction): (12.7.3) is a quantitative measure of the relative degree of order of equilibrium and nonequilibrium states, and (12.7.2) equates the mean energies of the equilibrium (more chaotic) and nonequilibrium (more ordered) states. In this manner, the Gibbs theorem is turned into a usable criterion of self-organization.

So, the Lyapunov functional (12.7.3) is a a quantitative measure of the change in the relative order of state of open system as the parameter a_f increases. The functional is zero when $a_f = 0$ and the distribution (12.5.12) coincides with the equilibrium Maxwell distribution.

The derivative $d\tilde{T}/da_f$, based on (12.7.2), gives the rate of transition from chaos to order as the controlling parameter a_f increases. For illustration let us use the Gaussian distribution (12.6.2); then for states above critical we have

$$\frac{3}{2}\kappa\tilde{T} = \frac{a_f - \gamma}{\beta},$$

(12.7.4)

and therefore the rate of transition from chaotic to more ordered motion in this range is positive and constant.

12.8. "Physical Chaos". General Formulation of S–Theorem

The example discussed in the preceding section is quite helpful, and yet it represents just one particular case. As a matter of fact, the state of equilibrium in the study of open systems can far not always be used as the reference point for comparing the relative order of different nonequilibrium states. This is especially true when the analysis is based on experimental data — for instance, on cardiograms or other time realizations of processes in medical and biological applications.

Let us return to Sect. 1.6.1 of the Introduction where we introduced the concept of "physical chaos", not to be confused with chaos at equilibrium. The latter is an absolute entity, because at equilibrium, in accordance with Gibbs theorem, the entropy (and hence the chaoticity) are at maximum.

By contrast, "physical chaos" is a relative concept. Let us explain this in detail.

We use the same notation as in Sect. 12.4. Let $a = a_0$ be the value of controlling parameter which by definition corresponds to the state of physical chaos. This state may be nonequilibrium. We compare the degree of order of this state which that of a different nonequilibrium state which corresponds to another value of controlling parameter, $a = a_0 + \Delta a$. If the degree of order of this other state is higher, then our choice of the first state as the state of physical chaos had been correct. Otherwise we have to look for a more chaotic initial state of the system.

How then do we construct the distribution function for a nonequilibrium state chosen for the "reference point" in comparing the degree of order of different states? There are at least two options.

1. The stationary solution for any value of Δa is found by solving the model kinetic equation. The state with $a = a_0$ we take for the state of physical chaos, and reduce the corresponding distribution function to the form of canonical Gibbs distribution:

$$f_0(A, T_{\text{eff}}, a_0) = \exp \frac{F_{\text{eff}}(T_{\text{eff}}, a_0) - H_{\text{eff}}(A, a_0)}{\kappa T_{\text{eff}}}, \quad \int f_0 dA = 1, \tag{12.8.1}$$

where we have used the notation for effective free energy and effective temperature.

Observe that this distribution only formally coincides with the equilibrium distribution (12.3.1), because all characteristics now relate to the nonequilibrium state. The choice of this state for the state of physical chaos is only justified if the nonequilibrium state with $a = a_0 + \Delta a$ is more ordered.

Naturally, in general the mean value of the effective Hamilton function will not be the same for the two selected states. To compare the relative degree of order of the two states we must carry out renormalization of the distribution function (12.8.1):

$$f_0(A, T_{\text{eff}}, a_0) \Rightarrow \tilde{f}_0(A, \tilde{T}_{\text{eff}}, a_0, \Delta a) = \exp \frac{\tilde{F}_{\text{eff}} - H_{\text{eff}}(A, a_0)}{\tilde{T}(\Delta a)}, \quad \int \tilde{f}_0 dA = 1, \tag{12.8.2}$$

The renormalized temperature $\tilde{T}$ as function of Δa is found from the assumption that the mean effective Hamilton function is the same for the two states in question:

$$\int H_{\text{eff}}(A, a_0) \tilde{f}_0(A, \tilde{T}_{\text{eff}}, a_0, \Delta a) dA = \int H_{\text{eff}}(A, a_0) f(A, a_0 + \Delta a) dA. \tag{12.8.3}$$

By solving this equation, we find the dependence of the renormalized effective temperature on the increment of controlling parameter

$$\tilde{T}_{\text{eff}}(\Delta a) \text{ at "initial condition" } \tilde{T}_{\text{eff}}(\Delta a)\big|_{\Delta a=0}. \tag{12.8.4}$$

By $\tilde{S}_0$ we denote the entropy of the state of physical chaos with renormalized temperature. Then, taking (12.8.3) into account, we get the following expression for the relevant Lyapunov functional:

$$\Lambda_S(\Delta a) = \tilde{S}_0 - S = \kappa \int \ln \frac{f(A, a_0 + \Delta a)}{\tilde{f}_0(A, \tilde{T}_{\text{eff}}, a_0, \Delta a)} f(A, a_0 + \Delta a) dA \geq 0. \tag{12.8.5}$$

In this way, we have two results. If the solution of (12.8.3) is such that

$$\tilde{T}_{\text{eff}}(\Delta a) > T_{\text{eff}}, \tag{12.8.6}$$

then for equalizing the mean effective energies one has to increase the temperature of physical chaos. This share will be converted in the other nonequilibrium state into the energy of the more ordered state.

If the solution of (12.8.3) is such as indicated above, our choice of the state of physical chaos is justified. Quantitatively, the change in the degree of order upon transition $a_0 \Rightarrow a_0 + \Delta a$ is given by (12.8.5).

We see that this criterion of the relative degree of order of nonequilibrium states of open systems includes a built-in test of the correct choice of the state with $a = a_0$ as the state of physical chaos.

Nevertheless, since the definition of physical chaos is a matter of convention, it will be expedient to double-check our conclusions. Namely, for the state of physical chaos we now take the state with $a = a_0 + \Delta a$ and carry out renormalization of distribution $f(A, a_0 + \Delta a)$ assuming that the mean value of $\langle H_{\mathrm{eff}}(A, a_0 + \Delta a) \rangle$ is constant. If we find therewith that $\tilde{T}_{\mathrm{eff}} < T_{\mathrm{eff}}$, then the conclusion that the state with $a_0 + \Delta a$ is more ordered is correct. If, on the other hand, we find that $\tilde{T}_{\mathrm{eff}} > T_{\mathrm{eff}}$, then it would be natural to assume that more ordered is the state for which the increase in the effective temperature after renormalization is smaller.

If there are several controlling parameters, the search for the most ordered state can be optimized. We shall discuss this point in chapters to follow.

First calculations of the relative degree of order based on the above criterion were carried out by the author a decade ago (see Klimontovich 1983, 1984). The review of seminal papers can be found in (Ebeling – Klimontovich 1984). It was in these papers that this criterion has been christened "S–theorem".

This name emphasizes the fact that S–theorem may serve as criterion of self-organization, since it allows measuring the relative degree of order of different nonequilibrium states of open systems.

2. Consider now the case when the form of effective Hamilton function is established directly on the basis of the time realization $A(t, a)$ (12.4.13).

As indicated at the end of Sect. 12.4, a long enough time realization of some process can be used for reconstructing the distribution function $f(A, a)$, which then allows one to define the effective Hamilton function by formula (12.4.15). The effective Hamilton function is so called because after renormalization to the specified mean value of H_{eff} the renormalized distribution function $\tilde{f}_0$, like in (12.8.2), assumes the form of canonical Gibbs distribution. The only difference is that in the initial distribution (12.4.15) the free energy is zero, and the respective effective temperature is unity.

Further we proceed as before, replacing the "initial condition" for the solution of (12.8.3) by

$$\tilde{T}_{\mathrm{eff}}(\Delta a)\big|_{\Delta a = 0} = 1 \ .$$

Instead of time realizations one may also use space realizations $A(r,a)$ which take the form of linear developments of two- or three-dimensional structures. The above criterion can be used for comparing the relative degree of order of complex spatial distributions. This allows one to investigate the evolution (degradation or self-organization) of spatial structures as the controlling parameter is varied.

12.9. Criterion of Self-Organization Based on Analysis of Time Spectra

In the study of relative degree of order of different states of open systems, it is sometimes more convenient to use the appropriate time spectra in place of time realizations:

$$I_0(\omega,a_0) \, , \; I(\omega,a_0 + \Delta a) \tag{12.9.1}$$

at different values of the controlling parameters. Then one gets two more options for employment of the above criterion of self-organization.

First, these spectra can be used for finding the appropriate distribution functions of intensity I:

$$f_0(I,a_0) \, , \; f(I,a_0 + \Delta a) \, , \; \int f_0 dI = \int f dI = 1 \; . \tag{12.9.2}$$

The effective Hamilton function is now defined by

$$f_0(I,a_0) = \exp\bigl(-H_{\mathrm{eff}}(I,a_0)\bigr) \; . \tag{12.9.3}$$

In this way, the state with $a = a_0$ is taken for the state of physical chaos. Further calculations follow the above-described scheme.

Secondly, the spectra (12.9.1) can be used for introducing normalized distribution functions of frequency values for any spectral interval $\omega_1 \le \omega \le \omega_2$:

$$f_0(\omega,a_0) = \frac{I_0(\omega,a_0)}{\int_{\omega_2}^{\omega_1} I_0 d\omega} \, , \; f(\omega,a_0 + \Delta a) = \frac{I(\omega,a_0 + \Delta a)}{\int_{\omega_2}^{\omega_1} I d\omega} \; . \tag{12.9.4}$$

Accordingly, the effective Hamilton function is defined by

$$f_0(\omega,a_0) = \exp\bigl(-H_{\mathrm{eff}}(\omega,a_0)\bigr) \; . \tag{12.9.5}$$

Further calculations are again in accordance with the above scheme.

In practice, the last method of comparing the relative degree of order of nonequilibrium states is often the most convenient (see, for example, Vasilyev – Klimontovich 1993; Anishchenko – Saponin – Anishchenko 1993).

Observe finally that thermodynamics of nonequilibrium irreversible processes is a most extensive branch of the modern statistical theory, and many of the arguments developed in this book belong to this domain of the statistical theory of open systems. In the next chapter we shall use the example of Boltzmann gas to demonstrate the feasibility of unified description of kinetic and hydrodynamic processes. This discussion will pave the way for developing a unified kinetic description of nonequilibrium processes in open active systems.

CHAPTER 13

UNIFIED DESCRIPTION OF KINETIC AND HYDRODYNAMIC PROCESSES

Chapters 6 and 11 were devoted to the two possible methods of description of nonequilibrium processes in a rarefied gas. The first of these is based on the kinetic Boltzmann equation, and the second on the equations of gas dynamics. The latter were derived from the approximate solution of the kinetic equation using perturbation theory in small Knudsen number. As a result, we constructed a closed theory for describing nonequilibrium processes on all scales.

It might seem that the job is done and we are ready to crack concrete problems with the aid of equations now available to us. Unfortunately, the situation is not that simple, and there are important obstacles that still have to be overcome.

We started to discuss these issues in the Introduction. At that point, however, we were only able to make some preliminary remarks, while now we have built up good basis for a detailed study (Klimontovich V; 1990c, d; 1992a, c; 1993b, c). Both kinetic and gasdynamic descriptions are carried out within the framework of a model of continuous medium. This is a common feature of the two approaches to the description of nonequilibrium processes. At the same time, the "continuous medium" in case of kinetic description is not the same as in case of gasdynamic description. We have already said that in Sect. 1.1.3, 1.1.4. This fact is manifested in that the structure of "continuous media" is different; in other words, we use different definitions of physically infinitesimal scales, and different concepts of what we call a "point" of continuous medium.

Now we are going to consider this matter on a constructive level. As a result, we shall construct a generalized kinetic equation which will serve as a basis for unified description of nonequilibrium processes on both kinetic and hydrodynamic scales. The transition to equations of gas dynamics will be accomplished without resorting to perturbation theory in Knudsen number, which will enable us to develop a unified kinetic description of nonequilibrium processes even in active media. Nonequilibrium phase transitions in such media may lead to appearance of various dissipative structures in space and time. It is sequences of nonequilibrium phase transitions that govern the processes of self-organization.

So, we have drawn up an ambitious program. Let us get down to business.

13.1. Necessity of Unified Description of Kinetic and Hydrodynamic Processes

When we go over from kinetic equation to equations of hydrodynamics using the above technique, we tacitly replace the fine-grained definition of continuous medium with a coarser-grained "gasdynamic" definition. It would be natural to ask whether this replacement is physically justified.

Recall that this transition was based on the approximate solution of the kinetic Boltzmann equation, using perturbation theory in small Knudsen number Kn. In the zero approximation the solution takes on the form of local Maxwell distribution (11.2.4). Gasdynamic functions ρ, u, T in these equations satisfy the Euler equations of gas dynamics neglecting the dissipative processes. The dissipative terms are taken into account in the first approximation of perturbation theory.

It follows that the zero approximation (the local Maxwell distribution) corresponds to local thermodynamic equilibrium. This means that only the reversible processes are taken into account. When, however, we go over to the first order in perturbation theory, the entire time symmetry is changed, and the equations become irreversible.

In this sense our theory is not regular. The irregularity of the theory is also manifested in the fact that the zero limit of Knudsen number is, strictly speaking, physically not feasible.

Indeed, the kinetic Boltzmann equation itself is derived from the Liouville equation by perturbation theory in small density parameter $\varepsilon = nr_0^3$ (see Ch. 7). Taking into account the definitions of Kn and ε, Knudsen number can be represented as

$$\mathrm{Kn} = \frac{l}{L} \sim \frac{1}{Lnr_0^2} \sim \frac{r_0}{L\varepsilon} \ . \tag{13.1.1}$$

We see that, given the parameters r_0 and L (which are defined by the size of the atoms and the characteristic scale of the system), the zero limit of Knudsen number corresponds to the infinitely large density parameter ε. This, however, is not compatible with the concept of rarefied Boltzmann gas.

Knudsen number can be small, but it cannot go to zero. Because of this, the choice of the principal approximation of conventional perturbation theory for Boltzmann equation is not justified from physical point of view. Anticipating the events, let us point out that this setback is removed when the generalized kinetic equation is used in place of the Boltzmann equation.

Let us continue with our comments.

When we go over from Boltzmann equation to equations of transfer for the densities of particles, momentum and energy, the new equations include (in addition to ρ, u, T) "excessive" functions π_{ij}, q, which are expressed via the distribution function by (11.1.2), (11.1.4). By convention, the former is called the tensor of viscous stress, and the latter the thermal flux.

Unfortunately, these terms are not quite to the point. Let us illustrate this with the example of thermal flux.

In the first approximation in Knudsen number the thus defined thermal flux is proportional to the temperature gradient (see (11.3.8)), in compliance with the empirical Fourier law. The coefficient of heat conductivity λ is therewith linked with the coefficient of temperature conductivity χ via the value of heat capacity at constant pressure c_p. This linkage, however, cannot be universal: in fact, the use of either c_p or c_v is dictated by the nature of the process. There also are other uncertainties.

At small Knudsen numbers the state is close to local equilibrium. The state of local equilibrium, however, is characterized not only by the temperature, but also by the density. Why then the transfer of heat (that is, the transfer of chaotic motion) should be characterized by the temperature gradient alone? We may even ask whether the use of temperature as one of the main characteristics in equations of gas dynamics is justified at all.

As a matter of fact, from the standpoint of kinetic theory the temperature is defined as the mean kinetic energy per particle. For this reason the fluctuations of the temperature are not small: the expression for the relative intensity of fluctuations does not include the $1/N$ factor (cf. (4.2.5) for the relative fluctuations of energy).

It follows that the temperature in equations of gas dynamics is not a thermodynamic function in the conventional sense. This becomes clear also from the fact that the mean and the most probable values of the kinetic energy in Maxwell distribution are

$$\langle E \rangle \equiv \left\langle \frac{p^2}{2m} \right\rangle = \frac{3}{2}\kappa T \ , \quad E_{\text{m.p.}} = \frac{1}{2}\kappa T \ , \tag{13.1.2}$$

and so the thermodynamic condition (4.3.3) does not hold (the corresponding limit is of the order of one).

We must conclude therefore that the mean kinetic energy, taken for the temperature in gas dynamics, is not a genuine thermodynamic characteristic. The question about the role of the higher moments of energy in the problem of heat transfer remains open.

Let us now return to the source of all these difficulties.

They all stem from the definition of thermal flux (11.1.5), for which we admittedly did not have sufficient reasons. Indeed, this expression characterizes the relative flux of kinetic energy of relative motion. Of course, this does to some extent characterize the transfer of chaotic motion (the transfer of heat). At the same time, we cannot ascertain that the transfer of heat is completely defined by this expression. After all, the deficiency of statistical description (the degree of chaoticity) is characterized by the entropy. It would be expedient therefore to link the thermal flux to the flux of entropy. How can this be accomplished?

We see that we are dealing with a superposition of two inconsistencies: first we take (11.1.5) for the definition of thermal flux without good enough reason, and then formally employ perturbation theory for deriving closed equations in gasdynamic functions ρ, u, T, which gets us into a physical quandary.

From arguments developed above we may conclude that the traditional two-level (kinetic vs. gasdynamic) approach to nonequilibrium processes in rarefied gases is not totally consistent. That is why we are in quest of the unified description of kinetic and gasdynamic processes in rarefied gases, and we start with the physical definition of "continuous medium".

13.2. Physical Definition of Continuous Medium

Physical definition of continuous medium is based on the concept of physically infinitesimal volume which is regarded as the "point" of continuous medium. By definition, the dimensions of the point are small compared with the characteristic scale of the problem; at the same time, the number of particles within the point (that is, within physically infinitesimal volume) is large. This implies that it is not possible to give a definition of the point size that would be valid for any level of description (whether kinetic, gasdynamic, or thermodynamic). For this reason we divide the task into three parts. First we shall give a concrete definition of physically infinitesimal scales in the kinetic theory. Then we define physically infinitesimal scales in the gasdynamic theory, and finally try to reconcile these two definitions and prepare the ground for a unified description of kinetic and gasdynamic processes. This will allow us to revise the traditional approach to the derivation of the kinetic equation with due account for the structure of "continuous medium", and eventually obtain the generalized kinetic equation. Every generalization has to be paid for, but we shall see that the game is worth the candles.

13.2.1. *Physically Infinitesimal Scales in Kinetic Theory*

Let us return to Sect. 1.1.3 of the Introduction, this time taking advantage of our knowledge of Boltzmann's kinetic theory.

The characteristic relaxation scales in the kinetic theory are the free path time τ and free path length l. Physically infinitesimal scales which would satisfy the general requirements of (1.1.11) can be selected from the following considerations.

On the average, each of the particles of the system experiences a collision within the time τ. Characteristic for the germination of irreversibility, however, is a much smaller time τ_{ph}, taken by a collision of one of N_{ph} particles within the "point" of continuous medium. It is this time that we have used in (1.1.16) for the physically infinitesimal time interval in the kinetic description. As a result, we arrived at relations (1.1.17) which define the relevant quantities. We see that the general requirements (1.1.11) are satisfied because the density parameter (1.1.12) is small.

These definitions signify the extent to which smoothing of molecular motion is possible upon transition to the approximation of continuous medium. In a gas at normal conditions we have $\varepsilon \sim 10^{-4}$, and so the "point" of continuous medium contains approximately 100 particles.

Since we have now a concrete estimate for the uncertainty of the position of gas particles, it is possible to define the Gibbs ensemble for nonequilibrium processes.

Smoothing of the microscopic characteristics over the physically infinitesimal volume (or the corresponding time interval τ_{ph}) transforms the initial reversible equations of Hamiltonian mechanics into the irreversible equations of statistical theory. In this chapter we are going to establish the concrete form of such equations.

Now we assume that the irreversibility originates on the physically infinitesimal scales τ_{ph}, l_{ph}. This assumption will be confirmed later also by comparing these scales with the characteristic scales of mixing in phase space r, p, which is the consequence of dynamic instability of motion of atoms of rarefied gas.

13.2.2. Physically Infinitesimal Scales in Gasdynamic Description

Return now to Sect. 1.1.4 of the Introduction. In case of gasdynamic description, the physically infinitesimal scales of time and length are governed by (1.1.19). The implication is that the "point" of continuous medium retains some trace of the relaxation process characteristic of the gasdynamic level of description. The relevant time interval τ_{ph}^{G} can be defined by equation similar to (1.1.16) in the kinetic theory. For relaxation processes of diffusion type, however, it would be more natural to begin with defining the number of particles N_{ph} by (1.1.20), and then proceed to obtain the desired relations (1.1.21) for the gasdynamic level of description.

As we see, the physically infinitesimal scales are functions of the characteristic scale L, or of the corresponding number of particles $N = nL^3$. Accordingly, the definition of physically infinitesimal scales depends crucially on the adopted level of description. Because of this, the definition of Gibbs ensemble for nonequilibrium processes will also depend on the level of description.

13.2.3. *Reconciliation of Kinetic and Gasdynamic Definitions of Physically Infinitesimal Scales*

Obviously, the "point" of continuous medium in gasdynamic description is "bolder" than in kinetic description; in other words,

$$V_{ph}^G \geq V_{ph} \, . \tag{13.2.1}$$

The "equals" sign here corresponds to the tiniest "point" possible in gasdynamic description. The size of this smallest possible "point" we denote by L_{min}, and the corresponding Knudsen number is

$$Kn_{min} = \frac{L_{min}}{L} \, . \tag{13.2.2}$$

From definitions (1.1.17), (1.1.21) it follows that inequality (13.2.1) can be rewritten in terms of two dimensionless parameters, Kn and ε:

$$\frac{V_{ph}^G}{V_{ph}} \sim \varepsilon^{3/10} Kn^{6/5} \leq 1 \, , \tag{13.2.3}$$

where the "equals" sign corresponds to the largest Knudsen number at which it is still possible to reconcile descriptions of kinetic and gasdynamic processes, and to the corresponding minimum point size:

$$Kn_{max} \sim \left(N_{ph}\right)^{1/2} , \quad L_{min} \sim \left(N_{ph}\right)^{1/2} l_{ph} \sim \frac{l}{\left(N_{ph}\right)^{1/2}} \, . \tag{13.2.4}$$

We see that L_{min} is much less than the free path length, being at the same time much greater than the scale l_{ph} relating to the kinetic description. It is this two-way inequality that opens the possibility of unified description of kinetic and gasdynamic processes also when Knudsen number is of the order of unity.

The relation for the relevant time interval is

$$\left(\tau_{ph}^G\right)_{min} \sim \frac{L_{min}^2}{D} \sim \tau_{ph} \, ; \tag{13.2.5}$$

that is, the minimum gasdynamic time scale is of the same order of magnitude as the physically infinitesimal time in kinetic description. This result will be used in the derivation of generalized kinetic equation.

So, the physically infinitesimal volume for the unified description of kinetic and gasdynamic processes (the "point size") is about L_{min}^3. From the values of N_{ph}, L_{min} as defined by (1.1.17), (13.2.4) we can estimate the number of particles N_{min} within such a "point":

$$N_{min} \sim \varepsilon^{-5/4} . \tag{13.2.6}$$

At normal conditions, when $\varepsilon \approx 10^{-4}$, the "point" contains about 10^5 particles.

13.2.4. Smoothing Over Physically Infinitesimal Volume.
Physical Knudsen Parameter

The particular form of smoothing function $F(\rho)$ depends on a number of factors, including the nature of external (mean) fields. We shall use the simplest smoothing function in the form of Gaussian distribution with mean mixing proportional to the force:

$$F(\rho) = \left(2\pi L_{min}^3\right)^{-3/2} \exp\left(-\frac{\left(\rho - \frac{b}{m}F\tau_{ph}\right)^2}{2L_{min}^2}\right) , \tag{13.2.7}$$

$$\langle\rho\rangle = \frac{b}{m}F\tau_{ph} .$$

The dispersion of this distribution is determined by the point size L_{min}; b is the mobility under the force F; $\langle\rho\rangle$ is the corresponding mean displacement over the time τ_{ph}.

For the function of dynamic variables we select the microscopic phase density (2.4.1). In place of r, ρ it will be more convenient now to use the "continuous-medium" variables R, v, and define the smoothing operation by

$$\tilde{N}(R,v,t) = \int N(R-\rho,v,t)F(\rho)d\rho . \tag{13.2.8}$$

For the operation of smoothing (which is required for the transition to the approximation of continuous medium) we shall use a dimensionless small parameter defined as the ratio of point size to the characteristic parameter of problem L. We call it the *physical Knudsen parameter* and denote by K_{ph}.

At the kinetic level of description we have $L \sim l$, and therefore

$$K_{ph} = \frac{l_{ph}}{l} \sim \frac{1}{N_{ph}} \; . \qquad\qquad (13.2.9)$$

The smallness of this parameter is ensured by condition $N_{ph} \gg 1$ which follows from the definition of point on the kinetic level of description.

For gasdynamic region the physical Knudsen parameter is

$$K_{ph}^G = \frac{l_{ph}^G}{L} \sim \frac{1}{\left(N_{ph}^G\right)^{1/2}} \; . \qquad\qquad (13.2.10)$$

The smallness of this parameter is again implied by the definition of point of continuous medium, this time on the gasdynamic level.

For the unified description of kinetic and gasdynamic processes it would be natural to define the physical Knudsen parameter as

$$K_{ph} = \frac{L_{min}}{L} \sim \frac{Kn}{\left(N_{ph}\right)^{1/2}} \; ; \; Kn_{max} \sim \left(N_{ph}\right)^{1/2} \sim \varepsilon^{-1/4} \; . \qquad\qquad (13.2.11)$$

The second relation here delimits the applicability of unified description in terms of the largest possible value of Knudsen number.

Now we are ready to embark upon derivation of the generalized kinetic equation for unified description of kinetic and gasdynamic processes. We shall use two possible averaging procedures for dynamic equations: smoothing with function (13.2.7) over physically infinitesimal volume, and averaging with respect to the above-defined nonequilibrium Gibbs ensemble.

13.3. Generalized Kinetic Equation for Unified Description of Kinetic and Gasdynamic Processes

For the initial reversible dynamic equation it is more convenient now to use (2.4.14) for the microscopic phase density in place of the Liouville equation. As the first step towards the irreversible kinetic equation, we introduce the relaxation term

$$-\frac{1}{\tau_{ph}}\Big[N(R,v,t)-\tilde{N}(R',v',t)\Big] ,$$
(13.3.1)

which describes the "adjustment" of microscopic distribution of particles to the corresponding smoothed distribution over the time τ_{ph}.

Recall that, according to (1.5.1), this time is of the same order of magnitude as the minimal characteristic time of dynamic instability of motion of particles of rarefied gas. Since dynamic instability leads to mixing in the phase space, the relaxation term describes the commencing transition to irreversibility.

As a result, instead of the initial reversible dynamic equation (2.4.14) we now have the following dissipative dynamic equation:

$$\frac{\partial N}{\partial t}+v\frac{\partial N}{\partial R}+\frac{F^{m}(R,t)}{m}\frac{\partial N}{\partial v}=-\frac{1}{\tau_{ph}}\Big[N(R,v,t)-\tilde{N}(R',v',t)\Big] ,$$
(13.3.2)

$$F^{m}(R,t)=F_{0}-\frac{\partial}{\partial R}\int\Phi\big(|R-R'|\big)N(R,v,t)dR'dv' .$$

Of course, these equations may be used for deriving the kinetic Boltzmann equation; smoothing will then be carried out over the physically infinitesimal volume V_{ph} with l_{ph} from (1.1.11). These scales define also the relevant nonequilibrium Gibbs ensemble. The first moment of random function $N(x,t)$ is then related to the one-point distribution $f(x,t)$ by equation (4.5.5), which in our current notation takes on the form

$$nf(R,v,t)=\big\langle N(R,v,t)\big\rangle .$$
(13.3.3)

Naturally, after smoothing of all terms in (13.3.2) over the volume V_{ph} the right-hand side of this equation goes to zero.

To construct the desired generalized kinetic equation we extend the volume of smoothing by defining the point size via the scale L_{min} (see (13.2.4)). Now we carry out averaging over the above-defined Gibbs ensemble, use identity (5.6.1), and thus obtain from (13.3.2) the following equation:

$$\frac{\partial f}{\partial t}+v\frac{\partial f}{\partial R}+\frac{F(R,t)}{m}\frac{\partial f}{\partial v}=I_{(v)}(R,v,t)+I_{(R)}(R,v,t) ,$$
(13.3.4)

$$F(R,t)=F_{0}-n\frac{\partial}{\partial R}\int\Phi\big(|R-R'|\big)f(R,v,t)dR'dv' .$$

Let us compare this with (5.6.2).

The equation for the distribution function now contains two terms instead of one. The first is similar to (5.6.3), and in our current notation is

$$I_{(v)}(R,v,t) = -\frac{1}{nm}\frac{\partial}{\partial v}\langle \delta F(R,t)\delta N(R,v,t)\rangle \ . \tag{13.3.5}$$

Like in (5.6.3), this term depends on the correlator of fluctuations of force and phase density. Recall that (5.6.3) coincides with the corresponding term (5.7.7) in the first equation of the BBGKY chain. This is why in case of rarefied gas it is reduced to Boltzmann collision integral.

This time, however, the right-hand side contains yet another term, which is defined by the relaxation term in the dynamic equation (13.3.2) and takes on the following form after averaging over the Gibbs ensemble:

$$I_{(R)}(R,v,t) = -\frac{1}{\tau_{\text{ph}}}\left(f(R,v,t) - \tilde{f}(R,v,t)\right) \ . \tag{13.3.6}$$

The smoothed distribution function is defined by expression similar to (13.3.1). Naturally, smoothing over the volume V_{ph} leaves this expression intact.

Since the "width" of the distribution function is now defined by the point size L_{min}, we may expand (13.3.6) in the physical Knudsen parameter (13.2.11). If the force is a function of coordinate, the nonlocality will enter the expansion also with the mean displacement. Retaining only the main terms in the expansion, we come to the following expression for the additional dissipative term:

$$-\frac{1}{\tau_{\text{ph}}}\left(f(R,v,t) - \tilde{f}(R,v,t)\right) = \frac{\partial}{\partial R}\left(\frac{L_{\text{min}}^2}{\tau_{\text{ph}}}\frac{\partial f}{\partial R}\right) - \frac{\partial}{\partial R}\left(\frac{b}{m}Ff\right) \ . \tag{13.3.7}$$

In order to get the final formula for the new "collision integral", we take into account the definitions of the relevant physically infinitesimal scales. Using (1.1.17), (13.2.4), we find

$$\frac{L_{\text{min}}^2}{\tau_{\text{ph}}} \approx \frac{l^2}{\tau} = D = b\frac{\kappa T}{m} \ , \tag{13.3.8}$$

where D is the coefficient of diffusion which is expressed in terms of mobility b and temperature T by the Einstein relation. For nonequilibrium processes the coefficient of diffusion D depends on the local temperature,

$$D(R,t) = \frac{b}{m} \kappa T(R,t) \,, \quad b = \tau \,. \tag{13.3.9}$$

Obviously, such generalization of the equilibrium Einstein relation is somewhat straining a point. Further on we shall discuss this issue.

Eventually we come to the final expression for the additional "collision integral":

$$I_{(R)}(R,v,t) = \frac{\partial}{\partial R}\left[D\frac{\partial f}{\partial R} - \frac{b}{m}F(R)f \right] \,, \quad D = \int D(R,t)\frac{dR}{V} \,. \tag{13.3.10}$$

Recall that here D is one of the three kinetic coefficients (kinematic viscosity v, temperature conductivity χ, and self-diffusion D) which are assumed to have one and the same value. The difference between them can be taken into account by using a more sophisticated smoothing function.

Now return to the first dissipative term. In principle, we may use it directly in Boltzmann's form. It is possible, however, to replace it with a simpler expression.

Indeed, in accordance with the above pattern of development of irreversibility in a rarefied gas, the characteristic scales of the kinetic Boltzmann theory τ, l (free path time and length) are not only individual, but also collective characteristics. For example, $\tau = N_{min}\tau_{ph}$ characterizes the time interval within which all N_{min} particles within a "point" of continuous medium will participate in collisions. The individual characteristic is the physically infinitesimal time τ_{ph} taken by the collision of one of the particles (on the average, of course) within a "point".

With due account for the "collective nature" of paired collisions, Boltzmann's collision integral can be replaced by a simpler expression, averaged over the angles in the space of velocities for individual paired collisions and similar to the Fokker – Planck expression in the theory of Brownian motion (see Ch. 15):

$$I_{(v)}(R,v,t) = \frac{\partial}{\partial v}\left[D_{(v)}(R,t)\frac{\partial f}{\partial v} \right] + \frac{\partial}{\partial v}\left[\frac{1}{\tau}(v - u(R,t))f \right] \,. \tag{13.3.11}$$

Here $D_{(v)}$ is the coefficient of diffusion in the space of velocities, linked with the local temperature by the Einstein relation (cf. (13.3.9)):

$$D_{(v)}(R,t) = \frac{1}{\tau}\frac{\kappa T(R,t)}{m} = \frac{1}{\tau}\frac{n}{\rho}\int \frac{m\delta v^2}{2}\,f\,dv \,. \tag{13.3.12}$$

In (13.3.9) – (13.3.12), τ is the mean "collision time" of the selected particle.

We see that the kinetic equation now contains two dissipative terms. One of them is defined by redistribution of particles in the space of velocities. This redistribution, like in case of Boltzmann approximation, is caused by paired collisions particles of rarefied gas. The potential of interaction of colliding particles is taken into account through the collision time τ.

The dissipation is again caused by paired collisions. However, in writing the simplified expression for the relevant collision integral we took into account the "collective manifestations" of Boltzmann's parameters τ, l.

Important for the forthcoming discussion is the fact that the collision integral $I_{(v)}(R,t)$ has the same properties as Boltzmann's collision integral:

$$\int \varphi(v) I_{(v)}(R,t) dv \begin{cases} = 0 & \text{if } \varphi = 1, \ v, \ v^2 \\ \geq 0 & \text{if } \varphi = -\kappa \ln f(R,v,t) \end{cases} \tag{13.3.13}$$

The "upper" property is useful for derivation of equations of gas dynamics in functions ρ, u, T, and the "lower" for the proof of the corresponding generalized H–theorem. In the proof of H–theorem we also use the following equation:

$$\int \ln f_{\text{loc}} \ I_{(v)}(R,v,t) dv = 0 \ , \tag{13.3.14}$$

where we have used the "upper" property of (13.3.13), and f_{loc} is the local Maxwell distribution (11.3.3).

At equilibrium, the generalized kinetic equation is satisfied by the Maxwell – Boltzmann distribution, since the additional collision integral (13.3.10) vanishes upon substitution of Boltzmann distribution.

The following is also important for further applications.

Replacement of Boltzmann collision integral with a simpler expression averaged with respect to angles is possible because there are many particles within a point of continuous medium (see (13.2.6)). On the same ground we may specify a class of distribution functions which are isotropic with respect to relative velocity:

$$f(R,v,t) = f\left(R, |v - u(R,t)|, t\right) \ . \tag{13.3.15}$$

This class includes, in particular, the local Maxwell distribution. Sometimes the mean velocity is zero, and the distribution with respect to velocities is isotropic. Such is the case, for instance, in the state of equilibrium.

13.4. H–Theorem for Generalized Kinetic Equation. Entropy Flux and Entropy Production. Lyapunov Functionals

Now we rewrite (13.3.4) in explicit form with due account for the above expressions for "collision integrals":

$$\frac{\partial f}{\partial t} + v\frac{\partial f}{\partial R} + \frac{F(R)}{m}\frac{\partial f}{\partial v} = \frac{\partial}{\partial v}\left[D_{(v)}(R,t)\frac{\partial f}{\partial v}\right] + \frac{\partial}{\partial v}\left[v - u(R,t)f\right]$$

$$+ \frac{\partial}{\partial R}\left[D\frac{\partial f}{\partial R} - \frac{b}{m}F(R)f\right], \quad b = \tau . \tag{13.4.1}$$

For establishing the form of equations of entropy balance and Lyapunov functionals it is important that the structure of dissipative terms in the kinetic equations is quite different.

Indeed, the properties of the first term, which takes care of dissipation due to the redistribution of particles' velocities, are the same as those of Boltzmann collision integral. This implies that we may use the same boundary conditions for the distribution function as in Boltzmann's kinetic theory:

$$f(R,v,t)\big|_{v_i=\pm\infty} = 0 , \quad f(R,v,t)\big|_{R_i=\pm\infty} = 0 , \quad i = x,y,z . \tag{13.4.2}$$

The first condition states that the probability of infinitely large values of kinetic energy of particles is zero, and the second prevents escape of particles to infinity. In particular, by this condition the convective flow of particles is zero:

$$nm\int vf(R,v,t)dv\Big|_{R_i=\pm\infty} = \rho u\big|_{R_i=\pm\infty} = 0 . \tag{13.4.3}$$

The properties of the second dissipative term are not so general. Since the second term includes spatial diffusion and mobility of particles under the force F, we need additional boundary conditions. In particular, the additional boundary conditions for a closed system are:

$$\rho u\big|_S - D\frac{\partial\rho}{\partial R}\Big|_S = 0 , \quad F(R,t) = 0 , \tag{13.4.4}$$

that is, the total flux of matter normal to the surface is zero, and the mean force on a particle is zero.

13.4.1. H–Theorem

In Sect. 6.4 we gave the proof of H–theorem based on the kinetic Boltzmann equation. Let us now rewrite the definition of density of local entropy (6.4.3):

$$S(R,t) \equiv \frac{\rho(R,t)}{m} s(R,t) = -\kappa n \int \ln\big(nf(R,v,t)\big) f(R,v,t) dv \ . \tag{13.4.5}$$

Here we also introduce the entropy per one particle, which enters equation (11.4.2) of the second law of thermodynamics for local equilibrium. Using this definition of entropy, we derive the equation of balance of density of local entropy, which can be written in the following form:

$$\frac{\partial \rho s}{\partial t} + \frac{\partial}{\partial R}\left[\left(\rho u - D\frac{\partial \rho}{\partial R}\right)s\right] = \frac{\partial}{\partial R}\left(D\rho\frac{\partial s}{\partial R}\right) + \sigma(R,t) \ . \tag{13.4.6}$$

At this point some explanations are due.

The flux of entropy is a sum of convection and diffusion terms:

$$j_S(R,t) = j_{\text{con}}(R,t) + j_{\text{dif}}(R,t) \ , \tag{13.4.7}$$

whereof the first is proportional to the entropy, and the second to the entropy gradient:

$$j_{\text{con}} = \left(\rho u - D\frac{\partial \rho}{\partial R}\right)s \ , \quad j_{\text{dif}} = -D\rho\frac{\partial s}{\partial R} \ . \tag{13.4.8}$$

The convective transfer of entropy is proportional to the flux of matter and the entropy. The flux of matter is built up of convection and diffusion terms. In establishing the structure of j_{con} we used the constraint (13.3.15) on the class of distribution functions. The diffusion term j_{dif} will be used in Sect. 13.5 for the definition of thermal flux.

Let us now look at the expression for entropy production, which for the kinetic Boltzmann equation is given by (6.4.5). Now we must replace the Boltzmann collision integral with expression (13.3.11), and transform with due account for the properties (13.3.13). To show expressly that this part of entropy production is positive, we add a zero term to it:

$$n\int \ln f_{\text{loc}}(R,v,t) \ I_{(v)}(R,v,t) dv = 0 \ , \tag{13.4.9}$$

where f_{loc} is the local Maxwell distribution.

The second part in entropy production is defined by the additional dissipative term (13.3.10), which is also positive. For the sake of symmetry we add a zero term $\partial \ln f_0 / \partial R$, where f_0 is the equilibrium Maxwell distribution. All in all, the production of entropy is given by

$$\sigma(R,t) = \kappa n \int \left[D_{(v)} f \left(\frac{\partial}{\partial v} \ln \frac{f}{f_{loc}} \right)^2 + D f \left(\frac{\partial}{\partial R} \ln \frac{f}{f_0} \right)^2 \right] dv \geq 0 . \tag{13.4.10}$$

This expression is a sum of two positive contributions. The first is determined by the redistribution of particles' velocities and is proportional to the coefficient of diffusion $D_{(v)}$. The second is determined by the redistribution of particles in space and is proportional to the coefficient of diffusion D.

13.4.2. State of Local Equilibrium

At local equilibrium the first term vanishes, and entropy production is wholly determined by redistribution of particles in space. Using the local Maxwell distribution, we can reduce (13.4.10) to

$$\sigma(R,t) = \frac{\kappa}{M} \left[D\rho \left(\frac{\text{grad}\,\rho}{\rho} \right)^2 + v\rho \frac{m}{\kappa T} \left(\frac{\partial u_i}{\partial R_j} \right)^2 + \frac{3}{2} \rho \chi \left(\frac{\text{grad}\,T}{T} \right)^2 \right] \geq 0 . \tag{13.4.11}$$

Let us compare this with (11.4.4) obtained with the aid of Boltzmann kinetic equation for entropy production in case of local equilibrium.

First, we now have an additional positive term which is determined by self-diffusion of particles. Observe that for slow processes, when the pressure is practically constant (see Ch. 19), the equation of state $p = (\rho/m)\kappa T$ implies that

$$-T \,\text{grad}\,\rho = \rho \,\text{grad}\,T \quad \text{if} \quad p = \text{const} \tag{13.4.12}$$

(assuming that $D = \chi$). If so, the first and the third terms on the right-hand side of (13.4.11) can be merged together, and the resulting term will be the same as the second "temperature" term on the right-hand side of (11.4.4), which is proportional to heat capacity at constant pressure c_p.

Now let us look at the terms which depend on velocity derivatives.

Recall that in transition from the kinetic Boltzmann equation to gasdynamic equations we defined the tensor of viscous stress π_{ij} by (11.1.2), (11.1.3). In the zero approximation in Knudsen number (at local equilibrium) the tensor of viscous stress is zero, and in the first approximation it is given by (11.3.6). A similar expression (12.2.4) is used in phenomenological hydrodynamics, with appropriate argumentation.

A different expression for the tensor of viscous stress follows from the generalized kinetic equation:

$$\pi_{ij} = -\rho v \frac{\partial u_i}{\partial R_j} \; . \tag{13.4.13}$$

For incompressible liquids this difference is not significant, since it does not tell on the Navier – Stokes equation. For compressible liquids there are some consequences. In this connection let us once again point out the following.

The symmetry of tensor of viscous stress in gas dynamics is postulated in its definition (11.1.3), and is preserved when perturbation theory in Knudsen parameter is applied. In hydrodynamics the requirement for the symmetry of tensor of viscous stress is also to a certain extent artificial.

Finally, let us quote the expression for the vector of entropy flux at local equilibrium. With due account for (13.4.8), we use the local Maxwell distribution to find the appropriate formula for the entropy:

$$S(R,t) \equiv \rho s = \rho \left[\frac{3}{2} \frac{\kappa}{m} \left(\ln \frac{2 \pi \kappa T}{m} + 1 \right) - \frac{\kappa}{m} \ln \frac{\rho}{m} \right] . \tag{13.4.14}$$

In contrast with the conventional expression (see (11.4.4)), the convection term here accounts for entropy transfer by self-diffusion.

The diffusion contribution to the flux of entropy is

$$j_{\text{dif}} = -D\rho \frac{\partial s(R,t)}{\partial R} = -\frac{\kappa}{m} D\rho \frac{\text{grad} \rho}{\rho} + \frac{3}{2} \frac{\kappa}{m} \chi\rho \frac{\text{grad} T}{T} \tag{13.4.15}$$

and contains the gradients of both density and temperature. For slow processes, when the pressure is practically constant, equation (13.4.12) holds and the result coincides with the corresponding term in the equation of entropy balance (11.4.4).

13.4.3. *Lyapunov Functional Λ_S for Closed System*

For a closed system, when conditions (13.4.2) – (13.4.4) hold, the energy of rarefied gas is the kinetic energy of its particles. Like in case of kinetic Boltzmann equation, the mean kinetic energy is conserved in the course of evolution towards equilibrium. Because of this, H–theorem can be expressed in terms of Lyapunov functional Λ_S by two inequalities (1.6.3), (1.6.4), or, in our current notation,

$$\Lambda_S = S_0 - S(t) = \kappa \int \ln \frac{f(R,v,t)}{f_0(v)} f(R,v,t) dv \frac{dR}{V} \geq 0 ,$$

(13.4.16)

$$\frac{d\Lambda_S}{dt} = -\int \sigma(R,t) dR \leq 0 , \quad \left\langle \frac{p^2}{2m} \right\rangle = \text{const} .$$

Entropy production is given by (13.4.10).

Since it is the mean energy rather than the energy itself that is conserved in the course of evolution towards equilibrium, fluctuations of energy are possible. We discussed this at the end of Sect. 6.4.

13.4.4. Non-Closed System. Lyapunov Functional Λ_F

Assume now that the mean force F is a nonzero potential force. By $f_0(R,v)$ we denote the equilibrium solution of kinetic equation (13.4.1), which is the Maxwell – Boltzmann distribution:

$$f_0(R,v) = \exp \frac{F_0 - H(R,v)}{\kappa T} , \quad H = \frac{p^2}{2m} + U(R) .$$

(13.4.17)

By definition (12.4.4), we introduce the free energy of nonequilibrium state

$$F(t) = \int H(R,v) f(R,v,t) dv \frac{dR}{V} - T \int S(R,t) \frac{dR}{V} ,$$

(13.4.18)

and the appropriate Lyapunov functional (cf. (12.4.5))

$$\Lambda_F(t) = F(t) - F_0 = \kappa T n \int \ln \frac{f(R,v,t)}{f(R,v)} f(R,v,t) dv \frac{dR}{V} \geq 0 .$$

(13.4.19)

In order to study the time evolution of this functional we use once again the kinetic equation (13.4.1). Boundary condition (13.4.3) is now replaced by a more general condition

$$j|_S = \left(\rho u - D\frac{\partial \rho}{\partial R} + \frac{b}{m}F\rho\right)\Bigg|_S = 0 \; , \tag{13.4.20}$$

which states that the total flux of matter across the boundary of the system is zero. Of course, other boundary conditions are possible.

Using (13.4.1) and (13.4.20) we find the time derivative of Lyapunov functional Λ_F:

$$\frac{d\Lambda_F}{dt} = -T \int \sigma(R,t)\frac{dR}{V} \leq 0 \; , \tag{13.4.21}$$

where $\sigma(R,t)$ is the entropy production as defined by (13.4.10). In place of Maxwell distribution $f_0(v)$, however, we must now use Maxwell – Boltzmann distribution (13.4.17).

Along with the functional $\Lambda_F(t)$, we can also define the local Lyapunov functional

$$\Lambda_F(R,t) \equiv \rho\lambda_F = \kappa Tn\int \ln\frac{f(R,v,t)}{f_0(R,v)}f(R,v,t)dv \geq 0 \; . \tag{13.4.22}$$

With the aid of kinetic equation we find the following equation of balance for this functional:

$$\frac{\partial}{\partial t}(\rho\lambda_F) + \frac{\partial}{\partial R}\left[\left(\rho u - D\frac{\partial \rho}{\partial R} + \frac{b}{m}Ff\right)\lambda_F\right] = \frac{\partial}{\partial R}\left(D\rho\frac{\partial}{\partial R}\lambda_F\right) - T\sigma(R,t) \; . \tag{13.4.23}$$

The structure of this equation is similar to that of the equation of entropy balance (13.4.6); the difference is that the flux of Lyapunov functional contains an additional term proportional to the force, and that Maxwell – Boltzmann distribution is used in place of Maxwell distribution in (13.4.10).

By integration over a sufficiently large volume one can go over from the local equation of balance to equation (13.4.21). Boundary condition (13.4.20) must then be supplemented by an additional condition: the nonequilibrium state of the system must occur within a limited volume. In other words, the system must be placed in a thermostat, and is able to exchange energy with the outside world (that is why we call it an open system).

So, according to the criterion based on Lyapunov functional the state of equilibrium corresponds to the minimum of free energy and is stable. In the parlance of Sect. 12.4, the adopted definition of nonequilibrium free energy may be classified as "thermodynamical". In the same section we also noted that another Lyapunov functional (Λ_S) is more adequate for comparing the relative degree of order of different states (in our present context, for the process of time evolution). In case of the generalized kinetic equation, however, the mean value of Hamilton function is not conserved in the course of time evolution towards the state of equilibrium. Because of this, one has to carry out renormalization to the specified value of mean energy before using the entropy-based functional as measure of the relative degree of order. This will naturally change the structure of the initial kinetic equation. An example of such renormalization will be given in the chapter devoted to Brownian motion in nonlinear systems.

13.5. Definition of Thermal Flux for Arbitrary Knudsen Numbers

Recall that "thermal flux" is one of the "excessive" functions (11.1.5) in equations of transfer. As indicated in Sect. 13.1, this definition is not quite consistent. To wit, function (11.1.5) is the mean value of transfer of relative kinetic energy, and does not account for the entire transport of heat.

Because of this, equation (11.3.8) (obtained by perturbation theory and expressing the empirical Fourier law) is not totally satisfactory. Now the generalized kinetic equation offers an opportunity to overcome these difficulties.

Since entropy at $F = 0$ is the measure of chaoticity, it would be natural to express thermal flux in terms of entropy gradient. This approach is also justified by the fact that the flux of entropy is given by (13.4.7) for arbitrary Knudsen numbers. The second contribution, accounting for the diffusion of entropy, can be used for defining the heat flux:

$$q(R,t) = -D\rho T \frac{\partial s}{\partial R} \ , \quad \rho s = -\kappa n \int \ln f(R,v,t) f(R,v,t) dv \ . \tag{13.5.1}$$

We see that in general one has to know the solution of the kinetic equation in order to define the thermal flux.

The situation is much simplified for the region of local equilibrium, when local Maxwell distribution holds. Then the local entropy is given by (13.4.14), and thermal flux is (assuming that $D = \chi$)

$$q(R,t) = \frac{\kappa}{m} DT \frac{\partial \rho}{\partial R} - c_v \rho \chi \frac{\partial T}{\partial R} \ , \quad c_v = \frac{3}{2} \frac{\kappa}{m} \tag{13.5.2}$$

We see that thermal flux in the state of local equilibrium depends in general not only on the velocity gradient, but also on the gradient of temperature. This complies with the overall structure of the second law of thermodynamics for local equilibrium:

$$dQ = Tds + pd\left(\frac{1}{\rho}\right), \quad \varepsilon = \frac{3}{2}\frac{\kappa T}{m}, \quad p = \frac{\rho}{m}\kappa T . \tag{13.5.3}$$

A simplifying assumption is required for the transition from (13.5.2) to the Fourier law. In gasdynamic region we single out the domain of slow processes (low-frequency spectral range). As we shall see, the pressure in this domain is practically constant. Relation (13.4.12) between the gradients of density and temperature holds under this constraint. Then expression (13.5.2) for thermal flux can be reduced to

$$q(R,t) = -\lambda\frac{\partial T}{\partial R}, \quad \lambda = c_p\rho\chi , \quad c_p = \frac{5}{2}\frac{\kappa}{m} . \tag{13.5.4}$$

Thus we have arrived at the Fourier formula, whereat the coefficient of thermal conductivity depends on heat capacity at constant pressure.

Observe that in case of incompressible medium, when the density is constant, the coefficient of thermal conductivity λ in Fourier law depends on heat capacity at constant volume c_v.

Finally, for the adiabatic process, when the temperature and pressure in (13.4.14) are linked by

$$T\rho^{1-\gamma} = T\rho^{-\frac{2}{3}} = \text{const} , \quad \gamma = \frac{c_p}{c_v} , \tag{13.5.5}$$

the following relation holds for the gradients of temperature and density:

$$\frac{1}{T}\frac{\partial T}{\partial R} = \frac{2}{3\rho}\frac{\partial \rho}{\partial R} , \quad \text{and therefore} \quad q(R,t) = 0 . \tag{13.5.6}$$

So, from (13.5.6), (13.5.2) it follows that for a locally adiabatic process the thermal flux is zero at every point of the system. This is a direct consequence of the inclusion of self-diffusion into equations of gas dynamics.

Recall that these particular results only hold under condition of local equilibrium. In the general case, the thermal flux is expressed by (13.5.1) via the gradient of entropy (per one particle). This expression only holds when $F(R,t) = 0$. It is only then that the mean energy is conserved in the course of evolution, and the difference in entropies of equilibrium and nonequilibrium states is a Lyapunov functional.

In the presence of external field, when $F(R,t) \neq 0$, the relevant Lyapunov functional is expressed by (13.4.19), (13.4.10) in terms of the difference in free energies of equilibrium and nonequilibrium states, and the equation of balance for the local functional has the form of (13.4.23). Function $\sigma(R,t)$ is given, as before, by (13.4.10). In the presence of external field f_0 is the equilibrium Maxwell – Boltzmann distribution, and f_{loc} the corresponding local distribution.

13.6. Thermal Flux in External Field

In the presence of external field it is natural to express thermal flux in terms of the gradient of the local functional $\lambda_F(R,t)$. To see the difference between such definition of thermal flux and (13.4.8), let us consider a particular example.

Assume that the mean velocity $u(R,t)$ is zero, and the temperature T is constant. This means we are dealing with thermal flux in a thermostat, and the thermal flux at local equilibrium is completely determined by the spatial distribution of heated ($T \neq 0$) particles.

Under these conditions, the class of distribution functions (13.3.15) is narrowed to

$$f(R,v,t) = f(R,|v|,t) \;, \tag{13.6.1}$$

and from the generalized kinetic equation (13.4.1) follows a closed equation for the density $\rho(R,t) = mn\int f(R,|v|,t)dv$,

$$\frac{\partial \rho}{\partial t} = \frac{\partial}{\partial R} D\left[\frac{\partial \rho}{\partial R} - \frac{F}{\kappa T}\rho\right] \equiv \frac{\partial}{\partial R}\left[D\rho\frac{\partial}{\partial R}\ln\frac{\rho}{\rho_0}\right], \tag{13.6.2}$$

which is the equation of self-diffusion in the potential field $F = -\operatorname{grad}U(R)$. In writing the second form of the right-hand side we have noted that the Boltzmann distribution ρ_0 is the equilibrium solution of the diffusion equation.

In this approximation the Lyapunov functional (13.4.22) is reduced to

$$\Lambda_F(R,t) \equiv \rho(R,t)\lambda_F(R,t) = \kappa T \ln\frac{\rho(R,t)}{\rho_0(R)}\rho(R,t) \;. \tag{13.6.3}$$

Like in Sect. 13.4 (see (13.4.19)), sign-constant is only the integral characteristic $\int \Lambda_F(R,t)(dR/V)$. The corresponding equation of balance is

$$\frac{\partial \rho \lambda_F}{\partial t} + \frac{\partial}{\partial R}\left\{\left[-D\left(\frac{\partial \rho}{\partial R} - \frac{F}{\kappa T}\rho\right)\right]\lambda_F\right\} = \frac{\partial}{\partial R}\left(D\rho\frac{\partial}{\partial R}\lambda_F\right) - \sigma_F ,$$

(13.6.4)

$$\sigma_F = \kappa T D\rho\left(\frac{\partial}{\partial R}\ln\frac{\rho}{\rho_0}\right)^2 \geq 0 .$$

We see that the diffusion flow of the local characteristic λ_F is expressed via the difference of gradients of entropies of equilibrium and nonequilibrium states. Eventually, we arrive at the following definition of thermal flux:

$$q(R,t) = D\rho\,\mathrm{grad}\,\lambda_F(R,t) = -TD\rho\,\mathrm{grad}\left[s(R,t) - s_0(R)\right] .$$

(13.6.5)

In the absence of external field, this expression coincides in form with the general definition (13.5.1) based on the equation of entropy balance (13.4.6). We see that in the presence of external field the thermal flux depends on the relative spatial distributions for nonequilibrium and equilibrium states.

Recall that currently we are dealing with a particular case, when the mean velocity is zero and the temperature is constant. The local entropy is completely determined by the distribution of density. Taking (13.6.3) into account, we come to the following expression:

$$q(R,t) = TD\rho\,\mathrm{grad}\left[\rho(R,t) - \rho_0(R)\right] ,$$

(13.6.6)

which is defined by the difference between the distributions of the density for nonequilibrium and equilibrium states. The transfer of heat in this example takes place at constant temperature and is therefore not governed by the Fourier law. Instead, it is completely determined by the transport of "heated" particles from the region where they are more numerous to the region where they are less numerous than at the state of equilibrium.

This non-conventional definition of thermal flux opens new possibilities for the theory of heat transfer. Some of these will be considered in the sections to follow, devoted to the theory of heat transfer in active media.

Let us now examine the changes in equations of gas dynamics when we derive them from the generalized kinetic equation (13.3.1). The important fact is that we no longer have to rely on perturbation theory in small Knudsen number.

CHAPTER 14

TRANSITION FROM GENERALIZED KINETIC EQUATION TO EQUATIONS OF GAS DYNAMICS

In Ch. 11 we accomplished a transition from the kinetic Boltzmann equation to equations of gas dynamics, using perturbation theory in small Knudsen number. In the first approximation in perturbation theory we obtained a closed set of equations in gasdynamic functions $\rho(R,t)$, $u(R,t)$, $T(R,t)$, which is certainly an elegant and important result.

At the same time, we encountered quite a few problems concerning both the general structure of equations of gas dynamics and the structure of individual terms.

In particular, this applies to the definitions of tensor of viscous stress and the vector of thermal flux. Taking advantage of the equation of entropy balance (13.4.6) we were able to give the definition of thermal flux in terms of entropy gradient which is more general than the conventional definition. We also questioned the validity of the choice of thermodynamic functions, including the temperature. For that matter, the fluctuations of gasdynamic temperature (the mean kinetic energy over the distribution $f(R,v,t)$ are not small. What is then the role of the higher moments?

In this chapter we shall construct equations of gas dynamics on the basis of generalized kinetic equation (13.4.1), without having to use perturbation theory in small Knudsen number. This automatically removes many of the above difficulties. We shall have to deal, however, with certain new problems, associated, for instance, with self-diffusion in gas dynamics (Klimontovich V; 1992a, c; 1993b, c).

14.1. Equations of Gas Dynamics with Self-Diffusion

In place of kinetic Boltzmann equation we now start with the generalized kinetic equation (13.4.1). Recall that representation (13.3.11) of the "collision integral" $I_{(v)}(R,v,t)$ is based on the fact that the number of particles within a "point" is $N_{min} \sim \varepsilon^{5/4} \gg 1$. For the same reason we restricted the class of distribution functions under consideration $f(R,v,t)$ by condition (see (13.3.5))

$$f(R,p,t) = f\big(R, m|v - u(R,t)|, t\big) . \tag{14.1.1}$$

235

Under this condition, the closed set of dissipative gasdynamic equations in functions ρ, u, T follows from the generalized kinetic equation without using perturbation theory in small Knudsen number.

Indeed, we only have to substitute distribution (14.1.1) into the kinetic equation (13.4.1) and go over to equations in the moments of the distribution function. This brings us directly to the desired closed set of equations in functions ρ, u, T, which is conveniently represented as a set of equations of balance of density ρ, density of momentum ρu, and density of total (hydrodynamic and internal) kinetic energy:

$$\frac{\partial \rho}{\partial t} + \frac{\partial}{\partial R} j(R,t) = 0 \; , \quad j = \rho u - D\frac{\partial \rho}{\partial R} + \frac{b}{m} F\rho \; ; \tag{14.1.2}$$

$$\frac{\partial \rho u_i}{\partial t} + \frac{\partial}{\partial R_j}\left(j_j u_i\right) = -\frac{\partial p}{\partial R_i} + \rho v\frac{\partial^2 u}{\partial R^2} + \frac{\rho}{m} F_i \; , \quad p = \frac{\rho}{m}\kappa T \; ; \tag{14.1.3}$$

$$\frac{\partial}{\partial t}\left(\frac{\rho u^2}{2} + \frac{p}{m}\varepsilon(R,t)\right) + \frac{\partial}{\partial R_i}\left[j_i\left(\frac{u^2}{2} + \varepsilon\right) + u_i p - \rho v u_j\frac{\partial u_i}{\partial R_j} - c_v\rho\chi\frac{\partial T}{\partial R_i}\right] = \frac{\rho}{m} Fu \; , \tag{14.1.4}$$

$$\varepsilon(R,t) = \frac{\rho}{m}\frac{3}{2}\kappa T \; .$$

Let us compare these with the equations of transfer (11.1.6), (11.1.7), (11.1.10).

Observe first of all that these transfer equations do not contain the "excessive" functions $\pi_{ij}(R,t)$, $q(R,t)$, and comprise a closed set. As a matter of fact, $\pi_{ij}(R,t)$ and $q(R,t)$ are identically zero for the selected class of distribution functions:

$$\pi_{ij} = mn\int\left(\delta v_i \delta v_j - \frac{1}{3}\delta_{ij}\delta v^2\right)f(R,|v-u|,t)dv = 0 \; , \tag{14.1.5}$$

$$q = n\int \delta v\frac{m\delta v^2}{2} f(R,|v-u|,t)dv = 0 \; , \quad \delta v = v - u \; , \tag{14.1.6}$$

since the integrands in both cases are odd functions of relative velocity δv.

Note that in case of transfer equations based on the kinetic Boltzmann equation such relations only hold in the zero approximation in Knudsen number, when the solution of kinetic equation is the local Maxwell distribution. The set of equations in functions ρ, u, T does not contain dissipative terms and is known as Euler equations.

The situation becomes quite different when the generalized kinetic equation is used: owing to the presence of the second dissipative term ("collision integral" $I_{(R)}(R,v,t)$) in equation (13.4.1), equations (14.1.2) – (14.1.4) form a closed set of equations in gasdynamic functions.

Now let us draw a parallel between each equation of the set (14.1.2) – (14.1.4) and their counterparts (11.1.6), (11.3.10), (11.3.11) based on the kinetic Boltzmann equation in the first approximation of perturbation theory in Knudsen number.

First of all we reduce our new equations of gas dynamics to a more convenient form. Equation of continuity (14.1.2) remains unchanged. Equation of transfer of density of momentum is transformed with the aid of (14.1.2) to the equation in velocity

$$\frac{\partial u}{\partial t} + \left(\frac{j}{\rho}\text{grad}\right)u = -\frac{1}{\rho}\text{grad}\, p - v\Delta u + \frac{1}{m}F \; . \tag{14.1.7}$$

The flux of matter $j(R,t)$ is once again defined by expression which enters equation of continuity (14.1.2).

Equation of transfer of density of energy is transformed with the aid of (14.1.2) and (14.1.7) to the equation in the pressure of rarefied gas:

$$\frac{\partial p}{\partial t} + \frac{j}{\rho}\text{grad}\,p + \frac{5}{2}p\,\text{div}\,u = p\frac{\partial}{\partial R}\left[\frac{D}{\rho}\frac{\partial \rho}{\partial R} - \frac{b}{m}F\right] + \frac{2}{3}\lambda\frac{\partial^2 T}{\partial R^2} + \frac{2}{3}\rho v\left(\frac{\partial u_i}{\partial R_j}\right)^2, \tag{14.1.8}$$

where λ is the coefficient of heat conductivity,

$$\lambda = \frac{3}{2}\frac{\kappa}{m}\rho\chi = c_v\rho\chi \; .$$

Now we are ready to begin our juxtaposition.

First we look at the continuity equations (14.1.2) and (11.1.6).

In (11.1.6) the flux of matter was completely defined by the product of density and the hydrodynamic velocity, $j = \rho u$. For that matter, this relation serves as the definition of hydrodynamic velocity. Now the situation is different.

To wit, the flux of matter in equation of continuity (14.1.2) is a sum of three parts: convection flow ρu, flow due to self-diffusion, and flow due to mobility under action of the mean force.

Equation for velocity (14.1.7) differs from the corresponding Navier – Stokes equation (11.3.10) in two respects: (1) the flux of momentum in the second term on the left-hand side is now defined by the combined contribution of the three parts indicated above, and (2) the tensor of viscous stress is now given by a simpler nonsymmetrical expression

(13.4.13), which corresponds to hydrodynamic flow of continuous medium with due account for boundary conditions.

In the approximation of incompressible fluid this circumstance does not affect the form of Navier – Stokes equation. It is important, however, for the equation of entropy balance (see below).

Recall finally that the additional "collision integral" in the kinetic equation (13.4.1) has been derived under assumption that all three kinetic coefficients are equal, $D = v = \chi$. This allows us to use different symbols for kinetic coefficients in the terms corresponding to different flows. Such are the coefficients D (in self-diffusion term) and v (in the term of viscous friction) in (14.1.7).

Now let us compare equations for the pressure (11.3.11) and (14.1.8). We see as many as four distinctions: (1) in place of u on the left-hand side there is a new term j/ρ defined by the total flux of matter; (2) the right-hand side contains also dissipative terms defined by self-diffusion and mobility of atoms of the gas; (3) the tensor of viscous stress is given not by (11.3.6), but by a simpler expression (13.4.13). As indicated above, this difference is insignificant for the Navier – Stokes equation, but becomes important in the equation of entropy balance; (4) the coefficient of thermal conductivity in (11.3.11) was proportional to heat capacity at constant pressure c_p, whereas in (14.1.8) it is proportional to heat capacity at constant volume c_v. This is because self-diffusion is taken into account.

Recall that the set of equations of gas dynamics (14.1.2) – (14.1.4) only holds for distribution functions which satisfy condition (14.1.1), which ensures elimination of "excessive" functions π_{ij}, q from equations for functions ρ, u, T.

For anisotropic media, when the smoothing function (13.2.7) is defined by a more general distribution, the class of distribution functions will be broader:

$$f(R,v,t) = f\left(R, |v_x - u_x|, |v_y - u_y|, |v_z - u_z|, t\right) .$$

                                                                                (14.1.9)

Distributions of this kind are encountered, for instance, in the theory of electron-ion plasma in the presence of magnetic field. Here we shall confine ourselves to considering only the above-defined class of distribution functions.

We would like to emphasize that the selected class of distribution functions is nevertheless very fertile, since it is equivalent to the infinite number of moments. Apart from the selected functions ρ, u, T, which are the lowest moments of the distribution, it includes the infinite set of higher moments of relative velocity.

Within the framework of the selected class of functions, let us now consider the simpler examples of gasdynamic equations.

14.2. Diffusion Stage of Relaxation Towards Equilibrium

Relaxation towards equilibrium occurs in several stages; in case of rarefied gas one may distinguish primarily the kinetic and the gasdynamic stages. The first is characterized by the free path time τ, and the second by the diffusion relaxation time $\tau_D = L^2/D$ (recall that D is one of the three kinetic coefficients). In the generalized kinetic equation we assumed that $D = v = \chi$. In equations of gas dynamics this restriction can be lifted, and kinetic coefficients in dissipative terms may each have its own value.

At the gasdynamic stage of evolution towards equilibrium it will be also expedient to separate the fast (sonic) and the slow (diffusion) processes. It is the diffusion processes that are characteristic of the final stage of relaxation towards equilibrium. Let us illustrate this statement by showing that the nonlinear set of gasdynamic equations for states close to equilibrium reduces to a set of three independent diffusion equations for the processes of viscous friction, self-diffusion, and heat conduction.

With subscript "0" we mark off the equilibrium values of gasdynamic functions, $\rho_0(R)$, T_0, $u_0 = 0$. Deviations from the equilibrium state we denote by $\delta\rho$, δT, $\delta u = u$ (since $u_0 = 0$). For states close to equilibrium the deviations are small, and we may retain only the linear terms in equations (14.1.1), (14.1.6), (14.1.7). Using equations of state, we come to a closed linear set of equations in functions $\delta\rho$, u, δT:

$$\frac{\partial \delta\rho}{\partial t} + \rho_0 \operatorname{div} u = \frac{\partial}{\partial R}\left[D\frac{\partial \delta\rho}{\partial R} - \frac{b}{m} F \delta\rho \right], \quad D = b\frac{\kappa T_0}{m}, \quad b = \tau; \tag{14.2.1}$$

$$\rho_0 \frac{\partial u}{\partial t} = -\frac{\kappa T_0}{m} \operatorname{grad} \delta\rho - \frac{\kappa \rho_0}{m} \operatorname{grad} \delta T + \rho_0 v \Delta \delta u; \tag{14.2.2}$$

$$\frac{\partial}{\partial t}\left(\frac{\kappa T_0}{m} \delta\rho + \frac{\kappa \rho_0}{m} \delta T \right) + \frac{5}{3}\rho_0 \operatorname{div} u = \frac{\kappa T_0}{m}\left[\frac{\partial}{\partial R}\left(D\frac{\partial \delta\rho}{\partial R} - \frac{b}{m} F \delta\rho \right) \right] + \frac{\kappa \rho_0}{m} \chi \Delta \delta T. \tag{14.2.3}$$

In these equations we have taken into account the fact that $\rho_0(R)$ at equilibrium is the Boltzmann distribution.

The set of variables includes vector $u(R,t)$. Recall that any vector $a(R,t)$ can be represented as the sum of rotational (superscript "$\perp$") and potential (superscript "$\parallel$") components:

$$a = a^\perp + a^\parallel, \quad \operatorname{div} a^\perp = 0, \quad \operatorname{curl} a^\parallel = 0. \tag{14.2.4}$$

Both components can be expressed in terms of vector $a(R,t)$: for example,

$$a^{\perp}(R,t) = a(R,t) + \frac{1}{4\pi} \int \frac{\text{grad}_{R'} \, \text{div}_{R'} \, a(R',t)}{|R - R'|} dR' \; . \tag{14.2.5}$$

Taking advantage of this fact, we represent the vector of velocity as

$$u(R,t) = u^{\parallel} + u^{\perp} \; , \quad \text{curl}\, u^{\parallel} = 0 \; , \quad \text{div}\, u^{\perp} = 0. \tag{14.2.6}$$

Since the transverse components of gradients are zero, from equation of motion (14.2.2) we find the following equation for the curl velocity field:

$$\frac{\partial u^{\perp}}{\partial t} = v\Delta u^{\perp} \; , \quad \text{div}\, u^{\perp} = 0 \; . \tag{14.2.7}$$

In this way, at the final stage of relaxation towards equilibrium the curl velocity component relaxes to its zero value by diffusion law.

Now we turn to equations in functions $\delta\rho$, δT, and $u^{\parallel}$. They coincide with equations (14.2.2) – (14.4.4), where we replace the total velocity vector with its potential component, $u(R,t) \Rightarrow u^{\parallel}(R,t)$. Let us single out those equations which describe the slowest processes.

Denote by T, L the characteristic scales of the problem under consideration — for example, the size of channel and one of the diffusion times τ_v, τ_{χ}, or τ_D. Let v_s be the speed of sound. Then we may define yet another characteristic time L/v_s taken by the sound wave to travel over distance L. For slow processes this time is much less than the diffusion time,

$$\frac{L}{v_s} \ll T \sim \frac{L^2}{D} \; , \; \text{and} \; \; \text{Kn} = \frac{l}{L} \ll 1 \; \text{since} \; D \sim v_s l \; . \tag{14.2.8}$$

This inequality is therefore equivalent to the condition of smallness of Knudsen number. In the study of wave and fluctuation processes the same condition can be expressed in terms of frequencies and wave numbers of the relevant Fourier components,

$$\omega \sim Dk^2 \ll kv_s \; , \; \text{or} \; \; \text{Kn} = kl \ll 1 \; . \tag{14.2.9}$$

We see that the characteristic length is the wavelength $1/k$.

In a similar way we may also characterize the fast processes:

$$T \sim \frac{L}{v_s} << \frac{L^2}{D} = \tau_D \; ; \; \text{for Fourier components} \; \omega \sim kv_s >> vk^2 \; . \qquad (14.2.10)$$

We see that in case of fast processes the characteristic region lies near the sonic resonance. In this region the damping of the wave is small.

On the strength of these definitions we may divide the system of equations in functions $\delta\rho$, δT, $u^{\parallel}$ into two subsystems which describe, respectively, the slow and the fast processes. Let us start with equation (14.2.1).

For comparison, we are going to express separate terms via the relative fluctuation of density $\delta\rho/\rho_0$. Under conditions (14.2.8), the first and the third terms in (14.2.1) are of the same order of magnitude. From juxtaposition of the second and the first terms on the left-hand side we find that

$$\frac{u^{\parallel}}{v_T} \sim \frac{L}{v_T T} \frac{\delta\rho}{\rho_0} \sim \frac{D}{Lv_T} \frac{\delta\rho}{\rho_0} \sim \text{Kn} \frac{\delta\rho}{\rho_0} << \frac{\delta\rho}{\rho_0} \; , \qquad (14.2.11)$$

where we have noted that the speed of sound is of the order of thermal velocity.

This estimate indicates that the second term on the right-hand side of (14.2.1) is infinitesimal of higher order. Keeping the remaining terms, we come to equation of self-diffusion

$$\frac{\partial \delta\rho}{\partial t} = \frac{\partial}{\partial R}\left[D\frac{\partial \delta\rho}{\partial R} - \frac{b}{m}F\delta\rho \right], \quad D = b\frac{\kappa T_0}{m} \; , \quad b = \tau \; . \qquad (14.2.12)$$

Now we look at equation (14.2.3). By virtue of (4.2.11) we may drop the term with $\text{div}\,u(R,t)$. The first terms on the left and right-hand sides cancel out on the strength of the diffusion equation. As a result, we come to equation of heat conduction

$$\frac{\partial T}{\partial t} = \chi\Delta T \; . \qquad (14.2.13)$$

It follows that at the last stage of relaxation towards equilibrium the slow (as defined by conditions (14.2.8), (14.2.9)) dissipative processes are described by three independent diffusion equations in functions $\delta\rho$, δT, and $u^{\parallel}$ with three independent kinetic coefficients D, χ, v. In general, of course, the three coefficients may have different values.

Let us compare these equations with those derived from conventional equations of gas dynamics which do not take self-diffusion into account. The equation for diffusion of the curl component of velocity coincides with the diffusion equation (14.2.7). The equation for density $\delta\rho$ retains its form of diffusion equation, but the "coefficient of diffusion" is now represented by the temperature conductivity χ, whereas the equation of heat conduction itself remains intact:

$$\frac{\partial\delta\rho}{\partial t} = \chi\Delta\delta\rho \; , \quad \frac{\partial\delta T}{\partial t} = \chi\Delta\delta T \; . \tag{14.2.14}$$

Now the question is why the coefficient of temperature conductivity determines the diffusion of characteristics as different as the temperature and the density. Is this associated with the incompleteness of equations of gas dynamics neglecting self-diffusion, and the resulting inconsistency of (14.2.14)? To answer these questions we turn to equations of gas dynamics (11.1.6), (11.3.10), (11.3.11) obtained in Ch. 11 on the basis of the kinetic Boltzmann equation.

Equations (14.2.1) – (14.2.3) hold for small deviations from equilibrium, given also that $D = 0$, $b = 0$. Hence

$$\frac{\partial\delta\rho}{\partial t} + \rho_0 \operatorname{div} u = 0 \; , \quad \rho_0 \frac{\partial u}{\partial t} = -\operatorname{grad} \delta p + \rho_0 v\Delta\delta u \; ,$$

$$\tag{14.2.15}$$

$$\frac{\partial\delta\rho}{\partial t} + \frac{5}{3}\rho_0 \operatorname{div} u = \frac{2}{3}\lambda\Delta\delta T \; , \quad \delta p = \frac{\kappa T_0}{m}\delta\rho + \frac{\kappa\rho_0}{m}\delta T \; , \quad \lambda = c_p\rho_0\chi \; ,$$

where δp is the deviation of pressure from its equilibrium value. It is important that the coefficient of heat conductivity is defined via heat capacity at constant pressure.

Observe that now we cannot take advantage of the smallness of Knudsen parameter (see conditions (14.2.8)) and drop the terms containing the potential component of velocity $u^{\|}$. Indeed, the first equation then would have implied that $\delta\rho = 0$, and hence $\delta T = 0$. It follows that the procedure used for obtaining equations (14.2.14) is inconsistent from the standpoint of perturbation theory in small Knudsen number.

Namely, we first assume that the deviation of pressure is $\delta p(R,t) = 0$, and therefore

$$\frac{\operatorname{grad}\rho}{\rho} = -\frac{\operatorname{grad} T}{T} \tag{14.2.16}$$

for all times (all frequencies). Next we use the first equation to express $\operatorname{div} u$ in terms of the derivative of $\delta\rho$ (notwithstanding the fact that the terms are infinitesimals of different order). As a result, we come to the equation of heat conduction (14.2.14). By virtue of (14.2.16), the same equation holds for $\delta\rho$.

There is no reason to claim that such derivation of equations (14.2.14) is consistent.

This applies, in particular, to the use of equation (14.2.17). For slow processes the change in the pressure is small; this does now mean, however, that the change in the pressure is zero at every instant of time evolution, or for every frequency of the spectrum. We shall discuss this matter later in connection with the calculation of fluctuations of gasdynamic functions.

Now let us look into the relaxation of fast wave dissipative processes.

14.3. Wave Excitations in Gas Dynamics

We go back to equations (14.2.2) – (14.2.4) which describe the time evolution of small deviations of gasdynamic functions from their equilibrium values. This time we are interested in the fast processes as defined by conditions (14.2.11). In particular, these conditions imply that in the non-resonance region the dissipative terms are small:

$$\frac{vk^2}{\omega} \sim \frac{vk^2}{kv_s} \sim lk \sim Kn \ll 1 . \tag{14.3.1}$$

We assume that the medium is homogeneous, and the force is $F = 0$.

Since the equations are linear in $\delta\rho$, δT, u, it will be convenient to deal with the Fourier components of these functions. The relevant equations have the following form (dropping subscript "0" at ρ, T):

$$\left(-i\omega + Dk^2\right)\delta\rho + i\rho(k\delta u) = 0 ,$$

$$\left(-i\omega + vk^2\right)i\rho(ku) = k^2\delta p \equiv k^2\left(\frac{\kappa\rho}{m}\delta T + \frac{\kappa T}{m}\delta\rho\right) , \tag{14.3.2}$$

$$-i\delta p + pDk^2\frac{\delta\rho}{\rho} + \frac{5}{3}i\rho(ku) + \frac{p}{T}\chi k^2\delta T = 0 .$$

By assumption (to be confirmed later), the deviation from equilibrium is a weakly damped sound wave. Under this assumption the local thermodynamic process is adiabatic ($\delta s = 0$), and therefore the deviations of density, temperature and pressure are linked by the following relations:

$$\delta T = \frac{p}{c_v\rho}\frac{\delta\rho}{\rho} , \quad \delta p = \frac{c_p}{c_v}\frac{\kappa T}{m}\delta\rho = v_s^2\delta\rho , \quad \delta T = \frac{1}{c_p\rho}\delta\rho . \tag{14.3.3}$$

Using these relations for transforming the small dissipative terms, we bring the last two equations in (14.3.2) to the form

$$\left(-i\omega + \frac{c_p}{c_v}Dk^2 + \frac{\kappa}{mc_p}\chi k^2\right)\delta p + \frac{5}{3}pi(ku) = 0 \ ,$$

$$(14.3.4)$$

$$\left(-i\omega + \nu k^2\right)pi(ku) = k^2\delta p \ .$$

Using the second equation to eliminate function (ku), and retaining only the linear dissipative terms, we arrive at the following equation:

$$\left(-\omega^2 + v_s^2k^2 - i2\omega\gamma\right)\delta p = 0 \ ,$$

$$(14.3.5)$$

where the speed of sound and the coefficient of damping are defined by

$$v_s^2 = \frac{c_p}{c_v}\frac{\kappa T}{m} \ , \quad \gamma = \frac{1}{2}\left[\nu + \frac{c_p}{c_v}D + \frac{\kappa}{mc_p}\right]k^2 \ .$$

$$(14.3.6)$$

We see that the coefficient of damping of sound wave is defined by a combination of three dissipative coefficients when self-diffusion is taken into consideration. Recall that usually the coefficient of damping in gas dynamics is expressed in terms of two dissipative coefficients:

$$\gamma = \frac{1}{2}\left[\frac{4}{3}\nu + \left(\frac{c_p}{c_v} - 1\right)\chi\right]k^2 \ .$$

The two expressions coincide when the kinetic coefficients have the same value: then $\gamma = Dk^2$.

The effects of self-diffusion on the structure of weak shock wave were considered in (Klimontovich 1990c, d) on the basis of the same equations of gas dynamics as derived in this chapter. Self-diffusion was found to be quite important for the distribution of entropy in the transition region.

We shall return to the effects of self-diffusion in connection with the fluctuations of gasdynamic functions. Such fluctuations are definitive for Brownian motion and molecular light scattering. The critical growth of these fluctuations when external parameters are changed is a harbinger of instability, a sign of the incipient structural changes in the system under consideration.

From derivation of equations of gas dynamics (14.1.2) – (14.1.4) it follows that the primary cause of all three diffusion processes is the additional "collision integral" $I_{(R)}$ in the generalized kinetic equation (13.4.1). This "collision integral" accounts mainly for the dynamic instability of motion of atoms and the resulting mixing in the phase space.

Now let us consider another important special case of the general set of equations of hydrodynamics, the approximation of incompressible gas. We shall see that the use of generalized kinetic equation once again leads to constructive results.

14.4. Navier – Stokes Equation for "Incompressible" Gas

Incompressibility can be expressed by the condition of constant density

$$\int f(R, |v - u|, t) dv = \text{const} .\qquad (14.4.1)$$

We also assume that the temperature of the gas is constant,

$$\frac{m}{3} \int (v - u)^2 f(R, |v - u|, t) dv = \text{const} .\qquad (14.4.2)$$

Then the continuity equation (14.1.2) takes on the form

$$\text{div}\, u = 0 .\qquad (14.4.3)$$

This equation implies that the field of velocity is rotational.

Recall (see (14.2.5)) that any vector $a(R, t)$ can be represented as the sum of rotational (superscript "$\perp$") and potential (superscript "$\parallel$") components. Both components can be expressed in terms of vector $a(R, t)$ (see (14.2.6)).

Let us find the equation for the field of rotational velocity $u^{\perp}(R, t)$.

After separation of rotational component the pressure term in (14.1.3) drops out. Then

$$\frac{\partial u^{\perp}}{\partial t} + \left[\left(u^{\perp}\, \text{grad} \right) u^{\perp} \right]^{\perp} = v \Delta u^{\perp} + \frac{1}{m} F^{\perp} , \quad \text{div}\, u^{\perp} = 0 .\qquad (14.4.4)$$

Here we have separated the rotational component of the force, and the rotational component of vector $\left(u^{\perp}\, \text{grad} \right) u^{\perp}$, which in general, of course, is not a rotational vector.

Equation (14.4.4) is sufficient for finding the velocity field given the initial and boundary conditions. Useful is also another form of this equation.

Let us transform the second term on the left-hand side of (14.4.4), using the general definition (14.2.6) for the rotational vector component when $a = \left(u^\perp \operatorname{grad}\right)u^\perp$. We also note that the second term on the right-hand side of (14.2.6) can be represented as gradient of some scalar function,

$$\operatorname{grad}_R\left[\frac{1}{4\pi}\int\frac{\operatorname{div}_{R'}\left[\left(u^\perp \operatorname{grad}_{R'}\right)u^\perp(R',t)\right]}{|R-R'|}dR'\right] \equiv \frac{1}{\rho}\operatorname{grad} p_G \; , \tag{14.4.5}$$

where p_G is the "*gasdynamic pressure*" which ensures elimination of nonrotational component from the equation for velocity $u^\perp$. The constant factor $1/\rho$ is included into the right-hand side so that p_G should have the dimensions of pressure. Thermodynamic pressure under the given conditions (constant density and temperature) does not enter the equation for velocity.

Observe that relation (14.4.5) may be regarded as an equation for finding the pressure p_G given the velocity field. By applying the divergence operator, (14.4.5) can be reduced to Poisson equation

$$\Delta p_G = -\rho \operatorname{div}_R\left(u^\perp \operatorname{grad}_R\right)u^\perp(R,t) = -\rho\frac{\partial u_i^\perp}{\partial R_j}\frac{\partial u_j^\perp}{\partial R_i} \; . \tag{14.4.6}$$

In the second equation here we have taken advantage of the fact that $\operatorname{div} u^\perp = 0$.

As a result, the second term on the left-hand side on (14.4.4) becomes

$$\left[\left(u^\perp \operatorname{grad}\right)u^\perp\right]^\perp = \left(u^\perp \operatorname{grad}\right)u^\perp + \frac{1}{\rho}\operatorname{grad} p_G \; . \tag{14.4.7}$$

and the equation for velocity assumes the standard form of Navier – Stokes equation for incompressible fluid,

$$\frac{\partial u^\perp}{\partial t} + \left(u^\perp \operatorname{grad}\right)u^\perp = -\frac{1}{\rho}\operatorname{grad} p_G + \Delta u^\perp + F^\perp \; . \tag{14.4.8}$$

Together with (14.4.6) (or (14.4.5)) this equation forms a closed set for the rotational velocity field $u^\perp(R,t)$ and "gasdynamic pressure" $p_G(R,t)$.

Of course, $F^\perp$ is not always a potential force. An example will be considered in connection with the convective motion in gas dynamics.

Expedient is also another form of Navier – Stokes equation. Using the vector identity

$$(a \operatorname{grad})a \equiv \frac{1}{2}\operatorname{grad} a^2 - [a \operatorname{curl} a] ,$$ (14.4.9)

and applying curl_R to the Navier – Stokes equation, we arrive at the equation for the curl of velocity

$$\frac{\partial}{\partial t}\operatorname{curl}_R u^\perp = \operatorname{curl}_R \left[u^\perp \operatorname{curl}_R u^\perp \right] + \nu \Delta \operatorname{curl}_R u^\perp + \operatorname{curl}_R F^\perp .$$ (14.4.10)

This equation does not contain "gasdynamic pressure" p_G, and is therefore a closed differential equation in velocity $u^\perp$.

Recall that the boundary condition for the Navier – Stokes equation states that $u^\perp = 0$ on solid surface, and implies molecular adhesion of adjacent liquid layer with the solid surface.

14.5. Equation of Entropy Balance for "Incompressible" Gas

So, for the case of "incompressible" gas we have a closed dynamic dissipative equation for the curl velocity field. This is the Navier – Stokes equation, the basic equation of hydrodynamics of incompressible fluid. The dissipation in this equation comes from viscous friction. From thermodynamic standpoint, however, the Navier – Stokes equation is not quite consistent, since it does not agree with the equation of entropy balance as formulated above.

Recall that so far we have obtained two equations of entropy balance, which correspond, respectively, to the general kinetic description and the approximation of local equilibrium. The first is (13.4.6); it has been derived from the general kinetic equation under condition $F = 0$. In approximation of local equilibrium the production of entropy is given by (13.4.11), being defined by three dissipative processes.

From (13.4.11) it follows that entropy production is nonzero also for incompressible gas at arbitrary constant temperature,

$$\sigma_G(R,t) = \frac{m}{T} \nu\rho \left(\frac{\partial u_i^\perp}{\partial R_j} \right)^2 ,$$ (14.5.1)

and is completely defined by the solution of the Navier – Stokes equation for rotational field (subscript "G" is there to emphasize this circumstance). How then do we define the flux of entropy and the entropy itself in case of "incompressible" gas?

We know that one of the main assumptions of thermodynamics of nonequilibrium processes is the assumption of local equilibrium. For rarefied gas, the entropy and the diffusion flow of entropy are given by (13.4.14), (13.4.15). According to these formulas, for "incompressible" gas at constant temperature the entropy is constant, and the diffusion flow of entropy is zero. Under these conditions, the convective entropy flux (13.4.8) is zero too.

Thus, in approximation of local equilibrium for "incompressible" gas at constant temperature we come to a contradiction with the second law of thermodynamics: the production of entropy is nonzero, whereas the thermodynamic entropy is constant.

Is it possible to overcome this difficulty within the framework of "incompressible" gas model, or one has to modify the model by, for instance, taking the temperature change into account? We shall see that thermal motion must be taken into consideration. Let us illustrate this point with the example of thermal convection in rarefied gas.

14.6. Thermal Convection in Rarefied Gas

14.6.1. Generalized Kinetic Equation for Description of Thermal Convection

Now we use the generalized kinetic equation (13.4.1) for describing thermal convection in "incompressible" rarefied gas, when F is the force of gravity

$$F = m\vec{g} \equiv mg\vec{\gamma} \,, \tag{14.6.1}$$

where

$$\vec{g} = (0,0,-g) \,, \text{ and } \vec{\gamma} = (0,0,1) \,. \tag{14.6.2}$$

We thus assume that z–axis is directed vertically towards the surface of the Earth.

Once again we select a class of functions; this time, however, we assume from the outset that the velocity field is rotational, and hence

$$f(R,v,t) = f\left(R, \left|v - u^{\perp}(R,t)\right|, t\right) \,. \tag{14.6.3}$$

Accordingly, the condition of incompressibility (14.4.1) now becomes

$$\int f\left(R, \left|v - u^{\perp}(R,t)\right|, t\right) dv = \text{const} \,, \quad \text{div} \, u^{\perp}(R,t) = 0 \,. \tag{14.6.4}$$

This implies that we are considering the zero approximation in small parameter

$$\frac{mgH}{\kappa T} \; ,$$

(14.6.5)

where H is the characteristic thickness of gas layer. In other words, we neglect the change in density in the field of gravity (such as the barometric law in the state of equilibrium). In the field of gravity, in case of inhomogeneous temperature distribution, there is also the Archimedes' force (buoyancy). The buoyancy is proportional to the relative variation of density due to the difference between the local temperature $T(R,t)$ in the layer and the temperature of "thermostat" T_0. To avoid contradiction with the condition of incompressibility (14.6.4), one can express the buoyancy via the coefficient of isothermal compressibility.

Observe finally that the equation of motion (14.4.8) includes only the rotational component of the force. Because of this, the force which does not depend on the temperature difference drops out, and we have

$$F^{\perp}(R,t) = \left[\beta\left(T(R,t) - T_0\right)mg\vec{\gamma}\right]^{\perp} \; .$$

(14.6.6)

To emphasize the role of the temperature difference, we have retained here the term with T_0 which is actually of little consequence for the forthcoming discussion. Recall that the coefficient of thermal expansion is

$$\beta = -\frac{1}{\rho_0}\left(\frac{\partial\rho}{\partial T}\right)_p \; , \quad \text{or, for rarefied gas,} \quad \beta = \frac{1}{T_0} \; .$$

(14.6.7)

Eventually, the generalized kinetic equation (13.4.1) becomes

$$\frac{\partial f}{\partial t} + v\frac{\partial f}{\partial R} + \left[\beta(T(R,t) - T_0)mg\vec{\gamma}\right]^{\perp}\frac{\partial f}{\partial v}$$

$$= \frac{\partial}{\partial v}\left[D_{(v)}(R,t)\frac{\partial f}{\partial v}\right] + \frac{\partial}{\partial v}\left[v - u^{\perp}(R,t)f\right] + \frac{\partial}{\partial R}\left[D\frac{\partial f}{\partial R}\right] \equiv I_{(v)} + I_{(R)} \; .$$

(14.6.8)

This equation, with due account for the specific properties of the distribution function, can be used for the kinetic description of convective motion.

14.6.2. *Gasdynamic Equations for Convective Motion*

The closed set of equations (14.1.2), (14.1.7), (14.1.8) in gasdynamic functions ρ, u, T (or pressure p) has been obtained under condition (13.3.15) from the generalized kinetic equation (13.4.1) without using perturbation theory. Let us now derive a simpler set of equations which would correspond to the kinetic equation (14.6.8).

Under conditions (14.6.3), (14.6.4), the continuity equation (14.1.2) reduces to

$$\rho = \text{const} , \quad \text{div}\, u^{\perp}(R,t) = 0 . \tag{14.6.9}$$

Equation of motion (14.1.7) once again takes on the form (14.4.8). This time, however, we may specify the form of rotational component of the force, and write the desired equation for the rotational velocity as

$$\frac{\partial u^{\perp}}{\partial t} + \left(u^{\perp}\,\text{grad}\right)u^{\perp} = -\frac{1}{\rho}\,\text{grad}\, p_G + \nu\Delta u^{\perp} + \left[\beta\big(T(R,t)-T_0\big)g\vec{\gamma}\right]^{\perp} ,$$

$$\tag{14.6.10}$$

$$\Delta p_G = -\rho \frac{\partial u_i^{\perp}}{\partial R_j}\frac{\partial u_j^{\perp}}{\partial R_i} .$$

These equations must be supplemented by equation for the temperature, which follows from (14.1.8) under condition of constant density:

$$\frac{\partial T}{\partial t} + \left(u^{\perp}\,\text{grad}\right)T = \chi\frac{\partial^2 T}{\partial R^2} + \frac{2}{3}\frac{m}{\kappa}\,\nu\left(\frac{\partial u_i^{\perp}}{\partial R_j}\right)^2 . \tag{14.6.11}$$

Now we have a closed set of equations for rotational velocity and temperature; they are called the Boussinesq equations. We have obtained them from the kinetic equation (14.6.8) without using perturbation theory in Knudsen number.

Note that the second of these equations is nonlinear. The nonlinear term on the left-hand side characterizes the linkage between the rotational velocity field with the distribution of temperature. We shall see that this term is important for the instability of convective motion which develops when the temperature gradient is sufficiently large. The condition of instability will be considered in the next section.

The nonlinear term on the right-hand side may be regarded as the source of heat determined by the rotational motion. This term is proportional to squared Mach number $M = u/v_T$, and is therefore negligible when flow velocity is much less than the speed of sound. This allows using a simpler equation for the temperature for the description of low-velocity convective motion.

14.7. Condition of Instability of Convective Motion

Consider convective motion of "incompressible" fluid in a horizontal layer of thickness H heated from below. Let T_1 be the temperature of the lower surface, and T_0 the temperature of the upper surface of the layer; the z–axis is directed vertically upwards.

We start with the simplest case, when gravity is neglected, and there is no hydrodynamic motion, $u = 0$. Then the Boussinesq equations reduce to

$$\frac{d^2T}{dz^2} = 0 \; , \quad T(z=0) = T_1 \; , \quad T(z=H) = T_0 \; , \quad u^\perp = 0 \; , \tag{14.7.1}$$

whence follows the expression for the temperature profile

$$T(z) = T_1 - \frac{T_1 - T_0}{H} z \equiv T_1 - \frac{\Delta T}{H} z \; , \quad 0 \le z \le H \; . \tag{14.7.2}$$

By superscript "1" we mark off the quantities which characterize small deviations from this particular solution:

$$u^\perp = \left(u^1\right)^\perp \; , \quad T = T^1 + T_1 - \frac{\Delta T}{H} z \; . \tag{14.7.3}$$

Now we substitute these expressions into the Boussinesq equations and retain only the linear terms in small deviations. We also note that the rotational component of gradient-wise force is zero, and that the gradient of "hydrodynamic pressure" p_G in the linear approximation with respect to velocity is zero (see (14.4.5)). As a result, we get the following set of linear equations in small deviations (superscript "1" is dropped):

$$\frac{\partial u^\perp}{\partial t} = \nu \Delta u^\perp + \left[\beta T(R,t)g\vec{\gamma}\right]^\perp \; ,$$

$$\tag{14.7.4}$$

$$\frac{\partial T}{\partial t} - \frac{\Delta T}{H}\left(u^\perp \vec{\gamma}\right) = \chi \Delta T \; .$$

Recall that earlier we considered the diffusion relaxation of density, velocity and temperature towards equilibrium. This process was described by the set of three independent diffusion equations (14.2.8), (14.2.13), (14.2.14). The present situation is different in two respects. First, we are now dealing with "incompressible" gas, when the density is constant, and the diffusion equation (14.2.13) therefore becomes an identity.

Secondly, we are considering relaxation towards a nonequilibrium state which is characterized by a temperature gradient. We shall see that such nonequilibrium state in the field of gravity is stable only if the temperature gradient is small enough. When the temperature gradient exceeds a certain critical value, the nonequilibrium state with the temperature profile (14.7.4) becomes unstable. Hydrodynamic motion arises which causes redistribution of the temperature; this is thermal convection.

For analyzing the stability of stationary nonequilibrium solution (14.7.3) we shall use the set of equations (14.7.3) for small deviations from equilibrium. This set is supplemented by the following boundary conditions:

$$u(z=0) = u(z=H) = 0 \ , \quad T(z=0) = T(z=H) = 0 \ . \tag{14.7.5}$$

The first of these implies that the layer in question is confined between solid walls, on which the velocity of viscous flow goes to zero. The second condition means that the the confining planes have, respectively, the temperatures of the "heater" and the "cooler", which do not depend on the process within the layer.

Since the set of equations is linear, for the purpose of studying the stability in layer unrestricted in the directions x and y the solution (with due account for the boundary conditions) can be presented in the form

$$u_z(x,y,z) = \mathrm{Re}\, u_n(k_x, k_y, \lambda) \exp(-\lambda t + ik_x x + ik_y y) \sin\frac{\pi n}{H} z \ ,$$

$$\tag{14.7.6}$$

$$T(x,y,z) = \mathrm{Re}\, T_n(k_x, k_y) \exp(-\lambda t + ik_x x + ik_y y) \sin\frac{\pi n}{H} z \ .$$

Here we have taken into account that the force in the equation for velocity $u^\perp$ is directed along the z-axis, and the equation for temperature includes only one velocity component $u_z^\perp$. Because of this, the convective motion involves only one component of the velocity, whereas the other two relax according to the diffusion law.

In order to obtain equations in complex amplitudes u_n and T_n, we substitute (14.7.6) into (14.7.4) and use the following notation:

$$k^2 = k_\perp^2 + k_n^2 \ , \quad \text{where} \quad k_\perp^2 = k_x^2 + k_y^2 \ , \quad \text{and} \quad k_n^2 = \left(\frac{\pi n}{H}\right)^2 \ . \tag{14.7.7}$$

As a result, we get the set of algebraic equations

$$\left(\lambda - vk^2\right)u_n\left(k_x,k_y\right)+\beta g\frac{k_\perp^2}{k^2}T_n\left(k_x,k_y\right)=0 \ ,$$

$$(14.7.8)$$

$$\frac{\Delta T}{H}u_n\left(k_x,k_y\right)+\left(\lambda - \chi k^2\right)T_n\left(k_x,k_y\right)=0 \ .$$

From condition of solvability of this set, we find a quadratic equation in λ. The structure of solution of this equation determines the nature of time evolution of small deviations from the stationary nonequilibrium solution.

To reduce this solution to a form suitable for analysis, we introduce two dimensionless parameters, Prandtl and Rayleigh numbers as defined by

$$P = \frac{v}{\chi} \ , \ \ Ra = \frac{g\beta H^4}{v\chi}\frac{\Delta T}{H} \ .$$

$$(14.7.9)$$

Then the solution of quadratic equation in λ can be represented as

$$\lambda^{(\pm)} = \frac{v+\chi}{2}k^2 \pm \sqrt{\frac{(v-\chi)^2}{4}k^4 + v\chi\,Ra\frac{k_\perp^2}{H^4 k^2}} \ .$$

$$(14.7.10)$$

When Rayleigh number is positive (that is, the heating is from below, $\Delta T = T_1 - T_0 \geq 0$), both roots of equation are real. The solution $\lambda^{(+)}$ is positive for any $\Delta T \geq 0$, whereas $\lambda^{(-)}$ may be positive and negative. The null of $\lambda^{(-)}$ occurs at a certain "critical" value of Rayleigh number. From condition $\lambda = 0$ we find that

$$Ra_{cr}(k) = \frac{k^6 H^4}{k_\perp^2} \equiv \frac{\left(a^2 +(\pi n)^2\right)^3}{a^2} \ , \ \ \text{where} \ a^2 = k_\perp^2 H^2 \ .$$

$$(14.7.11)$$

We see that the critical Rayleigh number depends on the number of harmonic n along the z–axis, and on the value of a defined in terms of the thickness of layer H and the wave number $k_\perp$ which characterizes the size of the cell of spatial structure which results from the developing instability of the initial stationary state (see below). We can also find the minimum value of the critical Rayleigh number,

$$[Ra_{cr}]_{min} = \frac{27}{4}\pi^4 \ , \ \text{if} \ n = 1 \ \text{and} \ a = \frac{\pi}{\sqrt{2}} \ .$$

$$(14.7.12)$$

Given the definition of critical Rayleigh number (14.7.11), expression (14.7.10) can be rewritten as

$$\lambda^{(\pm)} = \frac{v+\chi}{2} k^2 \left(1 \pm \sqrt{1 + 4\frac{v\chi}{v+\chi}\frac{\text{Ra} - \text{Ra}_{cr}}{\text{Ra}_{cr}}} \right). \tag{14.7.13}$$

Now we can trace the evolution of $\lambda^{(\pm)}$ as Rayleigh number increases (with increasing temperature difference ΔT). When Rayleigh number is zero, we have $\lambda^{(+)} = vk^2$ and $\lambda^{(-)} = \chi k^2$. This implies that relaxation of velocity and temperature occurs independently according to diffusion law. In the neighborhood of critical Rayleigh number,

$$\lambda^{(+)} = (v + \chi)k^2 , \quad \lambda^{(-)} = \frac{v\chi}{v+\chi}\frac{\text{Ra}_{cr}(k) - \text{Ra}}{\text{Ra}_{cr}(k)} . \tag{14.7.14}$$

We see that at the critical point $\lambda^{(-)}$ changes its sign and becomes negative. As a result, a small deviation from the equilibrium state (14.7.2) grows exponentially, and the initial state becomes therefore unstable.

The development of this instability leads to appearance of stationary spatial honeycomb structure called Benard cells. This is just a visible sign of orderly convective motion. By contrast to molecular convection, the arrangement of opposing "hot" and "cold" flows of matter is regular. To wit, the warmer gas goes upwards in the middle of the cells, and the colder gas goes downwards at the periphery of each cell. The process resembles well regulated street traffic. The shape of Benard cells may be different, but usually they are hexagonal. Their diameter is about the same as the thickness of the layer, and does not depend on the horizontal dimensions of the layer.

In this way, a spatially homogeneous stationary distribution is spontaneously replaced by a stationary inhomogeneous distribution of velocity and temperature when the temperature difference exceeds a certain critical value. This is one example of a dissipative structure.

Intuitively it is clear that we are dealing with a process of self-organization: the development of convective instability leads to appearance of a more ordered state. Sometimes it is not so easy to distinguish between self-organization and degradation, and one has to rely on criteria of relative degree of order of states of open systems (criteria of self-organization).

Of course, the evolution towards a new state can only be described by nonlinear equations. The linear approximation only gives indications that instability develops which may lead to transition to a new state. The basics of nonlinear theory of convective motion will be discussed in the chapter devoted to the theory of Brownian motion in active open systems.

Finally, let us discuss the limits of applicability of equations of gas dynamics for description of nonequilibrium processes.

14.8. Kinetic and Gasdynamic Description of Heat Transfer

In this chapter the equations of gas dynamics were obtained on the basis of generalized kinetic equation (13.4.1) without using perturbation theory in Knudsen number. The only constraint is imposed on the class of distribution functions (14.1.1). Obviously, this class includes the local Maxwell distribution, which is completely characterized by three gasdynamic functions which satisfy the same set of equations of gas dynamics (14.1.2) – (14.1.4). What is the role of the higher even moments of relative velocity δv which can be obtained from distribution functions of type (14.1.1)?

Since this question remains in force when the hydrodynamic velocity $u(R,t)$ is zero, we may further restrict the class of functions by condition

$$f(R,v,t) = f(R,|v|,t) \ . \tag{14.8.1}$$

This allows us to introduce the distribution function of energy values

$$f(R,E,t) = \int \delta\left(E - \frac{mv^2}{2} \right) f(R,v,t)dv \ , \tag{14.8.2}$$

and then use (13.4.1) to obtain a closed kinetic equation

$$\frac{\partial f(R,E,t)}{\partial t} = I(R,E,t) + \chi\frac{\partial^2 f}{\partial R^2} \tag{14.8.3}$$

with "collision integral"

$$I(R,E,t) = 2\frac{\partial}{\partial E}\left[D\left(E\frac{\partial f}{\partial E} - \frac{1}{2} \right) f \right] + 2\frac{\partial}{\partial E}\left(\frac{1}{\tau}Ef \right) , \tag{14.8.4}$$

where

$$D = \frac{1}{\tau}\frac{2}{3}\langle E \rangle_{R,t}$$

is the "energy" diffusion coefficient. The spatial diffusion coefficient is χ, since we are concerned with heat transfer equations. According to (13.3.13), the "collision integral"

$I(R,E,t)$ has the following property:

$$\int E \ I(R,E,t) \ dE = 0 \ . \tag{14.8.5}$$

Using this property and the kinetic equation (14.8.3), we find the equation for mean energy (equation of heat conduction):

$$\frac{\partial \langle E \rangle}{\partial t} = \chi \frac{\partial^2 \langle E \rangle}{\partial R^2} \ , \ \ \langle E \rangle_{R,t} = \frac{3}{2} \kappa T(R,t) \ . \tag{14.8.6}$$

We can also write the relevant equation for dispersion of kinetic energy (dispersion of temperature):

$$\frac{\partial}{\partial t} \left\langle (\delta E)^2 \right\rangle = \frac{4}{\tau} \left(\frac{2}{3} \langle E \rangle^2 - \left\langle (\delta E)^2 \right\rangle \right) + \chi \frac{\partial^2 \left\langle (\delta E)^2 \right\rangle}{\partial R^2} + 2 \chi \left(\frac{\partial \langle E \rangle}{\partial R} \right)^2 \ . \tag{14.8.7}$$

We see that energy fluctuations are not small for all values of Knudsen number. This is natural, since the energy itself and the higher moments are defined by the one-particle distribution, and therefore the "$1/N$ factor" equals to one. For this reason, the relative temperature fluctuations are also of the order of unity.

This implies that the temperature is not a thermodynamic characteristic in the Gibbsian sense. In this connection we may raise a number of questions.

First of all, why is it that the temperature (the first moment of kinetic energy in the one-particle distribution) should be more important than the higher moments? The answer is seemingly obvious: the equation of heat conduction follows from the kinetic equation (14.8.3) at arbitrary Knudsen numbers and is closed. However, even for small Knudsen numbers (that is, at local equilibrium), the solution of this equation does not define the flux of heat. As stated above, even at local equilibrium the thermal flux depends not only on the temperature gradient by Fourier's law, but also on the gradient of density (see (13.5.1)). This only applies to slow processes, when the pressure is practically constant.

As follows from (14.8.7), the source in equation for temperature fluctuations depends on squared temperature gradient. This is natural, since this source is proportional to the relevant contribution to entropy production. This source gains importance as Knudsen number increases. Thermal flux is then defined in terms of entropy gradient, and so the contribution of temperature fluctuations to thermal flux can no longer be neglected. A consistent description of thermal processes can then only be based on the kinetic equation for distribution function of kinetic energy values (instant temperature).

We shall more than once return to the problem of heat transfer in the description of nonequilibrium processes in active media. First, however, we have to deal with Brownian motion in nonlinear systems, which will form the basis for construction of generalized kinetic equations for active media. In doing this we shall rely heavily on the above results of statistical theory of nonequilibrium processes in rarefied gases.

CHAPTER 15

NONLINEAR BROWNIAN MOTION

Brownian motion is a familiar classroom demonstration. This phenomenon was discovered as early as 1827 by a British botanist called Robert Brown who was the first to report incessant chaotic movement of pollen particles suspended in liquid.

The cause of Brownian motion was understood much later. Theoretical foundations have only been laid at the turn of this century in the classical papers of Albert Einstein, Marian Smoluchowski and Paul Langevin. Jerky motion of suspended particle struck at random by surrounding atoms is a visualization of the atomic structure of "continuous medium". The main results of this theory were soon confirmed experimentally by Jean Perrin and Theodor Svedberg.

The theory of Brownian motion today is one of the main chapters of statistical theory of open systems. This is not surprising if we recall the following.

By contrast to Boltzmann's theory, where the elementary objects are atoms, the theory of Brownian motion considers small but macroscopic particles which experience the force of friction when they move in the medium. For this reason, the problem of irreversibility must be taken into consideration straight from the outset.

The kinetic theory of gases is based on the reversible equations of motion of atoms. However, because the motion of atoms is dynamically unstable, it is practically impossible to describe nonequilibrium processes on the basis of reversible equations. This is why all "working" equations of the statistical theory are irreversible — such as the kinetic Boltzmann equation, equations of gas dynamics, and the generalized kinetic equation. Then the distribution function $f(r,p,t)$ and the corresponding gasdynamic functions are macroscopic characteristics; they are the first moments of the relevant dynamic functions.

As shown in Ch. 9 and 10, a more comprehensive description can be accomplished by taking the fluctuations of macroscopic functions into account. They are actually regarded as "Brownian particles" engaged in unceasing irregular dance.

In this way, a nineteenth-century saloon curiosity has become one of the fundamental concepts of the modern statistical theory of open systems.

In this context, much of the material presented so far in this book belongs in fact to the theory of Brownian motion. We have, however, at least two important reasons for going back to the principles of this theory.

First, we must learn how to build the theory of nonequilibrium processes in open systems without relying on microscopic models. Only then we shall be able to describe the more complicated physical phenomena, and especially the nonequilibrium processes in chemistry, biology, sociology.

Secondly, the theory of Brownian motion opens numerous new practical applications of the statistical theory of processes in open systems.

We start with the simplest case when the equation of motion of Brownian particle is linear. Further on we shall analyze Brownian motion in passive and active nonlinear systems.

15.1. Two Ways of Describing Brownian Motion

15.1.1. *Langevin Equation*

Let Brownian particle be a sphere of radius a and mass M. By v we denote the velocity of particle relative to liquid. The sphere moving in liquid is acted upon by the force of friction, which at constant velocity v is given by Stokes' formula

$$\vec{F} = -M\gamma v \; , \quad \gamma = \frac{6\pi a}{M}\eta \; , \quad \eta = \rho v \; , \tag{15.1.1}$$

where the coefficient of friction γ is proportional to the dynamic viscosity η.

The equation of motion with only the force of friction is not sufficient for describing Brownian motion (such equation corresponds to the approximation of continuous medium). To account for the atomic structure of the medium, Langevin introduced an additional force $F_L \equiv My(t)$,

$$\frac{dr}{dt} = v \; , \quad \frac{dp}{dt} + \gamma p = F_0 + My(t) \; , \quad F_0 = -\mathrm{grad}\, U \; , \tag{15.1.2}$$

where F_0 is the external force (for instance, gravity).

This equation of motion includes three forces: the Stokes' force (same as in the approximation of continuous medium), the external force, and the Langevin force. The latter is a random function of time and reflects the existence of atomic structure of the medium.

Assume that the medium is at equilibrium, and the external force is zero. Then all directions of the random force are equivalent, and its mean value is therefore zero. Now what is the structure of the second moment $\langle y_i(t)y_j(t') \rangle$? We may assume that the characteristic time of correlation of the values of Langevin force τ_{cor}^{L} is much less than the relaxation time due to viscous friction $\tau_{rel} = 1/\gamma$, that is,

$$\tau_{cor}^{L} \ll \tau_{rel} = \frac{1}{\gamma} .$$ (15.1.3)

In the zero approximation in this parameter the correlation time is taken to be zero. Such random source is said to be delta-correlated. It would be natural to assume further that there is no correlation between different components of the Langevin source, since there is no preferential direction. As a result, we obtain expressions for the two moments,

$$\langle y_i(t) \rangle = 0 , \quad \langle y_i(t)y_j(t') \rangle = 2D \delta_{ij} \delta(t - t') ,$$ (15.1.4)

where $2D$ is the intensity of Langevin source (the mean intensity of random kicks from the side of atoms of the medium). Factor two is introduced for convenience, so that D in the kinetic equation below could be regarded as a coefficient of diffusion.

Inequality (15.1.3) allows us to regard $y(t)$ as a Gaussian random process. Then for the statistical description of the process it is sufficient to know the first two moments.

Equations of motion with Langevin source are not yet closed, because the intensity of noise D is unknown. It can be defined from condition of statistical equilibrium between Brownian particles and the surrounding medium. This is the so-called Einstein relation

$$D = \gamma \frac{\kappa T}{M} ,$$ (15.1.5)

which links the intensity of Langevin source with the dissipative factor γ and the temperature T. Historically, this formula is the first example of fluctuation-dissipation relation.

15.1.2. Fokker – Planck Equation

Langevin equation is a stochastic differential equation — that is, a differential equation which contains both deterministic and random forces. Random can be not only the forces, but also any other parameters of the differential equation. Statistical characteristics of Brownian motion can be found by solving these equations.

There is an alternative approach to the description of Brownian motion, based on the solution of kinetic equation for the one-particle distribution $f(r,v,t)$ of Brownian particles in six-dimensional phase space (r,v). This equation is similar to Boltzmann kinetic equation for rarefied gas; in the theory of Brownian motion the kinetic equation is usually referred to as the Kramers equation, or (more commonly) the Fokker – Planck equation.

For the kinetic description of Brownian motion one has to define the ensemble of noninteracting Brownian particles (the appropriate Gibbs ensemble). Then, instead of following the movement of individual particles, we are dealing with their distribution in six-dimensional phase space, or, in other words, with a "continuous medium" of noninteracting Brownian particles. Drawing an analogy with hydrodynamics, the Langevin description corresponds to the method of Lagrange, and the kinetic description to the method of Euler.

The Fokker – Planck equation can be established in different ways, one of which is based on the Langevin equation. For example, the kinetic equation which corresponds to equations (15.1.2) has the following form:

$$\frac{\partial f}{\partial t} + v\frac{\partial f}{\partial r} - \frac{1}{M}\frac{\partial U}{\partial r}\frac{\partial f}{\partial v} = D\frac{\partial^2 f}{\partial v^2} + \frac{\partial}{\partial v}(\gamma v f) \equiv I_{F-P} \ . \tag{15.1.6}$$

Brownian particles move in the liquid which by itself is at equilibrium. If, by assumption, equilibrium can be established between the Brownian particles and the medium, then the equilibrium solution of Fokker – Planck equation is the Maxwell – Boltzmann distribution

$$f(r,v) = C\exp\left(-\frac{\frac{Mv^2}{2} + U(r)}{\kappa T}\right) \ , \quad \int f(r,v)drdv = 1 \ . \tag{15.1.7}$$

Substitution of this distribution into the Fokker – Planck equation results in the Einstein relation (15.1.5).

15.2. Brownian Motion in Medium with Nonlinear Friction. Three Forms of Fokker – Planck Equation

Consider now the case when the coefficient of friction is a function of velocity,

$$\gamma(v) = \gamma(-v) \ ; \quad \text{for example,} \quad \gamma(v) = \gamma\left(1 + \alpha v^2\right) \ . \tag{15.2.1}$$

This will obviously complicate the Langevin equations. First of all, we should anticipate that the intensity of noise will also be a function of velocity $D = D(v)$, as an implication of fluctuation dissipation relation in case of nonlinear friction. This gives rise to an additional "stochastic force", proportional to the derivative of $D(v)$. The Langevin equations then become

$$\frac{dr}{dt} = v , \quad \frac{dv}{dt} + \gamma(v)v + \frac{1}{M}\frac{\partial U}{\partial r} + a\frac{\partial D}{\partial v} = \sqrt{D(v)}\,y(t) . \tag{15.2.2}$$

Here we have already separated the intensity of random source from the Langevin force, and so the moments of random function $y(t)$ are now defined as

$$\langle y(t)\rangle = 0 , \quad \langle y_i(t)y_j(t')\rangle = 2\delta_{ij}\delta(t - t') . \tag{15.2.3}$$

Now we are worse off than before: in place of one unknown constant D we have an unknown function $D(v)$ and unknown coefficient a.

Assume, as before, that there exists equilibrium between the Brownian particles and the medium. Moreover, we assume that the structure of Einstein's fluctuation-dissipation relation remains the same. Then (15.1.5) becomes

$$D(v) = \gamma(v)\frac{\kappa T}{m} . \tag{15.2.4}$$

These assumptions will be later supported with concrete examples. In particular, we shall demonstrate that the kinetic Fokker – Planck equation can be constructed for a mixture of "heavy" and "light" rarefied gases. This equation describes evolution towards the state of equilibrium, whereas the nonlinear coefficients of diffusion and friction are linked by a generalized Einstein relation.

Now let us proceed from the Langevin equations to the corresponding kinetic Fokker – Planck equation for the distribution function $f(r,v,t)$.

Like we did in the kinetic theory of gases, we define the microscopic phase density in 6-dimensional phase space (r,v) as

$$N(r,v,t) = \sum_{1\le i\le N}\delta(r - r_i(t))\delta(v - v_i(t)) . \tag{15.2.5}$$

This time, however, functions $r_i(t)$, $v_i(t)$ satisfy the irreversible stochastic Langevin equations rather than the reversible Hamilton equations. Since the system of Brownian particles may be considered as an "ideal gas", in place of phase density (15.2.5) we may use the one-particle dynamic distribution

$$f^{(d)}(r,v,t) = \delta(r - r(t))\delta(v - v(t)) , \quad \int f^{(d)}(r,v,t)\,drdv = 1 . \tag{15.2.6}$$

Functions $r(t)$, $v(t)$ satisfy the Langevin equations (15.2.2). Given this, the equation for the dynamic distribution can be written as the continuity equation

$$\frac{\partial f^{(\mathrm{d})}}{\partial t} + v \frac{\partial f^{(\mathrm{d})}}{\partial r} - \frac{1}{M} \frac{\partial U}{\partial r} \frac{\partial f^{(\mathrm{d})}}{\partial v} = \frac{\partial}{\partial v} \left[\left(\gamma(v)v + a \frac{\partial D}{\partial v} \right) f^{(\mathrm{d})} \right] - \frac{\partial}{\partial v} \left[\sqrt{D(v)} y(t) f^{(\mathrm{d})} \right].$$

(15.2.7)

The sought-for statistical distribution is the first moment:

$$f(r,v,t) = \left\langle f^{(\mathrm{d})} \right\rangle, \quad \text{and the fluctuation} \quad \delta f = f^{(\mathrm{d})} - f .$$

(15.2.8)

The equation for the distribution function f follows from (15.2.7), which after averaging over the Gibbs ensemble takes on the form

$$\frac{\partial f}{\partial t} + v \frac{\partial f}{\partial r} - \frac{1}{M} \frac{\partial U}{\partial r} \frac{\partial f}{\partial v} = \frac{\partial}{\partial v} \left[\left(\gamma(v)v + a \frac{\partial D}{\partial v} \right) f \right] - \frac{\partial}{\partial v} \left[\sqrt{D(v)} \langle y(t) \delta f \rangle \right].$$

(15.2.9)

This equation is not closed, since it includes, along with the distribution function f, the correlator $\langle y(t) \delta f \rangle$ (we have noted that the mean value of $\langle y(t) \rangle$ is zero).

To obtain a closed equation, one must express the unknown correlator in terms of the distribution function f. Now this task is simpler than it was in case of Boltzmann's kinetic equation, since we start with the dissipative Langevin equations rather than the reversible Hamilton equations. We proceed as follows (see Klyatskin – Tatarskii 1973; Klimontovich II – V, 1990a, 1991).

Using equations for functions $f^{(\mathrm{d})}, f$ we construct equation for fluctuation δf. For calculating the correlator $\langle y(t) \delta f \rangle$ it is sufficient to know the solution for δf on small time intervals of the order of $\tau_{\mathrm{cor}}^{\mathrm{L}}$. This allows us to keep only the term with delta-correlated source $y(t)$ in the equation for δf. As a result, the equation for fluctuation of the distribution function assumes the form

$$\frac{\partial \delta f}{\partial t} = -\frac{\partial}{\partial v} \left[\sqrt{D(v)} y(t) f(r,v,t) \right],$$

(15.2.10)

whence follows the desired solution

$$\delta f(r,v,t) = -\frac{\partial}{\partial v} \left[\sqrt{D(v)} \int_0^\infty y(t-\tau) f(r,v,t-\tau) d\tau \right].$$

(15.2.11)

Substituting this solution into the last term of equation (15.2.9) and taking into account the structure of correlator $\langle y(t)y(t')\rangle$, we get

$$-\frac{\partial}{\partial v}\left[\sqrt{D(v)}\langle y\delta f\rangle\right] = \frac{\partial}{\partial v}\sqrt{D(v)}\frac{\partial}{\partial v}\left[\sqrt{D(v)}f\right]$$

$$= \frac{\partial}{\partial v}\left[D(v)\frac{\partial f}{\partial v}\right] + \frac{\partial}{\partial v}\left[\frac{1}{2}\frac{\partial D}{\partial v}f\right] . \tag{15.2.12}$$

Substituting this into the right-hand side of (15.2.9), we obtain the kinetic Fokker – Planck equation:

$$\frac{\partial f}{\partial t} + v\frac{\partial f}{\partial r} - \frac{1}{M}\frac{\partial U}{\partial r}\frac{\partial f}{\partial v} = \frac{\partial}{\partial v}\left[D(v)\frac{\partial f}{\partial v}\right] + \frac{\partial}{\partial v}\left[\left(\gamma(v)v + \left(a + \frac{1}{2}\right)\frac{\partial D}{\partial v}\right)f\right] . \tag{15.2.13}$$

Like the initial Langevin equations, the Fokker – Planck equation (15.2.13) is not closed because it still contains an unknown function $D(v)$ and unknown coefficient a. To make further progress, we take advantage of the statistical equilibrium between the Brownian particles and the surrounding medium.

Under this assumption, the equilibrium solution of Fokker – Planck equation must have the form of Maxwell – Boltzmann distribution (15.1.7). Substituting this distribution into (15.2.13) for the state of equilibrium we come to the equation

$$D(v) = \left[\left(a + \frac{1}{2}\right)\frac{v}{v^2}\frac{\partial D(v)}{\partial v} + \gamma(v)\right]\frac{\kappa T}{M} , \tag{15.2.14}$$

which links the source intensity $D(v)$ with the nonlinear dissipative coefficient $\gamma(v)$, but the unknown coefficient a is still there.

If the coefficient of nonlinear friction is known, this equation may be regarded as a differential equation with respect to the intensity of random source $D(v)$ — that is, as a differential fluctuation-dissipation relation. From statistical theory, however, it follows that such linkage is not differential (rather than that, the fluctuation factor is proportional to the dissipation factor). This implies that for the case in question the fluctuation-dissipation relation must reduce to the generalized Einstein relation (15.2.4), and hence the coefficient a in the equations of Langevin and Fokker – Planck is $a = -\frac{1}{2}$. On the strength of these arguments we come to the so-called "kinetic form" of Fokker – Planck equation for Brownian motion in medium with nonlinear friction:

$$\frac{\partial f}{\partial t} + v\frac{\partial f}{\partial r} - \frac{1}{M}\frac{\partial U}{\partial r}\frac{\partial f}{\partial v} = \frac{\partial}{\partial v}\left[D(v)\frac{\partial f}{\partial v}\right] + \frac{\partial}{\partial v}\left[\gamma(v)vf\right] = 0 . \tag{15.2.15}$$

The corresponding Langevin equations are:

$$\frac{dr}{dt} = v \ , \quad \frac{dv}{dt} + \gamma(v)v + \frac{1}{M}\frac{\partial U}{\partial r} - \frac{1}{2}\frac{\partial D}{\partial v} = \sqrt{D(v)}y(t) \ . \tag{15.2.16}$$

Now we can summarize the results.

Although the Langevin equations are nonlinear with respect to dissipation, the kinetic Fokker – Planck equation is linear with respect to the distribution function. Nonlinearity of the medium is taken into account by regarding the coefficients D and γ as functions of velocity.

Along with the "kinetic" Fokker – Planck equation (15.2.15) (K–form), other representations of Langevin equations and the corresponding Fokker – Planck equations can be found in literature. The most important of these are Ito's (I–form) and Stratonovich's (S–form) representations.

The difference between Ito's and Stratonovich's approaches consists mainly in different treatment of stochastic integrals which come up in the solutions of nonlinear stochastic Langevin equations (see Gardiner 1983, 1986; Van Kampen 1981, 1991; Horstemke – Lefever 1984, 1987; Haken 1983, 1985). Even though these representations are based on similar stochastic equations (such as (15.2.2) with $a = 0$), they result in different Fokker – Planck equations:

$$\frac{\partial f}{\partial t} + v\frac{\partial f}{\partial r} - \frac{1}{M}\frac{\partial U}{\partial r}\frac{\partial f}{\partial v} = \frac{\partial^2}{\partial v^2}[D(v)f] + \frac{\partial}{\partial v}[\gamma(v)vf] = 0 \tag{15.2.17}$$

(I–form), and

$$\frac{\partial f}{\partial t} + v\frac{\partial f}{\partial r} - \frac{1}{M}\frac{\partial U}{\partial r}\frac{\partial f}{\partial v} = \frac{\partial}{\partial v}\left\{(D(v))^{1/2}\frac{\partial}{\partial v}\left[(D(v))^{1/2}f\right]\right\} + \frac{\partial}{\partial v}[\gamma(v)vf] = 0 \tag{15.2.18}$$

(S–form).

Let us compare the above three forms of Fokker – Planck equation with equation (15.2.9). We see that I–form corresponds to $a = \frac{1}{2}$, S–form to $a = 0$, and K–form to $a = -\frac{1}{2}$, and it is only in the last case that equation (15.2.14) coincides with the generalized Einstein relation.

Equation (15.2.14) can be formally reduced to the form of Einstein relation (15.2.4) by introducing the effective coefficient of friction:

$$D(v) = \gamma_{\text{eff}}\frac{\kappa T}{M} \ , \quad \gamma_{\text{eff}} = \gamma(v) + \left(a + \frac{1}{2}\right)\frac{v}{v^2}\frac{\partial D(v)}{\partial v}\frac{\kappa T}{M} \ . \tag{15.2.19}$$

Using this definition, one can also reduce the general Fokker – Planck equation (15.2.13) to the "kinetic" form.

This, however, is just a formal trick. The question is why there are three different Fokker – Planck equations which correspond to the same nonlinear dynamic system with a given dissipative coefficient. The problem of choice remains open.

Three different Fokker – Planck equations give rise to three different forms of stationary distributions:

$$f(v) = \frac{C}{[D(v)]^{\nu}} \exp\left(-\int_{0}^{v} \frac{\gamma(v')}{D(v')} d\frac{v'^2}{2}\right), \quad \int f(v)dv = 1 , \tag{15.2.20}$$

where $\nu = 1$ for I–form, $\nu = \frac{1}{2}$ for S–form, and $\nu = 0$ for K–form. Only in the last case the stationary solution depends on the ratio of fluctuation and dissipation factors, in agreement with the general structure of fluctuation-dissipation relations.

Let us return to the Langevin equation (15.2.2) with arbitrary a. The Langevin force depends on v, and therefore its mean value is nonzero. The correlator is defined as above. As a result we obtain the expression

$$\left\langle \sqrt{D(v)}y(t) \right\rangle = \frac{1}{2}\left\langle \frac{\partial D(v)}{\partial v} \right\rangle , \tag{15.2.21}$$

which does not depend on a.

Recall that the dissipation function $\gamma(v)$ is even. Function $D(v)$ is also even. If we assume that function $U(r)$ is even too, then at the state of equilibrium all terms upon averaging independently go to zero.

In the next section we are going to give physical examples which support the choice of Fokker – Planck equation in the "kinetic" form (15.2.15). Later we shall also discuss the inverse transition from Fokker – Planck equation to Langevin equation. This is important because in the context of statistical theory it is the kinetic equations that are more justified, or, so to say, "primary". At the same time, the stochastic Langevin equations are better suited for practical applications.

15.3. Fokker – Planck Equation for Boltzmann Gas

Once again we consider rarefied gas, the nonequilibrium processes in which can be described on the basis of kinetic Boltzmann equation. Assume now that the gas contains atoms of a heavier gas. The heavier atoms may be naturally regarded as Brownian particles in the medium represented by the main gas. We denote the distribution function of impurity atoms by $f_{\text{imp}}(r,p,t)$, and proceed to find the appropriate kinetic equation.

We start with the set of Boltzmann equations for distribution functions of light and heavy atoms. Assume that the concentration of heavy atoms is small enough to make the collisions between them entirely unimportant. In the collision integral of light and heavy atoms we carry out expansion in small parameter $|p-p'|/Mv_T$ (ratio of the change in momentum of light atom to the momentum of impurity atom). As a result, we come to the Fokker – Planck equation

$$\frac{\partial f_{\text{imp}}}{\partial t} + v\frac{\partial f_{\text{imp}}}{\partial r} + F\frac{\partial f_{\text{imp}}}{\partial p} = \frac{\partial}{\partial p_i}D_{ij}(p)\frac{\partial f_{\text{imp}}}{\partial p_j} + \frac{\partial}{\partial p_i}\Big(A_i(p)f_{\text{imp}}\Big) . \tag{15.3.1}$$

In structure, this equation corresponds to (15.2.15); the difference is that now the diffusion is characterized by tensor $D_{ij}(p)$, and the dissipation by vector $A_i(p)$.

There are two possible ways of expressing these functions:
(1) The tensor of diffusion and vector of dissipation are expressed directly via the distribution function of the gas of light particles, which satisfies the relevant Boltzmann equation, and
(2) Using the fluctuation representation of Boltzmann collision integral (Klimontovich 1982a; IV, V), one can express the tensor of diffusion in terms of spectral density of fluctuations of the *potential of scattering δU*, and the vector of dissipation in terms of the imaginary part of the relevant susceptibility $\alpha(\omega,k,p)$:

$$D_{ij}(p) = \frac{1}{16\pi^3}\int \delta(\omega - kv)(\delta U \delta U)_{\omega,k,p}k_i k_j \, d\omega dk , \tag{15.3.2}$$

$$A_i(p) = \frac{1}{8\pi^3}\int \delta(\omega - kv)k_i \operatorname{Im}\alpha(\omega,k,p)d\omega dk . \tag{15.3.3}$$

Both functions in the integrands are, in turn, expressed via the distribution function of the light particles which satisfies the Boltzmann equation. At equilibrium, when the solution of Boltzmann equation is the Maxwell distribution, these functions are linked by the fluctuation-dissipation relation

$$(\delta U \delta U)_{\omega,k,p} = \frac{2}{\omega}\operatorname{Im}\alpha(\omega,k,p)\kappa T . \tag{15.3.4}$$

This relation ensures that functions $D_{ij}(p)$, $A_i(p)$ satisfy Einstein's formula

$$\frac{v_i D_{ij} v_j}{v^2} = D(v) = \gamma(v)\kappa T ,$$

$$\gamma(v) = \frac{1}{8\pi^3} \int \delta(\omega - kv)\frac{kv}{v^2}\operatorname{Im}\alpha(\omega,k,p)d\omega dk .$$

(15.3.5)

The concrete dependence of coefficient of friction on velocity via the function $\alpha(\omega,k,p)$ is defined by the potential of interaction of gas particles.

This example is given for illustration, as an argument in favor of the "kinetic" form of Fokker – Planck equation (15.2.15). Other examples can be found in (Klimontovich 1990a, V). Here we just mark the following.

Fluctuation representation of Boltzmann's collision integral offers a possibility of consistent generalization of the kinetic equation so as to take into account both the strong paired collisions (like those in Boltzmann gas) and the weak collective interactions between particles. The latter is especially important for systems of particles with Coulombian interaction. A typical example of such system is electron-ion plasma.

The problem of choosing the form of Fokker – Planck equations will be discussed further in Ch. 16, and in the chapters concerned with the kinetic description of nonequilibrium processes in active media. We shall see that the K–form is the most advantageous.

Observe once again that the coefficient of diffusion and the intensity of the Langevin force reflect the existence of the atomic structure of the medium. Atomic structure is the source of intrinsic "natural" noise. Of course, the intensity of external noise from various sources may be much greater. Nevertheless, the role of even the low-intensity natural sources is quite important, as will be illustrated with numerous examples. We shall also see that the intensity of natural noise increases dramatically when the system approaches all kinds of critical points (points of equilibrium or nonequilibrium phase transitions). As a rule, the growth of fluctuations is a harbinger of the forthcoming structural change.

If the intensity of noise is completely determined by fluctuations due to the existence of atomic structure of the medium, the corresponding Fokker – Planck equation describes time evolution towards equilibrium. In the presence of external sources, the stationary solution of Fokker – Planck equation may differ considerably from the equilibrium solution. When the external source is switched off, however, the system must go over to the equilibrium state.

Other forms of kinetic equations are also extensively used in the theory of Brownian motion.

15.4. Smoluchowski Equation. Master Equation

In physical (Rytov 1976; Lifshitz – Pitayevsky 1979; Stratonovich 1961, 1985; Van Kampen 1986; Gardiner 1986) and mathematical (Gikhman – Skorokhod 1974) literature the Fokker – Planck equation is often based on the so-called Smolukhovski equation (also known as Chapman – Kolmogorov equation). This equation may be interpreted as the condition of consistency of distribution functions of different orders.

Denote by x an arbitrary set of variables, and by $f(x,t)$, $f(x,t,x',t')$ the distribution functions for, respectively, the given time t, and the two consecutive instants t, t'. We use two identities,

$$f(x,t) = \int f(x,t,x',t')dx' \equiv \int f(x,t|x',t')f(x',t')dx' ,$$

$$\int f(x,t)dx = 1 .$$

(15.4.1)

The first of these is the condition of consistency, and the second gives the definition of conditional distribution function referring to two different time instants. This distribution function is called the *probability of transition*. We denote it as

$$p(x,t,x',t') \equiv f(x,t|x',t')$$

and use the normalization condition

$$\int f(x,t|x',t')dx = \int p(x,t,x',t')dx = 1 .$$

Then (15.4.1) can be rewritten as

$$f(x,t) = \int p(x,t,x',t')f(x',t')dx' .$$

(15.4.2)

Substituting into the right-hand side the value of $f(x',t')$ expressed via the distribution $f(x_0,t_0)$ at an earlier time, we obtain the integral relation which includes the intermediate point x':

$$f(x,t) = \int p(x,t,x',t')p(x',t',x_0,t_0)f(x_0,t_0)dx'dx_0 , \quad p(x,t,x',t') > 0 .$$

(15.4.3)

This can be used for obtaining a closed equation for transition probabilities. Into the left-hand side of (15.4.3) we substitute (15.4.2) with $x',t' \Rightarrow x_0,t_0$. Since the equation obtained in this way holds for arbitrary $f(x_0,t_0)$, we may equate the integrands. As a result, we arrive at the Smoluchowski equation

$$p(x,t,x_0,t_0) = \int p(x,t,x',t')p(x',t',x_0,t_0)dx' \ . \tag{15.4.4}$$

In order to return to (15.4.3), we have to multiply both sides by $f(x_0,t_0)$ and carry out integration with respect to x_0. It is also possible to go back from (15.4.3) to (15.4.2) and (15.4.1).

It follows that the above relations are exact, since our transitions between equations did not involve any simplifying assumptions. In particular, the integral relation (15.4.2) links the distribution functions $f(x,t)$, $f(x',t')$ via the probability of transition. However, this is not yet a kinetic equation, since the transition probability is unknown. We may only argue that it satisfies the Smoluchowski equation (15.4.4) and the normalization condition. To obtain a closed kinetic equation we need additional information about the system, like we did when deriving the kinetic Boltzmann equation for rarefied gas. How do we proceed now?

First we shall go over from the exact relation (15.4.2) to the so-called master equation which is simpler. For this we assume that there are two characteristic time scales, "the fast" and "the slow". Recall that we used a similar assumption in the derivation of the kinetic Boltzmann equation.

Assume that the distribution function $f(x,t)$ changes slowly with time. As before, the characteristic relaxation time we denote by τ_{rel}; the characteristic correlation time for the "fast" process is τ_{cor}. Let the probability of transition depend only on the "fast" time. Moreover, in the zero approximation with respect to τ_{rel}/τ_{cor} the process may be regarded as stationary.

Given this, equation (15.4.2) can be rewritten as (with the replacement $t,t' \Rightarrow t+\Delta t, t$, where $\Delta t = t - t'$)

$$f(x,t+\Delta t) = \int p(x,x',\Delta t)f(x',t)dx' \ . \tag{15.4.5}$$

Now we expand the left-hand side in Δt and retain the first two terms. Assume that there exists the limit

$$\lim_{\Delta t \to 0} \frac{1}{\Delta t} p(x,x',\Delta t) = W(x,x') \ , \tag{15.4.6}$$

which determines the rate of change of the transition probability. We also note that the probability of transition has the following properties:

$$\int p(x',t',x,t)dx' = 1 \ , \quad p(x',x,\Delta t) = \Delta t W(x',x) \ . \tag{15.4.7}$$

The first of these follows from the normalization condition of distribution functions $f(x,t)$, $f(x',t')$ in (15.4.1). In the second we use the limit (15.4.6) and retain only the main term in the expansion in Δt. Taking advantage of these properties, we transform the first term in the expansion in Δt on the left-hand side of (15.4.5):

$$f(x,t) = \int p(x',x,\Delta t)dx' f(x,t) = \Delta t \int W(x',x)dx' f(x,t) \ . \tag{15.4.8}$$

Now we use (15.4.6), (15.4.8) to go over from (15.4.5) to the desired master equation

$$\frac{\partial f(x,t)}{\partial t} = \int [W(x,x')f(x',t) - W(x',x)f(x,t)]dx' \ , \quad W(x,x') > 0 \ . \tag{15.4.9}$$

This equation is not yet closed, since the structure of functions W is not defined. Observe that many of the known kinetic equations of the statistical theory of nonequilibrium processes, including the kinetic Boltzmann equation, can be reduced to this form. The master equation, however, is nonlinear, because the probabilities W themselves depend on the distribution function $f(x,t)$. For spatially homogeneous Boltzmann gas the variable x is a vector, $x = p$.

So, we have made the first step towards particularizing equation (15.4.2). This has been made possible by the simplifying assumption that there exist two different time scales, and that the process is stationary on small time intervals. Then the probability of transition $p(x,t,x',t')$ is replaced by the simpler functions $W(x,x')$, $W(x',x)$, which do not explicitly depend on the time. The resulting kinetic equation is irreversible because small-scale correlations are neglected. This will be confirmed by the analysis of the relevant equation of entropy balance.

15.5. Two Ways of Transition from Master Equation to Fokker – Planck Equation

For future discussion it will be convenient to use a different form of master equation in place of (15.4.9). We represent the probability of transition as the sum of symmetrical and antisymmetrical parts:

$$W(x',x) = W^s(x,x') + W^a(x,x') ,$$

$$W^s(x,x') = W^s(x',x) , \quad W^a(x,x') = -W^a(x',x) .$$

(15.5.1)

Then (15.4.9) can be rewritten in the form

$$\frac{\partial f(x,t)}{\partial t} = \int \left[W^s(x,x')(f(x',t) - f(x,t)) - W^a(x,x')(f(x',t) + f(x,t)) \right] dx' .$$

(15.5.2)

It is interesting that all major kinetic equations (including the kinetic Boltzmann equation) can be reduced to this form (Klimontovich 1982a, IV, V). Reduced to the same form can also be the quantum kinetic equations for a plasma and for a system of atoms interacting with electromagnetic field. In such cases, however, the master equations are nonlinear, because the probabilities of transition themselves depend on the distribution functions. The transition to linear equations is only possible for Brownian motion, when the statistical properties of the medium are known, and the interaction between Brownian particles can be neglected. We shall return to this issue in the chapters devoted to the quantum kinetic theory.

As follows from (15.5.2), the transition probabilities in the stationary state are linked by

$$W^s(x,x') = W^a(x,x') \frac{f^{(st)}(x') + f^{(st)}(x)}{f^{(st)}(x') - f^{(st)}(x)} ,$$

(15.5.3)

which is a fluctuation-dissipation relation. For that matter, the symmetrical function $W^s(x,x')$ defines the coefficient of diffusion in the corresponding Fokker – Planck equation, and the antisymmetrical function $W^a(x,x')$ the coefficient of friction.

Now let us make the transition from the master equation (15.5.2) to the Fokker – Planck equation. We shall see that this transition is not unambiguous, and may lead to different forms of the desired equation.

The kinetic form of Fokker – Planck equation. In place of x, x' we introduce new variables Δx, x,

$$W^{s,a}(x,x') = W^{s,a}\left(x - x', \frac{x+x'}{2} \right) = W^{s,a}\left(\Delta x, x - \frac{\Delta x}{2} \right) , \quad \Delta x = x - x' ,$$

(15.5.4)

and rewrite the master equation as

$$\frac{\partial f(x,t)}{\partial t} = \int \left[W^s_{\Delta x, x-\Delta x/2}(f(x-\Delta x,t) - f(x,t)) \right.$$

$$\left. - W^a_{\Delta x, x-\Delta x/2}(f(x-\Delta x,t) + f(x,t)) \right] d(\Delta x) . \qquad (15.5.5)$$

Assume (!) that $\Delta x \ll x$, but the dependence of functions $W^{s,a}$ on the first argument Δx is not weak. For the sake of simplicity we consider a one-dimensional case when x is a scalar quantity, and carry out expansion in small variation $\Delta x \, \partial/\partial x$. Making use of equations

$$\int W^s_{\Delta x, x} \Delta x \, d(\Delta x) = 0 ,$$

$$\int W^a_{\Delta x, x} d(\Delta x) = 0 , \qquad (15.5.6)$$

we transform the first and the second terms in (15.5.5):

$$\frac{1}{2} \int (\Delta x)^2 \frac{\partial}{\partial x} \left(W^s_{\Delta x, x} \frac{\partial f}{\partial x} \right) d(\Delta x) = \frac{\partial}{\partial x} \left(D(x) \frac{\partial f}{\partial x} \right) ,$$

$$\int \Delta x \frac{\partial}{\partial x} \left(W^a_{\Delta x, x} f \right) d(\Delta x) = \frac{\partial}{\partial x} (A(x) f) , \qquad (15.5.7)$$

where the coefficients of diffusion and friction are

$$D(x) = \frac{1}{2} \int (\Delta x)^2 W^s_{\Delta x, x} d(\Delta x) ,$$

$$A(x) = \int \Delta x W^a_{\Delta x, x} d(\Delta x) . \qquad (15.5.8)$$

As a result, we come to the kinetic form of Fokker – Planck equation:

$$\frac{\partial f(x,t)}{\partial t} = \frac{\partial}{\partial x} \left(D(x) \frac{\partial f}{\partial x} \right) + \frac{\partial}{\partial x} (A(x) f) . \qquad (15.5.9)$$

In order to obtain Fokker – Planck equation in the forms of Ito and Stratonovich, similar to (15.2.17), (15.2.18), we must make the following replacement in (15.5.9):

$$A(x) \Rightarrow A(x) + \left(a + \frac{1}{2}\right)\frac{dD(x)}{dx} , \tag{15.5.10}$$

which implies redefinition of either the coefficient of diffusion, or the coefficient of friction. Here $a = \frac{1}{2}$ for Ito's representation, and $a = 0$ for Stratonovich's representation.

We see that the coefficients of diffusion and friction are completely defined in terms of, respectively, the even and the odd parts of transition probability only for the kinetic representation of Fokker – Planck equation. This is yet another argument in favor of the K–form of kinetic equation.

How do we obtain different forms of Fokker – Planck equation from one and the same master equation? For illustration we shall construct the Fokker – Planck equation in I–form.

We return to the initial equation (15.4.9) and introduce new variables in a fashion less symmetrical than in (15.5.4): $x, x' \Rightarrow x' - x$, $x' = \Delta x$, $x + \Delta x$, where $\Delta x = x' - x$. Accordingly, the first argument now is the difference in the former second and first arguments. The second argument remains the same. The transition probabilities are then redefined as follows:

$$W(x, x') \Rightarrow W(x' - x, x') = W(\Delta x, x + \Delta x) , \quad W(x', x) \Rightarrow W(-\Delta x, x) . \tag{15.5.11}$$

As a result, the master equation (15.4.9) becomes

$$\frac{\partial f(x,t)}{\partial t} = \int \left[W(\Delta x, x + \Delta x)f(x + \Delta x, t) - W(-\Delta x, x)f(x,t)\right]d\Delta x . \tag{15.5.12}$$

The terms of zero order cancel out because

$$\int W(\Delta x, x)d\Delta x = \int W(-\Delta x, x)d\Delta x \tag{15.5.13}$$

(this can be easily proved using definition (15.5.1) of the transition probability).

As a result, we come to the Fokker – Planck equation in I–form:

$$\frac{\partial f(x,t)}{\partial t} = \frac{\partial^2}{\partial x^2}\left[D(x)f(x,t)\right] + \frac{\partial}{\partial x}\left[A(x)f\right] , \tag{15.5.14}$$

where the coefficients of diffusion and friction are defined by

$$D(x) = \frac{1}{2} \int (\Delta x)^2 W(\Delta x, x) d\Delta x \ ,$$

$$(15.5.15)$$

$$A(x) = \int \Delta x W(\Delta x, x) d\Delta x \ ,$$

which are close in form to (15.5.8).

So we see that different expansions in Δx lead to different forms of Fokker – Planck equation. As we have already noted in Sect. 15.2, the choice between them has to be made from additional physical considerations. Now we are going to continue discussing this point.

Stationary solution of Fokker – Planck equation. As indicated above, the replacement (15.5.10) allows one to obtain all three forms of the Fokker – Planck equation from (15.5.9). For the one-dimensional case the general stationary solution of these equations can be written in the form

$$f(x) = \frac{C}{(D(x))^{\nu/2}} \exp\left[-\int\limits_0^x \frac{A(x')}{D(x')} dx' \right], \quad \int f dx = 1 \ ,$$

$$(15.5.16)$$

where $\nu = 2$ for I–form, $\nu = 1$ for S–form, and $\nu = 0$ for K–form of the Fokker – Planck equation. Only in the last case the structure of the stationary solution is the simplest and is completely defined by the ratio of the fluctuation factor $D(x)$ and the dissipation factor $A(x)$.

Elsewhere we have considered the stationary solution of Fokker – Planck equation for Brownian motion in the medium with nonlinear friction (see (15.2.20)). At equilibrium, the coefficients $D(v)$, $\gamma(v)$ satisfy the Einstein relation (15.2.4), and we come to the Maxwell distribution. When the master equation (15.4.9) (or (15.5.2)) is used, the situation is in general more complicated.

Master equations are used not only for systems in thermostat, when the motion of Brownian particles occurs in the medium which is at equilibrium. They also describe relaxation in media which are in a stationary but not equilibrium state. In such situation one might question the validity of Einstein relation. We shall see, however, that its use is both possible and necessary.

Assume once again that the generalized coordinate x in the master equation can be interpreted as the velocity of "Brownian particle". Expressed in terms of this velocity can be, for instance, the electric current in a self-oscillatory system (Van der Pol oscillator). Then the coefficients of diffusion $D(v)$ and friction $A(v)$ will depend also on the coefficient of feedback a_f. Owing to the presence of feedback, the stationary state will be other than the state of equilibrium. From (15.5.15) we find that

$$f(v,a_f) = \frac{C}{(D(v,a_f))^{1/2}} \exp\left[-\int_0^v \frac{A(v',a_f)}{D(v',a_f)} dv'\right], \quad \int f dv = 1, \quad (15.5.17)$$

with the above values of v for the three forms of Fokker – Planck equation.

It would be natural to assume that in the absence of feedback the coefficients

$$D(v, a_f = 0) = D(v), \quad A(v, a_f = 0) = A(v) \equiv \gamma(v)v \quad (15.5.18)$$

satisfy the Einstein relation (15.2.4), and the distribution $f(v)$ coincides with the Maxwell distribution. The latter only occurs for the K–form of the Fokker – Planck equation — that is, when $v = 0$.

This again brings us to the conclusion that the kinetic form of the Fokker – Planck equation is preferable from the standpoint of statistical theory. This conclusion will be corroborated with numerous concrete examples in the chapters to follow.

Now we are going to study the master equation for a system of atoms interacting with electromagnetic field. Here the atoms act as Brownian particles, and the fluctuation electromagnetic field as the medium.

15.6. Master Equation for System of Atoms in Electromagnetic Field

The kinetic theory of atoms and field has much advanced in recent years to match the progress in quantum electronics. This theory stems from the classical Einstein's paper of 1916, in which he formulated the first equation of balance for atoms at rest and equilibrium field, and introduced the coefficients of induced and spontaneous emission (Einstein's coefficients).

Consider the most simple model when the atoms are at rest and are homogeneously distributed in space. Then the state of atoms is characterized by the distribution function f_n for the values of internal energy of atom E_n. In this way, the state of atoms is defined by a discrete set of variables n. For the case in question the kinetic equation can be written in the form (Klimontovich III, V)

$$\frac{\partial f_n}{\partial t} = \sum_m \left[B_m^n \frac{(\delta E \delta E)_{\omega_{nm}}}{4\pi^2} (f_m - f_n) - \frac{1}{2} A_m^n (f_m + f_n) \right] = I_n, \quad (15.6.1)$$

where

$$B_m^n = \frac{4\pi^2 |d_{nm}|^2}{3\hbar^2} \ , \quad A_m^n = \frac{4|d_{nm}|^2}{3\hbar c^3} \omega_{nm}^3 \ , \quad \frac{1}{2} A_m^n \equiv \gamma_m^n \tag{15.6.2}$$

are Einstein's coefficients which are found in the course of derivation of the kinetic equation. For the system of atoms in equilibrium electromagnetic field the spectral density of fluctuations of electric field is

$$\frac{1}{4\pi^2} (\delta E \delta E)_\omega \equiv \rho_\omega = \frac{\omega^2}{\pi^2 c^3} \frac{1}{2} \hbar \omega \coth \frac{\hbar \omega}{2\kappa T} \equiv \frac{\omega^2}{\pi^2 c^3} \kappa T_\omega \ , \quad \omega = \omega_{nm} \ , \tag{15.6.3}$$

where ρ_ω is Planck's distribution for the mean energy of equilibrium electromagnetic radiation including the zero-point energy.

At equilibrium, the solution of kinetic equation (15.6.1) is the Gibbs – Boltzmann distribution

$$f_n = \frac{1}{Z} \exp\left(-\frac{E_n}{\kappa T}\right), \quad Z = \sum_n \exp\left(-\frac{E_n}{\kappa T}\right). \tag{15.6.4}$$

It is important that the "collision integral" includes only the spectral density at the transition frequencies ω_{nm}. This corresponds to the approximation of infinitesimally narrow resonances, when the spectral density of fluctuations is represented by a set of infinitesimally narrow spectral lines. The essence of this approximation is discussed in detail in (Klimontovich 1987e, V); it corresponds to the so-called collisionless approximation in the calculation of small-scale fluctuations, the exclusion of which leads to irreversible kinetic equations.

Let us compare the kinetic equation (15.6.1) with master equation (15.5.2). This will allow us to particularize the expressions for transition probabilities $W^{s,a}$ for the case in question:

$$W_{nm}^s = B_m^n \frac{(\delta E \delta E)_{\omega_{nm}}}{4\pi^2} \ , \quad W_{nm}^a = \frac{1}{2} A_m^n \equiv \gamma_m^n \ . \tag{15.6.5}$$

Using (15.6.2), (15.6.3), (15.6.5), we find the connection between transition probabilities W^s, W^a:

$$W_{nm}^s = W_{nm}^a \coth \frac{\hbar \omega_{nm}}{2\kappa T} \ . \tag{15.6.6}$$

In case of the system of atoms in equilibrium electromagnetic field, this equation links the fluctuation characteristic W_{nm}^s to the dissipation characteristic W_{nm}^a, and particularizes the general fluctuation-dissipation relation (FDR) (15.5.3) which holds good also for nonequilibrium stationary states.

Observe that the argument of coth is not the current frequency of the spectrum, but rather the transition frequency ω_{nm}. This structure of FDR is typical of quantum systems (Klimontovich 1987e, V).

Let us quote the expressions for coefficients of diffusion and friction (in the general case they are defined by (15.5.8)). We go over from continuous variables x, x' to the corresponding discrete variables n, m and make substitution similar to (15.5.4). Then the expressions (15.6.5) for the probabilities of transition take on the form

$$W_{n,m}^{s,a} = W_{n-m,(n+m)/2}^{s,a} \equiv W_{\Delta_{nm},n-\Delta_{nm}/2}^{s,a} \, , \quad \Delta_{nm} = n - m \, . \tag{15.6.7}$$

We assume, like we did in case of continuous variables, that the dependence on Δ_{nm} is strong, but at the same time it is possible to carry out expansion in Δ_{nm} in the argument $n - \Delta_{nm}/2$. As a result, we come to the following expressions for the local (n-dependent) coefficients of diffusion and friction:

$$D_n = \frac{1}{2} \sum_{\Delta_{nm}} (\Delta_{nm})^2 W_{\Delta_{nm},n}^s \, , \quad A_n = \sum_{\Delta_{nm}} \Delta_{nm} W_{\Delta_{nm},n}^a \, , \tag{15.6.8}$$

which are similar to (15.5.8) above. In the next section these results will be particularized for the case of quantum atom oscillator.

To end this section, we quote the equation of balance of mean energy, which follows from the kinetic equation (15.6.1):

$$\frac{d\langle E \rangle}{dt} = \sum_{nm} \gamma_m^n (f_n - f_m) \left[\kappa T \omega_{nm} - \frac{1}{2} \hbar \omega_{nm} \frac{f_m + f_n}{f_m - f_n} \right], \quad \langle E \rangle = \sum_n E_n f_n(t) \, . \tag{15.6.9}$$

At equilibrium the right-hand side of equation of balance is zero, and the mean energy is defined by Planck's formula.

So we have succeeded in particularizing the expressions for the transition probabilities W^s, W^a for the system of atoms and field. As follows from (15.6.3) – (15.6.6), they are defined in terms of atomic characteristics: the matrix element of dipole moment d_{nm}, transition frequency ω_{nm}, and the field temperature — that is, the temperature of the medium where the Brownian motion of atoms takes place. To refine our results even further, we must consider a particular model of the atom.

15.7. Brownian Motion of Quantum Atoms Oscillators

15.7.1. *Master Equation*

Let us return to the kinetic equation (15.6.1), and consider the atom as a one-dimensional quantum oscillator. Such "atom oscillator" (that is, Brownian particle) can be visualized as a small but macroscopic electric circuit. By ω_0 we denote the eigenfrequency of oscillator. The square of matrix element can be represented as

$$|d_{nm}|^2 \Rightarrow |x_{nm}|^2 = \frac{\hbar}{m\omega_0}\left(\frac{m}{2}\delta_{m-1,n} + \frac{m+1}{2}\delta_{m+1,n}\right). \tag{15.7.1}$$

Then we come to the following expression for the "collision integral" I_n in the kinetic equation (15.6.1):

$$I_n = \gamma(\omega_0)\left\{\coth\frac{\hbar\omega_0}{2\kappa T}\left[\frac{n+1}{2}(f_{n+1} - f_n) + \frac{n}{2}(f_{n-1} - f_n)\right]\right.$$

$$\left. -\left[-\frac{n+1}{2}(f_{n+1} + f_n) + \frac{n}{2}(f_{n-1} + f_n)\right]\right\}; \quad \omega_0 = \omega_{n+1,n}, \tag{15.7.2}$$

where

$$\gamma(\omega_0) = \frac{2e^2\omega_0^2}{3mc^2} \tag{15.7.3}$$

is the coefficient of radiation friction.

Now we can write the equation of balance of mean energy:

$$\langle E \rangle = \sum_n \left(n + \frac{1}{2}\right)\hbar\omega_0 f_n. \tag{15.7.4}$$

Two forms of equation in the mean energy $\langle E \rangle$ are useful,

$$\frac{d\langle E \rangle}{dt} = \gamma(\omega_0)\left(\kappa T_{\omega_0} - \langle E \rangle\right) \equiv D_{(E)} - \gamma(\omega_0)\langle E \rangle, \tag{15.7.5}$$

where we have used the definition of the coefficient of diffusion

$$D_{(E)} = \gamma(\omega_0)\kappa T_{\omega_0} \ , \quad \kappa T_{\omega_0} = \frac{1}{2}\hbar\omega_0 \coth\frac{\hbar\omega_0}{2\kappa T} \ ; \qquad (15.7.6)$$

subscript (E) indicates that $D_{(E)}$ is the coefficient of diffusion with respect to energy values. This formula may be regarded as the quantum generalization of the classical Einstein relation (15.1.5).

Now let us look at expressions (15.6.8) which define the local coefficients of diffusion and friction in the zero approximation with respect to Δ_{nm}/n. For quantum atom oscillator they can be rewritten as

$$D_n = \gamma(\omega_0)\frac{1}{2}\coth\frac{\hbar\omega_0}{2\kappa T}n \equiv \frac{D_{(E)}}{\hbar\omega_0}n \ , \quad A_n = \gamma(\omega_0)n \ , \qquad (15.7.7)$$

where we have used definitions (15.7.3), (15.7.6) for the coefficients of radiation friction and diffusion. We see that the local coefficients of diffusion and friction are linear functions of n. Making use of (15.7.7), we can reduce the quantum "collision integral" (15.7.2) to a more convenient form:

$$I_n(t) = \left[D_{n+1}(f_{n+1} - f_n) - D_n(f_n - f_{n-1})\right]$$

$$+ \frac{1}{2}\left[A_{n+1}(f_{n+1} + f_n) - A_n(f_n + f_{n-1})\right] \ . \qquad (15.7.8)$$

We see that the right-hand side contains two induced contributions, which are proportional to the relevant coefficients of diffusion. Their signs are determined by the relative population of the adjacent levels. The two last terms are proportional to the relevant coefficients of friction. The first of these is positive, since it corresponds to the increase in population at the expense of the higher level. The second is negative and corresponds to the escape to the lower level.

At equilibrium, the diffusion and the dissipation terms cancel out pairwise by virtue of fluctuation-dissipation relation (15.7.6).

Observe that for quantum atom oscillator the matrix elements are defined by (15.7.1), which implies that the transition probabilities in variable n can only change by ± 1 in the course of time evolution as described by the master equation with "collision integral" (15.7.2). Processes of this kind are commonly known as *one-step processes*. Accordingly, the quantum kinetic equation with "collision integral" (15.7.2) or (15.7.8) is an example of master equation for one-step process. Since the local coefficients of diffusion and friction (15.7.7) are linear functions of n, we are dealing here with a *linear one-step process*.

The general structure of master equations for one-step processes will be discussed in the next section. We shall once again see the ambiguity of such equations. Prior to that, however, we shall prove that our current master equation corresponds to the canonical form of Fokker – Planck equation.

15.7.2. Fokker – Planck Equation

Transition to Fokker – Planck equation is based on the expansion in the inverse quantum number $1/n$, which implies that from the discrete spectrum of oscillator energy values we go over to the continuous spectrum. Since $E = n\hbar\omega_0$ for large n, we come to the following equation for the distribution function of energy:

$$\frac{\partial f(E,t)}{\partial t} = \frac{\partial}{\partial E}\left(D_{(E)}E\frac{\partial f}{\partial E}\right) + \frac{\partial}{\partial E}(\gamma E f) , \int_0^\infty f dE = 1 . \tag{15.7.9}$$

The equilibrium solution of the Fokker – Planck equation (15.7.9) is the Boltzmann distribution with the quantum temperature (15.7.6):

$$f(E)\frac{1}{\kappa T_{\omega_0}}\exp\left(-\frac{E}{\kappa T_{\omega_0}}\right) , \langle E\rangle = \kappa T_{\omega_0} . \tag{15.7.10}$$

Examples of such equations for electric circuit can be found in (Klimontovich 1987e, V); the emf is then given by

$$\left(\varepsilon^2\right)_\omega = 2R\kappa T_{\omega_0} , \tag{15.7.11}$$

which in the classical approximation coincides with the known Nyquist formula.

15.8. Master Equations for One-Step Processes

15.8.1. Traditional Definition of Transition Probability

For description of one-step processes we return to the master equation (15.4.9) and introduce discrete variable n in place of continuous variable x. The relevant equation is

$$\frac{\partial f_n}{\partial t} = \sum_{n'} \left[W_{nn'} f_{n'}(t) - W_{n'n} f_n(t) \right] , \quad \sum_{n=0}^{\infty} f_n(t) = 1 .$$ (15.8.1)

The following expression is used traditionally for the transition probability (Van Kampen 1981; Gardiner 1983)

$$W_{nn'} = g_{n'} \delta_{n,n'+1} + r_{n'} \delta_{n,n'-1} .$$ (15.8.2)

It is assumed therefore that one event corresponds either to emergence (birth) at state n, or to disappearance from state n (death). This parlance is used in the theory of populations. In semiconductor theory one refers to g_n as the *coefficient of generation*, and to r_n as the *coefficient of recombination*.

Substitution of (15.8.2) into (15.8.1) results in the following master equation:

$$\frac{\partial f_n}{\partial t} = g_{n-1} f_{n-1} + r_{n+1} f_{n+1} - \left(g_n + r_n \right) f_n , \quad \sum_{n=0}^{\infty} f_n = 1$$ (15.8.3)

(we assume that n varies from zero to infinity). Equation (15.8.3) must be supplemented by "boundary conditions"

$$r_0 = 0 , \quad g_{-1} = 0 , \quad \text{and therefore} \quad \frac{\partial f_0}{\partial t} = r_1 f_1 - g_0 f_0 .$$ (15.8.4)

These conditions forbid escape (recombination) from the lowest state and generation from states with negative numbers n. The local coefficients of diffusion and friction are defined by combinations of r_n, g_n:

$$D_n = \frac{1}{2}(r_n + g_n), \quad A_n = r_n - g_n .$$ (15.8.5)

Master equation (15.8.3) is widely used for describing most diverse processes (Van Kampen 1981; Gardiner 1983): radioactive decay, shot noise, processes of chemical kinetics, and various "predator – prey" (Volterra) systems. This equation works well whenever the main feature of the phenomenon in question is the competition between "birth" and "death", or "ionization" and "recombination". As a rule, such cases are dominated by nonequilibrium (although perhaps stationary) states, while the equilibrium state retreats backstage. In such situations the use of functions g_n, r_n as the main characteristics is perfectly justified.

Equilibrium, however, remains a fundamental concept. It is this state that is stable when the controlling factors are switched off, and which corresponds to the highest degree of chaos. It would be natural to demand therefore that master equations should describe, in particular, the processes of evolution towards equilibrium. Such equations may be more conveniently formulated not in terms of functions g_n, r_n, but rather in terms of the local coefficients of diffusion and friction related to the latter by (15.8.5). We have seen this in case of Brownian motion of quantum atoms oscillators in equilibrium electromagnetic field.

To confirm this point, let us consider some implications of the master equation (15.8.3).

Equation for the first moment of distribution f_n is

$$\frac{d\langle n \rangle}{dt} = -\left(\langle r_n \rangle - \langle g_n \rangle\right) = -\langle A_n \rangle .$$

(15.8.6)

Accordingly, the relaxation of the first moment is determined by the mean value of the local coefficient of friction. In case of linear one-step process, when

$$D_n = Dn , \quad A_n = (r - g)n \equiv \gamma n ,$$

(15.8.7)

equation (15.8.6) becomes

$$\frac{d\langle n \rangle}{dt} = -\gamma \langle n \rangle .$$

(15.8.8)

We see that the value of $\langle n \rangle$ relaxes towards zero. If, by way of example, we use formulas (15.8.8) for quantum atom oscillator, equation for the mean energy becomes

$$\frac{d\langle E \rangle}{dt} = -\gamma(\omega_0)\langle E \rangle , \quad \gamma(\omega_0) = \frac{2e^2\omega_0^2}{3mc^3} .$$

(15.8.9)

As opposed to (15.7.5), which follows from the master equation with collision integral (15.7.2) or (15.7.8), equation (15.8.9) does not describe relaxation towards equilibrium with the thermostat.

This disadvantage of the master equation (15.8.3) is manifested also in the structure of the corresponding Fokker – Planck equation. Indeed, expansion in $1/n$ results in Fokker – Planck equation in Ito's form

$$\frac{\partial f(n,t)}{\partial t} = \frac{\partial^2}{\partial n^2}\left[D_n f(n,t)\right] + \frac{\partial}{\partial n}\left[A_n f(n,t)\right] .$$

(15.8.10)

The stationary solution of this equation has the structure (15.5.16) with $v = 2$, and does not agree with the Einstein formula.

We see that the traditional definition of transition probability (15.8.2) for one-step processes leads to a number of results which contradict the main assumptions of the statistical theory. Our immediate task will therefore consist in trying to define the probability of transition in such a way as to overcome these difficulties.

15.8.2. Non-Traditional Definition of Transition Probability

We use the general form of master equation (15.5.2) and define the symmetrical and antisymmetrical parts of transition probability as (Klimontovich 1992b)

$$W_{nn'}^{s} = D_{n+1}\delta_{n+1,n'} + D_n\delta_{n-1,n'} = W_{n'n}^{s} \, ,$$

$$(15.8.11)$$

$$W_{nn'}^{a} = -\frac{1}{2}\left[A_{n+1}\delta_{n+1,n'} - A_n\delta_{n-1,n'}\right] = -W_{n'n}^{a} \, .$$

Substituting these expressions into (15.5.2), we get the master equation for one-step processes which is different from (15.8.3). The "collision integral" is now defined by (15.7.8), and the local coefficients of diffusion and friction are, as before, linked with the coefficients of generation and recombination by (15.8.5).

Recall that for the linear one-step process, when the coefficients of diffusion and friction are defined by (15.7.7), we come to the kinetic equation for quantum atoms oscillators in equilibrium electromagnetic field.

Expression (15.7.8) for the collision integral has a clear-cut physical sense. The diffusion is responsible for the "induced" transitions, and the signs of the respective terms depends on the relative population of the adjacent states.

The traditional definition of probability of transition allows one to separate the symmetrical and antisymmetrical parts $W_{nn'}^{s,a}$ which are expressed via the coefficients of diffusion and friction. The resulting expressions, however, defy straightforward physical interpretation.

Let us discuss some properties of the non-traditional master equation for one-step processes. For the state of equilibrium, from (15.7.8) follows the relation between the local coefficients of diffusion and friction

$$D_n = \frac{1}{2} A_n \frac{f_n + f_{n+1}}{f_n - f_{n+1}} \, ,$$

$$(15.8.12)$$

which is a fluctuation-dissipation relation. Since the coefficients of diffusion and friction are positive, the higher level at equilibrium is less populated. For the linear one-step process, when formulas (15.7.7) hold good, the Boltzmann distribution

$$f_n = \frac{1}{Z} \exp\left[-\frac{E_n}{\kappa T_{\omega_0}}\right], \quad \sum_n f_n = 1 \tag{15.8.13}$$

follows from (15.8.12). The distribution with respect to n is, therefore, exponential.

There are other implications of the master equation obtained in this section. The equation for the mean value of $\langle n \rangle$ differs from (15.8.6) and has the form

$$\frac{d\langle n \rangle}{dt} = \frac{1}{2}\left[(g_{n+1} - r_{n-1}) - (r_n - g_n)\right]. \tag{15.8.14}$$

To understand the physical meaning of this difference, let us consider the approximation of large n, when we can go over to continuous variable. Retaining only the main terms, we arrive at the following equation:

$$\frac{d\langle n \rangle}{dt} = \left\langle \frac{dD_n}{dn} \right\rangle - (\langle r_n \rangle - \langle g_n \rangle) \equiv \left\langle \frac{dD_n}{dn} \right\rangle - \langle A_n \rangle. \tag{15.8.15}$$

This equation differs from (15.8.6) in that it contains an additional term which is determined by the diffusion. For the linear one-step process, when the coefficients of diffusion and friction are defined by (15.7.7), from (15.8.15) follows the equation of balance of mean energy of quantum atoms oscillators in equilibrium electromagnetic field,

$$\frac{d\langle E \rangle}{dt} = \gamma(\omega_0)\left[\kappa T_{\omega_0} - \langle E \rangle\right], \tag{15.8.16}$$

which, by contrast to (15.8.9), does describe the process of relaxation towards the equilibrium value of mean energy.

Finally, let us consider the corresponding Fokker – Planck equation. Unlike (15.8.10), now it has the canonical form

$$\frac{\partial f_n}{\partial t} = \frac{\partial}{\partial n}\left[D_n \frac{\partial f_n}{\partial n}\right] + \frac{\partial}{\partial n}\left[A_n f_n\right]. \tag{15.8.17}$$

Because of this, the equilibrium solution is completely defined by the ratio of fluctuation and dissipation factors,

$$f_n = C \exp\left[-\int_0^n \frac{A_{n'}}{D_{n'}} dn'\right].$$

(15.8.18)

For the linear one-step process, when the local coefficients of diffusion and friction are defined by (15.7.7), from (15.8.17) follows the Fokker – Planck equation (15.7.9) for the distribution function of energy, and the Boltzmann distribution (15.7.10) follows from (15.8.18).

In the sections to follow we shall illustrate the difference between alternative descriptions of stochastic processes with concrete examples. First, however, we are going to explore the possibility of transition from the Fokker – Planck equation for the distribution $f(r,v,t)$ in the phase space of coordinates and velocities to the Einstein – Smoluchowski equation for a simpler distribution $f(r,t)$. We shall see that this task is very similar to the problem of transition from the kinetic equation to equations of gas dynamics, and is associated with similar difficulties (Klimontovich 1992c, 1993b). These difficulties can and will be overcome in the theory of Brownian motion by going over to the description of nonequilibrium processes based on the generalized kinetic equations (Ch. 17). We shall see that these equations work especially well in the case of Brownian motion in nonlinear active media.

The generalized kinetic equations will help us to draw the limits of applicability of reaction-diffusion equations, such as the known Ginzburg – Landau equation. It will be possible to go beyond the limitations of these equations so as to obtain information concerning the higher moments (which is important in the neighborhood of critical points), and to construct a more consistent theory of large-scale (also known as coarse-grained) fluctuations (kinetic, hydrodynamic, reaction-diffusion fluctuations).

15.9. Spatial Diffusion. Einstein – Smoluchowski Equation

So far in the Langevin equations and in the kinetic equations we have assumed that the medium which hosts Brownian motion is practically unlimited. Now, in addition to the internal parameters D, γ, we introduce the external parameter, the characteristic size of the system L. This gives rise to a new parameter of time, the diffusion time

$$\tau_D = \frac{L^2}{D_{(r)}},$$

(15.9.1)

where $D_{(r)}$ is the coefficient of spatial diffusion. Earlier for the description of Brownian motion we used the coefficient of diffusion in the space of velocities $D \equiv D_{(v)}$. Now the meaning of coefficient D in equations (15.5.9), (15.5.14) depends on the interpretation of the generalized coordinates x.

When the Langevin equations (15.2.2) are used, the correlation time of the source is $\tau_{cor} = 0$. Nonzero are two parameters of time, $\tau_{rel} = 1/\gamma$ and τ_D. If the diffusion time is much greater than the relaxation time, it would be natural to anticipate the feasibility of transition from the Fokker – Planck equation for the distribution $f(r,v,t)$ to the Einstein – Smoluchowski equation for a simpler distribution $f(r,t)$. There are two ways of doing this.

15.9.1. Spatial Diffusion. Langevin Method

We return to the Langevin equations (15.2.2) for Brownian particle, and assume that the external field is absent ($U = 0$) and that the diffusion is slow, $\tau_D \gg \tau_{rel}$. Then it would be natural to neglect the velocity derivative dv/dt in the Langevin equations as small compared to γv. Eliminating the velocity, we come to the Langevin equation in the coordinate

$$\frac{dr}{dt} = \frac{y(t)}{\gamma} \equiv y_r(t) ,$$ (15.9.2)

where $y_r(t)$ is the Langevin source which determines the displacement of Brownian particle kicked about by the atoms of the medium. The moments of this source are

$$\langle y_r(t) \rangle = 0 , \quad \langle y_r(t) y_r(t') \rangle = 3 \cdot 2 D_r \delta(t - t') , \quad D_{(r)} = \frac{\kappa T}{M\gamma} .$$ (15.9.3)

The last of these defines the coefficient of diffusion in conventional space.

From the Langevin equation (15.9.2) one can go over to the relevant equation for the distribution function $f(r,t)$. Following the guidelines set forth in Sect. 15.2, we come to the Einstein – Smoluchowski equation in $f(r,t)$:

$$\frac{\partial f}{\partial r} = D_r \Delta_r f , \quad \int f(r,t) \frac{dr}{V} = 1 , \quad n(r,t) = N f(r,t) .$$ (15.9.4)

It is natural that this equation coincides with the familiar equation of diffusion. We have also defined the density of Brownian particles $n(r,t)$. The analytical solution of this equation is well known. We shall only quote for the moments of displacement of Brownian particle $r - r_0$:

$$\langle r - r_0 \rangle = 0 \ , \ \ \langle (r - r_0)^2 \rangle = 3 \cdot 2 D_{(r)} (t - t_0) \ . \tag{15.9.5}$$

The second equation states that the mean square displacement of Brownian particle is proportional to the time (Einstein's formula).

15.9.2. Diffusion of Brownian Particle in External Field

When the external force $F = -\operatorname{grad} U(r)$ is taken into account in the description of slow processes, equation (15.9.2) includes a new term and becomes

$$\frac{dr}{dt} = -\frac{1}{M\gamma} \frac{dU}{dr} + y_r(t) \ . \tag{15.9.6}$$

The moments of the Langevin source are, as before, given by (15.9.3). Transition to the kinetic equation follows the familiar scheme. As a result, we come to a more general Einstein – Smoluchowski equation

$$\frac{\partial f(r,t)}{\partial t} = D_{(r)} \frac{\partial^2 f}{\partial r^2} - \frac{\partial}{\partial r} \left[\frac{F(r)}{M\gamma} f \right], \ \ D_{(r)} = \frac{\kappa T}{M\gamma} \ , \ \ F(r) = -\frac{\partial U}{\partial r} \ , \tag{15.9.7}$$

which is also known as the Kramers equation. Equilibrium solution is the Boltzmann distribution

$$f(r,t) = \frac{\exp\left(-\dfrac{U(r)}{\kappa T}\right)}{\displaystyle\int \exp\left(-\dfrac{U(r')}{\kappa T}\right) dr'} \ . \tag{15.9.8}$$

So, we have obtained the equation for the distribution function $f(r,t)$ for two cases: (1) free Brownian particles, and (2) particles in external field. In both cases we started with the Langevin equation under the condition of slow motion $dv/dt \ll \gamma v$. The meaning of this condition may be quite different, however, depending on the form of the potential $U(r)$. Let us look at this matter more closely.

Recall that we considered the free motion of particles under condition

$$\tau_D = \frac{L^2}{D_{(r)}} \gg \tau_{\rm rel} = \frac{1}{\gamma} \ , \tag{15.9.9}$$

which contains the squared size of the system and therefore holds as long as the system is large enough.

In the presence of external field the situation becomes much different if the field restricts the movement of particles. By r_0 we denote the characteristic size of the region, and consider two examples which are important for our subsequent discussion.

1. *Harmonic oscillator*: $F = -M\omega_0^2 r$ is the elastic force. Then the equilibrium solution (15.9.8) coincides with the Gaussian distribution with the potential

$$U(r) = \frac{M\omega_0^2 r^2}{2} , \quad r_0^2 \sim \langle r^2 \rangle = \frac{\kappa T}{M\omega_0^2} . \tag{15.9.10}$$

Now r_0 acts as L, and condition (15.9.9) becomes

$$\tau_D = \frac{\gamma}{\omega_0^2} \gg \frac{1}{\gamma} , \quad \text{and therefore} \quad \gamma \gg \omega_0 . \tag{15.9.11}$$

We see that the Kramers equation (15.9.7) holds for a bounded Brownian particle (harmonic oscillator) only when the damping is strong (overdamped oscillator). This situation is of certain practical interest — for instance, in connection with the Brownian motion of fragments of polymer molecules. More interesting, however, is the opposite extreme, when the damping is weak. The two extreme cases have been extensively studied elsewhere (see, for example, Risken 1984; Hanggi 1990).

2. *Brownian particle as bistable element*: the elastic force is nonlinear, and the potential is given by

$$U(r) = \frac{M\omega_0^2 r^2}{2}\left(-a + \frac{b}{2}r^2\right) , \quad a = a_f - 1 , \quad b > 0 . \tag{15.9.12}$$

The coefficient a_f characterizes the action of effective field — for instance, the Lorenz field in a dielectric (Klimontovich III, IV; Andreev – Emelyanov – Ilyinskii 1988, 1993). When a_f is large enough, so that $a > 0$, the coefficient of elasticity becomes negative, and the system becomes bistable. As in case of self-oscillatory systems (open active systems), the coefficient a_f can be referred to as the coefficient of feedback.

As a rule, feedback is the property of the medium in which Brownian motion takes place. The coefficient of nonlinearity b may have different nature; we shall distinguish two possible cases.

(a) Nonlinearity is the property of an individual element of the system (Brownian particle), rather than the property of the medium. Then the equilibrium solution of (15.9.7) is represented by the Boltzmann distribution (15.9.8) with the potential (15.9.12). If $a < 0$, the Boltzmann distribution has one maximum at $x = 0$. The behavior at large x is determined by the term with coefficient b. If $a > 0$, the Boltzmann distribution has two maxima, and the system is bistable.

(b) Both the coefficient of feedback a_f and the coefficient b depend on the characteristics of the medium. In the absence of feedback ($a_f = 0$) we come to the Boltzmann distribution for harmonic oscillator. This implies that at $a_f = 0$ the thermostat has the highest possible symmetry — in other words, the system is in the most chaotic state.

From condition of existence of Boltzmann distribution in the medium with nonlinear elasticity one can define the coefficient of diffusion as function of coordinates,

$$D_{(r)}(r) = D_{(r)}\left(1 + br^2\right), \quad D_{(r)} = \frac{\kappa T}{M\gamma} \, . \tag{15.9.13}$$

Given this, the equilibrium distribution (for any value of a_f) becomes

$$f(r) = C \exp\left(-\frac{U_{eff}(r)}{\kappa T}\right), \quad U_{eff}(r) = \frac{M\omega_0^2}{2}\left[r^2 - \frac{a_f}{b}\ln\left(1 + br^2\right)\right], \tag{15.9.14}$$

where U_{eff} is the effective potential which takes into account the lowering of thermostat symmetry as the coefficient a_f increases. The change may be caused, for instance, by the change in temperature or density.

Eventually we come to the Einstein – Smoluchowski equation with the variable coefficient of spatial diffusion:

$$\frac{\partial f(r,t)}{\partial t} = \frac{\partial}{\partial r}\left[D_{(r)}(r)\frac{\partial f}{\partial r}\right] - \frac{\partial}{\partial r}\left[\frac{F(r)}{M\gamma}f\right],$$

$$\tag{15.9.15}$$

$$F(r) = -\frac{\partial U}{\partial r} = -M\omega_0^2 r\left(1 - a_f + br^2\right) \, .$$

The change in the symmetry of the thermostat can be taken into account in a different way. Namely, the coefficient of diffusion remains the same, whereas the force $F(r)$ is replaced by the relevant effective force

$$F(r) \Rightarrow F_{\text{eff}}(r) = -\frac{\partial U_{\text{eff}}}{\partial r} = -M\omega_0^2 r \left(1 - \frac{a_f}{1 + br^2}\right).$$ (15.9.16)

As a result, we get the following Einstein – Smoluchowski equation:

$$\frac{\partial f(r,t)}{\partial t} = D_{(r)} \frac{\partial^2 f}{\partial r^2} - \frac{\partial}{\partial r}\left[\frac{F_{\text{eff}}}{M\gamma} f\right], \quad D_{(r)} = \frac{\kappa T}{M\gamma},$$ (15.9.17)

the equilibrium solution whereof is, as before, distribution (15.9.14).

We see that in the presence of nonlinear potential (15.9.12) the thermostat (surrounding medium) may act on the Brownian particles in different ways. In the first case one may speak of the *linear thermostat*, since the nonlinearity is the property of each individual Brownian particle. The surrounding medium only affects the characteristic frequency: $\omega_0^2 \Rightarrow \omega_0^2(1 - a_f)$. The value of $a_f = 1$ corresponds to the bifurcation point (the appearance of "soft mode"). In the second case one may refer to a *nonlinear thermostat*, since the nonlinearity of the force acting on Brownian particle is due to the surrounding medium.

15.9.3. Stationary Distributions in "Linear" and "Nonlinear" Thermostats

The above two cases correspond to different Einstein – Smoluchowski equations, and, as a consequence, to different stationary solutions, Boltzmann distribution with the potential (15.9.12) and distribution (15.9.14). It is only the latter that at $a_f = 0$ coincides with the Boltzmann distribution for Brownian motion of harmonic oscillators. The distinction between these two distributions is manifested, in particular, by the different behavior at large values of r, at the tails of the distributions.

For that matter, for the linear thermostat the fall-off of the distribution at large values of r is controlled by the nonlinear factor $\exp(-br^4)$ and is therefore much faster than in case of Boltzmann distribution with the harmonic oscillator potential (15.9.10). By contrast, the distribution (15.9.14), as a_f increases (that is, as the symmetry becomes lower), falls off at large values of r much slower than the Boltzmann distribution for linear oscillator.

Let us consider some other characteristics using the example of one-dimensional motion (replacing r by x).

(1) The locations of maxima coincide:

$$x_{\text{max}} = 0 \text{ if } a_f < 1, \text{ and } x_{\text{max}} = \pm\sqrt{\frac{a_f - 1}{b}} \text{ if } a_f > 1.$$ (15.9.18)

(2) The ratios of distribution functions at $x = x_{max}$ and $x = 0$ show that the relative depth of the pit for the symmetrical bistable potential is less in case of nonlinear thermostat. Because of this, the barrier is surpassed more easily when the nonlinearity is collective. This conclusion is confirmed by calculations of the corresponding dispersions. For example, in the domain of Gaussian approximation the ratio of dispersions is

$$\frac{\left\langle (\delta x)^2 \right\rangle_{NL}}{\left\langle (\delta x)^2 \right\rangle_L} = a_f \, , \quad a_f > 1 \, . \tag{15.9.19}$$

We see that the form of Einstein – Smoluchowski equation depends on the nature of interaction between the particle and the medium.

The Einstein – Smoluchowski equation has been obtained on the basis of the Langevin equation (15.9.6), which follows in its turn from the more general set of Langevin equations (15.2.2) for position and velocity of Brownian particle. As simplifying assumptions we used either the condition of slow spatial diffusion of free particle ($\tau_D \gg \gamma^{-1}$), or the condition of strong damping ($\gamma \gg \omega_0$). The first condition is always satisfied for large enough systems, since the diffusion time is proportional to the square of the characteristic size of the system L. The second condition is based on the internal parameters, and is far not universal. What can be done if inequality $\gamma \gg \omega_0$ does not hold?

There also are other important questions. How do we describe, for instance, spatial diffusion when the coefficient of friction depends on velocity and we have to deal with dissipative nonlinearity? All these questions are part of the general problem concerning the relationship between the kinetic and the hydrodynamic description of Brownian motion. To begin with, let us consider the possibility of hydrodynamic description for the simplest model of Brownian motion. This will help us later to find solutions for the more complicated cases.

15.10. Hydrodynamic Description of Brownian Motion

Let us go back to the Fokker – Planck equation (15.1.6) for the distribution function $f(r,v,t)$. The temperature of thermostat enters this equation via the coefficient of diffusion $D_{(v)}$. We are going to use the known scheme of transition from the kinetic Boltzmann equation to the equations of gas dynamics (Chapman – Cowling 1970; Grad 1949; Klimontovich I, IV; Silin 1971; Zhdanov 1982; Schram 1991).

In the kinetic description of Brownian motion we are actually dealing with a two-component continuous medium. One of the components is the medium which represents the thermostat. This medium may be nonequilibrium. The second component of our exemplary continuous medium consists of noninteracting Brownian particles. Naturally, a more general case is also possible when the interaction of Brownian particles is taken into account. The first results in this direction were obtained by Chandrasekhar (1943).

So, we are considering a two-component continuous medium comprised of the thermostat with temperature T and "continuous medium" of noninteracting particles. The thermostat is assumed to be linear in the above sense. Then Brownian motion is described by the kinetic equation (15.1.6).

Hydrodynamic functions for Brownian particles are $\rho_B(r,t)$, $u_B(r,t)$, $T_B(r,t)$; we shall retain subscript "B" only at the temperature to distinguish it from the temperature of thermostat. The continuity equation for the density of particles is

$$\frac{\partial \rho}{\partial t} + \frac{\partial \rho u}{\partial r} = 0 \ . \tag{15.10.1}$$

The second is the equation for the density of momentum,

$$\frac{\partial \rho u_i}{\partial t} + \frac{\partial \rho u_i u_j}{\partial r_j} = -\frac{\partial p}{\partial r_i} - \frac{\partial \pi_{ij}}{\partial r_j} + \frac{\rho}{m} F_i(r) - \gamma \rho u_i \ , \quad p = \frac{\rho}{m} \kappa T_B \ , \tag{15.10.2}$$

where p is the pressure of Brownian particles, and π_{ij} is the so far unknown "tensor of viscous stress". Finally, we have the equation for the density of kinetic energy,

$$\frac{\partial}{\partial t}\left[\frac{\rho u^2}{2} + \frac{3}{2}\frac{\rho}{m}\kappa T_B \right] + \frac{\partial}{\partial r_i}\left[u_i\left(\frac{\rho u^2}{2} + \frac{3}{2}\frac{\rho}{m}\kappa T_B + p \right) + \pi_{ij} u_j + q_i \right]$$

$$= 3\gamma\rho\left[\frac{\kappa T}{m} - \left(\frac{u^2}{3} - \frac{\kappa T_B}{m} \right) \right] + \rho F u \ , \tag{15.10.3}$$

where q_i is the so far unknown "vector of thermal flux". The last term on the right-hand side of (15.10.2) and the first term on the right-hand side of (15.10.3) are the moments of "collision integral" in the Fokker – Planck equation.

Distribution function $f(r,t)$ in the Einstein – Smoluchowski equation is linked with the density of Brownian particles by equation $\rho(r,t) = mnf$. Accordingly, in order to find the desired equation we must eliminate all "excessive" functions u, T_B, π_{ij}, q_i from equations of hydrodynamics of Brownian particles. This can only be done by perturbation theory, under the assumption that the diffusion process described by Einstein – Smoluchowski equation is the slowest in time and the smoothest in coordinates. In case of diffusion of free particles, the first condition is expressed by inequality $\tau_D \gg \gamma^{-1}$, and the second by $Kn = (v_T/\gamma)/L \ll 1$. In other words, the gradients of hydrodynamic functions are assumed to be small.

In the zero approximation with respect to these parameters, equation (15.10.3) implies that the temperature of Brownian particles is the same as the temperature of thermostat,

$$T_B = T \ . \tag{15.10.4}$$

In case of free Brownian particles ($F = 0$), the main terms on the right-hand side of (15.10.2) are the first and the last. Then, with due account for (15.10.4), we find that

$$\rho u = -\frac{\kappa T}{\gamma m}\frac{\partial \rho}{\partial r} = -D_{(r)}\frac{\partial \rho}{\partial r} \ . \tag{15.10.5}$$

Substituting this expression into the continuity equation (15.10.1), we arrive at the Einstein – Smoluchowski equation in function $\rho(r,t)$, and hence in the distribution function $f(r,t)$.

In the presence of external force, given that the process is slow ($\gamma \gg \omega_0$ in case of oscillator) and the Knudsen number is small, we proceed likewise. Equation (15.10.4) still holds, since the correction is proportional to the gradient and is therefore small. In equation for the density of momentum we have a new term which is proportional to the gradient of potential, and is only small given that the potential is smooth enough (again, $\gamma \gg \omega_0$ in case of oscillator). Then (15.10.5) becomes

$$\rho u = -D_{(r)}\frac{\partial \rho}{\partial r} + \frac{\rho}{M\gamma}F(r) \ . \tag{15.10.6}$$

Substitution of this expression into the continuity equation (15.10.1) results in the Einstein – Smoluchowski equation (15.9.7).

Now a brief summary. The construction of Einstein – Smoluchowski equation in this section has been based on the kinetic Fokker – Planck equation (15.1.6) for the distribution function of Brownian particles $f(r,v,t)$ in six-dimensional phase space of coordinates and velocities. The system of noninteracting Brownian particles is thus regarded as a continuous medium interacting with the thermostat. Interaction with the thermostat is characterized by the coefficient of friction γ and the coefficient of diffusion D. Both these coefficients in (15.1.6) are constant, which means that the Brownian motion is linear. The force of friction is nonzero also when the thermostat is regarded as "continuous medium". The coefficient of diffusion is an integral characteristic of the atomic structure of the surrounding medium (atomic structure of thermostat).

Transition to the Einstein – Smoluchowski equation has been carried out in two stages. First we made the transition to equations in hydrodynamic characteristics of Brownian particles. The resulting set of equations (15.10.1) – (15.10.3) is not closed, being obtained without any simplifying assumptions.

The closed diffusion equation (15.9.7) is obtained under several important constraints: smoothness of the potential (in particular, condition $\gamma >> \omega_0$), and smallness of Knudsen number. These constraints, together with the assumption that γ and D are constant (linear approximation of Brownian motion), narrow the applicability of the Einstein – Smoluchowski equation. Because of this, we again face the problem of construction of the generalized kinetic equation for unified description of Brownian motion on the kinetic and hydrodynamic (diffusion) scales. This problem will be considered in Ch. 17. In the last section of the present chapter we shall apply the general results of thermodynamics of irreversible processes to the theory of Brownian motion.

15.11. Evolution of Free Energy and Entropy at Brownian Motion. Lyapunov Functionals Λ_F, Λ_S

15.11.1. Master Equation. H–Theorem

By $f_0(x,t)$ we denote the stationary solution of master equation (15.4.9), and represent it as the canonical Gibbs distribution, like we did in Sect. 12.4. The effective Hamilton function can be defined in a number of ways; for instance, distribution (12.4.11) involves both the effective free energy F_{eff} and the effective temperature T_{eff}. In general, these characteristics are nonequilibrium; they are linked by the normalization condition.

The intensity of Langevin source (see Ch. 16), which also defines the coefficient of diffusion, may act as the effective temperature. When the general form of master equation (15.4.9) is used, however, there is no explicit information about the structure of stationary solution and the generally nonlinear coefficient of diffusion. As a result, a clear-cut definition of the effective temperature is not feasible.

There are two possible definitions of the nonequilibrium free energy of the stationary state.

On the one hand, we may formally put the effective temperature equal to unity, and represent the stationary distribution as

$$f_0(x) = \exp[F_{eff} - H_{eff}(x)] , \quad F_{eff} = -\int \exp[-H_{eff}(x)]dx . \qquad (15.11.1)$$

This representation amounts to the inclusion of the effective temperature into the definition of free energy and effective energy H_{eff}.

On the other hand, we may define the effective Hamilton function by equation similar to (12.4.15). This implies that

$$T_{\text{eff}} = 1 , \quad F_{\text{eff}} = 0 . \tag{15.11.2}$$

This representation is especially suitable when the information is derived directly from the experiment — for instance, from time realizations of the process $x(t,a)$. We have used this approach in the formulation of criterion of the relative degree of order of states of open systems in the form of S–theorem (Sect. 1.6.6, 12.7).

Now we return to the master equation (15.4.9) and represent its stationary solution in the form (15.11.1), introducing thus the free energy of nonequilibrium stationary state. A "thermodynamic relation" links this quantity with the mean effective energy and the corresponding entropy (cf. (12.4.6)):

$$F_{\text{eff}} = \int H_{\text{eff}}(x) f_0(x) dx - \int \ln f_0(x) \ f_0(x) dx . \tag{15.11.3}$$

In a similar way we define the nonequilibrium free energy for the process of time evolution (cf. (12.4.8)):

$$F(t) = \int H_{\text{eff}}(x) f(x,t) dx - \int \ln f(x,t) \ f(x,t) dx . \tag{15.11.4}$$

The difference in the thus defined free energies is reduced to (cf. (12.4.9)):

$$\Lambda_F = F(t) - F_0 = \int \ln \frac{f(x,t)}{f_0(x)} f(x,t) dx \geq 0 . \tag{15.11.5}$$

We see that the free energy is at minimum in the stationary state. Now let us show that the free energy monotonically decreases in the course of time evolution towards equilibrium as described by the master equation (15.4.9). This will ensure fulfilment of condition (12.4.10), and prove that Λ_F is a Lyapunov functional.

We differentiate (15.11.5) with respect to time, and use master equation (15.4.9). Taking the normalization condition into account, we get

$$\frac{d\Lambda_F}{dt} = \int \ln \frac{f(x,t)}{f_0(x)} [W_{xx'} f(x',t) - W_{x'x} f(x,t)] dx dx'$$

$$\equiv \int W_{xx'} f_0(x') \left[\frac{f(x',t)}{f_0(x')} \ln \frac{f(x,t)}{f_0(x)} - \frac{f(x',t)}{f_0(x')} \ln \frac{f(x',t)}{f_0(x')} \right] dx dx' . \tag{15.11.6}$$

To determine the sign of the integrand we take advantage of the equation

$$\int W_{xx'} f_0(x') \left[\frac{f(x,t)}{f_0(x)} - \frac{f(x',t)}{f_0(x')} \right] dx dx' = 0 ,$$ (15.11.7)

where we have noted that $f_0(x)$ is the stationary solution, and therefore

$$W_{xx'} f_0(x') - W_{x'x} f_0(x) = 0 .$$ (15.11.8)

By virtue of (15.11.7.), we may rewrite (15.11.6) as

$$\frac{d\Lambda_F}{dt} = \int W_{xx'} f(x') \left[-a' \ln \frac{a}{a'} - a + a' \right] dx dx' \le 0 ,$$ (15.11.9)

where we have used the notation

$$a = \frac{f(x,t)}{f_0(x)} , \quad a' = \frac{f(x',t)}{f_0(x')} ,$$ (15.11.10)

and the textbook inequality

$$\ln \frac{a}{a'} \ge 1 - \frac{a'}{a} .$$

Inequalities (15.11.5) and (15.11.9) prove that the difference in nonequilibrium free energies Λ_F is a Lyapunov functional. The process of evolution leads towards the stationary state which, according to this criterion, is stable.

The same result was interpreted by Van Kampen (1981) as the *increase in entropy*. In this connection we would like to recall the following (see Sect. 14.4).

The difference in the free energies and the corresponding Lyapunov functional can serve as the measure of how far the current state is removed from equilibrium. This information, however, is not sufficient for concluding whether or not the evolution under consideration is a process of self-organization. To answer this question, we must introduce another Lyapunov functional,

$$\Lambda_S = S_0 - \tilde{S}(t) = \int \ln \frac{\tilde{f}(x,t)}{f_0(x)} \tilde{f}(x,t) dx \ge 0 , \quad \frac{d\Lambda_S}{dt} \le 0 ,$$ (15.11.11)

where $\tilde{f}(x,t)$ is the renormalized distribution. Renormalization is based on the assumption that the mean effective Hamilton function H_{eff} remains constant in the course of evolution. Lyapunov functional Λ_S can be used as the measure of relative order of states in the course of evolution towards equilibrium. Inequalities (15.11.11) express Boltzmann's H–theorem for processes described by the master equation (15.4.9).

Various criteria of self-organization in open systems will be discussed further in Ch. 21.

15.11.2. *Fokker – Planck Equation. H–Theorem*

Now we turn to Fokker – Planck equation (15.5.9), which follows either from the master equation (15.4.9) or from the equivalent equation (15.5.2). The stationary solution is then given by (15.5.16) with $v = 0$, which can be represented in the form (cf. (15.11.1))

$$f_0(x) = \exp[F_{\text{eff}} - H_{\text{eff}}(x)] \, , \quad H_{\text{eff}}(x) = \int \frac{A(x')}{D(x')} dx' \, . \tag{15.11.12}$$

Once again we introduce the Lyapunov functional Λ_F by formula (15.11.5). Its time derivative is found from (15.5.9),

$$\frac{d\Lambda_F}{dt} = -\int D(x) f(x,t) \left(\frac{\partial}{\partial x} \ln \frac{f(x,t)}{f_0(x)} \right)^2 dx \le 0 \, . \tag{15.11.13}$$

We see that in case of Fokker – Planck equation the free energy decreases monotonically in the course of time evolution towards the equilibrium distribution (15.11.12). Naturally, the Lyapunov functional Λ_F in this case also shows how far the system is from equilibrium, but is not an indicator of self-organization. In order to obtain a criterion of self-organization, one must again carry out renormalization to the specified value of mean effective energy. We shall return to this issue in Ch. 21.

15.11.3. *Einstein – Smoluchowski Equation. H–Theorem*

We return to Einstein – Smoluchowski equation (15.9.17). The nonlinear thermostat acts on Brownian particle through the effective force. The stationary solution in the thermostat coincides with the equilibrium solution and is given by (15.9.14). When the coefficient of feedback is zero, $a_f = 0$, the solution is the Boltzmann distribution for harmonic oscillator.

The Lyapunov functional Λ_F is defined by expression similar to (15.11.5):

$$\Lambda_F = F(t) - F_0 = \kappa T \int \ln\frac{f(r,t)}{f_0(r)}\, f(r,t)dr \ge 0 \; . \tag{15.11.14}$$

The time derivative is found with the aid of (15.9.17),

$$\frac{d\Lambda_F}{dt} = -\kappa T \int D_{(r)} f(r,t)\left(\frac{\partial}{\partial r}\ln\frac{f(r,t)}{f_0(r)}\right)^2 dr \le 0 \; . \tag{15.11.15}$$

The Lyapunov functional Λ_F again indicates how far the system is from equilibrium. The criterion of self-organization must be based on a functional defined in terms of the entropy difference.

So, we have demonstrated the feasibility of extending the thermodynamic concept of free energy to a large class of equations which describe Brownian motion in nonlinear media. Notwithstanding a certain artificiality of such definition, Lyapunov functionals Λ_F are very helpful in dealing with many problems, some of which will be discussed in the next chapter.

CHAPTER 16

EXAMPLES OF NONLINEAR BROWNIAN MOTION

In the preceding chapter we discussed methods used for describing nonlinear Brownian motion. Now we are going to illustrate the efficiency of the general theory with concrete examples.

There exists a vast body of literature on the theory of Brownian motion. As indicated in Ch. 15, the foundations of this theory have been laid in the classical papers of Albert Einstein, Paul Langevin and Marian Smoluchowski. The following list is only intended to give the reader some notion of the creators and the scope of this fundamental theory: Pontryagin – Andronov – Vitt 1933; Davydov 1936; Leontovich 1983; Chandrasekhar 1943; Stratonovich 1961, 1967, 1985, 1993; Lifshitz – Pitayevskii 1979; Rytov 1976; Malakhov 1968; Tikhonov 1966; Klyatskin – Tatarskii 1973; Lax 1968; Klimontovich – Kovalev – Landa 1972; Klimontovich III – V, 1991, 1992b; Van Kampen 1981; Akhmanov – D'yakonov – Chirkin 1981; Risken 1984; Haken 1983; Gardiner 1983; Horstemke – Lefever 1984; Zel'dovich – Mikhailov 1987; Mikhailov 1989; Hanggi – Talkner – Borkovec 1990; Moss 1992.

For this chapter we have selected only those examples of nonlinear Brownian motion which are of interest from the standpoint of the general theory (this, of course, does not diminish their own importance). Many of these examples illustrate the difference between traditional and non-traditional approaches to the description of Brownian motion.

16.1. Brownian Motion in Self-Oscillatory Systems. Van der Pol Oscillator

Van der Pol oscillator is a classical example of electrical self-oscillatory system. It contains a linear electrical oscillatory circuit to which amplified feedback is applied.

Neglecting fluctuations, the process in oscillator can be described by a set of nonlinear dynamic dissipative equations in charge and current. The "electron structure" of the flow of electric charge is taken into account by inclusion of the source of random emf $\mathcal{E}$ into the dynamic equations (the Langevin source).

Consider an oscillator with soft excitation, when the linear component of the coefficient

300

of friction changes sign when the coefficient of feedback becomes sufficiently large. By electromechanical analogy, we obtain the following set of Langevin equations:

$$\frac{dx}{dt} = v, \quad \frac{dv}{dt} + \left(-\alpha + \beta v^2\right)v + \omega_0^2 x + \frac{1}{2}\frac{dD}{dv} = \sqrt{D(v)}y(t), \tag{16.1.1}$$

where α_f is the coefficient of feedback, γ and β are the coefficients of linear and nonlinear friction, $\alpha = \alpha_f - \gamma$.

Equations (16.1.1) are similar to the Langevin equations (15.2.16) for Brownian particle in the medium with nonlinear friction. This time, however, the steady component of friction may change its sign. Following the guidelines of Sect. 15.2, we go from (16.1.1) to the corresponding Fokker – Planck equation in K–form,

$$\frac{\partial f}{\partial t} + v\frac{\partial f}{\partial r} - \omega_0^2 \frac{\partial f}{\partial v} = \frac{\partial}{\partial v}\left[D(v)\frac{\partial f}{\partial v}\right] + \frac{\partial}{\partial v}\left[\left(-\alpha + \beta v^2\right)vf\right], \tag{16.1.2}$$

which is similar to (15.2.15).

Exact solution of equations for self-oscillations cannot be obtained even in the dynamic regime described by equations (16.1.1) without the Langevin source. It is far more difficult to solve the corresponding Langevin and Fokker – Planck equations. The equations can be considerably simplified, however, when the dissipative parameters are much smaller than the frequency of oscillations,

$$|\alpha|, \ \alpha_f, \ \delta\langle v^2\rangle << \omega_0. \tag{16.1.3}$$

To put it differently, this means that all relaxation times are large compared with the period of oscillations. Then the equations under consideration can be made much simpler by carrying out averaging over the oscillation period $2\pi/\omega_0$. Mathematical aspects of perturbation theory have been worked out in (Krylov – Bogolyubov 1934; Mitropolskii 1971).

Now let us consider an alternative way of describing Brownian motion in Van der Pol oscillator.

We begin with the definition of the coefficient of nonlinear diffusion $D(v)$ (see Sect. 15.2). Recall that the Einstein relation (15.2.4) was established on the basis of the Fokker – Planck equation (15.2.15) in K–form and the condition of existence of equilibrium solution (15.1.7).

Now the situation is different, because the system is open owing to the presence of feedback. For this reason the equilibrium solution only exists under the additional condition $\alpha_f = 0$:

$$f(x,v,\alpha_f=0)=C\exp\left[-\frac{H(x,v)}{\kappa T}\right], \quad H(x,v)=\frac{Mv^2}{2}+\frac{M\omega_0^2}{2}, \tag{16.1.4}$$

where $H(x,v)$ is the Hamilton function of linear oscillatory circuit. By electromechanical analogy, mass M corresponds to inductance L, coordinate x to the charge, etc.

Now we substitute this distribution into (16.1.2) and find that

$$D(v)=\left(\gamma+\beta v^2\right)\frac{\kappa T}{M}=D\left(1+\frac{\beta}{\gamma}v^2\right). \tag{16.1.5}$$

We have once again come to the Einstein relation (15.2.4).

Assume now that the coefficient of feedback is nonzero. Then, putting the "collision integral" in (16.1.2) to zero, we obtain the corresponding stationary solution, which is again represented as the canonical Gibbs distribution:

$$f(x,v,\alpha_f)=\exp\left[-\frac{H_{\text{eff}}}{\kappa T}\right], \quad H_{\text{eff}}=H(x,v)-\frac{\alpha_f}{2\beta}\ln\left(1+\frac{\beta}{\gamma}v^2\right). \tag{16.1.6}$$

If $\alpha_f=0$, the effective Hamilton function coincides with $H(x,v)$ as defined by (16.1.4).

We see that this stationary solution does not satisfy equation (16.1.2), since it does not bring the left-hand side to zero. The situation can be formally improved by redefining the velocity of Brownian particle:

$$v\Rightarrow v_{\text{eff}}=\frac{\partial H_{\text{eff}}}{\partial Mv}=v-\alpha_f\frac{1}{\gamma+\beta v^2}v. \tag{16.1.7}$$

In case of linear oscillator there is no need to renormalize also the coordinate. This becomes necessary, however, when Brownian motion occurs in the presence of both dissipative and nondissipative nonlinearities. A system of this kind will be discussed in Ch. 17. Recall that nondissipative nonlinearity has already been considered in Sect. 15.9 in connection with Brownian motion in bistable elements. Combination of two types of nonlinearities gives rise to many new regimes of Brownian motion, since both equilibrium and nonequilibrium "phase transitions" may take place simultaneously in systems of this kind. Equilibrium phase transitions may, for instance, have a considerable effect on the rates of chemical reactions. One may even speak of a new type of chemical catalysis, when the rate of chemical reaction is controlled by the amount of feedback.

So, along with equation (16.1.2), one can use a different (model!) Fokker – Planck equation:

$$\frac{\partial f}{\partial t} + \frac{\partial H_{eff}}{\partial Mv}\frac{\partial f}{\partial x} - \omega_0^2 x\frac{\partial f}{\partial v} = \frac{\partial}{\partial v}\left[D(v)\frac{\partial f}{\partial v}\right] + \frac{\partial}{\partial v}\left[\left(-\alpha + \beta v^2\right)vf\right], \tag{16.1.8}$$

whose stationary solution is given by (16.1.6).

So, by redefining the velocity according to (16.1.7), we have dramatically changed the structure of the Fokker – Planck equation. The question now is whether this is just a contrivance for simplifying the "established" equation (16.1.2), or there are physical arguments in favor of such redefinition. In the next chapter we shall give evidence supporting the last statement, but now we return to the solution (16.1.6).

Recall the problem of the distribution of velocities of gas particles in the presence of heat sources (Sect. 12.5). Note that after integration with respect to x the distribution (16.1.6) formally coincides with the distribution of particle velocities (12.5.12). The only difference is that the distribution (12.5.12) is three-dimensional.

In Sect. 12.5 we have also analyzed a one-dimensional velocity distribution, which is the projection of three-dimensional distribution on to the plane $v_y = 0$, $v_z = 0$. This one-dimensional distribution coincides with (16.1.6) — of course, with a different interpretation of the process and its parameters.

This allows us to use the ready-made results (12.5.13) – (12.5.20), with some comments on the specific features of our current problem. Distribution (16.1.6) holds for all values of the coefficient of feedback, and defines, in particular, the fluctuations of velocity both at the threshold of generation and much above the critical value.

Using the criterion of S–theorem (see, for example, Sect. 12.7), one can calculate the relative degree of order of different regimes of generation which correspond to different values of α_f. For the reference point it would be natural to take the state of equilibrium with the distribution (16.1.4), which corresponds to $\alpha_f = 0$. Equation similar to (12.7.2),

$$\frac{3}{2}\kappa\tilde{T}(\alpha_f) = \int \frac{Mv^2}{2}C_0\exp\left[-\frac{H(x,v)}{\kappa T(\tilde{\alpha}_f)}\right]dxdv = \int \frac{Mv^2}{2}C\exp\left[-\frac{H_{eff}(x,v)\alpha_f}{\kappa T}\right]dxdv, \tag{16.1.9}$$

is used for renormalizing the equilibrium state to the preassigned value of the mean kinetic energy of oscillations in the state with $\alpha_f > 0$. This equation allows us to find the effective temperature as function of α_f,

$$\tilde{T} = \tilde{T}(\alpha_f) \text{ at "initial condition" } \tilde{T}(\alpha_f)\big|_{\alpha_f=0} = T . \tag{16.1.10}$$

Since the maximum entropy corresponds to the state of equilibrium given that the values of mean energy are the same, the degree of order for all states with $\alpha_f > 0$ is higher than that of the equilibrium state with $\alpha_f = 0$. Equation (16.1.9) also allows one to check whether the degree of order increases monotonically with increasing α_f.

Quantitative assessment of the relative degree of order of different states is based on the equation similar to (12.7.2). We shall discuss this issue in Sect. 16.6.

16.2. Van der Pol Oscillator. Symmetrized Nonlinearity

In some cases (for instance, in the theory of solid state lasers) the process of generation is described by equations in x and v which are symmetrical with respect to dissipative nonlinearity:

$$\frac{dx}{dt} + \frac{1}{2}(-\alpha + \beta E)x = v , \quad \frac{dv}{dt} + \frac{1}{2}(-\alpha + \beta E)v + \omega_0^2 x = 0 , \tag{16.2.1}$$

where

$$E = \frac{M}{2}\left(v^2 + \omega_0^2 x^2\right) \equiv H(x,v) \tag{16.2.2}$$

is the energy of oscillations.

An exact equation in E (without averaging over the period of oscillations) follows from (16.2.1):

$$\frac{dE}{dt} = (\alpha - \beta b E)E , \quad \alpha = \alpha_f - \gamma . \tag{16.2.3}$$

The solution of these equations can be expressed in terms of fixed values of energy E_0 and phase ϕ_0 (we assume that $M = 1$):

$$x(t) = \omega_0^{-1}\sqrt{2E(t)}\cos(\omega_0 t + \phi_0) ,$$

$$v(t) = -\sqrt{2E(t)}\sin(\omega_0 t + \phi_0) ; \tag{16.2.4}$$

$$E(t) = \frac{E_0 \dfrac{\alpha}{\beta}}{E_0 - \left(E_0 - \dfrac{\alpha}{\beta}\right)e^{-\alpha t}} ; \quad \text{and} \quad E(t) = \frac{E_0}{1 + E_0 \beta t} \quad \text{if } \alpha = 0 . \tag{16.2.5}$$

The character of time evolution depends on the sign of α. At $\alpha < 0$ the system is at rest with $E = 0$; at $\alpha > 0$ steady oscillations are established with frequency ω_0 and energy of the limiting cycle $E = a/b$. Thus, the value of $\alpha = 0$ corresponds to the bifurcation point.

The same expression also defines the relaxation time as function of the deviation from the bifurcation point. At a finite distance from the point of bifurcation, the system approaches either of the two stationary states according to exponential law with the relaxation time $\tau_{rel} \sim 1/|\alpha|$. Closer to the bifurcation point the exponential dependence is replaced by power law. In the limit of $\alpha = 0$ we have $E(t) \propto 1/t$, and the dependence on E_0 disappears.

Fluctuations are taken into consideration by introducing Langevin sources into these equations, and establishing the appropriate form of the Fokker – Planck equation. Like in Ch. 15, we give preference to the K–form of the Fokker – Planck equation.

We shall use the same equations as those developed for describing Brownian motion with nonlinear friction. By analogy with (15.2.16), we write the Langevin equation for the energy as

$$\frac{dE}{dt} + (-\alpha + \beta E)E - \frac{1}{2}\frac{d}{dE}[D(E)E] = [D(E)E]^{\frac{1}{2}}y(t) . \tag{16.2.6}$$

The moments of the Langevin source $y(t)$ are given by

$$\langle y(t)\rangle = 0 , \quad \langle y(t)y(t')\rangle = 2\delta(t-t') . \tag{16.2.7}$$

The only difference is that in the definition of the coefficient of diffusion we factor out the quantity E, which comes up in the transition to polar coordinates and remains in case of constant diffusion.

Transition to the equation for distribution function $f(E,t)$ follows the guidelines set forth in Sect. 15.2, and results in

$$\frac{\partial f(E,t)}{\partial t} = \frac{\partial}{\partial E}\left[D(E)E\frac{\partial f}{\partial E}\right] + \frac{\partial}{\partial E}[(-\alpha + \beta E)Ef] . \tag{16.2.8}$$

We can also write the equation for the mean energy,

$$\frac{d\langle E\rangle}{dt} + \langle(-\alpha + \beta E)E\rangle = \left\langle\frac{dD(E)E}{dE}\right\rangle , \tag{16.2.9}$$

which is obviously not closed.

Function $D(E)$ in (16.2.8) is defined from condition that in the absence of feedback the solution should be the Boltzmann distribution. Hence

$$D(E) = D\left(1 + \frac{\beta}{\gamma}E\right), \quad D = \gamma\kappa T \ . \tag{16.2.10}$$

The stationary solution of (16.2.8) is again represented as the canonical Gibbs distribution

$$f_0(E) = C\exp\left[-\frac{H_{\text{eff}}(E, \alpha_f)}{\kappa T}\right] \tag{16.2.11}$$

with the effective Hamilton function

$$H_{\text{eff}}(E, \alpha_f) = E - \frac{\alpha_f}{\beta}\ln\left(1 + \frac{\beta E}{\gamma}\right) \ .$$

At $\alpha_f = 0$ this solution coincides with the Boltzmann distribution.

Diffusion $D(E)$ reflects the "atomicity" of charge transfer. Often employed is the approximation of predetermined noise: $D(E) = D$. In this approximation the Einstein relation $D = \gamma\kappa T$ does not hold, and so the stationary solution is

$$f(E, \alpha) = \exp\frac{F - H_{\text{eff}}}{D/\gamma} \ , \quad H_{\text{eff}}(E) = -\alpha E + \frac{1}{2}\beta E^2 \ , \tag{16.2.12}$$

where F and D/γ are, respectively, the effective free energy and the effective temperature.

Let us compare the stationary distributions corresponding to the different definitions of Langevin source (and hence to the different definitions of coefficient of diffusion).

When the coefficient of diffusion is a function of energy, the distribution function falls off more slowly at high energies. The "tails" of distributions are therefore different. At the same time, the locations of maxima coincide:

$$E_{\text{max}} = \alpha/\beta \ \text{if} \ \alpha > 0 \ , \quad \text{and} \ E_{\text{max}} = 0 \ \text{if} \ \alpha = 0 \ . \tag{16.2.13}$$

It is interesting to compare the two distributions in the domain of Gaussian approximation. For the region of well-developed generation we may rewrite (16.2.12) as

$$f = \sqrt{\frac{\beta}{2\pi D}}\exp\left[-\frac{\left(E - \frac{\alpha}{\beta}\right)^2}{\frac{2D}{\beta}}\right]; \quad \varepsilon = \frac{D\beta}{\alpha^2} \ll 1 \ , \tag{16.2.14}$$

where ε is a dimensionless parameter which is small in the regime of well-developed generation. Then the value of the distribution function at $E = 0$ can be assumed to be zero. Then the expression for the relative dispersion of energy is

$$\frac{\left\langle (\delta E)^2 \right\rangle}{\langle E \rangle^2} = \frac{D\beta}{\alpha^2} \equiv \varepsilon \ll 1 .$$ (16.2.15)

To bring (16.2.11) to the form of Gaussian distribution, we expand the exponent in $(E - E_{\max})$ and retain the first two derivatives. As a result, we arrive at the distribution similar to (16.2.14); the relative dispersion, however, is now given by

$$\frac{\left\langle (\delta E)^2 \right\rangle}{\langle E \rangle^2} = \frac{D\beta}{\alpha^2} \frac{\gamma + \alpha}{\gamma} \equiv \varepsilon \frac{\gamma + \alpha}{\gamma} \ll 1 .$$ (16.2.16)

Recall that for natural fluctuations we have $D = \gamma \kappa T$.

We see that even in the Gaussian approximation the energy-dependent coefficient of diffusion alters the expression for the relative dispersion. Expression (16.2.16) only coincides with (16.2.15) when $\alpha \ll \gamma$. Accordingly, the applicability of expressions (16.2.14), (16.2.15) is restricted by two inequalities:

$$\varepsilon = \frac{D\beta}{\alpha^2} \ll 1 , \quad \alpha \ll \gamma .$$ (16.2.17)

This means that we are in the domain of well-developed generation, but not far from the threshold.

A similar two-way inequality defines the limits of applicability of Landau's theory of phase transitions (Landau – Lifshitz 1976; Patashinskii – Pokrovskii 1982). Later on we shall return to this problem.

16.3. Combined Action of Natural and External Noise

Of special practical interest is the case when the system experiences the action of both natural noise (via the function $D(E)$) and external noise. The following example shows that it is necessary to take the natural noise into account even when the external noise is strong (Klimontovich 1990a).

We return to the Langevin equation (16.2.6) and assume that the value of coefficient of feedback fluctuates under the action of external source,

$$\alpha_f \Rightarrow \alpha_f + \sqrt{\sigma} y(t) , \tag{16.3.1}$$

where σ is the intensity of parametric noise. The moments of random function $y(t)$ are given by (16.2.7). Then the Langevin equation becomes

$$\frac{dE}{dt} + (-\alpha + \beta E)E - \frac{1}{2}\frac{d}{dE}\left[\sqrt{D(E)E} + \sqrt{\sigma}E\right]^2 = \left[\sqrt{D(E)E} + \sqrt{\sigma}E\right]y(t) . \tag{16.3.2}$$

Following the scheme of Sect. 15.2, we find the corresponding Fokker – Planck equation:

$$\frac{\partial f(E,t)}{\partial t} = \frac{\partial}{\partial E}\left\{\left[\sqrt{D(E)E} + \sqrt{\sigma}E\right]^2 \frac{\partial f}{\partial E}\right\} + \frac{\partial}{\partial E}\left[(-\alpha + \beta E)Ef\right] , \tag{16.3.3}$$

which coincides with (16.2.8) if $\sigma = 0$.

The stationary solution of this equation is

$$f(E) = C\exp\left(\int_0^E \frac{\alpha - \beta E'}{\left[\sqrt{D(E')} + \sqrt{\sigma}E'\right]^2}dE'\right) , \quad D(E) = D\left(1 + \frac{\beta}{\gamma}E\right) , \tag{16.3.4}$$

and coincides with (16.2.12) if external noise is absent ($\sigma = 0$). If, on the contrary, the source of natural fluctuations is switched off ($D = 0$), the solution (16.3.4) becomes

$$f(E) = C\exp\left(\int_0^E \frac{\alpha - \beta E'}{\sigma E'}dE'\right) , \quad \int_0^\infty f(E)dE = 1 . \tag{16.3.5}$$

We see that the integral in the exponent contains logarithmic divergence at small values of E. At the same time, in the general expression (16.3.4) the denominator is finite even when $E = 0$ owing to the presence of natural noise, and the integral does not diverge. This proves that the inclusion of natural noise (even small) may alter dramatically the behavior of distribution and lead to a physically correct result. In this way, the known difficulty of calculation of fluctuations in oscillator with fluctuating parameters is overcome.

16.4. Symmetrized Oscillator. Distribution of Coordinates and Velocities

Let us now establish the more general Fokker – Planck equation for distribution function $f(x,v)$. For this purpose we introduce the appropriate Langevin sources into the dynamic equations (16.2.1). Recall that energy E is expressed in terms of x, v by (16.2.2). Once again we use the method of Sect. 15.2, getting as a result the following Fokker – Planck equation:

$$\frac{\partial f(x,v,t)}{\partial t} + v\frac{\partial f}{\partial x} - \omega_0^2 x\frac{\partial f}{\partial v} = \frac{1}{2}\left\{\frac{\partial}{\partial v}\left[D(E)\frac{\partial f}{\partial v}\right] + \frac{1}{\omega_0^2}\frac{\partial}{\partial x}\left[D(E)\frac{\partial f}{\partial x}\right]\right\}$$

$$+ \frac{1}{2}\left\{\frac{\partial}{\partial v}[(-\alpha + \beta E)vf] + \frac{\partial}{\partial x}[(-\alpha + \beta E)xf]\right\}. \qquad (16.4.1)$$

To go over from (16.4.1) to our former equation (16.2.8) we must use the relationship between distribution functions $f(E,t)$ and $f(x,v,t)$.

The nonlinear coefficient of diffusion is found from the condition of existence of Gibbs distribution for oscillator when $\alpha_f = 0$; the resulting expression is similar to (16.2.10):

$$D(E) = D\left(1 + \frac{\beta}{\gamma}E\right); \quad E = \frac{1}{2}\left(v^2 + \omega_0^2 x^2\right). \qquad (16.4.2)$$

The stationary solution of (16.4.1) is

$$f_0(x,v,\alpha_f) = C\exp\left(-\frac{H_{\text{eff}}(x,v)}{\kappa T}\right), \quad H_{\text{eff}} = E - \frac{\alpha_f}{\beta}\ln\left(1 + \frac{\beta}{\gamma}E\right), \qquad (16.4.3)$$

and coincides with the canonical Gibbs distribution when $\alpha_f = 0$.

16.5. H-Theorem for Van der Pol Oscillator

We return to Sect. 12.4, 15.11.1, 15.11.2, and represent distribution (16.4.3) in the form similar to (12.4.11):

$$f_0(x,v,\alpha_f) = \exp\frac{F_{\text{eff}} - H_{\text{eff}}(x,v,\alpha_f)}{\kappa T} \qquad (16.5.1)$$

(this time we are dealing with the true temperature of the thermostat rather than with the effective temperature). Like in (12.4.12) or (15.11.5), we introduce a functional defined as the difference between the free energy of the state at time t and the free energy of the stationary state:

$$\Lambda_F(t) = F(t) - F_{\text{eff}} = \kappa T \int \ln \frac{f(x,v,t)}{f_0(x,v,\alpha_{\text{f}})} f(x,v,t)dxdv \geq 0 \ . \tag{16.5.2}$$

With the aid of Fokker – Planck equation (16.4.1) we obtain the derivative of this functional, and find it to be similar to (15.11.13). Hence, Λ_F is a Lyapunov functional.

Let us quote the results for the simpler equation (16.2.8), which will be required in Sect. 16.10 for evaluation of the maximum permissible discrete time step in the numerical solution of Fokker – Planck equation. Lyapunov functional satisfies two inequalities:

$$\Lambda_F(t) = F(t) - F_0(T,\alpha_{\text{f}}) = \kappa T \int_0^\infty \ln \frac{f(E,t)}{f_0(E,\alpha_{\text{f}})} f(E,t)dE \geq 0 \ , \tag{16.5.3}$$

$$\frac{d\Lambda_F}{dt} = -\kappa T \int_0^\infty D(E)Ef\left(\frac{\partial}{\partial E}\ln\frac{f(E,t)}{f_0(E,\alpha_{\text{f}})}\right)^2 dE \equiv -\sigma_F \leq 0 \ , \tag{16.5.4}$$

where σ_F is the analog of entropy production. The stationary solution for f_0 is given by (16.2.11).

In this way, Lyapunov functional is defined as the difference in nonequilibrium free energies. As indicated elsewhere (see, for instance, Sect. 15.11.2), the functional Λ_F can serve as a measure of deviation from the stationary state. However, this information is in general not sufficient for deciding whether the evolution in question is a process of self-organization. This problem can only be solved with the aid of functional based on the difference in the entropies of the stationary and the current nonequilibrium states.

In order to define Lyapunov functional Λ_S in case of evolution described in terms of distribution function $f(E,t)$, we must go over to renormalized distribution $\tilde{f}(E,t)$. Renormalization is carried out under assumption that the mean effective Hamilton function (16.2.11) remains constant in the course of time evolution towards the stationary state. This condition may be expressed as

$$\int_0^\infty H_{\text{eff}}(E,\alpha_{\text{f}})\tilde{f}(E,t)dE = \int_0^\infty H_{\text{eff}}(E,\alpha_{\text{f}})f_0(E,\alpha_{\text{f}})dE \ . \tag{16.5.5}$$

Naturally, this condition is not satisfied in the course of time evolution towards the stationary state according to Fokker – Planck equation (16.2.8). In order to comply with this condition, we must introduce the new temperature $\tilde{T}$ which is a functional of the distribution $\tilde{f}(E,t)$:

$$\tilde{T}\{\tilde{f}\} = \frac{\int \frac{(-\alpha + \beta E)^2 E}{\gamma + \beta E} \tilde{f} dE}{\int \frac{d}{dE}[(-\alpha + \beta E)E]\tilde{f} dE} . \tag{16.5.6}$$

The renormalized distribution satisfies the equation which is nonlinear in $\tilde{f}$:

$$\frac{\partial \tilde{f}}{\partial t} = \frac{\partial}{\partial E}\left[\tilde{D}(E)E\frac{\partial \tilde{f}}{\partial E}\right] + \frac{\partial}{\partial E}[(-\alpha + \beta E)E\tilde{f}] , \quad \tilde{D}(E) = \kappa \tilde{T}\{\tilde{f}\}(\gamma + \beta E) . \tag{16.5.7}$$

The stationary solution of this equation coincides with (16.2.11), and the temperature $\tilde{T}$ is of course the same as the thermostat temperature T.

After renormalization we may introduce the functional defined by the difference in entropies,

$$\Lambda_S = S_0 - \tilde{S}(t) = \kappa \int_0^\infty \ln\frac{\tilde{f}(E,t)}{f_0(E,\alpha_f)}\tilde{f}(E,t)dE \geq 0 . \tag{16.5.8}$$

Using (16.5.7), one can prove that this functional decreases monotonically in the course of time evolution towards the stationary state:

$$\frac{d\Lambda_S}{dt} = -\kappa \int_0^\infty \tilde{D}(E)E\tilde{f}\left(\frac{\partial}{\partial E}\ln\frac{\tilde{f}(E,t)}{f_0(E,\alpha_f)}\right)^2 dE \equiv -\sigma \leq 0 . \tag{16.5.9}$$

The last two inequalities indicate that Λ_S is a Lyapunov functional. They also prove that the stationary state is stable at any value of the feedback parameter α_f.

In this way we have proved H–theorem for equation (16.5.7), which is similar to Boltzmann's H–theorem for rarefied gas. There is, however, one question.

Lyapunov functional Λ_S for Boltzmann equation is a natural characteristic, since the mean energy is conserved in the course of evolution towards equilibrium. Currently, however, the situation is quite different: to make condition (16.5.5) hold, we had to change the structure of the initial equation and to replace (16.2.8) by (16.5.7). It follows that we have proved the H–theorem for a different system. Is it of any value for the original problem? Or should we put up with the results for the functional Λ_F? Such questions are certainly not without reason. For the example under consideration, however, the Lyapunov functional Λ_S is actually relevant to the initial equation (16.2.8), since it includes, for instance, its stationary solution f_0. Because of this, the analysis of the relative degree of order based on H–theorem is useful. We shall illustrate this clearly in the next section with the example of evolution of stationary states as the parameter of feedback is varied.

16.6. Self-Organization in Van der Pol Oscillator. S–Theorem

In Sect. 1.6.6 and 12.7 – 12.9 we have formulated the criterion of the relative degree of order of states of open systems in the form of S–theorem. Now we shall apply this criterion for the particular case of evolution of stationary states of Van der Pol generator as the parameter of feedback (the controlling parameter) is varied.

So, let us return to the stationary solution (16.2.11) and write it out for three selected states:

1. *Feedback parameter is zero*, $\alpha_f = 0$. The stationary solution (16.2.11) is then the Boltzmann distribution:

$$f_{(1)} = \frac{1}{\kappa T} \exp\left(-\frac{E}{\kappa T}\right) .$$

(16.6.1)

2. *Generation threshold*, $\alpha_f = \gamma$. We also assume that the nonlinearity is small. Then we can expand (16.2.11) in $\kappa T \beta/\gamma$, and find that

$$f_{(2)} = \sqrt{\frac{2\beta}{\pi\gamma\kappa T}} \exp\left(-\frac{\beta E^2}{2\gamma\kappa T}\right) .$$

(16.6.2)

3. *Regime of well-developed generation*. In this case we may use the Gaussian distribution

$$f_{(3)} = \sqrt{\frac{1}{2\pi\langle(\delta E)^2\rangle}} \exp\left(-\frac{\left(E - \frac{\alpha}{\beta}\right)^2}{2\langle(\delta E)^2\rangle}\right) , \quad \langle(\delta E)^2\rangle = \frac{D}{\beta} .$$

(16.6.3)

The relative dispersion of energy is given by (16.2.16). From the above distributions we find the mean energy values,

$$\langle E\rangle_{(1)} = \kappa T , \quad \langle E\rangle_{(2)} = \sqrt{\frac{2}{\pi}\frac{\gamma\kappa T}{\beta}} , \quad \langle E\rangle_{(3)} = \frac{\alpha}{\beta} ,$$

(16.6.4)

and the corresponding entropy values:

$$S_{(1)} = \ln \kappa T + 1 , \quad S_{(2)} = \ln \sqrt{\frac{\pi \gamma \kappa T}{2\beta}} + \frac{1}{2} , \quad S_{(3)} = \ln \sqrt{\frac{2\pi \kappa T}{\beta}} (\gamma + \alpha) + \frac{1}{2} . \quad (16.6.5)$$

Now let us analyze the implications of these results.

First of all, we note that, under our assumption that $\kappa T \beta / \gamma \ll 1$, the entropy increases as the system moves towards the regime of well-developed generation:

$$S_{(1)} < S_{(2)} < S_{(3)} . \quad (16.6.6)$$

This could be interpreted as a decrease in the order of states as the generation develops. Intuitively it is clear, however, that the degree of order should increase. What is wrong?

To clear up the situation we shall compare the values of mean energy for the three selected states. From (16.6.4) it follows that

$$\langle E \rangle_{(1)} < \langle E \rangle_{(2)} < \langle E \rangle_{(3)} . \quad (16.6.7)$$

We see that the mean energy also increases as the generation develops. At the same time, we know that S–theorem must be applied to the states whose mean energy is the same. Accordingly, we have to use an appropriate procedure of renormalization.

In our current example it would be natural to choose state "1" (the state of equilibrium) for the state of "physical chaos". Since the distribution function is then given by (16.6.1), the effective Hamilton function coincides with the energy E:

$$H_{\text{eff}}(E, \alpha_f) = E . \quad (16.6.8)$$

In this way the renormalization is carried out at a preassigned value of mean energy.

In this connection we recall equation (16.1.9), used as the additional condition for renormalizing the velocity distribution so as to comply with the criterion of S–theorem. Now, with due account for (16.6.8), this equation takes on the form

$$\kappa \tilde{T}(\alpha_f) = \int_0^\infty E \tilde{f}_0(E, \alpha_f = 0) dE = \int_0^\infty E f(E, \alpha_f) dE \quad (16.6.9)$$

and allows us to find the effective temperature as a function of the controlling parameter α_f:

$\tilde{T} = \tilde{T}(\alpha_f)$ at "initial condition" $\tilde{T}(\alpha_f)\big|_{\alpha_f=0} = T$.   (16.6.10)

Taking advantage of this result, we now compare state "1" with state "2" at $\langle E \rangle_{(1)} = \langle E \rangle_{(2)}$, and then state "1" with state "3" at $\langle E \rangle_{(1)} = \langle E \rangle_{(3)}$. The renormalized values of temperature and difference in entropies are given by

$$\kappa \tilde{T}_{(1)} = \sqrt{\frac{2}{\pi} \frac{\gamma \kappa T}{\beta}} > \kappa T \ , \quad \tilde{S}_{(1)} - S_{(2)} = \ln\frac{2}{\pi} + \frac{1}{2} > 0 \qquad (16.6.11)$$

for the first pair of states, and by

$$\kappa \tilde{T}_{(1)} = \frac{\alpha}{\beta} > \sqrt{\frac{2}{\pi} \frac{\gamma \kappa T}{\beta}} > \kappa T \ , \quad \tilde{S}_{(1)} - S_{(3)} = \ln\sqrt{\frac{1}{2\pi\varepsilon} \frac{\gamma + \alpha}{\gamma}} > \ln\frac{2}{\pi} + \frac{1}{2} > 0 \qquad (16.6.12)$$

for the second pair. Recall that

$$\varepsilon = \frac{\beta \gamma \kappa T}{\alpha^2} << 1$$

is a characteristic small parameter for the regime of well-developed generation.

From these results for the three selected states it follows that the effective temperature (16.6.9) steadily grows, and the entropy steadily decreases as the feedback parameter α_f is increased. Thus, the disorder added in state "1" by "heating" it to the temperature $\tilde{T}$, transforms in states "2" and "3" into the more ordered motion, since the entropy of these states is less, given that their mean energy is the same. This allows us to conclude that the evolution towards well-developed generation is a process of self-organization. What is more, these results confirm that our choice of the feedback parameter α_f for the controlling parameter is correct.

Note that the S–theorem as a criterion of the relative degree of order was first formulated for the case of Van der Pol oscillator (Klimontovich 1983c).

Shannon entropy and "S–information". Let us return to expressions (16.6.5) which define the values of Shannon entropy (or Shannon information)

$$S \equiv I = -\int \ln f(E) \ f(E) dE \qquad (16.6.13)$$

for the three selected stationary states of Van der Pol oscillator. To draw comparison with the above results based on the criterion of S–theorem, let us compare the values of Shannon entropy for states "3" and "2":

$$S_{(3)} - S_{(2)} \equiv I_{(3)} - I_{(2)} = \ln\left(2\frac{\gamma + \alpha}{\gamma}\right) > 0 , \qquad (16.6.14)$$

which implies that the entropy and the information increase as the generation develops. If this result is to be interpreted as the increase in chaoticity, we challenge both our physical intuition and the criterion of S–theorem.

It is possible, however, to construe this result, in the spirit of information theory, as the information gain: the information about the system increases upon transition to the regime of well-developed generation (Haken 1988). The state of well-developed generation is thus regarded as more informative.

16.7. Oscillator with Inertial Nonlinearity

The dynamic motion in Van der Pol oscillator is two-dimensional, being described by a set of two ordinary differential equations of the first order (16.1.1) without Langevin source. In such systems only the *simple attractors* are possible: the state of rest and the limiting cycle.

Three-dimensional systems may also have the so-called *strange attractors*. This term refers to such regions of phase space where all trajectories are dynamically unstable, and the divergence of trajectories is exponential. Because of this, the Kolmogorov – Krylov – Sinai entropy (K–entropy) is positive (for the definition of K–entropy see Ch. 21).

Quite understandable is the desire to explore the more complicated oscillators, in which the dynamic processes are described by at least three differential equations of the first order. Such systems include, in particular, the so-called oscillators with inertial nonlinearity, first described by K.F. Teodorchik (see Landa 1980; Anishchenko 1990).

The oscillatory circuit of Teodorchik's oscillator includes a thermistor, which gives rise to inertial nonlinearity and, as a consequence, to new regimes of oscillations. This oscillator, however, does not involve strange attractors. In order to produce complex motion, a modified oscillator with inertial nonlinearity was invented (Anishchenko – Astakhov 1983), in which the inertial converter is placed in the feedback circuit. The nature of motion in the oscillator depends considerably on the asymmetry of the nonlinear characteristic of the inertial element. The most "universal" spectrum of bifurcations is observed in such oscillator when the half-wave detector is used.

As a matter of fact, the strange attractor was discovered in the classical paper of E. Lorenz (1963). As a mathematical object, however, it was first analyzed by D. Ruelle and F. Takens (1971).

In connection with the strange attractor it would be interesting to mark the following. "Strange" irregular solutions of nonlinear equations of the same type as Lorenz equations were independently discovered by A. Grasyuk and A. Orayevskii in 1964 (see Orayevskii 1986) in their studies of oscillations in a molecular oscillator. At that time their discovery failed to draw much attention, since complex irregular motions in molecular generators did not seem to be too important. By contrast, for Lorenz the appearance of complex chaotic regimes explained why a long-term weather forecast was practically impossible. The present-day status of the problem of predictability of complex behavior is reported in (Kravtsov 1993).

The Anishchenko – Astakhov oscillator with inertial nonlinearity proved to be a very convenient device for experimental study of complex ("chaotic") behavior in relatively simple dynamic systems. In particular, it was found that as the feedback parameter increases (with fixed value of the inertial parameter), the limiting cycle is surpassed by a sequence of period-doubling bifurcations in accordance with Feigenbaum's scheme (Feigenbaum 1983). Beyond Feigenbaum's critical point, the strange attractor appears with intricately interlaced regions of chaos and order. What is more, there was good agreement between the results of physical and numerical simulations. In recent years, the fluctuation processes ("Brownian motion") in such oscillators, due to both natural and external noise, have been thoroughly investigated (Anishchenko 1990).

Evaluation of the relative degree of order was based on the criterion of S–theorem (Anishchenko – Klimontovich 1984); the entropy was calculated by the above-described scheme. Renormalization of the distribution function was carried out at the given intensity of oscillations (condition similar to (16.6.9)); the range of feedback parameter extended up to the critical Feigenbaum's point.

Calculations indicate that the entropy (renormalized to the preassigned value of intensity of oscillations) decreases as the system approaches the critical point of transition into the domain of strange attractor, in the process of period doubling sequence. This points to the increasing degree of order. In other words, the sequence of period-doubling bifurcations may be regarded as a process of self-organization. More detailed analysis shows that the decrease in entropy is nonmonotonic in the neighborhood of bifurcation points.

For our future discussion it is important that both the sequence of period-doubling bifurcations, and the main features of bifurcations in the domain of strange attractor, can be described on the basis of so-called *logistic equation*

$$x_{n+1} = \alpha(1 - x_n)x_n , \quad 0 \le \alpha \le 4 , \quad 0 \le x \le 1 , \tag{16.7.1}$$

which describes a one-dimensional process in discrete time with unit step. It is also referred to as the *equation of sequence*, because it may describe a sequence of locations of traces of trajectory of the process in question on the secant plane.

In this way, the one-dimensional equation in discrete time emulates the important properties of the system of three differential equations. This becomes possible because the dimensionality of phase space filled up by the trajectory is two with a small fraction.

Logistic equations were first introduced for describing the behavior of biological objects. We shall see that equations of this kind offer the possibility of construction of two- and one-dimensional models of complex motion. This complex motion owes its existence not to the escape of trajectory from the plane into three dimensions, as in case of oscillator with inertial nonlinearity, but to the more sophisticated nonlinearity.

If $3 \leq \alpha \leq 4$, the logistic equation describes a very complicated motion which corresponds to the state of the so-called *dynamic chaos*. This state may be characterized by appropriate distribution functions. The most chaotic state occurs when $\alpha = 4$. In general, the distribution functions $f(x, \alpha)$ can only be found from numerical simulations.

For the most chaotic state ($\alpha = 4$) the distribution function was found analytically by Ulam and Neumann:

$$f(x) = \frac{1}{\pi}(x(1-x))^{-\frac{1}{2}} , \quad \int_0^1 f(x)dx = 1 . \tag{16.7.2}$$

This distribution can be obtained in the following way (Lichtenberg – Liebermann 1982; Neimark – Landa 1992; Klimontovich 1990, 1991). By substituting $x \Rightarrow (x'+1)/2$ at $\alpha = 4$, equation (16.7.1) is reduced to

$$x_{n+1} = 1 - 2x_n^2 , \quad -1 \leq x \leq 1 . \tag{16.7.3}$$

Then, with $x = -\cos \pi y$, (16.7.3) is reduced to the linear equation

$$y_{n+1} = 2y_n , \quad 0 \leq y \leq 1 . \tag{16.7.4}$$

Hence follows a uniform distribution for y: $f(y) = 1$ if $0 \leq y \leq 1$, and the distribution $f(x) = f(y)dy/dx$ coincides with (16.7.2).

16.8. Bifurcations of Energy of Limiting Cycle. Oscillators with Multistable Stationary States

As indicated above, the transition to "dynamic chaos" in oscillator with inertial nonlinearity can be qualitatively described on the basis of logistic equation.

Now we are going to show that the same equation can be used for constructing a model of generalized Van der Pol oscillator, in which a cascade of bifurcations of limiting cycle energy is possible as the parameter of feedback increases (Klimontovich 1985, V).

We start with introducing dimensionless variables. The choice of scales is of course not unambiguous. A natural time interval for an oscillator is its period of oscillations $T = 2\pi/\omega_0$. Accordingly, the dimensionless variables can be defined as follows:

$$t' = \omega_0 t , \quad \alpha' = \alpha/\omega_0 , \quad E' = \beta E/\omega_0 , \quad D' = \beta D/\omega_0^2 = \gamma'\kappa T' . \tag{16.8.1a}$$

In the second case, the time scale is based on the relaxation time $1/\gamma$:

$$t' = \gamma t , \quad \alpha' = \alpha/\gamma , \quad E' = \beta E/\gamma , \quad D' = \beta D/\gamma^2 = \kappa T\beta/\gamma = \kappa T' . \tag{16.8.1b}$$

The form of equation (16.2.3) for energy E' does not depend explicitly on the choice of scales used for transition to dimensionless variables (further on the prime is dropped):

$$\frac{dE}{dt} = (\alpha - E)E . \tag{16.8.2}$$

The distinction becomes important, however, when the differential equation is replaced by equation in differences. The basic time scale will then define the discrete time step; the choice of time scale also affects the calculation of fluctuations.

Let us generalize equation (16.8.2) in such a way as to obtain a sequence of bifurcations of limiting cycle energy with increasing feedback parameter. With this purpose we replace (16.8.2) by equation in discrete time with step $\Delta = 1$, which corresponds to the period of oscillations in case of dimensionless variables (16.8.1a), and to the relaxation time $1/\gamma$ in case of dimensionless variables (16.8.1b). As a result, we come to the logistic equation

$$E_{n+1} = (\alpha + 1)E_n - E_n^2 \equiv F(E_n) , \quad 0 \leq \alpha + 1 \leq 4 , \quad 0 \leq E \leq 4 , \tag{16.8.3}$$

which after kth iteration becomes

$$E_{n+k} = F^{(k)}(E_n) . \tag{16.8.4}$$

To return from (16.8.4) to the differential equation, we define the sequence of derivatives

$$\frac{E_{n+k} - E_n}{k} \Leftrightarrow \frac{dE}{dt} , \quad k = 1,2,\dots . \tag{16.8.5}$$

As a result, we come to the sequence of differential equations

$$\frac{dE}{dt} = \frac{1}{k}\left[F^{(k)}(E) - E\right], \quad k = 1, 2, \dots .$$

(16.8.6)

Let us consider a few first equations in this sequence. At $k = 1$ we return to the initial equation (16.8.2). At $k = 2$ we get a new equation

$$\frac{dE}{dt} = \frac{1}{2}(\alpha - E)E\left[E^2 - (\alpha + 2)E + \alpha + 2\right] .$$

(16.8.7)

This equation admits four stationary states:

$$E_1 = 0 \text{ at } \alpha \leq 0 ; \quad E_2 = \alpha \text{ at } 0 \leq \alpha \leq 2 ;$$

(16.8.8)

$$E_{3,4} = \frac{\alpha + 2}{2} \pm \sqrt{\left(\frac{\alpha + 2}{2}\right)^2 - (\alpha + 2)} \text{ at } \alpha \geq 2 .$$

We see that ramification of values of limiting cycle energy occurs at $\alpha = 2$, and we are dealing with a bistable state. In general, the number of stationary states in the range of values of α up to the critical Feigenbaum's point is 2^k. The possible values of the limiting cycle energy coincide with the values of energy at the stationary points of the logistic equation (16.8.3).

Equations (16.2.1) for functions $x(t)$, $v(t)$ are generalized in a similar way, by making the substitution

$$\alpha - E \Rightarrow \frac{1}{k}\frac{F^{(k)}(E) - E}{E} .$$

(16.8.9)

The solutions for $x(t)$, $v(t)$ are given, as before, by (16.2.4), whereas $E(t)$ must be found from equation (16.8.6).

Now let us consider the relevant Langevin and Fokker – Planck equations, which can be written by analogy with (16.2.6), (16.2.8). The first of these is

$$\frac{dE}{dt} = \frac{1}{k}\left[F^{(k)}(E) - E\right] + \frac{1}{2}\frac{d}{dE}\left[D_{(k)}(E)E\right] + \sqrt{D_{(k)}(E)E}\,y(t) .$$

(16.8.10)

The moments of the source $y(t)$ are given by (16.2.7). The corresponding Fokker – Planck equation can be represented as

$$\frac{\partial f}{\partial t} = \frac{\partial}{\partial E}\left[D_{(k)}(E)E\frac{\partial f}{\partial E}\right] + \frac{\partial}{\partial E}\left[-\frac{1}{k}\left(F^{(k)}(E) - E\right)f\right] . \tag{16.8.11}$$

The nonlinear coefficient of diffusion is again found under assumption that in the absence of feedback ($\alpha_f = 0$) the system occurs at equilibrium, and the solution is the Boltzmann distribution (16.6.1):

$$D_{(k)}(E) = \frac{1}{k}\left[E - F^{(k)}(E)\right]\Big|_{\alpha_f = 0} \; ; \; D_{(1)}(E) = \kappa T'(\gamma' + E) . \tag{16.8.12}$$

In dimensional variables (see (16.8.1a,b)) the expression for $D_{(1)}$ coincides with (16.2.10). Accordingly, equation (16.8.11) with $k = 1$ coincides with the Fokker – Planck equation (16.2.8).

The stationary solution of (16.8.11) is

$$f_0(E, \alpha_f) = C\exp\left[\frac{1}{k}\int_0^\infty \frac{F^{(k)}(E') - E'}{D_{(k)}(E')}dE'\right] . \tag{16.8.13}$$

Naturally, the structure of the distribution becomes more and more complicated as the number of iteration increases, because there are more and more maxima of the distribution. The locations of maxima are determined by the roots of equation

$$E_{max} = F^{(k)}(E_{max}) \tag{16.8.14}$$

and therefore coincide with the locations of stationary points of the logistic equation (16.8.4).

Let us illustrate these results with a simple example, with $k = 2$ and $D_{(k)} = D = \text{const}$. The constant coefficient of diffusion implies that the noise is external (fixed). At the same time, the noise is not parametric, and so there are no such difficulties as we encountered in Sect. 16.3.

Recall that the value of $\alpha = 2$ in the dynamic regime at $k = 2$ is the point of bifurcation, associated with ramification of the energy of the limiting cycle (see (16.8.8)).

For the sake of clarity we single out three particular cases.

1. *Region below the bifurcation point* ($\alpha < 2$). In Gaussian approximation the stationary solution (16.8.13) becomes

$$f_0(E) = \sqrt{\frac{1}{2\pi D}\frac{2-\alpha}{2}} \exp\left[-\frac{1}{2D}\frac{2-\alpha}{2}(E-\alpha)^2\right], \quad 0 \le \alpha < 2. \tag{16.8.15}$$

Here and below the lower limit $E = 0$ in the normalization conditions is replaced by $E = -\infty$, which is justified if $\varepsilon = D\beta/\alpha^2 \ll 1$. From this distribution it follows that on approaching the critical point the dispersion of fluctuations increases as $2/(2-\alpha)$ — that is, according to Curie's law for the nonequilibrium phase transition.

2. *Critical point* ($\alpha = 2$). The stationary solution (16.8.13) then is

$$f_0(E) = \frac{1}{\Gamma\left(\frac14\right)} \sqrt[4]{\frac{2}{D}} \exp\left[-\frac{(E-\alpha)^4}{8D}\right], \quad \alpha = 2. \tag{16.8.16}$$

The relative dispersion is proportional to $\sqrt{\varepsilon}$ rather than to ε.

3. *Region above the bifurcation point* ($\alpha > 2$). The distribution displays two maxima, and the partial Gaussian distributions are

$$f_{1,2} = \sqrt{\frac{\alpha-2}{2\pi D}} \exp\left[-\frac{\alpha-2}{2\pi D}(E-E_{1,2})^2\right], \quad E_{1,2} = 2 \pm \sqrt{\alpha-2}. \tag{16.8.17}$$

On approaching the critical point from above, the dispersion of fluctuations also increases according to Curie's law, but this time with the factor of $1/(\alpha-2)$.

To end this section, we shall demonstrate that the above transition across the critical point is, according to the criterion of S–theorem, a process of self-organization.

There is the entropy jump for the states above and below the critical point in the Gaussian approximation. By S_+ we denote the value of entropy below ($\alpha < 2$), and by S_- above the critical point. Then the difference in entropies in the Gaussian approximation is

$$S_- - S_+ = \ln\sqrt{2} > 0. \tag{16.8.18}$$

So, when the critical point is crossed as α_f increases, the entropy decreases. Since in the neighborhood of the critical point ($|\alpha-2| \ll 1$) the mean energies have the same values above and below, $\langle E \rangle_- = \langle E \rangle_+$, we may use the criterion of S–theorem, which indicates in this case that the state above the critical point is more ordered. This means that the transition into the region of bistability is a process of self-organization.

The role of the parameter of order in this example of nonequilibrium phase transition is played by the difference in the mean energies of the two branches,

$$\eta = \langle E_+ \rangle - \langle E_- \rangle = 2\sqrt{\alpha - 2} \ , \quad \alpha \geq \alpha_{cr} = 2 \ . \tag{16.8.19}$$

We see that at the critical point the parameter of order is zero.

These results have been obtained in the Gaussian approximation. Because of this, they are quite similar to those obtained within the framework of Landau's theory of phase transitions of the second kind.

16.9. Oscillators in Discrete Time. Bifurcations of Energy of Limiting Cycle and Period of Oscillations

Recall that the logistic equation (16.8.3) or its equivalent (16.7.1) can be used for describing in discrete time the transition to the state of dynamic chaos through a sequence of period-doubling bifurcations. Alternatively, the set of differential equations (16.8.6) can be employed for the description of bifurcations of energy of limiting cycle. For the stationary distribution (16.8.13), this sequence of bifurcations corresponds to the appearance of new maxima at those values of E_{max} which coincide with the locations of stationary points of the logistic equation at the kth step of iteration.

At large enough values of k one might expect to observe chaotic behavior in the arrangement of maxima of the stationary distribution in that range of values of α_f which, according to the logistic equation, corresponds to the domain of dynamic chaos.

Now let us find a generalized logistic equation which would simultaneously describe a superposition of two cascades of bifurcations: bifurcations of energy of limiting cycle and bifurcations of period doubling (Klimontovich 1985; Klimontovich – Chetverikov 1987).

Making the substitution $E_{n+k} \Rightarrow E_{n+1}$ we reduce the logistic equation (16.8.4) to the form

$$E_{n+1} = F^{(k)}(E_n) \ , \quad k = 1, 2, \dots \ . \tag{16.9.1}$$

As a result, we get a family of logistic equations. At $k = 1$ we return to the initial equation (16.8.3), whose solution is well known. The relevant bifurcation diagram is shown in Fig. 4a. The sequence of bifurcations of period doubling starts at $\alpha = 2$ and ends at the critical point at $\alpha_{cr} = 2.58$. The edge of the broadest window of order appears at $\alpha = \sqrt{8}$. The value of $\alpha = 3$ corresponds to the state of the most developed "dynamic chaos".

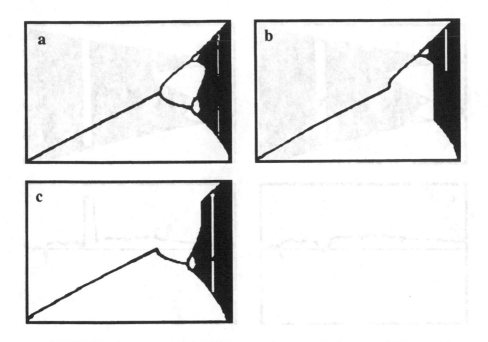

Figure 4: Bifurcation diagrams for the generalized logistic equation (16.9.1) for $k = 1$ (a), and $k = 2$ (b, c), for the values of controlling parameter in the range $0 \leq \alpha \leq (8)^{\frac{1}{2}}$. Diagrams (b) and (c) correspond to different initial conditions (see text).

With $k = 2$ the bifurcation pattern becomes much different. At $\alpha = 2$ we now have the bifurcation of energy of limiting cycle, and bistability arises. Depending on the initial conditions, the system goes to either the upper (Fig. 4b) or the lower (Fig. 4c) branch. The process of period doubling now only starts at $\alpha = \sqrt{6}$, when for the logistic equation the period would already have quadrupled. A "chaos-to-chaos" phase transition takes place at $\alpha = 2.6785$, which results in the state of "dynamic chaos" inherent in the logistic equation (this can be easily seen by superimposing Fig. 4b and Fig. 4c).

Recall that the differential equation (16.8.6) at $k = 2$ describes only the bifurcations of energy values; there are no bifurcations of period doubling.

Finally, let us consider the case of $k = 3$.

For the differential equation (16.8.6) the limiting cycle with $E = \alpha$ is now stable up to $\alpha = \sqrt{8}$, where three stationary states with different energies arise. The bifurcation diagram of the generalized logistic equation (16.9.1) is also changed dramatically. For comparison, the bifurcation diagrams for $k = 1$ and $k = 3$ are shown in Fig. 5a,b. Very clear is the difference between the states within the broadest window of order: Fig. 5a shows the states corresponding to all three energy values, whereas in Fig. 5b we only see that state

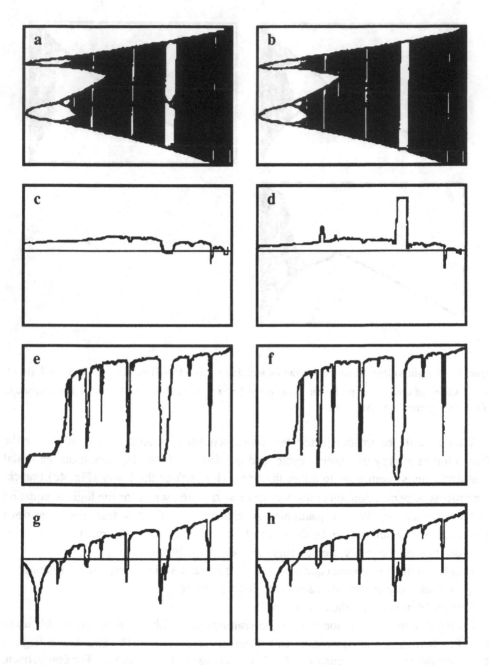

Figure 5: A set of characteristics of the solution of the generalized logistic equation (16.9.1) for $k = 1$ (left column), and $k = 3$ (right column), for the values of controlling parameter in the range $(6)^{1/2} \leq \alpha \leq 3$. For explanations see text.

corresponding to the largest value of energy. This means that only one of the three states at $k = 3$ has the region of attraction large enough to be observable.

To show the extent to which this difference is detected by various criteria of the relative degree of order, in Fig. 5 we have plotted α-dependence of the following quantities: the mean energy $\langle E \rangle$ (Fig. 5c,d), the Shannon entropy

$$S(\alpha) = -\int \ln f(E,\alpha)f(E,\alpha)dE \qquad (16.9.2)$$

(Fig. 5e,f), and the Lyapunov index (Fig. 5g,h) (see Ch. 21). The mean values are calculated from the distribution function $f(E,\alpha)$, constructed on the basis of time realization $E_n(\alpha)$.

These diagrams indicate that functions $\langle E \rangle$ and $S(\alpha)$ better reflect the difference of states in the windows of order than Lyapunov indices. We shall discuss this in detail in Ch. 21 which is devoted to comparison of various criteria of self-organization.

16.10. Criterion of Stability upon Transition to Discrete Time Based on H–Theorem

Let us return to the Fokker – Planck equation (16.2.8) for the distribution function of energy of oscillations. In particular, this equation describes the process of time evolution towards the stationary state. For numerical calculations this equation must be replaced by the appropriate equation in discrete time with step Δ. Denoting by $f(E,n)$ the corresponding distribution function, in place of (16.2.8) we get

$$\frac{f(E,n+\Delta)-f(E,n)}{\Delta} = \frac{\partial}{\partial E}\left[D(E)E\frac{\partial f(E,n)}{\partial E}\right] + \frac{\partial}{\partial E}[(-\alpha+\beta E)Ef(E,n)] . \qquad (16.10.1)$$

Observe that formally the stationary solutions of equations (16.2.8) and (16.10.1) in continuous and discrete time are given by the same expression (16.2.11). The problem is, however, whether these stationary solutions are stable. To answer this question we must analyze the behavior of small deviations from the stationary solution $f_0(E,\alpha_f)$:

$$f_1(E,t) = f(E,t) - f_0(E) , \quad f_1(E,n) = f(E,n) - f_0(E) . \qquad (16.10.2)$$

First we shall do this for the equation in continuous time. We make use of the H–theorem, which for the theory of Brownian motion is expressed by two inequalities (16.5.3), (16.5.4) in the Lyapunov functional Λ_F, defined in terms of the difference in nonequilibrium free energies of the current and the stationary states.

Substituting $f = f_0 + f_1$ into expression (16.5.3) for the difference of free energies, we carry out expansion in f_1 and retain only the main term of the series. Then for the Lyapunov functional Λ_F we find that

$$\frac{d}{dt}(F - F_0) = \frac{\kappa T}{2}\frac{d}{dt}\int \frac{1}{f_0}(f_1)^2 dE \le 0 \ . \tag{16.10.3}$$

Hence it follows that f_1 decreases with time, and the stationary solution of the Fokker – Planck equation (16.2.8) is stable.

Now let us consider the same problem in discrete time, when the Fokker – Planck equation has the form of (16.10.1).

Introduce a factor q which links the values of f_1 at two consecutive instants of discrete time with step Δ:

$$f_1(E, n + \Delta) = q f_1(E, n) \ . \tag{16.10.4}$$

Now we use definition (16.5.4) of "entropy production" $\sigma_F \ge 0$, which for small f_1 can be represented as

$$\sigma_F = \kappa T \int\limits_0^\infty \frac{DE}{f_0}\left(\frac{\partial f_1}{\partial E}\right)^2 dE \ge 0 \ . \tag{16.10.5}$$

The time derivative in (16.10.3) we replace by finite difference,

$$\frac{d}{dt}(F - F_0) = \kappa T \int \frac{f_1}{f_0}\frac{df_1}{dt} dE \Rightarrow \kappa T \int \frac{f_1(E, n)}{f_0}\frac{f_1(E, n + \Delta) - f_1(E, n)}{\Delta} dE$$

$$= \frac{\kappa T}{\Delta}(q - 1)\int \frac{f_1^2(E, n)}{f_0} dE = -\sigma_F \le 0 \ , \tag{16.10.6}$$

where q is as defined by (16.10.4).

From condition of stability, which requires that $|f_1|$ should decrease as the discrete time goes, it follows that $|q|$ should be less than one. Then from (16.10.6) follows the necessary and sufficient condition of stability

$$|q| = \left|1 - \Delta \frac{\sigma_F}{\kappa T \int \frac{1}{f_0} f_1^2 dE}\right| \leq 1 , \qquad (16.10.7)$$

which can be written in a more concise form in terms of Lyapunov functional Λ_F:

$$|q| = \left|1 - \frac{1}{2}\Delta \frac{\sigma_F}{\Lambda_F}\right| \leq 1 , \quad \Lambda_F = \frac{\kappa T}{2} \int \frac{1}{f_0} f_1^2 dE . \qquad (16.10.8)$$

We see that the criterion of stability of the stationary solution is expressed in terms of the general thermodynamic characteristics of nonequilibrium process: the analog of "entropy production" σ_F and the Lyapunov functional Λ_F.

From (16.10.8) we find that the region of stability is delimited by the two-way inequality

$$0 \leq \frac{1}{2}\Delta \frac{\sigma_F}{\Lambda_F} \leq 2 , \quad \Delta_{max} = 4\frac{\Lambda_F}{\sigma_F} . \qquad (16.10.9)$$

We see that it is always possible to make the step of discrete time small enough to ensure stability of the stationary distribution with respect to small deviation of arbitrary form.

To particularize this criterion of stability, let us specify the form of nonequilibrium distribution. We assume that the nonequilibrium distribution can be obtained from the stationary distribution by varying the parameter of feedback (the controlling parameter). Then the effective Hamilton function in (16.2.11) must be replaced: $H_{eff} \Rightarrow +\delta\alpha_f E$. In the simplest case, when the coefficient of diffusion is constant (see (16.2.12)), and the variation $\delta\alpha_f$ is small, the values of σ_F and Λ_F can be defined as

$$\sigma_F = \gamma(\delta\alpha_f)^2 \langle E \rangle , \quad \Lambda_F = (\delta\alpha_f)^2 \frac{\langle (\delta E)^2 \rangle}{2(\kappa T)} , \qquad (16.10.10)$$

and are thus expressed in terms of the mean energy and the dispersion of energy for the stationary distribution. The condition of stability then becomes

$$0 \leq \Delta\gamma\kappa T \frac{\langle E \rangle}{\langle (\delta E)^2 \rangle} \leq 2 . \qquad (16.10.11)$$

In particular, hence follows the value of the largest permissible time step

$$\Delta_{max} = \frac{2}{\gamma \kappa T} \frac{\langle (\delta E)^2 \rangle}{\langle E \rangle} \, . \tag{16.10.12}$$

Observe that Δ_{max} implicitly depends on the parameter of feedback via the mean energy and the dispersion. For the regime of well-developed generation these results can be reduced to a very simple form with the aid of (16.2.15):

$$0 \le \Delta \alpha \le 2 \, , \ \Delta_{max} = \frac{2}{\alpha} \, . \tag{16.10.13}$$

These expressions do not contain either the temperature or the coefficient of diffusion, because they pertain to the region of well-developed generation, which corresponds to the limit of $T \to 0$ when the dynamic distribution holds good. In our present case the distribution follows from (16.2.14) and has the form

$$f_0(E, \alpha_f) = \delta \left(E - \frac{\alpha}{\beta} \right), \ \langle E \rangle = \frac{\alpha}{\beta} \, . \tag{16.10.14}$$

These results can also be used in a different way. Namely, if we define the value of time step (for instance, by setting it equal to one, $\Delta = 1$), from (16.10.13) we find that the stationary solution is stable when

$$0 \le \alpha \le 2 \, , \ \alpha_{max} = 2 \, . \tag{16.10.15}$$

This condition coincides with the condition of stability of the limiting cycle, when the evolution in discrete time is described by the logistic equation (16.8.3). This is natural, because we started with the Fokker – Planck equation (16.2.8), from which we obtained equation (16.10.1) in discrete time.

Now we can redo all our calculations, and consider the more general Fokker – Planck equation (16.8.11) for the next level with $k = 2$. It corresponds to the stationary solution (16.8.13) with $k = 2$, which describes ramification of the energy of limiting cycle. Bifurcation occurs exactly at $\alpha = 2$, at the limit of stability as defined by (16.10.15).

By way of generalization, we may tackle the problem of stability of the most general stationary distribution (16.8.13) at arbitrary numbers k. The regions in which dynamic distributions hold good will be narrowed as k increases; because of this, the analysis of limits of stability of stationary states on the basis of H–theorem becomes even more important.

Recall that the final expressions (16.10.13) – (16.10.15) have been obtained for the region of well-developed generation. For comparison, let us consider another distinct state, the threshold of generation. This will show us how strong can be the dependence of limits of stability from the value of controlling parameter.

From distribution (16.6.2) it follows that the relative dispersion of energy is of the order of unity, and that $\langle E \rangle \sim \sqrt{D/\beta}$. Given this, from (16.10.12) we find the maximum permissible time step at the generation threshold: $(\Delta_{max})_{th} \sim 1/\sqrt{\beta D}$. Then

$$\frac{(\Delta_{max})_{th}}{\Delta_{max}} \sim \sqrt{\frac{\alpha^2}{\beta D}} \sim \sqrt{\frac{1}{\varepsilon}} \gg 1 , \tag{16.10.16}$$

where ε is the small parameter from the theory of well-developed generation.

We see that upon transition to discrete time the state of well-developed generation loses stability with much smaller values of Δ_{max} than does the state at the generation threshold. Comparing this conclusion with calculations of the relative degree of order of the same states on the basis of S–theorem, we find that the higher organized state is more vulnerable upon transition to discrete time. In other words, it has less "stability margin" than the more chaotic state at the threshold of generation. One might ask whether this is good or bad news. There is no categorical answer to this question. Sometimes the loss of stability may result in a higher organized state, which is good from the standpoint of self-organization. Such was the case with the ramification of values of energy of limiting cycle, and with the period-doubling bifurcations.

If, however, the loss of stability results in transition to a more chaotic state, this is bad as far as self-organization is concerned. In this respect *the higher organized state is more fragile*. We shall discuss more of this in Ch. 21 devoted to criteria of self-organization.

So far we have been considering examples based on various modifications of Van der Pol generator. Now we shall give some examples from other provinces.

16.11. Brownian Motion in Chemically Reacting Systems. Partially Ionized Plasma

Consider an example of Brownian motion when the "Brownian particles" are represented by the fluctuations of density of electrons, ions and atoms in partially ionized plasma. By "chemical reactions" we refer to the processes of ionization (for instance, electron impact ionization), and the processes of recombination involving three charged particles.

Theory of such fluctuations can be based on the set of kinetic equations for the distribution functions of electrons, ions and atoms (Klimontovich III, IV). The theory of kinetic fluctuations is built according to the scheme developed in Ch. 9, 10 above for Boltzmann gas: in particular, one may define the appropriate Langevin sources in the kinetic equations for the distributions of electrons, ions and atoms.

Relaxation towards the equilibrium or (in open systems) the stationary state proceeds, as a rule, in several stages. In many cases, local equilibrium with respect to translational degrees of freedom is achieved at the first stage, followed by the stage when local equilibrium with respect to internal degrees of freedom is established, and finally chemical equilibrium sets in.

Consider the state of plasma after completion of the first two stages of relaxation. Neglecting fluctuations, the evolution towards chemical equilibrium can be described on the basis of equations for the mean densities of electrons n_e, ions n_i, and atoms n_a. Of all possible processes of conversion of particles in a plasma we select just one, the process of ionization by electron impact. By α and β we denote the coefficients of ionization by electrons and recombination of electron and ion in the presence of electron. Then the dynamic equations can be written as

$$\frac{dn_e}{dt} = \alpha n_e n_a - \beta n_e^2 n_a \ , \quad n_e = n_i \ , \quad n_e + n_a = n \ . \tag{16.11.1}$$

The second equation states that the average numbers of electrons and ions are equal, and the third ensures that the total number of charged particles (for instance, electrons and atoms) is constant. At equilibrium, the so-called Saha's formula holds:

$$\frac{n_e n_i}{n_a} = \frac{\alpha}{\beta} \ , \quad \left[\frac{\alpha}{\beta} = \left(\frac{\mu \kappa T}{(2\pi\hbar)^2} \right)^{3/2} \frac{1}{Z} \right] . \tag{16.11.2}$$

In accordance with the scheme developed in Ch. 9, 10 (see Belyi – Klimontovich 1978), we define Langevin sources in the kinetic equations so as to take the atomic structure of "continuous medium" into account. This will allow us to find the relevant Langevin sources in the equations in random functions n_e, n_i, and n_a. The intensity of Langevin source, which acts as the coefficient of diffusion in the corresponding Fokker – Planck equation for the distribution function $f(n,t)$, is

$$D_{n_e} = \frac{1}{2V} \left[\alpha n_e n_a + \beta n_e^2 n_a \right] , \tag{16.11.3}$$

where V is the volume, and thus the particles are assumed to be uniformly distributed in space. Under this condition, two equations in (16.11.1) remain valid also when fluctuations are taken into account. Then (16.11.3) can be rewritten as

$$D_{n_e} = \frac{1}{2V}\left[\alpha n_e(n - n_e) + \beta n_e^3\right] \equiv \frac{1}{2}\left(r_{n_e} + g_{n_e}\right) , \qquad (16.11.4)$$

The coefficient of nonlinear friction follows from the dynamic equation in (16.11.1):

$$A_{n_e} = -\left[\alpha n_e(n - n_e) - \beta n_e^3\right] \equiv r_{n_e} - g_{n_e} , \qquad (16.11.5)$$

To establish correspondence with the structure of the master equation for one-step processes of Sect. 16.8.1, we have again introduced the coefficients of generation and recombination. Then, by analogy with (15.8.17), we may immediately write the Fokker – Planck equation for distribution function $f(n,t)$:

$$\frac{\partial f}{\partial t} = \frac{\partial}{\partial n_e}\left[D_{n_e}\frac{\partial f}{\partial n_e}\right] + \frac{\partial}{\partial n_e}\left[A_{n_e}f\right] , \quad \int f(n_e,t)dn_e = 1 . \qquad (16.11.6)$$

The stationary solution of this equation is

$$f(n_e) = C\exp\left[-\int\limits_0^{n_e} \frac{A(n_e')}{D(n_e')}dn_e'\right] . \qquad (16.11.7)$$

Hence it follows that the maximum of the distribution is defined by the solution of equation

$$A(n_{max}) = -\left[\alpha n_{max}(n - n_{max}) - \beta n_{max}^3\right] = 0 . \qquad (16.11.8)$$

The location of the maximum of the distribution satisfies Saha's formula.

In the Gaussian approximation the dispersion of $\delta n_e = n_e - n_{max}$ is given by

$$\left\langle(\delta n_e)^2\right\rangle = \frac{n_e}{V}\frac{n - n_e}{2n - n_e} . \qquad (16.11.9)$$

The dispersion vanishes in two extreme cases: when the degree of ionization is zero (the gas of neutral particles, $n_e = 0$), and when the plasma is completely ionized ($n_e = n$). This indicates that our source of fluctuations reflects the fact that the acts of chemical transformation are discrete; the dispersion is zero in the above two cases because chemical reactions do not take place.

For small deviations from equilibrium the fluctuation δn satisfies the linear Langevin equation with the coefficient of friction $\lambda = \alpha(2n - n_e)$. The expression for the diffusion coefficient can then be reduced to the standard form of fluctuation-dissipation relation:

$$D = \lambda \langle (\delta n_e)^2 \rangle , \quad \lambda = \alpha(2n - n_e) , \qquad (16.11.10)$$

whence it follows that the coefficient of diffusion also vanishes in the two extreme cases specified above.

Making use of the definitions (16.11.4), (16.11.5), (15.8.5) for the coefficients of generation and recombination, we may write the appropriate master equation. In the context of statistical theory, of the two possible definitions of transition probabilities (15.8.2) and (15.8.11) the latter is more "natural". It is only then that the K–form of the Fokker – Planck equation (16.11.6) follows from the master equation.

In case of master equation it is more natural to use the total numbers of particles rather than the densities:

$$N_e = Vn_e , \quad N_i = Vn_i , \quad N_a = Vn_a , \quad N = Vn . \qquad (16.11.11)$$

In these variables the dynamic equation (16.11.1) becomes

$$\frac{dN_e}{dt} = \alpha_V N_e N_a - \beta_V N_e^2 N_i , \quad \alpha_V = \frac{\alpha}{V} , \quad \beta_V = \frac{\beta}{V^2} . \qquad (16.11.12)$$

Let us also quote the expressions for the dispersion of the number of particles, and Saha's formula:

$$\langle (\delta N_e)^2 \rangle = (N_e)_{\text{max}} \frac{N - (N_e)_{\text{max}}}{2N - (N_e)_{\text{max}}} , \quad \frac{(N_e)_{\text{max}}^2}{N - (N_e)_{\text{max}}} = \frac{\alpha_V}{\beta_V} . \qquad (16.11.13)$$

In Ch. 18 we shall give the kinetic equation for describing the processes of ionization and recombination in the presence of spatial diffusion. Then the volume V will be replaced by the relevant physically infinitesimal volume V_{ph}.

In the presence of external source (for instance, photoionization in a given light field), it is possible to monitor the degree of order of partially ionized plasma as a function of, for instance, the strength of external field.

16.12. Malthus – Verhulst Process

The Malthus – Verhulst model was proposed many years ago for describing the survival of a population, for instance, of bacteria (see Van Kampen 1983; Horstemke – Lefever 1984).

Let N be the number of species in the population. If γ and a are, respectively, the rates of death and birth, and $\beta(N-1)$ is the rate of extinction because of competition between the species, then the process of time evolution at $N \gg 1$ can be described by the following differential equation:

$$\frac{dN}{dt} = (a - \gamma - \beta N)N \equiv (g_N - r_N)N \ . \tag{16.12.1}$$

Here in the definitions of coefficients g_N and r_N we have factored out the common multiplier N. Equation (16.12.1) is similar in structure to equation (16.2.3) for the energy of oscillations in Van der Pol oscillator. Observe that the coefficient of feedback α_f corresponds here to the birth rate a.

In this connection let us recall that the nonlinear coefficient of diffusion (16.2.10) in the Fokker – Planck equation (16.2.8) has been defined from condition that the stationary distribution in the absence of feedback ($\alpha_f = 0$) should coincide with the Boltzmann distribution. Following the same hypothesis, we must now define the nonlinear coefficient of diffusion as

$$D(N) = (\gamma + \beta N)T \ , \tag{16.12.2}$$

where T is the dimensionless counterpart of the temperature, which, as we shall see, can be taken for unity.

Alternatively, one may define the coefficient of diffusion in terms of the coefficients of ionization and recombination by formulas (15.8.5), (16.11.4). In this case, for equation (16.12.1) we have

$$g_N = aN \ , \quad r_N = (\gamma + \beta N) \ , \quad D = (\gamma + \beta N) - \frac{1}{2}a \ . \tag{16.12.3}$$

Thus defined, the coefficient of diffusion depends on the "feedback parameter". As the feedback parameter increases, the system displays a phase transition which results in a stationary state with a nonzero number N. However, the sign of the coefficient of diffusion may change. To avoid this complication, we shall stick to the definition (16.12.2) with $T = 1$. As before, we give preference to the K–form of the Fokker – Planck equation:

$$\frac{\partial f(N,t)}{\partial t} = \frac{\partial}{\partial N}\left[(\gamma + \beta N)N\frac{\partial f}{\partial N}\right] + \frac{\partial}{\partial N}[(-a + \gamma + \beta N)Nf] \ . \tag{16.12.4}$$

If the birth rate is zero, $a = 0$, the stationary distribution takes on the form

$$f_0(N) = \exp(-N) \ , \quad \int f_0 dN = 1 \ . \tag{16.12.5}$$

Since this description holds for $N \gg 1$, the exponential decrease means that the population is actually doomed to extinction.

By analogy with (16.2.11), the general stationary solution can be represented as

$$f_0(N,a) = C \exp\left[-H_{\mathrm{eff}}(N,a)\right] \ , \quad \text{where} \quad H_{\mathrm{eff}} = N - \frac{a}{\beta} \ln\left(1 + \frac{\beta}{\gamma} N\right) \tag{16.12.6}$$

is the effective Hamilton function. The maximum of the distribution is given by

$$N_{\max} = \frac{a - \gamma}{\beta} \ , \tag{16.12.7}$$

which expresses a quite natural requirement: birth rate must be high enough ($a > \gamma$) to ensure viability of the population.

If the relative dispersion of the number of species is small, we may go over from (16.12.6) to the Gaussian distribution by expanding in small deviation $\delta N = N - N_{\max}$. The condition of smallness of the relative dispersion is similar to (16.2.16):

$$\frac{\left\langle (\delta N)^2 \right\rangle}{N_{\max}^2} = \frac{\gamma}{a - \gamma} \frac{1}{N_{\max}} \ . \tag{16.12.8}$$

Accordingly, the Gaussian approximation is justified as long as $a \gg \gamma$, and the rate of extinction because of competition β is low enough.

An appropriate master equation must be used for describing processes with a small number of species. In this case also there are strong arguments which shift the balance in favor of the kinetic representation of master equation (see Sect. 15.8.1, 15.8.2).

Practical applications of the nonlinear theory of Brownian motion are numerous; they are discussed in specialized literature. Our task consisted in elucidating the important new aspects which will be necessary for further development of the theory. Next we are going to discuss the feasibility of unified kinetic description of fast and slow processes in Brownian motion. Then we shall not have to resort to the perturbation theory used in Sect. 15.9 and 15.10 for transition from Fokker – Planck to Einstein – Smoluchowski equation. This problem is similar to that discussed in Ch. 13.

CHAPTER 17

NONLINEAR BROWNIAN MOTION. UNIFIED DESCRIPTION OF KINETIC, HYDRODYNAMIC AND DIFFUSION PROCESSES

In both the kinetic theory and in the theory of Brownian motion we dealt with the transition from the kinetic description in six-dimensional phase space to a more coarse description in the space of three dimensions. In Ch. 11, for example, it was the transition from the Boltzmann equation to the comparatively simple equations of gas dynamics. Then, in the theory of Brownian motion, we went over from Fokker – Planck equation for the distribution function $f(r,v,t)$ to the simpler Einstein – Smoluchowski equation in the spatial distribution $f(r,t)$. As indicated in Sect. 15.10, this transition corresponds to the simplified hydrodynamic description of Brownian motion.

The transitions to simplified description were based on perturbation theory in Knudsen number in the gas theory, or in a similar small parameter in the theory of Brownian motion. In both cases we ran into serious difficulties, and had to consider the feasibility of unified description of kinetic and hydrodynamic processes without using perturbation theory in Knudsen number. In the present chapter we shall deal with the unified description of kinetic, hydrodynamic and diffusion processes in the theory of Brownian motion. Like in case of Boltzmann gas, this requires a more detailed account of the structure of "continuous medium" than this is usually done. The choice of physically infinitesimal scales will depend on the nature of forces acting on the Brownian particle.

Depending on the form of the potential, we distinguish two characteristic cases: motion in the "field of gravity", when the motion is delimited by the boundaries of the system, and motion in a field which constrains the movement of Brownian particles (an example is Brownian motion in the presence of elastic force). We shall begin with the latter case.

17.1. Generalized Kinetic Equation in Theory of Brownian Motion

Let Brownian particles be represented by linear harmonic oscillators. Their interaction with the thermostat gives rise to friction, which is proportional to the velocity multiplied by a constant coefficient γ. Then the Langevin equations have the form of (15.1.2) with $u(r) = M\omega_0^2 r^2/2$; they correspond to the Fokker – Planck equation (15.1.6) for the distribution function $f(r,v,t)$. The transition to the simpler equation (15.9.7) is only possible under the rather severe constraints (15.9.9), (15.9.11), which both express the requirement that the diffusion process should be slow. Here, however, we are dealing not with the diffusion of free Brownian particles within the volume of the system, but rather with the diffusion of oscillator with respect to variable r.

For a more complicated active system with nonlinear friction, like the electric circuit in oscillator, neither of the above methods of transition to Einstein – Smoluchowski equation (15.9.6) or (15.9.7) can be used. Moreover, it is not possible to satisfy condition (15.9.11) even for systems with linear friction (for instance, for an oscillator with weak friction).

In this connection we face the problem of constructing a more general equation for unified description of kinetic, hydrodynamic, and diffusion processes. We have done this for the Boltzmann equation in Ch. 13. With this purpose we had to particularize the structure of "continuous medium", which resulted in an additional term in the kinetic equation which takes care of the spatial diffusion of the distribution function. This term appears because of smoothing over the volume of the "point". Let us apply the same scheme to the case of Brownian motion.

By contrast to Boltzmann gas, the system currently under consideration consists of two components, one of which is the subsystem of noninteracting Brownian particles, and the other the thermostat. Interaction of Brownian particles with the thermostat is associated with two processes. One is friction, described in Sect. 15.1 by the Stokes' force. The force of friction can be calculated without taking the atomic structure of the thermostat into account; the atomic structure, however, must be included in the description of the process of diffusion in the phase space of both velocities and coordinates.

In case of Fokker – Planck equation (15.1.6) we deal with the diffusion in the space of velocities. Let us show that when the structure of "continuous medium" is taken into account, an additional term appears in the kinetic equation which accounts for the spatial diffusion.

We start with introducing physically infinitesimal scales for the medium which hosts Brownian motion ("thermostat"). The characteristic relaxation time with respect to velocities in the kinetic equation (15.1.6) is determined by the coefficient of friction γ, $\tau_{rel}^{(v)} \sim 1/\gamma$. By way of example, assume that the role of thermostat is played by some liquid, and the coefficient of friction γ is found from Stokes' formula (15.1.1). Then

$$\tau_{rel}^{(v)} \sim \frac{1}{\gamma} = \frac{M}{6\pi\eta a} \sim \frac{a^2}{v} \equiv \tau_D(L=a) \equiv \tau_{ph}^{(T)} \ . \tag{17.1.1}$$

Here and below D is the generalized coefficient of diffusion of liquid. For the diffusion of velocity of liquid we have $D = v$. From (17.1.1) we see that the relaxation time of velocity is of the same order as the time of diffusion by viscosity to a distance of the order of the size of Brownian particle a. This scale is the smallest in the hydrodynamics of the "medium", and therefore the relaxation time of velocity can be taken for the physically infinitesimal time scale of the "medium".

Let us define the corresponding displacement of Brownian particle:

$$L_{min}^{(B)} = V_{T_{(B)}} \tau_{ph}^{(T)} \equiv V_{T_{(B)}} \tau_{rel}^{(v)} \; ; \; \frac{L_{min}^{(B)}}{a} \sim \sqrt{\frac{r_0}{a}} << 1 \; , \tag{17.1.2}$$

where $V_{T_{(B)}}$ is the thermal velocity of Brownian particle; radius r_0 is of the order of the size of an atom. From (17.1.2) it follows that over the physically infinitesimal time interval of the "medium" the particle covers a distance which is much smaller than its own size.

The scale introduced in this way can be taken for the common physically infinitesimal scale for the two components, the medium and the system of noninteracting Brownian particle. This allows us to define the coefficient of spatial diffusion of Brownian particles as

$$D_{(B)} \sim \frac{\left(L_{min}^{(B)}\right)^2}{\tau_{ph}^{(T)}} \sim V_{T_{(B)}}^2 \tau_{ph}^{(T)} = \frac{\kappa T}{M \gamma} \; , \tag{17.1.3}$$

which actually is a known expression for the spatial diffusion. Observe that the above expressions are similar to (13.2.4), (13.2.5), (13.3.8) which we have used in Ch. 13 for deriving the generalized kinetic equation for unified description of kinetic and hydrodynamic processes in Boltzmann gas. Because of this, we may write the generalized kinetic equation for Brownian motion in the following form:

$$\frac{\partial f}{\partial t} + v \frac{\partial f}{\partial r} + \frac{1}{M} F(r,t) \frac{\partial f}{\partial v} = \frac{\partial}{\partial v} \left[D_{(v)}(r,t) \frac{\partial f}{\partial v} \right] + \frac{\partial}{\partial v} \left[\gamma(v - u(r,t)) f \right]$$

$$+ \frac{\partial}{\partial r} \left[D_{(B)}(r,t) \frac{\partial f}{\partial r} - \frac{F(r,t)}{\gamma M} f \right] . \tag{17.1.4}$$

In structure this equation is similar to the generalized kinetic equation (13.3.4); it is, however, linear with respect to the distribution function. Indeed, the coefficients of diffusion depend on the local temperature of the medium,

$$D_{(v)}(r,t) = \gamma \frac{\kappa T(r,t)}{M} \; , \; D_{(B)}(r,t) = \frac{\kappa T(r,t)}{\gamma M} \; ; \tag{17.1.5}$$

and $u(r,t)$ is the hydrodynamic velocity of the medium. If the medium which hosts Brownian particles is at equilibrium, then the temperature and the mean velocity are constant.

Assume that the force $F(r,t)$ consists in general of two components,

$$F(r,t) = F_0 - \frac{\partial}{\partial r} U(r,t) \ . \tag{17.1.6}$$

The first component is the external force (like gravity). The nature of the second term may vary; it may be an internal force when, for example, the Brownian particle is an oscillator or a bistable element, for which the potential $U(r)$ can be given by (15.9.12).

If the temperature and the mean velocity are constant, equation (17.1.4) corresponds to the following system of Langevin equations:

$$\frac{dr}{dt} = v + \frac{F(r,t)}{\gamma M} + z(t) \ , \ \langle z(t) \rangle = 0 \ , \ \langle z_i z_j \rangle_{t,t'} = 2D_{(B)}^{(r)} \delta_{ij} \delta(t - t') \ ,$$

$$\tag{17.1.7}$$

$$\frac{dv}{dt} + \gamma(v - u) = \frac{F}{M} + y(t) \ , \ \langle y(t) \rangle = 0 \ , \ \langle y_i y_j \rangle_{t,t'} = 2D_{(v)} \delta_{ij} \delta(t - t') \ .$$

In the first equation we take into account the variation of velocity due to mobility. In the second the friction is determined by the relative velocity. The coefficients of diffusion (the intensities of random kicks) are given by (17.1.5).

Let us particularize equation (17.1.4) for the case of Brownian motion of the system of noninteracting harmonic oscillators when the medium is at rest ($u = 0$) and the temperature T is constant:

$$\frac{\partial f}{\partial t} + v \frac{\partial f}{\partial r} + \omega_0^2 r \frac{\partial f}{\partial v} = \frac{\partial}{\partial v} \left[D_{(v)} \frac{\partial f}{\partial v} \right] + \frac{\partial}{\partial v} [\gamma v f] + \frac{\partial}{\partial r} \left[D_{(B)} \frac{\partial f}{\partial r} - \frac{\omega_0^2 r}{\gamma} f \right] \ . \tag{17.1.8}$$

The coefficients of diffusion are given by (17.1.5) with $T = \text{const}$.

The kinetic equations (17.1.4), (17.1.8) describe two relaxation processes, transition towards the distribution with respect to velocities (Maxwell distribution), and transition towards the distribution with respect to coordinates (Boltzmann distribution). Let us compare the two relaxation times for the case of Brownian motion of harmonic oscillator. From (17.1.8), given that the mean kinetic energy and the mean potential energy are determined by the temperature, we get the following estimates:

$$\tau_{\text{rel}}^{(v)} \sim \frac{1}{\gamma}, \quad \tau_{\text{rel}}^{(r)} \sim \frac{\gamma}{\omega_0^2} . \tag{17.1.9}$$

We see that, under condition (15.9.11), the distribution with respect to velocities is established much sooner than the distribution with respect to coordinates. This facilitates the transition from the Fokker – Planck equation to the simpler Einstein – Smoluchowsky equation (15.9.15). On the contrary, in case of weak damping the inequality is reversed, $\gamma \ll \omega_0$. Then, as we shall see in Sect. 17.3, the division of processes into the fast and the slow is much different.

Now let us return to the more general equation (17.1.4) which describes the motion of Brownian particles in a moving liquid, and consider the conditions of transition to the diffusion equation for Brownian particles.

17.2. Diffusion Description of Brownian Motion

Like we did in case of Boltzmann gas, we select the class of distribution functions such that

$$f(r,v,t) = f(r,|v - u|,t) . \tag{17.2.1}$$

For Boltzmann gas this constraint is justified because the number of particles within a "point" is very large, and the function of smoothing with respect to the volume of the "point" is isotropic. Now the function of smoothing is also isotropic; moreover, within the relaxation time the Brownian particle travels to a distance $L_{\text{min}}^{(B)}$ which is much smaller than the size a of the Brownian particle itself. By V_{min} we denote the volume which corresponds to this length; then the number of particles within this volume is

$$N_{\text{min}} \sim \frac{V_{\text{min}}}{r_0^3} \sim V_{\text{min}} n \sim \left(\frac{a}{r_0}\right)^{3/2} \gg 1 \tag{17.2.2}$$

(the last inequality holds because the Brownian particle is small but macroscopic).

Hydrodynamic variables pertaining to the system of Brownian particles are marked off with sub/superscript (B). From the kinetic equation (17.1.4) with condition (17.2.1), we obtain the following equation for the density $\rho_{(B)}$:

$$\frac{\partial \rho_{(B)}}{\partial t} + \frac{\partial}{\partial t}\left(\rho_{(B)} u\right) = \frac{\partial}{\partial r}\left[D_{(B)} \frac{\partial \rho_{(B)}}{\partial t} - \frac{1}{\gamma} \frac{F}{M} \rho_{(B)}\right] \equiv \frac{\partial j^{(B)}}{\partial r} . \tag{17.2.3}$$

If the "medium" is represented by incompressible fluid, this equation can be somewhat simplified,

$$\frac{\partial \rho_{(B)}}{\partial t} + u \frac{\partial \rho_{(B)}}{\partial t} = \frac{\partial}{\partial r}\left[D_{(B)} \frac{\partial \rho_{(B)}}{\partial t} - \frac{1}{\gamma} \frac{F}{M} \rho_{(B)} \right] \equiv \frac{\partial j^{(B)}}{\partial r} \quad . \tag{17.2.4}$$

These equations describe the diffusion of Brownian particles in the field of velocity $u(r,t)$ of the hydrodynamic flow of the "medium". Equations of this kind are commonly used for describing, for instance, the diffusion of impurity in laminar fluid flows. In turbulent flows the diffusion increases dramatically, and the above equations can only be used as the starting point for the description of the more complicated process of turbulent diffusion. We shall return to these issues in Ch. 22.

If the medium is homogeneous and stationary, we may set $u = 0$ and come to the Einstein – Smoluchowski equation (15.9.7). In this case the flux of density of Brownian particles in the medium is completely determined by two counterflows, of which one is shaped by diffusion, and the other by mobility under the action of force F. At equilibrium the two flows balance out, and the density assumes the Boltzmann distribution.

Recall that, according to (15.1.5), the coefficients of diffusion are proportional to the temperature of the medium. Because of this, equation (17.2.3) must in general be solved together with the complete set of equations of hydrodynamics. Observe that equations (17.2.3), (17.2.4) contain the hydrodynamic functions of the medium ρ, u, T, although out of the three hydrodynamic functions of the system of Brownian particles $\rho_{(B)}$, $u_{(B)}$, $T_{(B)}$ they only include the density. This implies that in this approximation the mean velocity and temperature of Brownian particles adjust themselves to the characteristics of the medium. In other words, the hydrodynamic motion of the medium completely entrains the Brownian particles, so that

$$T_{(B)} = T \ , \quad u_{(B)} = u \ . \tag{17.2.5}$$

This adjustment is only possible if the velocity relaxation time of Brownian particles is much less than the time of relaxation with respect to coordinates. It is this condition that justifies the choice of solution of the kinetic equation in the form of (17.2.1).

Naturally, kinetic equation (17.1.4) can be used for assessing the extent to which the "hypothesis of complete entrainment" is justified. This requires analyzing the equations for the higher moments of distribution function (17.2.1).

Finally, let us recall that we have already considered the hydrodynamic equations for Brownian particles in Ch. 15. There we started, however, with the Fokker – Planck equation (15.1.6) rather than with the generalized kinetic equation (17.1.4). We also assumed that the medium is stationary and occurs at constant temperature, and there was no need to mark off the hydrodynamic functions of Brownian particles with special subscript (B).

Let us compare the two continuity equations (15.10.1) and (17.2.3).

In the first of these, the flux of density of Brownian particles is completely determined by convective transfer ρu. Since the medium is at rest, velocity in this equation is the velocity of Brownian particles with respect to the medium, $u \equiv u_{(B)}$. In order to find the flux of Brownian particles we used the approximate solution (15.10.4), (15.10.5), which implies complete entrainment (in terms of temperature and relative velocity) of Brownian particles by the medium.

In (17.2.3), the same flow of matter $j^{(B)}$ is determined by the sum of three components: convective transfer of Brownian particles by the flow of fluid, diffusion of Brownian particles in the medium, and mobility of Brownian particles under the action of force F.

We have seen a similar difference between equations of gas dynamics obtained from the Boltzmann equation and from the generalized kinetic equation (13.3.4).

17.3. Two Models of Harmonic Oscillators

In case of Brownian motion in liquid, the Brownian particles are small macroscopic particles. We are not concerned with their internal structure. When we consider Brownian motion of oscillators, the situation seems to be as simple: one can imagine a weight on a spring immersed into a thermostat. Brownian motion of movable parts of measuring instruments is also performed in a concerted way by all particles of, say, the needle of galvanometer.

One must, however, be aware of the changes in the description of motion associated with the transition from "macroscopic" to "microscopic" Brownian particles. Let us consider two extreme situations. In the first case the Brownian particle is represented by a collection of free charged particles, like electrons in a plasma or in a conductor. In the second case the Brownian particle is built up of bound charge carriers, like quantum atoms oscillators. These concrete examples will be useful for the description of Brownian motion with nonlinear friction and nonlinear nondissipative force. Moreover, this discussion will pave the way for the study of Brownian motion in nonlinear active media.

17.3.1. Brownian Motion in Plasma Oscillatory Circuit

In 1928 Harry Nyquist proposed to use the appropriate Langevin equations for the calculation of fluctuations in electric circuit:

$$\frac{dq}{dt} = I \ , \ \ L\frac{dI}{dt} + RI + \frac{1}{C}q = \mathcal{E}(t) \ , \tag{17.3.1}$$

where q and I are the charge and the current; L, C and R are the inductance, capacitance and ohmic resistance. The role of the Langevin force is played by random emf $\mathcal{E}$, which reflects the existence of "atomic structure" of electric current and charge. The moments of random source are given by

$$\langle \mathcal{E}(t) \rangle = 0 \ , \ \ \langle \mathcal{E}(t)\mathcal{E}(t') \rangle = 2R\kappa T\delta(t - t') \ , \tag{17.3.2}$$

which are similar to (15.1.4). The second of these is called the Nyquist formula; it implies that the spectral density of random emf is

$$(\mathcal{E}\mathcal{E})_\omega = 2\kappa T \ , \tag{17.3.3}$$

and does not depend on frequency.

The langevin equations can be used in a standard way for finding the spectral density of fluctuations of charge and current. In particular, one can define the mean kinetic and potential energy of equilibrium fluctuations:

$$L\langle I^2 \rangle = \frac{\langle q^2 \rangle}{C} = \kappa T \ . \tag{17.3.4}$$

It follows that the electric circuit which acts as a "macroscopic" Brownian particle has only two degrees of freedom. This means that the motion of its constituent microparticles (electrons) is concerted. To elucidate this concerted action, let us consider the structure of individual elements of our electric circuit.

Assume that the electric circuit is a cylindrical conductor of length l and cross sectional area S. Then the values of L, C, R, and the proper frequency ω_0 are given by

$$L = \frac{nl}{e^2 nS} \ , \ \ R = \frac{ml}{e^2 nS}v \ , \ \ C = \frac{4\pi S}{l} \ , \ \ \omega_0^2 = \frac{4\pi e^2 n}{m} \ , \tag{17.3.5}$$

where v is the frequency of collisions, which defines the electric conductivity of the medium. The current, charge and emf can be expressed in terms of velocity, coordinate and force averaged over the volume of the conductor $V = Sl$:

$$I = enSv_V \ , \ \ q = enSx_V \ , \ \ \mathcal{E} = \frac{lm}{e}\frac{f_V}{m} \equiv \frac{lm}{e}y_V(t) \ . \tag{17.3.6}$$

The averaging procedure is defined in a standard way — like, for instance,

$$x_V(t) = \int x(R,t) \frac{dR}{V} .$$

(17.3.7)

Now we may rewrite Langevin equation (17.3.1) as

$$\frac{dx_V}{dt} = v_V , \quad \frac{dv_V}{dt} + \nu v_V + \omega_0^2 x_V = \frac{f_V}{m} \equiv y_V ;$$

(17.3.8)

whereas the moments of the Langevin force are

$$\langle y_V \rangle = 0 , \quad \langle y_V(t) y_V(t') \rangle = 2 \frac{D}{N} \delta(t - t') , \quad D = v \frac{\kappa T}{m} .$$

(17.3.9)

Let us compare (17.3.9) with (17.3.2). The noise intensity is now multiplied by $1/N$; this is natural, because the functions v_V, x_V now characterize the motion of a collective of N electrons on a per-electron basis. The definition of mean values is changed accordingly: for example,

$$L \langle I^2 \rangle = N m \langle v_V^2 \rangle = \kappa T , \quad \text{whence} \quad \langle v_V^2 \rangle = \frac{1}{N} \frac{\kappa T}{m} .$$

(17.3.10)

Now we may particularize the generalized Fokker – Planck equation (17.1.8) for the case under consideration. In variables I, q the equation for the distribution function $f(I,q,t)$ is

$$\frac{\partial f}{\partial t} + I \frac{\partial f}{\partial q} - \omega_0^2 q \frac{\partial f}{\partial I} = \frac{\partial}{\partial I} \left[D_{(I)} \frac{\partial f}{\partial I} \right] + \frac{\partial}{\partial I} [vIf] + \frac{\partial}{\partial q} \left[D_{(q)} \frac{\partial f}{\partial q} + \frac{\omega_0^2 q}{v} f \right] ,$$

(17.3.11)

where

$$D_{(I)} = v \frac{\kappa T}{L} , \quad D_{(q)} = \frac{\kappa T}{Lv} .$$

The equilibrium solution of this equation has the form

$$f_0(I,q) = C \exp \left[- \frac{H(I,q)}{\kappa T} \right] , \quad H = \frac{LI^2}{2} + \frac{q^2}{2C} .$$

(17.3.12)

Let us also write the corresponding equation in $f(v_V, x_V, t)$:

$$\frac{\partial f}{\partial t} + v\frac{\partial f}{\partial x} - \omega_0^2 x\frac{\partial f}{\partial v} = \frac{\partial}{\partial v}\left[D_{(v)}\frac{\partial f}{\partial v}\right] + \frac{\partial}{\partial v}[vvf] + \frac{\partial}{\partial x}\left[D_{(x)}\frac{\partial f}{\partial x} + \frac{\omega_0^2 x}{v}f\right],$$ (17.3.13)

where

$$D_{(v)} = v\frac{1}{N}\frac{\kappa T}{m} \ , \quad D_{(x)} = \frac{1}{N}\frac{\kappa T}{mv} \ ,$$

(we have dropped subscript V at x_V, v_V). The equilibrium solution of (17.3.13) also has the form of the canonical Gibbs distribution:

$$f_0(x_V, v_V) = C\exp\left[-\frac{H(x_V, v_V)}{\kappa T}\right], \quad H = \frac{Nm}{2}\left(v_V^2 + \omega_0^2 x_V^2\right) .$$ (17.3.14)

Factor N in the expression for the Hamilton function reflects the fact that velocity v_V and coordinate x_V are not the characteristics of an individual electron, but rather the collective variables of the system of N particles.

The proper frequency ω_0, as follows from its definition (15.3.5), is also a collective characteristic, since it is expressed in terms of not only the electron's charge e and mass m, but also the mean concentration of electrons n.

At equilibrium, the dispersions of variables x_V, v_V are given by (see also (17.3.10))

$$m\langle v_V^2\rangle = \frac{1}{N}\kappa T \ , \quad m\omega_0^2\langle x_V^2\rangle = \frac{1}{N}\kappa T \ .$$ (17.3.15)

These definitions include the factor of $1/N$, which is typical of the fluctuations of thermodynamic functions of a system (as opposed to the local thermodynamic functions in, for example, gas dynamics). In this connection let us recall that the mean dispersion of velocity for the one-particle Maxwell distribution does not contain the $1/N$ factor, and $m\langle v^2\rangle = \kappa T$ per degree of freedom. The situation is different for the N–particle Maxwell distribution which is the equilibrium solution of the kinetic Leontovich equation (9.1.1), (9.1.2). From (9.2.2) we may go over to the distribution of velocity

$$v_N = \frac{1}{N}\sum_{1\leq i\leq N}v_i \ ,$$ (17.3.16)

defined as the mean value for the collective of N particles. As a result, we come to the

Gaussian distribution

$$f(v_N) = \left(\frac{Nm}{2\pi\kappa T}\right)^{3/2} \exp\left(-N\frac{mv_N^2}{2\kappa T}\right). \tag{17.3.17}$$

By contrast to the Maxwell distribution, the dispersion of velocity is now smaller by a factor of N. The analogy with (17.3.14) is obvious.

Finally, let us estimate the ratio of the coefficient of damping v (frequency of "collisions") to the proper frequency ω_0. By definition (17.3.5), ω_0 coincides with the Langmuir frequency of electrons ω_e. Via the characteristic velocity (thermal velocity v_T or Fermi velocity v_F under the degeneracy condition), the latter is linked with the Debye radius — for instance, as $v_F = r_D \omega_e$. By l we denote the corresponding free path length; for example, $v_F = vl$. Then

$$\frac{v}{\omega_e} \sim \frac{r_D}{l} \sim \mu , \tag{17.3.18}$$

where μ is the so-called plasma parameter, which is small for rarefied plasma, and for nonideal plasma is of the order of unity. In both cases, condition $v \gg \omega_0$, which is necessary for the transition from the Fokker – Planck equation to the Einstein – Smoluchowski equation for a simpler distribution function, is not satisfied.

Now let us consider an alternative model, in which the electric circuit is represented by a piece of insulator. Then the microscopic elements of the medium in case of classical description can be regarded as atoms oscillators. Recall that the corresponding quantum kinetic equation has been discussed in Sect. 15.7.1. Its classical counterpart is the Fokker – Planck equation for distribution function of the values of energy of oscillations of atoms. Next we are going to consider Brownian motion of the collective of atoms within the sample. We shall also investigate the transition to the "one-particle" description.

17.3.2. Brownian Motion in Dielectric Oscillatory Circuit

Now we represent of "electric circuit" as a cylindrical chunk of dielectric of length L and cross-sectional area S. The dielectric is a system of N atoms oscillators with proper frequency ω_0. Assume that the displacement of electors in atoms is only possible in the direction x; R_i is the position of ith atom in the sample. Then, with due account for the radiation friction as defined by (15.7.3), equation for the local vector of polarization $P(R,t)$ can be written as

$$\frac{\partial^2 P(R,t)}{\partial t^2} + \gamma \frac{\partial P}{\partial t} + \omega_0^2 P = \frac{e^2 n}{m} E(R,t) \ . \tag{17.3.19}$$

Field E is created both by the exposure of the selected atom to the action from the side of other atoms of the sample, and by the fluctuation electromagnetic field. The inclusion of interaction between atoms calls for renormalization of γ and ω_0^2 (Klimontovich III, IV). We assume that such renormalization has already been carried out; then $E(R,t)$ is the equilibrium electromagnetic field.

Variable R enters equation (17.3.19) as a parameter. Because of this, by averaging with respect to the volume of the sample,

$$P_V = e n x_V \ , \quad x_V = \int x(R,t) \frac{dR}{V} \ , \tag{17.3.20}$$

we get Langevin equation for the volume-averaged displacement of electrons bound in atoms. This equation formally coincides with (17.3.8), we only have to replace the frequency of collisions ν by the coefficient of radiation friction:

$$\nu \Rightarrow \gamma = \frac{2e^2 \omega_0^2}{3mc^3} \ . \tag{17.3.21}$$

There is, however, an important distinction. Namely, the frequency ω_0 was formerly defined by (17.3.5), and was a collective characteristic. Now ω_0 is the proper frequency of an individual atom oscillator.

Equation for the distribution function $f(x_V, \nu_V, t)$ coincides (with the replacement (17.3.21)) with the generalized equation (17.3.13); the equilibrium solution is given by (17.3.14). Naturally, the dispersions of volume-averaged variables this time also contain the factor of $1/N$, and the corresponding energies are much less than thermal energy per degree of freedom.

Inasmuch as the system in question is a collection of individual atoms oscillators, a different level of description of Brownian motion is also possible. This problem will be treated in the following section.

17.4. Brownian Motion of Atoms Oscillators

Consider a gas of noninteracting atoms in equilibrium electromagnetic field. Following the Thomson model, we regard the atom as a sphere of radius r_0 which carries a uniformly distributed positive charge. The electron vibrates with respect to the center of the sphere.

The frequency of oscillations is found from the expression for the elastic force

$$eE = -\frac{e^2}{r^3}r = -m\omega_0^2 r \, , \quad \omega_0 = \sqrt{\frac{e^2}{mr^3}} \, . \tag{17.4.1}$$

For the sake of simplicity we consider one-dimensional oscillations; then at equilibrium the mean kinetic and potential energies are given by (cf. (15.7.10))

$$m\langle v^2 \rangle = m\omega_0^2 \langle x^2 \rangle = \kappa T_{\omega_0} = \frac{1}{2}\hbar\omega_0 \coth\frac{\hbar\omega_0}{2\kappa T} \, . \tag{17.4.2}$$

For atoms oscillators at room temperature the $\coth(\hbar\omega_0/2\kappa T)$ is close to one, and the amplitude is therefore defined by the zero-point energy $\hbar\omega_0/2$. Given this, from (17.4.1), (17.4.2) we find the following expressions for the amplitude, frequency and velocity of atom oscillator:

$$x_0 = \frac{\hbar^2}{me^2} \, , \quad \omega_0 = \frac{me^4}{\hbar^3} \, , \quad v_0 = \frac{e^2}{\hbar} \, , \tag{17.4.3}$$

which are known in quantum mechanics as Bohr's parameters.

Oscillations of electron in equilibrium electromagnetic field may be regarded as the motion of a "heavy" particle (electron) in the medium composed of "light" particles (photons). Indeed, the main role is played by the photons with frequency ω_0, and the mass of photon is much less than the mass of electron,

$$m_p = \frac{\hbar\omega_0}{c^2} \ll m \, . \tag{17.4.4}$$

Now we turn to the generalized Fokker – Planck equation (17.3.13). This time it will be the equation in the distribution function of microscopic variables x, v for atom oscillator in equilibrium electromagnetic field:

$$\frac{\partial f}{\partial t} + v\frac{\partial f}{\partial x} - \omega_0^2 x\frac{\partial f}{\partial v} = \frac{\partial}{\partial v}\left[D_{(v)}\frac{\partial f}{\partial v}\right] + \frac{\partial}{\partial v}[\gamma v f] + \frac{\partial}{\partial x}\left[D_{(x)}\frac{\partial f}{\partial x} + \frac{\omega_0^2 x}{\gamma}f\right] \, , \tag{17.4.5}$$

where the coefficients of friction and diffusion are given by

$$\gamma = \frac{2e^2\omega_0^2}{3mc^3} \, , \quad D_{(v)} = \gamma\frac{\kappa T_{\omega_0}}{m} \, , \quad D_{(x)} = \frac{\kappa T_{\omega_0}}{m\gamma} \, . \tag{17.4.6}$$

The quantum thermal factor is defined by Planck's formula (17.4.2). At room temperature in the optical frequency range it reduces to the zero-point energy (we have used this condition before).

Although equation (17.4.5) formally coincides with (17.3.13), it now defines the distribution of microscopic variables. Because of this, there is no $1/N$ factor in the expressions for the coefficients of diffusion. Like in case of transition to any irreversible equation, however, the dissipation is due to elimination of small-scale fluctuations, which do not figure in the final equation, being a kind of "hidden scale". To assess the magnitude of such hidden scales, let us get an estimate for the characteristic time scales of the kinetic equation. For this purpose we evaluate the diffusion times $\tau_{(v)}$, $\tau_{(x)}$ on the scales v_0, x_0 as defined by (17.4.3):

$$\tau_{(v)} \sim \frac{1}{\gamma} , \quad \tau_{(x)} \sim \frac{\gamma}{\omega_0^2} \equiv \frac{1}{\Gamma} \equiv \tau_e \equiv \frac{r_e}{c} , \quad r_e = \frac{e^2}{mc^2} , \tag{17.4.7}$$

where r_e is the classical electron radius, and τ_e is the corresponding "transit time". Given the definitions of γ and ω_0, we arrive at the following hierarchy of time scales:

$$\tau_{(x)} \ll \frac{2\pi}{\omega_0} \equiv T_0 \ll \tau_{(v)} , \quad \text{or} \quad \Gamma \equiv \frac{\omega_0^2}{\gamma} \gg \omega_0 \gg \gamma . \tag{17.4.8}$$

The "strength" of inequality "$\gg$" can be expressed in terms of the "fine structure constant" $\mu = e^2/\hbar c$:

$$\frac{\tau_{(x)}}{T_0} \sim \frac{T_0}{\tau_{(v)}} \sim \mu^3 , \quad \text{where} \quad \mu = \frac{e^2}{\hbar c} \cong \frac{1}{137} . \tag{17.4.9}$$

It follows that within the atomic limits the spatial diffusion is the fastest process. As a result, Boltzmann's distribution with respect to x is established on this scale within the time $\tau_{(x)}$, and can be used as the smoothing function for carrying out the transition to the smoother function $f(x,v,t)$. Marking off the function which satisfies (17.4.5) with tilde, we may define the operation of smoothing as

$$f(x,v,t) = \int \tilde{f}(x-x',v,t) F(x')dx' ;$$

$$\tag{17.4.10}$$

$$F(x) = \sqrt{2\pi x_0^2} \exp\left(-\frac{x^2}{2x_0^2}\right) , \quad x_0 = \sqrt{\frac{\hbar}{m\omega_0}} .$$

After smoothing, the second dissipative term in (17.4.5) drops out, and we arrive at the standard Fokker – Planck equation commonly used for describing the Brownian motion of harmonic oscillator:

$$\frac{\partial f}{\partial t} + v\frac{\partial f}{\partial x} + \omega_0^2 x\frac{\partial f}{\partial v} = \frac{\partial}{\partial v}\left[D_{(v)}\frac{\partial f}{\partial v}\right] + \frac{\partial}{\partial v}[\gamma v f] .$$ (17.4.11)

We see that in this equation the scale on which the spatial diffusion takes place within the volume of an atom is a "hidden parameter".

The second two-way inequality in (17.4.8) allows us to further simplify equation (17.4.11) by averaging over the period of oscillations (see, for example, Stratonovich 1961; Ch. 11 in Klimontovich IV):

$$\frac{\partial f}{\partial t} = \frac{1}{2}\left\{D_{(v)}\left[\frac{\partial^2 f}{\partial v^2} + \frac{1}{\omega_0^2}\frac{\partial^2 f}{\partial x^2}\right] + \frac{\partial}{\partial v}[\gamma v f] + \frac{\partial}{\partial x}[\gamma x f]\right\} .$$ (17.4.12)

This equation describes only the slowest process of relaxation over the time of the order of γ^{-1}.

We see that Brownian motion of atoms oscillators is possible with a varying degree of finesse. The most detailed information is contained in the generalized Fokker – Planck equation (17.4.5), where the negligibly small ("hidden") scales are less than τ_e, r_e (see (17.4.7)). Now we continue our study of nonlinear Brownian motion, but in Ch. 23 we shall return to the problem of "hidden parameters" in quantum mechanics.

17.5. Brownian Motion in Oscillator with Nonlinear Frequency. Van der Pol – Duffing System

Let us return to the Fokker – Planck equation (16.1.8) which describes Brownian motion in Van der Pol oscillator. The nonlinear coefficient of diffusion $D(v)$ has been found from condition of existence of equilibrium distribution (16.1.4) when the coefficient of feedback is zero. In the presence of feedback the stationary solution is represented by the canonical Gibbs distribution with the effective Hamilton function (16.1.6). This stationary solution satisfies the Fokker – Planck equation with the effective velocity as defined by (16.1.7).

Recall also that in the description of Brownian motion with bistable potential (15.9.12) the stationary distribution with respect to coordinates is represented by the Boltzmann distribution (16.9.14) with the appropriate effective potential.

We use the same reasoning which has brought us to the generalized Fokker – Planck equation (17.1.4), and write the corresponding equation for the case under consideration.

For this purpose we merge the effective Hamilton function (16.1.6) for Van der Pol oscillator with linear (at $\alpha_f = 0$) circuit, and the effective potential (15.9.14) for bistable element, into a unified effective Hamilton function:

$$H_{eff}(x,v,\alpha_f,a_f) = \frac{M}{2}\left[v^2 - \frac{\alpha_f}{\beta}\ln\left(1 + \frac{\beta}{\gamma}v^2\right)\right] + \frac{M\omega_0^2}{2}\left[x^2 - \frac{a_f}{b}\ln\left(1 + bx^2\right)\right]. \quad (17.5.1)$$

Oscillator with nonlinear frequency (with the potential (15.9.12)) is called the Duffing oscillator. The system under consideration combines the dissipative nonlinearity of Van der Pol oscillator with the nondissipative nonlinearity of Duffing oscillator; this is just one example of the numerous systems with combined nonlinearities.

Function (17.5.1) depends on two controlling parameters, the feedback parameter α_f, and the effective field parameter a_f. If both of these are zero, the effective Hamilton function coincides with the Hamilton function for linear oscillator (linear oscillatory circuit in the generator):

$$H_{eff}(x,v,\alpha_f = 0, a_f = 0) = \frac{Mv^2}{2} + \frac{M\omega_0^2 x^2}{2}. \quad (17.5.2)$$

The effective velocity (16.1.7) has been introduced for the Van der Pol oscillator via the derivative of the effective Hamilton function (16.1.6). When the oscillatory circuit includes a bistable element, we must introduce the appropriate effective coordinate. Accordingly, we once again use the following formal definitions:

$$v_{eff} = \frac{\partial H_{eff}}{\partial Mv} = \left(1 - \alpha_f \frac{1}{\gamma + \beta v^2}\right)v, \quad (17.5.3)$$

$$F_{eff} = -\frac{\partial H_{eff}}{\partial x} = -\left(1 - a_f \frac{1}{1 + bx^2}\right)M\omega_0^2 x. \quad (17.5.4)$$

As a result, in place of (17.1.4) we get the following generalized model Fokker – Planck equation:

$$\frac{\partial f}{\partial t} + v_{eff}\frac{\partial f}{\partial r} + \frac{1}{M}F_{eff}(x)\frac{\partial f}{\partial v} = \frac{\partial}{\partial v}\left[D_{(v)}(v)\frac{\partial f}{\partial v}\right] + \frac{\partial}{\partial v}\left[\left(-\alpha_f + \beta v^2\right)f\right]$$

$$+ \frac{\partial}{\partial x}\left[D_{(x)}(x)\frac{\partial f}{\partial x} - \frac{F}{\gamma M}f\right]. \quad (17.5.5)$$

where

$$D_{(v)} = \frac{\kappa T}{M}\left(\gamma + \beta v^2\right) , \quad D_{(x)} = \frac{\kappa T}{M\gamma}\left(1 + bx^2\right) .$$

The solution of this equation is given by the stationary distribution

$$f_0\left(x, v, \alpha_f, a_f\right) = C \exp\left(-\frac{H_{\text{eff}}\left(x, v, \alpha_f, a_f\right)}{\kappa T}\right) \qquad (17.5.6)$$

with the effective Hamilton function (17.5.1).

The process of relaxation towards a stationary state at arbitrary values of controlling parameters is very complicated. The simplest case corresponds to $\alpha_f = a_f = 0$; then we return to equation (17.1.8) which describes the Brownian motion of a system of harmonic oscillators in a thermostat.

If the values of controlling parameters are nonzero, the characteristic relaxation times may be divided into two groups:

$$\gamma^{-1} , \quad \left|\alpha_f\right|^{-1} , \quad \beta\left\langle v^2\right\rangle , \quad \tau_{(v)} \quad ; \quad \frac{\omega_0^2}{\gamma}\left(1, \left|\alpha_f\right|, \beta\left\langle r^2\right\rangle\right) , \quad \tau_{(r)} . \qquad (17.5.7)$$

This offers ample opportunities for the statistical description of how the dissipative nonlinear processes affect the nondissipative nonlinear processes, and the other way round. In the presence of feedback, for instance, the distribution with respect to velocities differs considerably from the equilibrium distribution. For that matter, the share of particles with large velocities in the stationary state with distribution (17.3.6) can be much larger than that at equilibrium. This circumstance may have a strong effect on the rate of transition of Brownian particle in bistable element from one possible state to the other, which brings us back to the classical problem of Brownian motion connected with reaching the boundaries or surmounting a potential barrier. Problems of this kind have been first treated in the classical papers of Pontryagin – Andronov – Vitt (1933) and Kramers (1940) (see also a review in Hanggi – Talkner – Borkovec 1990).

Now there is another possibility for varying the rate of transition ("reaction rate") by adjusting the parameter of feedback in oscillator. This is actually a new method of controlling the catalyzed chemical transformations. Also, the second controlling parameter a_f can be varied by adjusting the effective field. In this way the process of achieving the target can be optimized by manipulating the two parameters, one of which controls the dissipative nonlinearity, and the other the nondissipative nonlinearity. This task is rather sophisticated and has to be treated separately. Here we shall confine ourselves to a simple illustration.

17.6. Time of Crossing a Barrier

Let us look at equation (17.2.3) for the density of Brownian particles. This is the continuity equation, the first in the set of hydrodynamic equations for a spatial system of noninteracting particles. Equation (17.2.3) is not closed, since it includes, in addition to $\rho_{(B)}$, the flux of Brownian particles $j = \rho u$ (subscript (B) hereinafter is dropped).

Consider the following problem.

Assume that Brownian motion is one-dimensional and occurs in the field with potential (15.9.12). Then the force in (17.2.3) is

$$F(x) = -\frac{dU(x)}{dx} \ , \quad U(x) = \frac{M\omega_0^2 x^2}{2}\left(-a + \frac{b}{2}x^2\right) , \quad a = a_f - 1 \ , \quad b > 0 \ , \tag{17.6.1}$$

where a_f is the controlling parameter. If $a > 0$, the potential $U(x)$ has two minima at

$$x_{min} = \pm\frac{a}{b} \ , \quad U(x_{min}) = -\frac{M\omega_0^2}{2}\frac{a^2}{4b} \ , \quad U(x_{max} = 0) = 0 \ . \tag{17.6.2}$$

For convenience of notation, we use the distribution function $f(x,t)$ in place of density, and rewrite (17.2.3) in the form

$$\frac{\partial f(x,t)}{\partial t} + \frac{\partial j}{\partial x} = \frac{\partial}{\partial x}\left[D\frac{\partial f}{\partial x} - \frac{F(x)}{M\gamma}f\right] , \quad j = uf \ , \quad \int_{-\infty}^{+\infty} f(x,t)dx = 1 \ . \tag{17.6.3}$$

Two ways of solving the problem of the time of transition across a barrier are possible: by solving the time equation, and by solving the stationary equation with the appropriate "source" and "sink", which ensure a steady inflow of Brownian particles to the barrier. The second method is more comprehensible, and brings one faster to the desired goal.

So, assume that the distribution of particles is stationary but differs considerably from the equilibrium distribution: the source makes sure that all the particles are on the left-hand side of the barrier. After crossing the barrier all the particles disappear through the capacious "sink". This state of things is ensured by the following conditions:

$$\int_{-\infty}^{x_{max}} f(x)dx = 1 \ , \quad j(x) = j_0 \equiv \frac{1}{\tau_{tr}} \ , \tag{17.6.4}$$

where j_0 is the constant supply of particles (expressed, as follows from (17.6.3), in inverse time units), and τ_{tr} is the mean time of transition across the barrier. After carrying out one integration with respect to x, we get the following equation:

$$D\frac{df}{dx} + \frac{1}{M\gamma}\frac{dU(x)}{dx}f = \frac{1}{\tau_{\text{tr}}} , \quad D = \frac{\kappa T}{M\gamma} . \tag{17.6.5}$$

The solution of this equation can be expressed as

$$f(x) = \frac{1}{\tau_{\text{tr}}D}\exp\left(-\frac{U(x)}{\kappa T}\right)\int_x^0 \exp\frac{U(x')}{\kappa T}dx' , \quad x_{\max} = 0 , \tag{17.6.6}$$

whence, with due account for the normalization condition of (17.6.4) follows the "Einstein relation"

$$(\Delta x_{\text{eff}})^2 = D\tau_{\text{tr}} , \quad (\Delta x_{\text{eff}}^2) = \int_{-\infty}^0 \exp\left(-\frac{U(x)}{\kappa T}\right)\left(\int_x^0 \exp\frac{U(x')}{\kappa T}dx'\right)dx . \tag{17.6.7}$$

This relation establishes linkage between the coefficient of diffusion, the squared effective size of the region where the diffusion takes place, and the transition time (which acts as the effective diffusion time). The expression for the transition time is well known; for a high barrier, when the depth of pit is much greater than κT, it takes on the form obtained in the pioneering paper by Kramers (1940).

Expressions (17.6.7) permit finding the transition time at any value of the parameter which controls the height of the barrier. The relevant effective potential can be used in place of potential (17.6.1); in the one-dimensional case the effective potential is found from (15.9.15):

$$U_{\text{eff}}(x) = \frac{M\omega_0^2}{2}\left[x^2 - \frac{a_{\text{f}}}{b}\ln\left(1 + bx^2\right)\right] . \tag{17.6.8}$$

At $a_{\text{f}} = 0$ this is the potential of harmonic oscillator. For the effective potential the depth of pit is smaller (see (15.9.19), (15.9.20)), and therefore the time of transition across the barrier is less.

Let us consider the general expression (17.6.7) in two extreme cases.

1. *The controlling parameter is zero*, $a_{\text{f}} = 0$. Using (17.6.8) and the definition

$D = \kappa T / M\gamma$, we find the following estimate:

$$\left(\Delta x_{\text{eff}}\right)^2 \sim \frac{\kappa T}{M\omega_0^2} \ , \text{ and hence } \tau_{\text{tr}}^{(0)} \sim \frac{\gamma}{\omega_0^2} \ . \tag{17.6.9}$$

To understand the meaning of this result, we return to equation (17.6.3) and assess the diffusion time at $a_{\text{f}} = 0$ (that is, for the case of harmonic oscillator). Since the mean square displacement for Brownian motion in the thermostat is $\sim \kappa T / M\omega_0^2$, the diffusion time is

$$\tau_D \sim \frac{\gamma}{\omega_0^2} \ , \text{ and hence } \tau_D \sim \tau_{\text{tr}}^{(0)} \ . \tag{17.6.10}$$

This is natural, because the time of diffusion in the absence of barrier is the only characteristic time of relaxation.

2. *The barrier is tall*, $U(x_{\text{max}}) \gg \kappa T$. In this case, for the first exponent in (17.6.7) we may use the Gaussian distribution, and in the zero approximation in the relative dispersion go over to δ-function:

$$\exp\left(-\frac{U_{\text{eff}}(x)}{\kappa T}\right) \Rightarrow \exp\left[-\frac{(x - x_{\text{max}})^2}{2\sigma}\right] \Rightarrow \sqrt{2\pi\sigma_{\text{eff}}}\,\delta(x - x_{\text{max}}) \ ,$$

$$\tag{17.6.11}$$

$$x_{\text{max}} = \pm\sqrt{\frac{a_{\text{f}}}{b}} \ , \quad \sigma_{\text{eff}} = \frac{\kappa T}{M\omega_0^2} \ , \quad a_{\text{f}} \gg 1 \ .$$

Then (17.6.7) may be rewritten as

$$D\tau_{\text{tr}} = \sqrt{2\pi\sigma_{\text{eff}}} \int_{-x_{\text{max}}}^{0} \exp\frac{U(x')}{\kappa T} dx' \ , \quad x_{\text{max}} = \sqrt{\frac{a_{\text{f}}}{b}} \ . \tag{17.6.12}$$

This transition time is conveniently expressed via the time of transition $\tau_{\text{tr}}^{(0)}$ which corresponds to $a_{\text{f}} = 0$:

$$\tau_{\text{tr}} = \tau_{\text{tr}}^{(0)} \sqrt{\frac{2\pi M\omega_0^2}{\kappa T}} \int_{-x_{\text{max}}}^{0} \exp\frac{U(x')}{\kappa T} dx' \ , \tag{17.6.13}$$

where the integrand contains the effective potential as defined by (17.6.8). When the potential (17.6.1) is used, the expression for the dispersion in the Gaussian distribution is modified by a factor of $1/a_f$.

Recall finally that these results are based on the solution of Einstein – Smoluchowski equation. In case of nonlinear friction, the calculations must be based on the appropriate Fokker – Planck equation.

17.7. Mutual Influence of Equilibrium and Nonequilibrium Phase Transitions

A new rapidly developing branch of the theory of phase transitions is the theory of equilibrium and nonequilibrium phase transitions induced by external factors (for example, by laser irradiation or some kind of external noise) (Klimontovich III, IV; Horstemke – Lefever 1984; Andreev – Emel'yanov – Il'inskii 1988, 1993; Anishchenko 1990). Much attention has been paid recently to the problem of the so-called stochastic resonance.

This effect consists mainly in the following. When a bistable system experiences simultaneous action of deterministic signal and random noise, the response of the system is a nonmonotonic function of the intensity of external noise (see a review in Moss 1992). The specific feature of this problem is that the process is nonstationary, and requires solving the appropriate time-dependent Fokker – Planck equation.

Systems which may experience both equilibrium and nonequilibrium phase transitions are also under active investigation. The studies are focused on the mutual influence of the two essentially nonlinear stochastic processes.

In Sect. 15.5 we considered a system where the dissipative and the nondissipative nonlinearities are localized in two different elements of the electric circuit of Van der Pol – Duffing oscillator. Such processes can be described on the basis of the unified generalized Fokker – Planck equation for the distribution function of charge and current (coordinate and velocity) in the electric circuit with concentrated parameters. Bistability is ensured by, for instance, placing a ferroelectric into the capacitor of the oscillatory circuit. The nature of nondissipative nonlinearity is varied by changing the temperature in the critical neighborhood of phase transition of the second kind in the ferroelectric. Dissipative nonlinearity is created by applying feedback to the circuit.

To illustrate the diversity of interactions between the equilibrium and the nonequilibrium phase transitions, let us consider a few more examples of such systems.

17.7.1. *Effect of Phase Transition in Ferroelectric on Generation of Laser Radiation*

The active medium of a laser can be represented by ferroelectric crystal doped with optically active impurities. Because of the linkage between the lattice and the dope atoms, fluctuations of the lattice affect the process of generation. Since the intensity of fluctuations depends on how close the temperature of the sample T is to the critical point T_c, the nature of emission displays dramatic changes when the point of phase transition in ferroelectric crystal is approached. Calculations indicate (Klimontovich III, IV) that the coefficient of damping of electric polarization of the system of dope atoms γ on approaching the critical point increases as

$$\Delta\gamma \propto \frac{1}{(T-T_c)^2} \ . \tag{17.7.1}$$

The same law governs the critical value of inverse population of levels corresponding to the threshold of laser generation.

This result is based on the calculation of fluctuations in the linear approximation. In the immediate neighborhood of the critical point the growth is curbed by the dissipative nonlinearity which gains importance as the lattice displacement becomes sufficiently large. Fluctuations of lattice displacement at arbitrary temperatures can only be calculated by solving the corresponding Fokker – Planck equation for the distribution function of the values of crystalline lattice displacement.

17.7.2. Effect of Phase Transition in Liquid Crystal on Characteristics of Laser Radiation

The working medium of laser in (Kuroda – Kubota 1976) was represented by a solution of optically active dye atoms in nematic liquid crystal. A considerable lowering of generation threshold and reduction in wavelength of emitted radiation were observed as the critical point of phase transition to the nematic phase was approached.

This phenomenon is due to the effect of fluctuations of tensor of orientation ordering in liquid crystal on the optically active dye atoms (Klimontovich III, IV). Calculations have been based on the solution of the corresponding Langevin equation. At low frequencies, the intensity of fluctuations of tensor of orientation ordering increases in the neighborhood of the critical point as $(T-T_c)^{-1/2}$. These fluctuations result in an effective increase in squared modulus of matrix element of dipole moment of dye atoms. This in its turn causes lowering of the population threshold D_{th}. As the critical point is approached, the population threshold D_{th} varies at first as $(T-T_c)^{-1/2}$; in the immediate neighborhood of the critical point the rate of decrease is slowed down. The wavelength of laser radiation behaves in a similar fashion.

Examples of mutual influence of phase transitions are numerous. Interdependence of this kind can be used for stimulating chemical reactions and for controlling the rate of conversion by adjusting the values of controlling parameters. The availability of several controlling parameters gives more freedom in both controlling the time evolution and searching for the most organized stationary states.

17.8. Evolution Towards Stationary State in Systems with Two Controlling Parameters. H–Theorem

We return to the generalized Fokker – Planck equation (17.5.5) which contains two controlling parameters. Its stationary solution (17.5.6) can be represented as the canonical Gibbs distribution:

$$f_0(x,v,\alpha_f,a_f) = \exp\frac{F_{\text{eff}}(\alpha_f,a_f) - H_{\text{eff}}(x,v,\alpha_f,a_f)}{\kappa T} . \tag{17.8.1}$$

By contrast to (16.5.1), this solution contains two controlling parameters. The following "thermodynamic equality" holds for the nonequilibrium free energy of stationary state:

$$F_{\text{eff}} = \left\langle H_{\text{eff}}(x,v,\alpha_f,a_f) \right\rangle_0 - T\left\langle -\kappa \ln f_0(x,v,\alpha_f,a_f) \right\rangle_0 , \tag{17.8.2}$$

where $\langle ... \rangle_0$ denotes averaging over the stationary distribution. Now we use the appropriate definition of free energy of the nonstationary state, and introduce a functional defined as the difference in nonequilibrium free energies (see Sect. 13.4, 15.11):

$$\Lambda_F = F(t) - F_0 = \kappa T \int \ln\frac{f(x,v,t)}{f_0(x,v,\alpha_f,a_f)} f(x,v,t)dxdv \geq 0 . \tag{17.8.3}$$

With the aid of the kinetic equation, we find the time derivative of Λ_F,

$$\frac{d\Lambda_F}{dt} = \frac{d}{dt}[F(t) - F_0] = -\sigma_F \leq 0 , \tag{17.8.4}$$

where, like in (16.5.4) and (16.10.5) we use the notation for "entropy production" (decrease in the free energy σ_F):

$$\sigma_F = \kappa T \int \left[D_{(v)}(v)\left(\frac{\partial}{\partial v}\ln\frac{f}{f_0}\right)^2 + D_{(x)}(x)\left(\frac{\partial}{\partial x}\ln\frac{f}{f_0}\right)^2 \right] f dx dv \geq 0 . \tag{17.8.5}$$

The "equals" sign only holds for the stationary state. In the course of evolution towards the stationary state the nonequilibrium free energy decreases.

So, for the generalized Fokker – Planck equation we also have a Lyapunov functional Λ_F which is defined as the difference in nonequilibrium free energies. The stationary state is thus found to be stable. Like in Sect. 16.10, these inequalities can be used for defining the domain of stability of stationary state upon transition to the discrete time. The largest permissible step of discrete time will depend on two controlling parameters.

17.9. Optimization of Process of Evolution in Space of Controlling Parameters Based on S–Theorem

As pointed out in the Introduction and in Sect. 16.5, the criterion of S–theorem is not only designed for comparing the relative degree of order of nonequilibrium states of open systems, but can also be used for constructive purposes. Namely, it allows checking whether the choice of controlling parameters is correct. If there are several controlling parameters, this criterion permits optimizing the path towards the desired goal — for instance, to the most ordered or, alternatively, the most chaotic state.

Two examples of systems with two controlling parameters were considered in (Klimontovich – Bonitz 1988; Klimontovich IV): (1) Van der Pol oscillator with nonlinear characteristic, operating in soft or hard excitation regime depending on the values of controlling parameters, and (2) Van der Pol oscillator under external resonant action.

Since the system studied in Sect. 17.5, 17.8 also contains two controlling parameters a_f, α_f, one may use the criterion of S–theorem for optimizing the search for the most ordered stationary state of Van der Pol – Duffing oscillator.

As follows from (17.5.6), (17.5.1), the state which may be taken for the reference point in comparing the degree of chaoticity coincides with the equilibrium state when

$$a_f = 0, \ \alpha_f = 0 \ , \tag{17.9.1}$$

which corresponds to the canonical Gibbs distribution

$$f_0(x, v, \alpha_f = 0, a_f = 0) = C \exp\left[-\frac{H(x, v)}{\kappa T}\right] \tag{17.9.2}$$

with the Hamilton function (17.5.2).

For comparing the relative degree of states at different values of controlling parameters we must carry out renormalization of nonequilibrium distributions to the preassigned mean value of the Hamilton function (17.5.2). Recall that this has been done in Sect. 16.1 and 16.6 for the problem of Brownian motion in Van der Pol oscillator. Renormalization of distributions of velocity and energy required introducing the effective temperature defined from solutions of equations (16.1.9), (16.6.9). Let us write a similar equation for the more complicated Van der Pol – Duffing oscillator with two controlling parameters a_f, α_f:

$$\kappa\tilde{T}(\alpha_f, a_f) = \int H(x,v)C\exp\left[-\frac{H(x,v)}{\kappa\tilde{T}}\right]dxdv = \int H(x,v)f(x,v,\alpha_f,a_f)dxdv . \quad (17.9.3)$$

The nonequilibrium stationary distribution is given by (17.5.6). Under condition (17.9.1), we have $\tilde{T} = T$.

Now for the quantitative assessment of the relative degree of order of the equilibrium state with renormalized temperature $\tilde{T}$ and the selected nonequilibrium state, we use the distributions (17.9.2) with $T \Rightarrow \tilde{T}$ and (17.5.6) to construct a functional defined as the difference in the relevant entropies,

$$\Lambda_F(\alpha_f, a_f) = \tilde{S}(0,0) - S(\alpha_f, a_f) = \int \ln\frac{f(x,v,\alpha_f,a_f)}{\tilde{f}(x,v,0,0)} f(x,v,\alpha_f,a_f)dxdv \geq 0 . \quad (17.9.4)$$

Now let us sum up our results relating to the criterion of S–theorem. This criterion was formulated in Sect. 1.6.6, and was subsequently applied to various systems in Sect. 12.7, 12.8, 16.5, and in the present section. Given the availability of mathematical model of the system under consideration (for instance, the Fokker – Planck equation for Van der Pol or Van der Pol – Duffing oscillator), the use of this criterion is much simplified. Indeed, then we know the most chaotic "reference" state, which is the state of equilibrium. The additional requirement of constancy of the mean energy (like (16.1.9), (16.6.9), or (17.9.4)) serves just as the definition of the effective temperature. To find the dependence on one or, as in (17.9.4) two controlling parameters, one only needs to calculate the integrals which contain nonstationary distributions, like the last term in (17.9.4). It is not difficult to decide whether each consecutive state is more ordered or not, since any state in the sequence will be compared separately with the "reference" equilibrium state.

More complicated calculations are required when the criterion of S–theorem is applied to the time realizations (or other realizations of a process) obtained directly from experiment. In such cases there is no universal "reference state", and so one of the states to be compared is a priori taken for the state of "physical chaos". Eventually, however, the criterion of S–theorem allows one to check the correctness of such assumption. The distribution function of the state of "physical chaos" is used to define the effective Hamilton function H_{eff}. Renormalization of the distribution function of the state of "physical chaos" is based on the requirement that the mean values of the effective Hamilton function H_{eff} for the two states being compared should be the same. This condition is expressed as a transcendental equation which defines the effective temperature as a function of the controlling parameter. The form of solution of this transcendental equation allows one to judge whether the choice of a particular state for the state of "physical chaos" has been correct or not. Of course, since there is no common reference point, the nonequilibrium states have to be compared in pairs.

We shall discuss these issues in a more systematic way in Ch. 21, which deals specifically with the comparative study of various criteria of self-organization. In the next chapter we shall be concerned with the kinetic description of processes in active media.

CHAPTER 18

KINETIC THEORY OF ACTIVE MEDIA

In this chapter we discuss the use of kinetic methods in the description of nonequilibrium processes in active media. The concept of "active medium" embraces a wide scope of physical, chemical and biological systems. The important common property of active media consists in their capability of developing diverse space-time dissipative structures when the controlling parameters are varied. The term "dissipative structures" implies that dissipation plays a constructive role, as we have already said in the Introduction.

Active media may be conventionally divided into two classes.

The first class includes those media in which restructuring takes place as a result of phase transitions. The controlling parameters are represented by thermodynamic quantities, like the temperature at given pressure, or pressure at given temperature. A classical example of appearance of a new structure is a liquid-to-crystal phase transition, and, in general, all phase transitions of the second kind. The dissipation is important for both the emergence of the new phase, and for the relaxation of fluctuations; the "thermostat" serves as a source or a sink of energy. Structures which arise in systems of this kind are at equilibrium with the surrounding medium ("thermostat").

Active media whose stationary states differ considerably from the state of equilibrium belong to the second class; here the existence of a stationary state depends on the steady supply of energy. This class can be further divided into two subgroups. The first will include active media comprised of active elements (such as tiny oscillators), which by themselves are small but macroscopic active systems. Active elements may be represented by atoms with inverse population of levels, which constitute the working medium of generators of coherent radiation, lasers or masers. Coherent radiation is the result of collective action of individual active elements. Transition to laser regime of emission may be regarded as a nonequilibrium phase transition; the role of controlling parameter is played by the pumping which creates inverse population of atoms of laser medium.

The other subgroup includes systems consisting of passive elements; here the dissipative structures can only arise under nonequilibrium conditions. This is the case, for instance, with the transition from laminar to turbulent flow as the gradient of pressure increases. In Ch. 22 we shall demonstrate that turbulent flow is higher organized; this allows us to regard the transition from laminar to turbulent flow as a nonequilibrium phase transition; the elements of Reynolds tensor act as the parameters of order.

Another example is the emergence of a regular pattern of convective motion in liquid layer heated from below. The controlling parameter in this case is the temperature gradient.

Naturally, there is interaction between individual elements of active media. This interaction may take different forms; the simplest and the most common type of interaction occurs through spatial diffusion. If we refer to the changes in individual elements caused by variations in controlling parameter as "reactions", then the processes in active media in the presence of diffusion may be called reaction-diffusion processes. Our discussion of the kinetic theory of active media we begin with the study of the relevant equations.

First mathematical models of active media were proposed as early as forty years ago in the well known works of Wiener, Rosenblueth, Gel'fand and Tsetlin (see Romanovsky – Stepanova – Chernavsky 1984). A fundamental contribution to the study of active media is the theory of turbulence developed by Landau in 1944 (see Landau – Lifshitz 1986).

Today there exists extensive literature on various aspects of the theory of active media. Numerous results and references can be found, for example, in Vasiliev – Romanovsky – Yakhno 1987; Mikhailov 1990, 1994; Haken 1978, 1980; Kerner – Osipov 1994; Klimontovich III – V.

18.1. Kinetic Equations of Reaction Diffusion Type. Kinetic and Gasdynamic Description of Heat Transfer in Active Medium

The class of reaction-diffusion equations includes the famous Fisher – Kolmogorov – Petrovsky – Piskunov (FKPP) equations, introduced for describing the evolution of biological populations, the Turing equations (see Murray 1977), and the Ginzburg – Landau equation in its numerous modifications. The latter was intended in the first place for the description of superconductivity, and is widely used now for describing the nonequilibrium phase transitions and, in particular, the autowave processes (see Vasiliev – Romanovsky – Yakhno 1987).

So, the contemporary theory of active media is largely based on equations of reaction-diffusion type,

$$\frac{\partial X(R,t)}{\partial t} = F[X(R,t)] + \frac{\partial}{\partial R_i}\left[D_{ij}(X)\frac{\partial X}{\partial R_j} \right], \tag{18.1.1}$$

where $X(R,t)$ is the set of macroscopic functions which characterize the system under consideration, like concentrations of populations in biological systems, or concentrations of reactants in chemically reacting media. In physics they may be represented by the local densities of charge and current, the local density and temperature, etc. Functions $F(X)$ are nonlinear functions defined by the structure of the system in question; D_{ij} are the coefficients of spatial diffusion of the "elements" of active medium. Here and further we denote the radius-vector of points of the medium by R.

Recall that we have already encountered such equations in Sect. 14.8, devoted to the kinetic theory of heat transfer. Let us recapitulate the main results of our discussion. We started with the kinetic equation (14.8.3) for the distribution function of kinetic energy, which is a particular case of the generalized kinetic equation (13.4.1) for Boltzmann gas and also includes two dissipative terms. The first dissipative term depends on the redistribution of particles with respect to velocities, and the second describes the process of spatial diffusion of the relevant distribution function $f(E,R,t)$ (here and below E is the value of the kinetic energy, and R is the position in space).

From the kinetic equation (14.8.3) follows the equation in the mean kinetic energy (equation of heat conduction):

$$\frac{\partial \langle E \rangle}{\partial t} = \chi \frac{\partial^2 \langle E \rangle}{\partial R^2} \; , \quad \langle E \rangle_{R,t} = \frac{3}{2} \kappa T(R,t) \; . \tag{18.1.2}$$

From (14.8.3) also follows the equation for the dispersion of temperature (14.8.7). Let us rewrite the latter for the particular case when the temperature T is constant. Then it has the structure of a reaction-diffusion equation:

$$\frac{\partial}{\partial t} \left\langle (\delta E)^2 \right\rangle = \frac{4}{\tau} \left(\frac{2}{3} \langle E \rangle^2 - \left\langle (\delta E)^2 \right\rangle \right) + \chi \frac{\partial^2}{\partial R^2} \left\langle (\delta E)^2 \right\rangle \; . \tag{18.1.3}$$

The temperature conductivity χ here acts as the coefficient of diffusion; the reaction term in this example is linear. It has a very simple structure because the medium currently under consideration is passive.

Assume now that there is a nonlinear heat source. Then the constant relaxation time τ in the "collision integral" (14.8.4) will be replaced by a nonlinear relaxation coefficient $\gamma(E)$. By way of example, we define this function by (12.5.7), and once again find the nonlinear coefficient of diffusion from assumption that there exists the state of equilibrium with Maxwell distribution. Then

$$\gamma(E) = -a_s + \frac{1}{\tau} + bE \; , \quad D_{(v)}(E) = \left(\frac{1}{\tau} + bE \right) \kappa T_0 \; , \tag{18.1.4}$$

where a_s defines the heat source ("feedback"), and T_0 is the temperature of the thermostat.

Now equation (14.8.3) is replaced by the more general kinetic equation

$$\frac{\partial f(R,|v|,t)}{\partial t} + v\frac{\partial f}{\partial R} = \frac{\partial}{\partial v}\left[D_{(v)}(E)\frac{\partial f}{\partial v}\right] + \frac{\partial}{\partial v}\left[\gamma(E)vf\right] + \chi\frac{\partial^2 f}{\partial R^2} . \tag{18.1.5}$$

By contrast to equation (12.5.10), which has also been obtained from the kinetic Boltzmann equation, this equation is also linear with respect to distribution function f, but takes the spatial diffusion of the distribution function into account.

Linearity of equations (12.5.10), (18.1.5) with respect to to distribution function f is due to the fact that functions (12.5.7), (18.1.4) are defined not by the local value of kinetic energy $E(R,t) = \int Ef(E,R,t)dE$, but rather by its nonaveraged value. The replacement

$$E \Rightarrow E(R,t) = \int Ef(E,R,t)dE \tag{18.1.6}$$

in (12.5.10), (18.1.5) results in nonlinear kinetic equations which correspond to the "approximation of self-consistent field" with respect to the dissipative characteristics.

Juxtaposition of these two methods of kinetic description of nonequilibrium processes in active media constitutes a separate task. Here we just point out that the approximation of (12.5.7), (18.1.4) allows taking care of the additional correlations on the scale $r_{cor} > l_{ph}$ which are neglected in the approximation of self-consistent field.

Since the coefficients γ, D in (18.1.5) are functions of E, the equation for the mean energy (the temperature) is not closed. So we once again have to deal with the problem of closure for the chain of equations. The simplest solution is the approximation of first moments, when the distribution function $f(E,R,t)$ has the form

$$f(E,R,t) = \delta\big(E - E(R,t)\big) , \quad E(R,t) = \langle E\rangle_{R,t} . \tag{18.1.7}$$

As a result, we get the following equation for the mean energy:

$$\frac{\partial\langle E\rangle}{\partial t} = 2\left[\frac{3}{2}mD_{(v)}(\langle E\rangle) - \gamma(\langle E\rangle)\langle E\rangle\right] + \chi\frac{\partial^2\langle E\rangle}{\partial R^2} . \tag{18.1.8}$$

We see that in this approximation a reaction-diffusion equation of type (18.1.1) follows from the generalized kinetic equation (18.1.5). Equation (18.1.8), however, is more general, since, in addition to the spatial diffusion (heat conduction), it takes care of the internal diffusion $D_{(v)}(E)$. Because of this, by varying the controlling parameter a_s it is possible to pass smoothly the neighborhood of the bifurcation point $a_s = 1/\tau$, where the linear term in $\gamma(E)$ changes its sign. We shall demonstrate this in the next section for the case of active medium of bistable elements.

Equation (18.1.8) gives an example of nonlinear equation of heat conduction.

Equations of this kind are widely used as models of heat transfer (see, for example, Samarskii 1987). They, however, pertain to the one-moment approximation. Higher moments have to be taken into account, in particular, in the description of states close to the points of phase transitions. This, of course, does not exhaust the role of fluctuations.

In the approximation of second moments, in place of one equation (18.1.7) we obtain a set of equations of reaction-diffusion type, which considerably extends the capabilities of the model (Krinsky 1984; Kerner – Osipov 1983, 1984; Akhromeeva 1992).

Resorting to the perturbation theory in the moments, however, we meet with the same difficulties as were encountered in the transition from the kinetic Boltzmann equation to equations of gas dynamics. One complication is that the the states of active systems are, as a rule, far from equilibrium. This can be illustrated with the example of stationary solution of the kinetic equation (18.1.5).

Assume that the time evolution leads to the spatially homogeneous distribution of the particles of the system (12.5.12). This function describes the distribution with respect to velocities at arbitrary values of a_s, and may differ considerably from the Maxwell distribution. In this case the approximation in moments is not efficient enough, and one has to address the solution of the corresponding kinetic equation.

18.2. Manifestations of Structure of "Continuous Medium" in Processes of Time Relaxation

To begin with, let us briefly review the results of Ch. 1 and 13. In Boltzmann's kinetic theory the relaxation time is determined by the mean free path time τ of any of the system's particles. The physically infinitesimal time scale τ_{ph} comes up when the structure of "continuous medium" is taken into account. This scale defines the characteristic time interval for one of N_{ph} particles within a point of "continuous medium". This is an individual relaxation characteristic of the particle, whereas τ is a characteristic of the collective of N_{ph} particles confined within a "point"; this is the time within which each of the N_{ph} particles on the average experiences a collision.

Given all this, the relaxation time due to redistribution of velocities because of collisions in Boltzmann gas $\tau_{(v)}$, and the corresponding coefficient of diffusion $D_{(v)}$, can be represented as

$$\tau_{(v)} \equiv \tau \sim \tau_{ph} N_{ph} , \quad D_{(v)} \sim \frac{\kappa T}{m\tau} \sim \frac{1}{N_{ph}} \frac{\kappa T}{m\tau_{ph}} \equiv \frac{1}{N_{ph}} D . \tag{18.2.1}$$

In the last term we have singled out the "individual coefficient of diffusion", relating to the diffusion of one particle in the medium. It is this coefficient of diffusion that is used in the theory of Brownian motion, which is clear from the comparison between (18.2.1) and (17.3.9). The difference is that in Sect. 17.3 the entire sample containing N particles was regarded as a single particle, which is the standard approach in the theory of electric circuits with concentrated parameters. In the description of distributed systems, including active media, the smallest object is the "point" which contains N_{ph} particles; hence the factor of $1/N_{ph}$ in (18.2.1).

Now let us consider the relaxation characteristics of the second dissipative term in the generalized kinetic equation which describes spatial diffusion of the distribution function.

For a rarefied gas, the coefficient of spatial diffusion is

$$D_{(R)} \sim \frac{l^2}{\tau} \sim v_T l \ . \tag{18.2.2}$$

Hence it follows that the free path time can be defined by either of the two equivalent relations,

$$\tau = \frac{l}{v_T} \ , \quad \tau = \frac{l^2}{D_{(R)}} \ . \tag{18.2.3}$$

Thus, the free path time is the time of diffusion to the distance l; the latter, while being the conventional free path length, plays at the same time the role of the smallest "macroscopic" scale in the unified description of kinetic and hydrodynamic processes. Recall that the size of "point" is then defined by the value of L_{min} (see (13.2.4)).

Let us fix this smallest "macroscopic" length scale, on which the diffusion process is still possible, and pay attention to the fact that the free path time τ is a characteristic of the collective of N_{ph} particles contained within a "point" of continuous medium. Given the definition of physically infinitesimal scale τ_{ph} (1.1.17), the second expression in (18.2.3) can be rewritten as

$$\tau = N_{ph} \tau_{ph}, \text{ and hence } D_{(R)} = \frac{1}{N_{ph}} \frac{l^2}{\tau_{ph}} \ . \tag{18.2.4}$$

Here, like in (18.2.1), we have factored out $1/N_{ph}$. The second factor is the coefficient of diffusion which makes a particle travel to the minimum "macroscopic" distance l over the time of τ_{ph}.

So we see that in the analysis of nonequilibrium processes the existence of structure of "continuous medium" is revealed also in the form of dissipative characteristics. The latter depend on the number of particles within physically infinitesimal volume, regarded as a "point" of continuous medium. For every new system one has to redefine physically infinitesimal scales. For instance, we shall deal with the definition of physically infinitesimal scales for partially ionized plasma, which is an example of the system with chemical transformations. Chemical reactions are represented by the processes of ionization and recombination. In Ch. 23 a similar problem will be considered for the quantum system of atoms and electromagnetic field.

After this important aside, let us return to the main theme of the present chapter.

18.3. Medium of Bistable Elements. Kinetic Approach in Theory of Phase Transitions

18.3.1. Kinetic Equation

Let us return to the problem of Brownian motion of nonlinear atoms (Duffing oscillators). Then the dissipation is linear, but the elastic force is given by (17.5.4). In Sect. 15.5 the oscillator has been regarded as a small macroscopic object of mass M which performs Brownian motion in thermostat with temperature T. Now we consider a medium composed of Duffing oscillators; individual oscillators are linked only through spatial diffusion. As before, the size of the point of "continuous medium" is determined by physically infinitesimal volume V_{ph}.

Let, for instance, the fluctuation electromagnetic field once again act as the thermostat. Then γ is the coefficient of radiation friction; the corresponding relaxation time $\tau = 1/\gamma$ characterizes the time of damping of any one of N_{ph} oscillators within the physically infinitesimal volume — that is, within a "point" of continuous medium of atoms oscillators. Along with τ, one may also define a smaller time interval

$$\tau_{ph} = \frac{\tau}{N_{ph}} = \frac{1}{\gamma N_{ph}} , \tag{18.3.1}$$

which is an individual characteristics. In this respect the situation is similar to that discussed in Ch. 13 in connection with the derivation of the generalized kinetic equation (13.4.1) for Boltzmann gas.

Equation (13.4.1) contains two dissipative terms, defined, respectively, by the redistribution of particles with respect to velocities and with respect to R. The state of one-dimensional atoms oscillators is given by the distribution function of internal variables x, v, and the variables R, P which characterize the motion of oscillator as a whole. Let us confine ourselves here to considering a simper distribution $f(x,v,R,t)$. For establishing the kinetic equation we use the appropriate Langevin equations. The intensities of the sources in equations for x, v are determined by the coefficients of diffusion

$$D_{(v)} = \gamma \frac{\kappa T}{m} \ , \quad D_{(x)} = \frac{\kappa T}{m\gamma} \left(1 + bx^2 \right) \ . \tag{18.3.2}$$

In order to describe spatial diffusion of atoms oscillator as a whole, one must use the appropriate Langevin equation (cf. (15.9.2)),

$$\frac{dR}{dt} = Y(t) \ , \quad \langle Y \rangle = 0 \ , \quad \left\langle Y_i(t) Y_j(t') \right\rangle = 2D\delta_{ij}\delta(t - t') \ , \tag{18.3.3}$$

where D is the relevant coefficient of spatial diffusion.

Now we may use the three Langevin equations to find the Fokker – Planck equation for the distribution function of the complete set of variables

$$f(x, v, R, t) \ , \quad \int f(x, v, R, t) dx dv \frac{dR}{V} = 1 \ . \tag{18.3.4}$$

For the medium of Duffing oscillators the Fokker – Planck equation is

$$\frac{\partial f}{\partial t} + v \frac{\partial f}{\partial r} + \frac{1}{m} F_{\mathrm{eff}} \frac{\partial f}{\partial v} = \frac{\partial}{\partial v} \left[D_{(v)} \frac{\partial f}{\partial v} \right] + \frac{\partial}{\partial v} [\gamma v f]$$

$$+ \frac{\partial}{\partial x} \left[D_{(x)}(x) \frac{\partial f}{\partial x} - \frac{F(x)}{m\gamma} f \right] + D \frac{\partial^2 f}{\partial R^2} \ . \tag{18.3.5}$$

The effective force is given by (17.5.4); the coefficients of diffusion are given by (18.3.2). Here and above the force of friction in equation for the velocity is linear.

Possible is also a more coarse kinetic description, which corresponds to the approximation of "self-consistent field". Like in (18.1.6), this is achieved by replacing $x \Rightarrow x(R,t)$ in the nonlinear force and in the coefficient of diffusion. This replacement, however, will result in the loss of correlations with $\tau_{\mathrm{cor}} > \tau_{\mathrm{ph}}$.

Let us explore the feasibility of transition from (18.3.5) to equation for the simpler distribution function $f(x, R, t)$. This feat is similar to the transition from Fokker – Planck to Einstein – Smoluchowski equation. In case of Brownian motion of linear oscillator this can only be done under a rather stringent condition (15.9.10). For the nonlinear Duffing oscillator this constraint is much slackened if $|a_f - 1| \ll 1$, when the effective frequency of oscillations becomes much less than ω_0. If the potential $U(x)$ has two minima, slow transition across the barrier is possible. The frequency of this process is determined by the inverse transition time, $\omega_{\mathrm{tr}} = 1/\tau_{\mathrm{tr}}$, for which the condition $\gamma \gg \omega_{\mathrm{tr}}$ may hold. It is this circumstance that allows us to go over from (18.3.5) to the simpler equation

$$\frac{\partial f}{\partial t} = \frac{\partial}{\partial x}\left[D_{(x)}(x)\frac{\partial f}{\partial x} - \frac{F(x)}{m\gamma}f\right] + D\frac{\partial^2 f}{\partial R^2} , \quad \int f dx \frac{dR}{V} = 1 , \tag{18.3.6}$$

where $D_{(x)}$ is the same as in (18.3.2).

Let us demonstrate that in a particular case this equation can be reduced to an equation of reaction-diffusion type. For this purpose we consider a distribution for which the process is completely defined by the first moment:

$$f(x,R,t) = \delta(x - x(R,t)) , \quad \langle x \rangle = x(R,t) . \tag{18.3.7a}$$

Assume also that the diffusion with respect to internal variable x is negligibly small, $D_{(x)} = 0$. Equation in $x(R,t)$ follows from (18.3.6) and has the form of a Fisher – Kolmogorov – Petrovsky – Piskunov (FKPP) equation:

$$\frac{\partial x(R,t)}{\partial t} = F[x(R,t)] + D\frac{\partial^2 x}{\partial R^2} , \quad x(R,t) = \langle x \rangle_{R,t} . \tag{18.3.7b}$$

Recall that in case of Duffing oscillator the force is given by

$$F(x) = -m\omega_0^2 x\left(1 - a_f + bx^2\right) , \quad \text{where} \ x \equiv x(R,t) . \tag{18.3.8}$$

Equation (18.3.7b) with different functions $F(x)$ is widely used in theories of phase transitions and autowave processes; it is one of the basic equations of synergetics. In the sections to follow we shall discuss concrete examples to prove the point that the use of the more general kinetic equation (18.3.6) allows one to extract important additional information about the processes under study.

18.3.2. "Hydrodynamic Approximation" in Statistical Theory of Active Media

In the general case, when there are no simplifying assumptions, the kinetic equation (18.3.6) can only yield a chain of meshing equations in moments. Because of this, we are again faced with the problem of closure of infinite sequence of equations. Let us consider the simplest solution which yet allows taking the role of fluctuations into account.

To this end, we extend the concept of self-consistent approximation to include self-consistency not only with respect to x, but also with respect to $x^2 \equiv E$ — that is,

$$\left\langle x^3 \right\rangle \equiv \left\langle Ex \right\rangle \Rightarrow \left\langle E \right\rangle\!\left\langle x \right\rangle \;,\quad \left\langle x^4 \right\rangle \equiv \left\langle EE \right\rangle \Rightarrow \left\langle E \right\rangle\!\left\langle E \right\rangle \;. \tag{18.3.9}$$

In this approximation we obtain the following closed set of equations in functions $\langle x \rangle$ and $\langle x^2 \rangle$ (Klimontovich III – V):

$$\frac{\partial \langle x \rangle}{\partial t} = \Gamma(a - b\langle E \rangle)\langle x \rangle + D\frac{\partial^2 \langle x \rangle}{\partial R^2}\;,\quad \Gamma = \frac{\omega_0^2}{\gamma}\;,$$

$$\tag{18.3.10}$$

$$\frac{\partial \langle E \rangle}{\partial t} = 2\Big[\Gamma(a - b\langle E \rangle)\langle E \rangle + D_{(x)}\Big] + D\frac{\partial^2 \langle E \rangle}{\partial R^2}\;,\quad D_{(x)} = \frac{1}{N_{\mathrm{ph}}}\frac{\kappa T}{m\gamma}\;.$$

In writing these equations we have also made a simplifying assumption that the coefficient of diffusion does not depend on x, $D_{(x)} = \mathrm{const}$. The applicability of this approximation was discussed in Sect. 15.5.3.

Equations (18.3.9), (18.3.10) correspond to the "hydrodynamic approximation" for the kinetic equation (18.3.6); they make up a set of FKPP-type equations. In another extreme case the process can be describe by just one equation. This happens when the distribution is symmetrical with respect to x, so that $x(R,t) = \langle x \rangle = 0$. Then the first of the equations in (18.3.10) is identically zero, whereas the second reduces to

$$\frac{\partial E}{\partial t} = 2\Big[\Gamma(a - bE)E + D_{(x)}\Big] + D\frac{\partial^2 E}{\partial R^2}\;,\quad E \equiv \langle E \rangle = \left\langle (\delta x)^2 \right\rangle \;. \tag{18.3.11}$$

This equation also has the structure of FKPP equation. There is, however, an important distinction. As a matter of fact, all dissipation in the FKPP equation reduces to spatial diffusion with the coefficient D, while equation (18.3.11) includes, along with D, the coefficient of diffusion with respect to variable x. This is an obvious advantage of the kinetic equation (18.3.6).

The set of equations (18.3.10) corresponds to the distribution function

$$f(x,R,t) = \sqrt{\frac{1}{2\pi\sigma}}\,\exp\!\left[-\frac{(x - \langle x \rangle)^2}{2\sigma}\right]\;,\quad \sigma = \langle E \rangle - \langle x \rangle^2 \;. \tag{18.3.12}$$

Equations (18.3.10) are consistent only when the value of $\langle x \rangle$ is small enough.

Observe finally that the kinetic equation (18.3.6) can also be used for obtaining the equation in the zero-order moment in x — that is, for the function $f(R,t)$. The form of this distribution depends on the properties of the medium. If the medium is incompressible, then $f(R,t) = 1$. In the more general case we come to equation of self-diffusion,

$$\frac{\partial f}{\partial t} = D\frac{\partial^2 f}{\partial R^2} , \quad \int f(R,t)\frac{dR}{V} = 1 . \tag{18.3.13}$$

18.4. Lyapunov Functionals Λ_F, Λ_S. H–Theorem

Assume that the process of time evolution according to equation (18.3.6) leads to a spatially homogeneous state with the distribution $f_0(x)$. This distribution is known to us; its form is similar to (15.9.14):

$$f_0(x) = C\exp\left[-\frac{U_{\text{eff}}(x)}{\kappa T}\right], \quad U_{\text{eff}} = \frac{m\omega_0^2}{2}\left[x^2 - \frac{a_{\text{f}}}{b}\ln\left(1 + bx^2\right)\right] . \tag{18.4.1}$$

Let $f(x,R,t)$ be the distribution function for nonstationary states which satisfies equation (18.3.6). Then, in place of (15.11.13), we may introduce the local Lyapunov functional

$$\Lambda_F(R,t) = F(R,t) - F_0 = \kappa T\int \ln\frac{f(x,R,t)}{f_0(x)}f(x,R,t)dx , \tag{18.4.2}$$

for which the equation of balance follows from (18.3.6):

$$\frac{\partial\Lambda_F(R,t)}{\partial t} = D\frac{\partial^2\Lambda_F(R,t)}{\partial R^2} - \sigma_F(R,t) . \tag{18.4.3}$$

Here $-\sigma_F(R,t)$ is the "destruction" of free energy; it corresponds (up to the sign) to the production of entropy and is defined by

$$\sigma_F(R,t) = \kappa T\left\{\kappa T\int f\frac{\partial}{\partial x}\left(\ln\frac{f(x,R,t)}{f(x)}\right)^2 f(x,R,t)dx\right.$$

$$\left. + D\int f\frac{\partial}{\partial R}\left(\ln\frac{f(x,R,t)}{f(x)}\right)^2 f(x,R,t)dx\right\} \geq 0 . \tag{18.4.4}$$

Inequality (18.4.4) ensures that the stationary state is stable.

As indicated above, the Lyapunov functional Λ_F may characterize the deviation from the stable stationary state. It cannot, however, serve as a measure of the relative degree of order of states at different values of the controlling parameter a_{f}. In order to introduce

Lyapunov functional Λ_S as the difference in entropies, we must first redefine the distribution $f(x,R,t)$ so as to satisfy the normalization condition

$$\int U_{\text{eff}}(x,a_f)\tilde{f}(x,R,t)dx = \int U_{\text{eff}}(x,a_f)f_0(x,a_f)dx \ . \tag{18.4.5}$$

Since the right-hand side does not depend on R, the mean effective energy for the renormalized distribution must be the same at all points of the medium. This is certainly a stringent requirement.

If condition (18.4.5) is satisfied, we may introduce the Lyapunov functional as

$$\Lambda_S(R,t) = S_0 - S(R,t) = \kappa\int \ln\frac{\tilde{f}(x,R,t)}{f_0(x)}\tilde{f}(x,R,t)dx \geq 0 \ . \tag{18.4.6}$$

The corresponding equation of balance is

$$\frac{\partial\Lambda_S(R,t)}{\partial t} = D\frac{\partial^2\Lambda_F(R,t)}{\partial R^2} + \sigma(R,t) \ , \quad \sigma(R,t) \equiv \frac{\sigma_F(R,t)}{T} \ . \tag{18.4.7}$$

This equation is similar in structure to the equation of entropy balance for Boltzmann gas (13.4.6), derived from the generalized kinetic equation (13.4.1).

Now let us give a simple illustration.

Assume that the distribution $f(x,R,t)$ is completely defined by two functions which satisfy equations (18.3.10); then it has the form of (18.3.12). For the stationary state, from the second equation in (18.3.10) and (18.3.12) we find that

$$f_0(x) = \sqrt{\frac{1}{2\pi\langle E\rangle_0}}\exp\left(-\frac{x^2}{\langle E\rangle_0}\right), \quad \Gamma(a - b\langle E\rangle_0)\langle E\rangle_0 + D_{(x)} = 0 \ . \tag{18.4.8}$$

This equation allows us to find the mean value of $\langle E\rangle_0 \equiv \langle x^2\rangle_0$ for the stationary state at any value of the controlling parameter.

Return now to the additional condition (18.4.5), which becomes

$$\langle E\rangle_{R,t} \equiv \int x^2\tilde{f}(x,R,t)dx = \langle E\rangle_0 \ . \tag{18.4.9}$$

Accordingly, after renormalization it follows from (18.3.12) that

$$\tilde{f}(x,R,t) = \sqrt{\frac{1}{2\pi\tilde{\sigma}}} \exp\left(-\frac{(x-\langle x\rangle)^2}{2\tilde{\sigma}}\right) , \quad \tilde{\sigma} = \langle E\rangle_0 - \langle x\rangle^2 . \tag{18.4.10}$$

Substituting these expressions for the distribution functions into (18.4.6), we get

$$\Lambda_S(R,t) = \frac{1}{2}\ln\frac{\langle E\rangle_0}{\langle E\rangle_0 - \langle x(R,t)\rangle^2} \geq 0 , \quad \langle x\rangle^2 << \langle E\rangle_0 , \tag{18.4.11}$$

where the "equals" sign corresponds to the stationary state.

Now we have to write the equation in $\langle x\rangle$ in the same approximation. We turn to the first equation in (18.3.10), and, using equation (18.4.8) for $\langle E\rangle_0$ and taking advantage of the fact that $\langle E\rangle = \text{const}$, obtain the desired equation:

$$\frac{\partial\langle x\rangle}{\partial t} = -\gamma_{a_f}\langle x\rangle + D\frac{\partial^2\langle x\rangle}{\partial R^2} , \quad \gamma_{a_f} = \Gamma\big(-a + b\langle E\rangle_0\big) = \frac{D_{(x)}}{\langle E\rangle_0} . \tag{18.4.12}$$

We see that the evolution of function $\langle x\rangle$ is described by a linear equation which contains two dissipative terms. The first of these is determined by spatial diffusion with the coefficient D, and the second by the internal diffusion within the "point" of the medium. The second term depends considerably on the value of the controlling parameter. Let us quote the values of mean energy $\langle E\rangle_0$ as found by solving equation (18.4.8), and the coefficient of damping γ_{a_f} for a number of particular cases.

1. *The controlling parameter is zero*, and the coefficient of internal diffusion is small. Then

$$\langle E\rangle_0 = \frac{D_{(x)}}{\Gamma} = \frac{\kappa T}{m\omega_0^2} , \quad \gamma_{a_f} = \Gamma = \frac{\omega_0^2}{\gamma} . \tag{18.4.13}$$

The coefficient of damping does not depend on the structure of the "continuous medium", and is determined by the parameters of individual atom. For weakly damped oscillator the relaxation time is much less than the period of oscillations.

2. *Because of the field, the frequency of linear oscillator goes to zero, $a_f = 1$*, which corresponds to the point of bifurcation. Then

$$\langle E \rangle_0 = \sqrt{\frac{D_{(x)}}{b}} , \quad \gamma_{a_f} = \sqrt{D_{(x)}b} . \tag{18.4.14}$$

We see that at the bifurcation point the relaxation time depends on the structure of the "continuous medium".

3. *The system is far above the bifurcation point*, $a_f \gg 1$. Then the potential $U(x)$ and the corresponding effective potential $U_{\text{eff}}(x)$ have two minima; the mean energy and the relaxation time are given by

$$\langle E \rangle_0 = \frac{a_f}{b} , \quad \gamma_{a_f} = \frac{D_{(x)}}{a_f}b . \tag{18.4.15}$$

Now the relaxation time is proportional to $\langle E \rangle_0^{-1}$, and the process of relaxation is therefore slow.

We see that the character of time relaxation is changed considerably when the bifurcation point is crossed as the controlling parameter a_f increases. This is one of the manifestations of a phase transition taking place in the system. We shall return to this discussion in Sect. 18.6, devoted to the basics of the theory of equilibrium phase transitions of the second kind; the controlling parameter will be represented by the temperature.

18.5. Example of Generalized FKPP Equation

FKPP equation (18.3.7) has been obtained from the kinetic equation (18.3.6) under the condition $D_{(x)} = 0$. Let us consider a simple example which elucidates the role of this diffusion.

Assume that spatial diffusion with the coefficient D is one-dimensional and occurs in the direction ξ. Then the FKPP equation admits a stationary solution of the form

$$\langle x(\xi) \rangle = -\sqrt{\frac{a}{b}} \tanh\left(\sqrt{\frac{a}{2D}}\xi\right), \quad \langle x(\xi = \pm\infty) \rangle = \pm\sqrt{\frac{a}{b}} , \tag{18.5.1}$$

which describes the stationary transition from a positive value on the left to a negative value of the same magnitude on the right. The relative width of transition (the front width) is

$$\frac{\Delta\xi}{\langle x(\xi=\infty)\rangle}=\sqrt{\frac{2Db}{a^2}}\ ,\quad a=a_{\mathrm{f}}-1\ . \tag{18.5.2}$$

This solution only holds when $a>0$, above the bifurcation point. Moreover, it does not hold too close to the bifurcation point, since the front width exhibits unlimited growth.

In order to obtain a more general result, good for arbitrary values of the controlling parameter, one must use the kinetic equation (18.3.6) in place of the FKPP equation (Klimontovich 1991, V). After some straightforward algebra we come to an equation which differs from the FKPP equation by the replacement

$$a\Rightarrow a^*=\begin{cases}\dfrac{1}{2}a\left(1+\sqrt{\dfrac{2D_{(x)}b}{\pi a^2}}\right) & \text{if } a>0\ ,\\[4mm] \dfrac{1}{2}a\left(1-\sqrt{\dfrac{2D_{(x)}b}{\pi a^2}}\right) & \text{if } a<0\ .\end{cases} \tag{18.5.3}$$

Then the solution (18.5.1) (with the above replacement) depends on the coefficient of diffusion with respect to x. The relative front width now is

$$\frac{\Delta\xi}{\langle x(\xi=\infty)\rangle}=\sqrt{\frac{2Db}{a^{*2}}}\ . \tag{18.5.4}$$

This expression holds at any value of the controlling parameter; in particular, at the point of bifurcation ($a_{\mathrm{f}}=1$) we have

$$a^*=\frac{1}{2}\sqrt{D_{(x)}b}\ ,\quad\text{and}\quad \frac{\Delta\xi}{\langle x(\xi=\infty)\rangle}=\sqrt{\frac{D}{D_{(x)}}}\ . \tag{18.5.5}$$

Clearly, the use of kinetic equations in place of reaction-diffusion equations brings more details into the description of nonequilibrium processes in active media.

The term "active medium" refers here to systems with nondissipative instability. This is justified because, as the controlling parameter increases, the system recedes from the most chaotic equilibrium state. Less symmetrical and higher organized states arise when the point of bifurcation is crossed. The change of controlling parameter is brought about by the change in the acting field, like the Lorenz field in dielectrics. If the controlling parameter is determined by the temperature, we are dealing with a phase transitions of the second kind.

18.6. Temperature as Controlling Parameter at Phase Transitions

18.6.1. *Landau's Theory. Kinetic Approach*

Let us return to the kinetic equation (18.3.6) and assume that the characteristic time of spatial diffusion is much smaller than other relaxation times. Then equation (18.3.6) can be replaced by a simpler kinetic equation for the distribution function of variable x smoothed over the volume V:

$$\frac{\partial f(x_V,t)}{\partial t} = \frac{\partial}{\partial x_V}\left[D_{(x)}(x_V)\frac{\partial f}{\partial x_V} - \frac{F(x_V)}{m\gamma}f\right], \quad D_{(x)} = \frac{1}{N}\frac{\kappa T}{m\gamma}\left(1 + bx_V^2\right),$$

$$\tag{18.6.1}$$

$$F(x_V) = -m\omega_0^2\left[1 - a_f + bx_V^2\right]x_V .$$

The stationary solution of this equation is

$$f_0(x_V) = C\exp\left(-N\frac{U_{\text{eff}}(x_V)}{\kappa T}\right), \quad \int f(x_V,t)dx_V = 1 , \tag{18.6.2}$$

$$U_{\text{eff}} = \frac{m\omega_0^2}{2}\left[x_V^2 - \frac{a_f}{b}\ln\left(1 + bx_V^2\right)\right], \tag{18.6.3}$$

which corresponds to the distribution (18.4.1). Now, however, the volume-homogeneous distribution has presumably been established, and hence the factor of $1/N$ appears in the expression for $D_{(x)}$.

The controlling parameter a_f for the temperature-controlled phase transition depends on the temperature; it is useful to mark off on the temperature scale the point corresponding to the critical temperature T_c and take it for the reference point of the temperature. Since the controlling parameter a_f is dimensionless, there is yet another characteristic temperature parameter, which may be defined as the temperature half-width ΔT of the region of phase transition. Then the linkage between a_f and T is given by

$$a = 1 - a_f = \frac{T - T_c}{\Delta T} . \tag{18.6.4}$$

We assume that the medium is composed of nonlinear Duffing oscillators. Observe that this is only done for the sake of conspicuousness; all the main results are general.

We select three characteristic values of the controlling parameter:

1.
$$a_f = 0 , \quad T \geq T_c + \Delta T . \tag{18.6.5}$$

In this case the acting field, which induces the phase transition, is zero; the distribution (18.6.2) coincides with the Boltzmann distribution for harmonic oscillator.

2.
$$a_f = 1 , \quad T = T_c . \tag{18.6.6}$$

This is the critical point of phase transition.

3.
$$a_f \gg 1 , \quad T \leq T_c - \Delta T . \tag{18.6.7}$$

In this temperature range the phase transition has been accomplished.

The kinetic equation (18.6.1) may serve as basis for Landau's theory of phase transitions, provided that the spatial diffusion is the fastest relaxation process.

We restrict our discussion to those facts which will be useful for the statistical theory of open systems. Our next task consists in showing that a phase transition which takes place when the temperature falls below the critical point may be regarded as a process of self-organization. This view is supported, in particular, by calculations of entropy according to the criterion of S–theorem. The theory of phase transitions is presented, for example, in Landau – Lifshitz 1976; Stanley 1971; Patashinskii – Pokrovskii 1982; Klimontovich III, IV.

18.6.2. Relative Degree of Order at Phase Transitions. S–Theorem

Let us start by considering, along with (18.6.2), a simpler distribution for the critical region of phase transition. Assume that the following two inequalities hold:

$$bx_V^2 \ll 1 , \quad |a_f - 1| = \frac{|T - T_c|}{\Delta T} \ll 1 , \tag{18.6.8}$$

the second of which delimits the critical region, and the first will be explained later on. Then the distribution (18.6.2) assumes the form

$$f_0(x_V) = C \exp\left[-N \frac{m\omega_0^2\left[(1-a_f)x_V^2 + \frac{b}{2}x_V^4\right]}{2\kappa T_c}\right], \quad 1-a_f = \frac{T-T_c}{\Delta T} << 1 . \qquad (18.6.9)$$

Distributions (18.6.2), (18.6.9) at $T < T_c$ have two most probable values, corresponding to

$$(x_V)_{m.p.} = \pm\sqrt{\frac{|a|}{b}} . \qquad (18.6.10)$$

After Landau, we introduce the "parameter of order" η, for which we take that of the most probable values which is distinguished by the arbitrarily small external force, $\eta = \sqrt{|a|/b}$. We see that at the critical point the parameter of order is zero. As the critical point is approached, the susceptibility, which determines the change in the parameter of order under the action of weak external field, is known to vary according to Curie's law, $|T - T_c|^{-1}$. Using distributions (18.6.2), (18.6.9), one can calculate the fluctuations of the parameter of order $\delta\eta = x_V - \eta$.

Now we apply the criterion of S–theorem for assessing the relative order of states before and after crossing the critical point. The natural choice for the reference state is the state with $a_f = 0$; the acting field which determines restructuring of the potential (the appearance of two minima) is then absent.

Distribution (18.6.2) for the selected state coincides with the Boltzmann distribution for harmonic oscillator. This allows us to define H_{eff} as the potential energy of harmonic oscillator. As a result, the additional condition (12.8.3) or (16.5.6) can be rewritten as

$$\frac{\kappa\tilde{T}}{m\omega_0^2} = \int x_V^2 \tilde{f}_0(x_V)dx_V = \int x_V^2 f(x_V, a_f)dx_V , \qquad (18.6.11)$$

where the Boltzmann distribution with the temperature $\tilde{T}$ must be substituted into the first integral, and distribution (18.6.2) (or, for the critical region, (18.6.9)) into the second.

For illustration, let us select another state with $a_f >> 1$. For the two selected states the mean values of $\langle x_V^2 \rangle$ are:

$$(1) \ \langle x_V^2 \rangle = \frac{1}{N}\frac{\kappa T}{m\omega_0^2} , \quad \text{and} \ (2) \ \langle x_V^2 \rangle = \frac{T_c - T}{\Delta Tb} . \qquad (18.6.12)$$

The second expression is a consequence of the Gaussian distribution which follows from

(18.6.9) after expansion in small deviation from the most probable value $x_V - \eta$. The mean value and the dispersion are given by

$$\langle x_V \rangle \equiv \eta = \sqrt{\frac{a_f - 1}{b}}, \quad \sigma = \frac{1}{N} \frac{\kappa T_c}{2m\omega_0^2 (a_f - 1)}, \quad a_f - 1 = \frac{T_c - T}{\Delta T}. \tag{18.6.13}$$

The relative dispersion is small. We see that in this approximation the value of $a_f - 1$ satisfies the two inequalities of (18.6.8). The first of these indicates that $T - T_c$ is confined within the region ΔT, whereas the second states that T is not too close to T_c.

Making use of equation (18.6.11) and expressions (18.6.12) we find the renormalized temperature at the state of equilibrium:

$$\kappa \tilde{T} = N \frac{T_c - T}{\Delta T} \frac{m\omega_0^2}{b}. \tag{18.6.14}$$

Inequality

$$\kappa \tilde{T} \gg \kappa T \tag{18.6.15}$$

certainly holds because N is quite large. We see that the temperature of the equilibrium state must be increased considerably to make condition (18.6.11) hold good for the two selected states.

The difference in entropies of the two states in question is

$$S_{(1)} - S_{(2)} = \int \ln \frac{f_{(2)}}{f_{(1)}} f_{(2)} dx_V \sim \ln N \gg 0. \tag{18.6.16}$$

It follows that phase transition in a macroscopic system to a less symmetrical state is associated with a substantial decrease in the entropy. According to the criterion of S–theorem, we are thus dealing with the process of self-organization.

Observe that distribution (18.6.2) has been replaced by the less complicated distribution (18.6.9) only to simplify the analysis. Equation (18.6.11) can be used with the general distribution (18.6.2) for finding the renormalized temperature $\tilde{T}$ at any value of the controlling parameter (that is, at any value of the temperature, including T_c. This allows one to compare the relative order of states of the system under consideration.

To end this section, let us compare the relative degree of order in case of Brownian motion of an individual Duffing oscillator, when the equilibrium stated is defined by distribution (15.9.14), and one of the oscillators of the system considered in the present section. We assume that the masses and the values of controlling parameter are the same, and define a region below the critical point where Gaussian approximation can be used. In the second case the mean value and the dispersion of x_V are given by (18.6.13); in the first case in the expression for dispersion we must set $N = 1$. Since the mean energies in both cases are the same (to an accuracy of $1/N$), there is no need of renormalization, and the relative degree of chaoticity is determined by the difference in the entropies of the states "$N - 1$" and "N". For the Gaussian approximation we find that

$$S_{(N-1)} - S_{(N)} = \ln N \geq 0 . \tag{18.6.17}$$

This means that the degree of chaoticity for an individual particle is higher than that for the same particle in a collective. The number of "members" of the collective N plays the role of controlling parameter.

So far we have been considering media in which the dissipation is nonlinear. It is in such systems that phase transitions can be controlled by the temperature. Next we shall consider examples of active media where nonequilibrium phase transitions are feasible.

18.7. Medium of Bound Oscillators

In Sect. 16.2 – 16.4 we considered Brownian motion in oscillator with symmetrized nonlinearity. Assume that N such oscillators are elements of active medium. Then we may define the local value of the energy of oscillations,

$$E(R,t) , \quad E(t) = \int E(R,t) \frac{dR}{V} . \tag{18.7.1}$$

The relevant FKPP equation is

$$\frac{\partial E(R,t)}{\partial t} = (\alpha - \beta E(R,t))E(R,t) + D \frac{\partial^2 E}{\partial R^2} . \tag{18.7.2}$$

To make the description more comprehensive, we must take into account the structure of individual elements of the medium. With this purpose we use a more general equation for the distribution function $f(E,R,t)$:

$$\frac{\partial f}{\partial t} = \frac{\partial}{\partial E}\left[D_{(E)}(E)E\frac{\partial f}{\partial E}\right] + \frac{\partial}{\partial E}[(-\alpha + \beta E)Ef] + D\frac{\partial^2 f}{\partial R^2} , \tag{18.7.3a}$$

$$\int_0^\infty dE \int \frac{dR}{V} f(E,R,t) = 1 , \quad D_{(E)} = \gamma \kappa T \left(1 + \frac{b}{\gamma} E \right) . \tag{18.7.3b}$$

Like equation (18.3.6) for the medium composed of bistable elements, this equation contains two coefficients of diffusion. Let us consider two extreme cases.

1. *The noise within elements of the medium is negligibly small* ($D_{(E)} = 0$). Then (18.6.3) admits a particular solution

$$f(E,R,t) = \delta\big(E - E(R,t)\big) , \quad \langle E \rangle = \int_0^\infty E f dE = E(R,t) . \tag{18.7.4}$$

2. *The characteristic time of spatial diffusion is much less than the other relaxation times.* Then the spatially homogeneous distribution is established at the first stage of relaxation, and so we may go over to equation in a simpler distribution function $f(E,t)$, which formally coincides with (16.2.8).

So far it has been assumed that the oscillators are only linked through spatial diffusion. Imagine now that all oscillators which compose the medium have common feedback. The most complete description of such medium can be based on the kinetic equation for the "N–particle distribution function".

It is also possible, however, to single out the coherent motion of oscillators of the system. We consider the stationary state, and denote by $\mathcal{E}$ and E, respectively, the values of total energy and energy per particle, $\mathcal{E} = NE$. Now let us look at two particular cases.

1. *Feedback is zero*, and the system is at equilibrium (state "0"). Then we come to the Boltzmann distribution,

$$f_{(0)} = \frac{N}{\kappa T} \exp\left(-N \frac{E}{\kappa T} \right) , \quad \int_0^\infty f_{(0)}(E) dE = 1 . \tag{18.7.5}$$

2. *All oscillators occur in the state of well-developed generation* (state "g"). Then the distribution function is (cf. (16.6.3))

$$f_{(g)} = \sqrt{\frac{Nb}{2\pi\alpha\kappa T}} \exp\left[-N\frac{\left(E-\frac{\alpha}{\beta}\right)^2}{\frac{2\alpha\kappa T}{\beta}}\right].$$ (18.7.6)

Like in Sect. 16.6, we may once again compare the relative degree of order at equilibrium and in the state of well-developed generation. In place of (16.6.12) we now have

$$\tilde{S}_{(0)} - S_{(g)} = \ln\left(N\frac{1}{2\pi\varepsilon}\frac{\gamma+\alpha}{\gamma}\right) >> 0 \ , \ \ \varepsilon = \frac{\beta\gamma\kappa T}{\alpha^2} << 1 \ .$$ (18.7.7)

We see that, upon transition to the regime of well-developed generation, the increase in the degree of order is much greater (by $\ln N$) for a collective of oscillators than for an individual oscillator. This is a manifestation of the organizing role of the collective. This result is similar to (18.6.17) for equilibrium phase transition.

We may also demonstrate the organizing role of the collective in a different way. Like in Sect. 18.6, it is possible to compare the degree of order of an individual oscillator in the state of well-developed generation, and one of the oscillators of a collective at one and the same value of feedback parameter. Making use of distribution functions (16.6.3), (18.7.6), we find the corresponding entropies:

$$S_{(1)} = \ln\sqrt{\frac{2\pi D}{b}} + \frac{1}{2} \ , \ \ S_{(N)} = \ln\sqrt{\frac{2\pi D}{Nb}} + \frac{1}{2} \ , \ \ D = \gamma\kappa T \ .$$ (18.7.8)

SInce the mean values of the energy of oscillations are the same, there is no need for renormalization. The difference in these entropies defines the relative degree of order of the two states:

$$S_{(1)} - S_{(N)} = \ln\sqrt{N} \geq 0 \ , \ \ N \geq 1 \ .$$ (18.7.9)

We see that, like in (18.6.17), the number N once again acts as the controlling parameter.

The element of active medium has been represented here by the most simple Van der Pol oscillator with symmetrized nonlinearity. The use of more sophisticated elements, like Van der Pol – Duffing oscillators, oscillators with inertial nonlinearity, etc., offers immense possibilities for constructing models of active media. As our last example, let us consider a chemically reacting active medium.

18.8. Kinetic Description of Media with Chemical Reactions

As a chemically reacting medium we consider partially ionized plasma, the chemical reactions being represented, as in Sect. 16.11, by electron impact ionization of atoms and the relevant three-particle recombination.

Recall that the dynamic equation for the mean density of number of electrons has the form of (16.1.1). Coefficients of ionization and recombination are linked by Saha's formula. We also assume that the last two equations in (16.11.1) also hold good when fluctuations are taken into account. Then the relevant Langevin equation is (Belyi – Klimontovich 1978; Klimontovich IV):

$$\frac{dn_e}{dt} = \alpha\left(n - n_e - \frac{\beta}{\alpha}n_e^2\right)n_e + \frac{1}{2}\frac{dD(n_e)}{dn_e} + \sqrt{D(n_e)}y(t) . \qquad (18.8.1)$$

In place of (16.11.3), we use the following expression for the intensity of Langevin source (the coefficient of diffusion in the kinetic equation):

$$D(n_e) = \frac{\alpha}{2V_{ph}}n_e\left(n - n_e + \frac{\beta}{\alpha}n_e^2\right) , \qquad (18.8.2)$$

which contains the physically infinitesimal volume. The moments of Langevin source in (18.8.1) are given by

$$\langle y(t)\rangle = 0 , \quad \langle y(t)y(t')\rangle = 2\delta(t - t') . \qquad (18.8.3)$$

The expression for the relevant coefficient of friction follows from the dynamic equation (without Langevin source), and has the form of (16.11.5):

$$A_{n_e} = -\left[\alpha n_e(n - n_e) - \beta n_e^3\right] . \qquad (18.8.4)$$

For kinetic description of partially ionized plasma with due account for the processes of ionization and recombination, we introduce the distribution function of velocities of physically infinitesimal elements of the medium v, their positions R, and the value of n_e or the relevant number of particles within physically infinitesimal volume $N_e = n_e V_{ph}$,

$$f(R, v, n_e, t) , \quad \int f(R, v, n_e, t)dv\frac{dR}{V}dn_e = 1 . \qquad (18.8.5)$$

In this way, we expand the phase space for a more comprehensive kinetic description.. The generalized kinetic equation itself can be written as

$$\frac{\partial f}{\partial t} + v \frac{\partial f}{\partial R} + \frac{F}{m} \frac{\partial f}{\partial v} = I_{(v)} + I_{(R)} + I_{(n_e)} \ . \tag{18.8.6}$$

As compared with the generalized kinetic equation for Boltzmann gas, equation (18.8.6) contains an additional "collision integral" which takes care of the chemical transformations in the model under consideration.

So as to define the structure of this "collision integral", as well as the structure of $I_{(v)}$, one may rely on the regular methods of the kinetic theory (Klimontovich II – IV; Klimontovich – Kremp 1981; Klimontovich – Kremp – Kräft 1987). Elucidation of the form of $I_{(R)}$, however, requires taking into account the structure of the "continuous medium"; in other words, one has to define the physically infinitesimal scales for the case of partially ionized plasma.

This "consistent" derivation of the generalized equation is an extremely arduous task. The problem becomes even more complicated when we are dealing with open systems and describe states far from equilibrium. In case of medium comprised of Van der Pol – Duffing oscillators we saw that the structure of both "collision integrals" and the corresponding stationary solutions in open systems is quite different. This circumstance may greatly affect the course of chemical transformations.

Because of this, we would rather confine ourselves to just establishing the connection between the kinetic equation and the corresponding reaction-diffusion equation.

Assume that the time of transition to local equilibrium is much less than the other relaxation times. Then we may go over from (18.8.6) to equation in a simpler distribution function $f(R, n_e, t)$:

$$\frac{\partial f}{\partial t} = \frac{\partial}{\partial n_e} \left[D(n_e) \frac{\partial f}{\partial n_e} \right] + \frac{\partial}{\partial n_e} \left[A_{n_e} f \right] + D \frac{\partial^2 f}{\partial R^2} \ . \tag{18.8.7}$$

Let us consider two extreme cases for this equation.

1. *The coefficient of diffusion D is so large* that the diffusion time is much less than the characteristic times of ionization and recombination. Then the homogeneous distribution over the volume of plasma is established at the first stage of relaxation. This allows us to use characteristics averaged with respect to the volume V, and replace $V_{ph} \Rightarrow V$ in (18.8.2). As a result, we come back to equation (16.11.6) for the simpler distribution function $f(n_e, t)$, whose stationary solution (16.11.7) may be written explicitly as

$$f_0(n_e) = C \exp \int_0^{n_e} \frac{n - n'_e - \frac{\beta}{\alpha} n'^2_e}{\frac{2}{V} \left[n - n'_e + \frac{\beta}{\alpha} n'^2_e \right]} dn'_e \ , \quad \int_0^{\infty} f dn_e = 1 \ . \tag{18.8.8}$$

The maximum of this distribution is determined by the solution of equation

$$A(n_{max}) = -\left[\alpha n_{max} (n - n_{max}) - \beta n^3_{max} \right] = 0 \ . \tag{18.8.9}$$

This coincides with the stationary solution of the dynamic equation, and hence satisfies Saha's formula.

In the Gaussian approximation the dispersion is given by (16.11.9). As indicated earlier, the dispersion vanishes when the ionization is zero, and when the plasma is completely ionized. This implies that the source of fluctuations in the Langevin equation (18.8.1) characterizes the discrete nature of acts of ionization and recombination within physically infinitesimal volumes.

Physically infinitesimal scales for partially ionized plasma can be defined along the same guidelines as we have used in case of Boltzmann gas and completely ionized plasma (Klimontovich II – V, 1971). Naturally, the scales will depend on the degree of ionization, varying from those for the Boltzmann gas when the degree of ionization is zero, to the values already obtained for the completely ionized plasma.

2. *We neglect the internal structure of the point of "continuous medium"* $(D(n_e) = 0)$, and select the class of distribution functions

$$f(n_e, R, t) = \delta(n_e - n_e(R, t)) \ , \quad \langle n_e \rangle = n(R, t) \ . \tag{18.8.10}$$

For the "field variable" $n(R, t)$ we obtain the following equation:

$$\frac{\partial n_e(R, t)}{\partial t} = \alpha n_e(R, t) \left[n - n_e(R, t) - \frac{\beta}{\alpha} n^2_e(R, t) \right] + D \frac{\partial^2 n_e}{\partial R^2} \ . \tag{18.8.11}$$

In this way we have arrived at the equation of reaction-diffusion type, an FKPP equation. The "reaction term" here is defined by the right-hand side of the corresponding dynamic equation. Interaction between individual elements of the medium occurs only through spatial diffusion.

Equation (18.8.11) may be refined by taking the internal diffusion into account. To this end, we regard (16.8.10) as the approximation of the first moment. This brings us to the following equation:

$$\frac{\partial n_e(R,t)}{\partial t} = \alpha n_e(R,t)\left[n - n_e(R,t) - \frac{\beta}{\alpha}n_e^2(R,t)\right] + \frac{dD(n_e)}{dn_e} + D\frac{\partial^2 n_e}{\partial R^2} .$$ (18.8.12)

The inclusion of internal diffusion calls for redefinition of the coefficients of ionization and recombination. These additional contributions are important when there exist bifurcation points which result from imperfection of the plasma, and from external causes.

So, we have considered a number of exemplary kinetic equations for different media. Of course, the elements of the media may be more sophisticated, like oscillators with inertial nonlinearity or other oscillators discussed in Ch. 16. As indicated above, the elements of active media may be represented by biological objects. The interested reader may refer to abundant literature. Recall that the kinetic equations, like the dynamic dissipative equations for the distribution functions, can also be used for the calculation of large-scale (kinetic and hydrodynamic) fluctuations. This is the task of the next chapter.

CHAPTER 19

KINETIC THEORY OF FLUCTUATIONS IN ACTIVE MEDIA

The kinetic theory of fluctuations for Boltzmann gas has been developed in Ch. 9, 10. We employed the results of this theory for calculating the kinetic fluctuations in partially ionized plasma in Sect. 16.11, 18.8. Now we are going to calculate the equilibrium and especially the nonequilibrium fluctuations in active media.

Let us give a brief summary of the problem.

The physical basis behind the transition from the reversible equations of mechanics to the irreversible and approximative kinetic equations is the dynamic instability of motion of elements of the system (for instance, atoms in Boltzmann gas). The resulting mixing of trajectories of particles makes it possible to go over to the simpler equations for the macroscopic characteristics of continuous medium. This, as indicated in the Introduction, is the constructive aspect of the dynamic instability of motion of "particles" of the system.

By defining the physically infinitesimal scales of the medium in a way consistent with the characteristics of mixing, we have been able to separate the small-scale and the large-scale (or fine-grained and coarse-grained) fluctuations. Elimination of the small-scale fluctuations gives rise to "collision integrals" in the kinetic equations. Large-scale fluctuations are not taken into account at the kinetic stage of the theory. It is the large-scale fluctuations, however, that are responsible for such fundamental phenomena as Brownian motion (in the broadest sense), scattering of waves in various media, turbulent motions of diverse physical nature.

As demonstrated in Ch. 10, one of the possible methods of calculation of such fluctuations is based on the solution of the kinetic equation with appropriate Langevin source, whereas the intensity of random source is defined by the solution of the kinetic equation itself. This makes the problem of calculation of fluctuations a natural part of the more general kinetic theory. Its solution carries additional information about the system and brings us closer to the "complete knowledge" of the system.

We shall see that a consistent theory of large-scale fluctuations must be rely on the generalized kinetic equations which form the basis of the unified kinetic, hydrodynamic and diffusion theory of fluctuations. Much effort is required, however, to substantiate this claim.

19.1. Unified Description of Kinetic and Hydrodynamic Fluctuations

We know that there are two ways of calculating the fluctuations for the kinetic Boltzmann equation. One is based on the Leontovich equation for the many-particle distribution function. Then the intensity of the source which reflects the atomic structure of the medium is given by (9.7.7). The intensity is nonzero even at equilibrium, when it is defined by a simpler expression (9.7.8). The other is the Langevin method. As follows from (10.4.7), the intensity of Langevin source in the kinetic Boltzmann equation is given by the same expressions (9.7.7), (9.7.8).

Since these expressions are completely determined by the structure of the linearized collision "operator" and the solution of the kinetic equation, they actually are universal and can be used for calculating the fluctuations on the basis of generalized kinetic equation (13.3.4) with "collision integrals" (13.3.10), (13.3.11). In the explicit form this is equation (13.4.1).

Universality of formulas (9.7.7), (9.7.8) is, of course, not accidental. They are universal because universal is their cause, the atomic structure of "continuous medium".

The linearized collision operator (9.7.2) in (9.7.7) is determined by the structure of Boltzmann's collision integral. For the generalized kinetic equation it is now represented as a sum of two operators,

$$\delta \hat{I}_{(v,R)} = \delta \hat{I}_{(v)}(R,v,t) + \delta \hat{I}_{(R)}(R,v,t) \ . \tag{19.1.1}$$

The first operator is defined by the structure of the nonlinear "collision integral" (13.3.11). Its action on the arbitrary function $F(R,v,t)$ in linear approximation in F is described by

$$-\delta \hat{I}_{(v)} F(R,v,t) = -\frac{\partial}{\partial v}\left[D_{(v)}(R,t)\frac{\partial}{\partial v} F \right] - \frac{\partial}{\partial v}\left[\frac{1}{\tau}(v - u(R,t))F \right]$$

$$-\frac{\partial}{\partial v}\left[\delta \hat{D}_{(v)}(R,t)\frac{\partial}{\partial v} f \right] + \frac{\partial}{\partial v}\left[\frac{1}{\tau}\delta \hat{u}(R,t)f \right] \ . \tag{19.1.2}$$

The coefficient of diffusion $D_{(v)}$ and the mean velocity u are expressed in terms of the distribution function $f(R,v,t)$ by (13.3.12). The action of the corresponding operators on F is described by

$$\delta \hat{D}_{(v)} F = \frac{1}{\tau} \frac{n}{\rho} \int \frac{m \delta v^2}{2} F dv - D_{(v)} \frac{nF}{\rho} \,,$$

$$\tag{19.1.3}$$

$$\delta \hat{u} F = \frac{n}{\rho} \int v F dv - u \frac{nF}{\rho} \,.$$

Recall that the "collision integral" $I_{(v)}$, like Boltzmann's collision integral, has the properties (13.3.13). A similar property holds good for (19.1.2):

$$n \int \varphi(v) \delta I_{(v)} F(R,v,t) dv = 0 \quad \text{if} \quad \varphi(v) = 1. \; v, \; v^2 \,. \tag{19.1.4}$$

This property will be used in calculating the fluctuations of gasdynamic functions $\delta\rho(R,t)$, $\delta u(R,t)$, $\delta T(R,t)$.

The second "collision integral" (13.3.10) is linear, and therefore the action of the corresponding "linearized collision operator" on the arbitrary function $F(R,v,t)$ is defined by

$$\delta \hat{I}_{(R)} F(R,v,t) = - \left\{ \frac{\partial}{\partial R} \left[D \left(\frac{\partial}{\partial R} - \frac{F(R,t)}{\kappa T} \right) \right] \right\} F(R,v,t) \,. \tag{19.1.5}$$

The "minus" sign in the definitions of operators (19.1.2), (19.1.5) is introduced so as to emphasize the dissipative nature of the "collision integrals".

So, for the unified description of kinetic and hydrodynamic fluctuations, a Langevin source $y(R,v,t)$ must be introduced into the generalized kinetic equation (13.4.1). The moments of Langevin source are given by (10.4.2), whereas its intensity, instead of being defined by (10.5.13) (or (9.7.7)), now is given by the more general expression

$$A(R,v,R',v') = \left[\left(\delta \hat{I}_{(R,v)} + \delta \hat{I}_{(R',v')} \right) - \left(\delta \hat{I}_{(R,v)} + \delta \hat{I}_{(R',v')} \right)_0 \right]$$

$$\times \; n\delta(R - R')\delta(v - v') f(R,v,t) \,. \tag{19.1.6}$$

As before, subscript "0" indicates that the "collision operators" in parentheses only act on the regular functions, and do not act on the delta-function. Distribution function $f(R,v,t)$ satisfies the generalized kinetic equation (13.4.1) (the kinetic equation without the source of fluctuations).

19.2. Kinetic Fluctuations at Brownian Motion

In this section we shall discuss simple examples of application of the above general results, starting with the kinetic equations of the theory of Brownian motion.

19.2.1. *Langevin Source in Fokker – Planck Equation*

Let us return to the Fokker – Planck equation (15.1.6), where we introduce the Langevin source and write the equation in the fluctuation δf :

$$\left[\frac{\partial}{\partial t} + v \frac{\partial}{\partial r} + \frac{F}{M} \frac{\partial}{\partial v} + \delta \hat{I}_v^{(F-P)} \right] \delta f(r,v,t) = y(r,v,t) \ . \tag{19.2.1}$$

The moments of Langevin source are given by (10.4.2). This time, however, introduced into (10.5.13) or (9.7.7) must be the appropriate linearized "collision operator", whose structure is defined by the right-hand side of equation (15.1.6):

$$\delta \hat{I}_{F-P} = -\left[D \frac{\partial^2}{\partial v^2} + \frac{\partial}{\partial v} \gamma v \right] \cdots \ . \tag{19.2.2}$$

As a result, we obtain the following expression for the intensity of Langevin source:

$$A_{F-P} = 2 \frac{D}{n} \delta(r-r') \frac{\partial^2}{\partial v \partial v'} \left(\delta(v-v') f(r,v,t) \right) , \quad D = \gamma \frac{\kappa T}{M} \ . \tag{19.2.3}$$

Thus, the intensity of the source, which reflects the "atomic structure" of the system of Brownian particles, is determined by the solution of Fokker – Planck equation (15.1.6) and is proportional to the relevant coefficient of diffusion.

Let us discuss some implications of expression (19.2.3).

We define a volume within the system which is greater than or about equal to the physically infinitesimal volume, $\Delta V \geq V_{ph}$, and introduce coordinates and velocities of Brownian particles averaged with respect to this volume:

$$\delta r_{\Delta V} = \int r \delta f(r,v,t) \frac{dr dv}{\Delta V} \ , \quad \delta v_{\Delta V} = \int v \delta f(r,v,t) \frac{dr dv}{\Delta V} \ . \tag{16.2.4}$$

The equations for these variables follow from the Langevin equation (19.2.1):

$$\frac{d\delta r_{\Delta V}}{dt} = \delta v_{\Delta V} \ , \quad d\delta v_{\Delta V} + \mathcal{W}_{\Delta V} = y_{\Delta V}(t) \ . \tag{16.2.5}$$

Formally, these equations are similar to the Langevin equations for individual Brownian particle (15.1.2); this time, however, we are dealing with averaged nondeterministic (random) motion of a collective of $\Delta N = n\Delta V$ particles, rather than with the motion of a single particle. This fact is reflected in the structure of the intensity of Langevin source $y_{\Delta V}(t)$:

$$\left\langle y_{\Delta V}(t)y_{\Delta V}(t')\right\rangle_{ij} = 2\frac{D}{\Delta N}\delta_{ij}\delta(t-t') \ , \quad \Delta N = n\Delta V \ , \tag{16.2.6}$$

which is smaller by a factor of ΔN than that for a single Brownian particle.

In this connection let us recall the problem of Brownian motion of collective variables (charge and current) in oscillatory circuit. The collective nature of motion in equations (17.3.1), (17.3.2) is concealed because of the use of macroscopic variables, and only comes up when we go over to averaged characteristics related to one electron (equations (17.3.8), (17.3.9)).

Now we go over from equation (19.2.1) to the corresponding "hydrodynamic equations". This will allow us to introduce Langevin sources into equations (15.10.1) – (15.10.3) based on the Fokker – Planck equation (15.1.6). As a result, we shall also find the Langevin source in Einstein – Smoluchowski equation.

19.2.2. Langevin Source in Einstein – Smoluchowski Equation

In order to obtain the intensity of the source in the continuity equation (15.10.1), we must carry out integration with respect to v and v'. As a result, we find that the intensity of the source of fluctuations $y(R,t) = \int y(R,v,t)dv$ is zero.

This does not imply, however, that the fluctuations of density of Brownian particles are zero; we shall see that their source enters via the fluctuations of flow of matter ρu in equation (15.10.1). Let us prove this statement.

By $\delta j(R,t)$ we denote the Langevin source in equation of balance of density of momentum. Then

$$J_{\text{source}}(R,t) = Mn\int vy(R,v,t)dv \ , \tag{19.2.7}$$

and the corresponding correlator of fluctuations is

$$\langle J_i J_j \rangle^{\text{source}}_{r,t,r',t'} = 2Dmp\delta_{ij}\delta(r-r')\delta(t-t') \ . \tag{19.2.8}$$

Recall now that if the time of diffusion τ_D is much greater than the relaxation time $\tau_{\text{rel}} = 1/\gamma$, then it is possible to go over from equations of hydrodynamics of Brownian particles (15.10.1) – (15.10.3) to the closed equation in the density $\rho(R,t)$ (the Einstein – Smoluchowski equation (15.9.7) in the function $f(R,t) = \rho(R,t)/Mn$). The density of momentum is given by (15.10.6), the substitution whereof into (15.10.1) results in the Einstein – Smoluchowski equation.

With due account for fluctuations, we get the following equation with the source in place of (15.10.6):

$$\rho u = -D_{(R)}\left[\frac{\partial \rho}{\partial R} - \frac{F}{\kappa T}\rho\right] + J_{\text{source}} \ , \quad D_{(R)} = \frac{\kappa T}{M\gamma} \ . \tag{19.2.9}$$

Substitution of this expression into the continuity equation results in the Einstein – Smoluchowski equation (15.9.7) with the Langevin source:

$$\frac{\partial \rho}{\partial t} = \frac{\partial}{\partial R}\left[D_{(R)}\left(\frac{\partial \rho}{\partial R} - \frac{F}{\kappa T}\rho\right)\right] - \frac{\partial}{\partial R} J_{\text{source}} \ . \tag{19.2.10}$$

This equation involves the derivative of the source; of course, the source may be redefined as

$$y(R,t) = -\frac{\partial}{\partial R} J_{\text{source}} \ . \tag{19.2.11}$$

Making use of (19.2.8), we find the corresponding correlator:

$$\langle yy \rangle_{R,t,R',t'} = 2D_{(R)}\frac{\partial^2}{\partial R \partial R'} M\rho(R,t)\delta(R-R')\delta(t-t') \ . \tag{19.2.12}$$

Function $\rho(R,t)$ is found by solving equation (15.9.7).

The last result can also be obtained from the general expression (10.5.13) (or (9.7.7)) by substituting the appropriate expression for the linearized "collision operator",

$$\delta\hat{l}_{\text{E-S}} = -\frac{\partial}{\partial R}\left[D_{(R)}\left(\frac{\partial}{\partial R} - \frac{F}{\kappa T}\right)\right] \ , \tag{19.2.13}$$

which brings us back to (19.2.12).

Now let us calculate the fluctuations of density $\delta\rho$. Consider equation (19.2.10) and assume that $F = 0$. Then the mean density ρ for the steady state is constant, and the equation for the Fourier components of $\delta\rho$ can be written as

$$\left(-i\omega + D_{(R)}k^2\right)\delta\rho(\omega,k) = y(\omega,k) .$$ (19.2.14)

The expression for the spectral density of the source follows from (19.2.12):

$$(yy)_{\omega,k} = 2D_{(R)}k^2 M\rho .$$ (19.2.15)

The right-hand side does not depend on the frequency; accordingly, the source is "white noise" in time.

The expression for the space-time spectral density also follows from the above formulas:

$$(\delta\rho\delta\rho)_{\omega,k} = \frac{2D_{(R)}k^2}{\omega^2 + D_{(R)}k^2} M\rho .$$ (19.2.16)

Carrying out integration with respect to the frequency, we find the spatial spectral density,

$$(\delta\rho\delta\rho)_k = \frac{1}{2\pi}\int(\delta\rho\delta\rho)_{\omega,k}\,d\omega = M\rho .$$ (19.2.17)

We see that the spatial spectral density does not depend on the wave number; because of this, the dispersion of fluctuations

$$\left\langle(\delta\rho)^2\right\rangle = \frac{1}{(2\pi)^3}\int(\delta\rho\delta\rho)_k\,dk = \frac{1}{2\pi^2}\int_0^\infty(\delta\rho\delta\rho)_k k^2\,dk$$ (19.2.18)

is infinite (the integral diverges at large wave numbers).

The same conclusion may be derived in a different way. Using (19.2.14), we find the spatial correlation function

$$\left\langle\delta\rho\delta\rho\right\rangle_{R,R'} = \frac{1}{(2\pi)^3}\int(\delta\rho\delta\rho)_{\omega,k}\exp\left[ik(R - R')\right]dk = M\rho\delta(R - R') ,$$ (19.2.19)

whence it follows that the spatial correlation is only nonzero at $R - R'$, where its value is infinite. Accordingly, infinite is also the dispersion, which follows from (19.2.16) with $R = R'$.

Such physically unreasonable results are due to the neglect of the structure of the "continuous medium" of Brownian particles.

19.2.3. Time Correlations at Brownian Motion

Two groups of problems may be distinguished in the study of time evolution of the distribution of Brownian particles. The first includes the processes of time evolution of the given distribution $f(R_0, v_0, t_0)$ at $t = t_0$ towards the equilibrium distribution of Brownian particles (the Maxwell – Boltzmann distribution). This a typical Cauchy problem of mathematical physics for the Fokker – Planck and Einstein – Smoluchowski equations.

To the second group we attribute the problems of calculation of space-time correlations of fluctuations which are observed also when the equilibrium state has been achieved. Let us discuss a few simple problems of this kind.

Assume that $F = 0$; at equilibrium the particles are distributed homogeneously with the density ρ. By

$$\tau = t - t' , \quad r = R - R' \tag{19.2.20}$$

we denote the difference of times and points of observation. Fluctuation $\delta\rho(R, t)$ at equilibrium homogeneous state is a stationary homogeneous process, and the relevant correlations depend only on the absolute values of τ, r.

We return to (19.2.16) and carry out Fourier transform with respect to ω, getting as a result the following expression for time correlation of spatial Fourier components of fluctuations of density:

$$(\delta\rho\delta\rho)_{k,\tau} = \frac{1}{2\pi}\int(\delta\rho\delta\rho)_{\omega,k}\exp(i\omega\tau)d\omega = M\rho \, \exp\left(-D_{(R)}k^2\tau\right) . \tag{19.2.21}$$

We see that when k is fixed the correlation time is of the order of $1/Dk^2$.

Now, carrying out Fourier transform with respect to k, we find the space-time correlation of fluctuations of density of Brownian particles:

$$\langle\delta\rho\delta\rho\rangle_{r,\tau} = \frac{M\rho}{\left(4\pi D_{(R)}|\tau|\right)^{3/2}} \, \exp\left(-\frac{r^2}{4D_{(R)}|\tau|}\right) . \tag{19.2.22}$$

Let us consider some implications of this result.

The above expressions have been obtained without boundary conditions. They only hold for a boundless medium. When the size of the system L is finite, the admissible values of τ are restricted from above; namely, τ must be much less than the time of diffusion τ_D.

Consider now the time correlations at nearby points:

$$\frac{r^2}{D_{(R)}} << \tau << \frac{L^2}{D_{(R)}} \equiv \tau_D .$$

(19.2.23)

Given this inequality, from (19.2.22) we find that the time correlation function is

$$\langle \delta\rho\delta\rho \rangle_{r=0,\tau} = \frac{M\rho}{\left(4\pi D_{(R)}|\tau|\right)^{3/2}} \propto \frac{1}{|\tau|^{3/2}} .$$

(19.2.24)

Because of (19.2.23), here we may not set $\tau = 0$. In other words, it is not possible to derive the expression for the one-time correlator from (19.2.24). If transition to the limit $\tau \to 0$ is carried out in the initial expression (19.2.22), we come back again to (19.2.19) and to infinite dispersion.

To remove these difficulties, we define the physically infinitesimal volume and apply the appropriate smoothing procedure. For the smoothing function we choose Gaussian distribution with the dispersion $2l_{ph}^2$. Then in place of (19.2.19) we get

$$\langle \delta\rho\delta\rho \rangle_{r,\tau} = \frac{M\rho}{\left[4\pi\left(l_{ph}^2 + D_{(R)}|\tau|\right)\right]^{3/2}} \exp\left[-\frac{r^2}{4\left(l_{ph}^2 + D_{(R)}|\tau|\right)}\right] .$$

(19.2.25)

At $r = 0$ hence follows the finite expression for the one-time correlator (with a finite size of the "point"). If $D\tau >> l_{ph}^2$, we come back to expression (19.2.22).

Finally, with due account for the finite size of the "point", we obtain the following expression in place of (19.2.21):

$$(\delta\rho\delta\rho)_{k,\tau} = M\rho \, \exp\left[-\left(l_{ph}^2 + D_{(R)}|\tau|k^2\right)\right] .$$

(19.2.26)

At $\tau = 0$ hence follows the formula for the spatial spectral density of one-time correlator,

$$(\delta\rho\delta\rho)_{k,\tau=0} = M\rho \, \exp\left(-l_{ph}^2 k^2\right) .$$

(19.2.27)

Formerly it was "white noise" in the space of wave numbers; now the distribution with respect to wave numbers has a maximum at $k = 0$. The dispersion of this distribution is of the order of $1/l_{ph}^2$. This result will be used in the next chapter.

Let us now recapitulate the results of the present section.

In (19.2.23) we may now refine the left-hand inequality with due account for the size of the point and for definition (17.1.1), and rewrite it as

$$\frac{1}{\gamma} \sim \tau_{ph} = \frac{l_{ph}^2}{D_{(R)}} \ll \tau \ll \frac{L^2}{D_{(R)}} \equiv \tau_D . \tag{19.2.28}$$

In this way, for l_{ph} we take the size of Brownian particle.

This two-way inequality defines the range of τ for the diffusion description of fluctuations at Brownian motion. We see that in this range the time correlations at nearby points fall off slowly, according to power law $\propto 1/\tau^{3/2}$. Observe that in the kinetic region, where the distribution with respect to velocities is established, the time correlations fall off much faster, by exponential law with the correlation time $\tau_{cor} \sim 1/\gamma$.

So, as the system approaches the state of equilibrium, the process of damping of correlations slows down, switching over from exponential to power law. So far, however, we have been neglecting the finite size of the system. What happens when τ becomes much greater than the time of diffusion, and each Brownian particle within this period reaches the wall more than once (*multiple diffusion*). We shall see that in this case the fall-off of correlations is even more slow, and power law is replaced by logarithmic dependence. For large arguments this may be interpreted as the existence of *residual time correlations*, which keep the memory of the "atomic structure" of the "continuous medium" of Brownian particles.

Similar developments take place in the structure of time and space spectra. To see this more clearly, we rewrite (19.2.25) as

$$\langle \delta\rho \delta\rho \rangle_{r,\tau} = \frac{M\rho}{\left[4\pi \left(l_{ph}^2 + D_{(R)}|\tau| \right) \right]^{d/2}} \exp\left[-\frac{r^2}{4\left(l_{ph}^2 + D_{(R)}|\tau| \right)} \right] , \tag{19.2.29}$$

where d is the dimensionality of the system. Then, for close points and for τ in the range defined by (19.2.28), we have

$$\langle \delta\rho\delta\rho \rangle_{r=0,\,\tau} \propto \frac{1}{|\tau|^{d/2}} \,. \tag{19.2.30}$$

So as to single out the dependence on frequency, we use the Fourier transform of this function:

$$\int_0^\infty \cos\omega\tau \frac{1}{\tau^{d/2}} d\tau = \omega^{d/2-1} \int_0^\infty \cos x \frac{1}{x^{d/2}} dx \,. \tag{19.2.31}$$

Hence it follows that the time spectral density is

$$(\delta\rho\delta\rho)_{r=0,\,\omega} \propto \omega^{d/2-1} \,. \tag{19.2.32}$$

So, for τ in the range defined by (19.2.28), both the time correlations and the corresponding spectrum vary according to the power law. The powers of τ and ω depend on the dimensionality of the system, which normally is an integer, $d = 1,\ 2,\ 3$. There are, however, many examples of systems characterized by non-integer dimensionality (the so-called *fractal systems*). Some systems of this kind will be discussed in the next chapter.

Formulas (19.2.30), (19.2.32) for systems with fractal dimensions are usually written as

$$\langle \delta\rho\delta\rho \rangle_{r=0,\,\tau} \propto \frac{1}{|\tau|^{\beta}} \,, \quad (\delta\rho\delta\rho)_{r=0,\,\omega} \propto \omega^{\gamma} \,, \quad \gamma = \beta - 1 \,. \tag{19.2.33}$$

The relationship between the exponents for systems with fractal dimensions remains the same.

There is yet another exponent which characterizes the dispersion of position of Brownian particle at given value of τ,

$$\langle r^2 \rangle = 3 \cdot 2 D_{(R)} \tau^{\alpha} \,. \tag{19.3.34}$$

If the dimensionality of the system is expressed by whole numbers — that is,

$$d = 1,\ 2,\ 3,\ \text{and } \beta = \frac{d}{2} \,, \quad \gamma = \frac{d}{2} - 1 \,, \quad \alpha = 1 \,, \tag{19.3.35}$$

then the Brownian motion is referred to as "normal" or "classical". If the dimensionality is fractional, and $\alpha \neq 1$, we speak of "anomalous" Brownian motion. Examples of anomalous Brownian motion will be discussed in the next chapter.

Brownian motion may be anomalous also in a different sense. Recall that so far we have been mainly dealing with Brownian motion in unrestricted medium. Otherwise, the diffusion at $\tau \gg \tau_D$ will be multiple, and the volume traversed by a particle will be much greater than the actual volume of the system. In the limit, the volume of the system may be reduced to a point, which corresponds to dimension zero, $d = 0$. Expressions (19.2.30), (19.2.32) then become

$$\langle \delta\rho\delta\rho \rangle_{r=0,\tau} = \text{const} , \quad \left(\delta\rho\delta\rho \right)_{r=0,\omega} \propto \frac{1}{\omega} . \tag{19.2.36}$$

The first of these implies that there exist residual correlations. The magnitude of the constant characterizes the correlation which persists at arbitrarily large times τ.

The $1/\omega$ spectrum is practically universal; this spectrum is often referred to as flicker noise, since it was first discovered as flicker of the cathode in electron ray tube. This process is an example of anomalous Brownian motion, and will also be discussed in the next chapter.

Let us now embark upon calculation of fluctuations of hydrodynamic functions $\rho(R,t)$, $u(R,t)$, $T(R,t)$. We shall start with the traditional approach to calculation of hydrodynamic fluctuations based on Boltzmann's kinetic theory. Difficulties encountered on this path will be overcome by recalculating the fluctuations of hydrodynamic functions on the basis of generalized kinetic equation developed in Ch. 13.

19.3. Calculation of Hydrodynamic Fluctuations Based on Boltzmann's Kinetic Theory

19.3.1. General Properties of Langevin Source in Boltzmann Equation

In transition from the kinetic Boltzmann equation to equations of gas dynamics (Ch. 11) we have used the general properties (6.2.2) of Boltzmann's collision integral. Similar properties hold for the correlator of Langevin source in the kinetic Boltzmann equation.

The correlator is defined by (10.4.2), (10.4.7) (or (9.7.7)). Like we did when deriving equation (6.2.1), we multiply correlator (10.4.2) by two (this time arbitrary) functions of momenta p_1 and p_2. Denoting these functions by, respectively, $\varphi(p_1)$ and $\psi(p_2)$, we consider the following integral:

$$I(r_1,r_2) = \int \varphi(p_1)\psi(p_2)\langle yy \rangle_{r_1,p_1,r_2,p_2,t} dp_1 dp_2 . \tag{19.3.1}$$

Into the right-hand side we substitute the expression for the correlator of the source, and make use of the expression for the linearized collision operator (9.7.2). After some simple but lengthy algebra (see Klimontovich II), expression (19.3.1) is reduced to

$$
I(r_1, r_2, t, t') = \frac{n^2}{2} \int_0^\infty \rho d\rho \int_0^{2\pi} d\varphi \int dp_1 dp_2 |v_1 - v_2|
$$

$$
\times \left[\varphi(p_1) + \varphi(p_2) - \varphi(p_1') - \varphi(p_2') \right]
$$

$$
\times \left[\psi(p_1) + \psi(p_2) - \psi(p_1') - \psi(p_2') \right]
$$

$$
\times f(r_1, p_1, t) f(r_2, p_2, t) \delta(r_1 - r_2) \delta(t - t') . \tag{19.3.2}
$$

This expression goes to zero if one of the following equations is satisfied:

$$
\varphi(p_1) + \varphi(p_2) = \varphi(p_1') + \varphi(p_2') ,
$$

$$
\psi(p_1) + \psi(p_2) = \psi(p_1') + \psi(p_2') . \tag{19.3.3}
$$

Since energy and momentum are conserved at elastic collisions, we have

$$
I(r_1, r_2, t) = 0 \text{ if } \begin{cases} \varphi = 1, \ p, \ \dfrac{p^2}{2m} \text{ (arbitrary function } \psi) \\[3mm] \psi = 1, \ p, \ \dfrac{p^2}{2m} \text{ (arbitrary function } \varphi) \end{cases} \tag{19.3.4}
$$

and hence the correlator $\langle yy \rangle_{x_1, t, x_2, t'}$ exhibits properties which are similar to the properties of Boltzmann collision integral. This fact is very important for the solution of our next problem.

19.3.2. Langevin Sources in Equations of Gas Dynamics

Properties (19.3.4) of correlator of Langevin source imply that the Langevin source $y(r, p, t)$ itself satisfies the following condition:

$$
n \int \varphi(p) y(r, p, t) dp = 0 \text{ if } \varphi = 1, \ p, \ \frac{p^2}{2m} . \tag{19.3.5}
$$

Let us once again reproduce the Langevin equation (10.4.8) for the fluctuation of phase density or the corresponding fluctuation of the distribution function ($\delta N(x,t) = n\delta f(x,t)$):

$$\left(\frac{\partial}{\partial t} + v\frac{\partial}{\partial r} + F\frac{\partial}{\partial p} + \delta\hat{I}_p\right)\delta f(x,t) = y(x,t) ,$$

(19.3.6)

which we are going to use for obtaining the equations for fluctuations of gasdynamic functions. Here we confine ourselves to the state of equilibrium; then the mean values of ρ and T are constant, and the mean velocity is zero, $u = 0$. Fluctuations of gasdynamic functions are linked with δN by the following relations:

$$\delta\rho(r,t) = m\int \delta N dp , \quad \rho\delta u(r,t) = m\int v\delta N dp , \quad \frac{\rho}{m}\frac{3}{2}\delta T(r,t) = \int \frac{p^2}{2m}\delta N dp .$$

(19.3.7)

Because of the properties (19.3.5), the Langevin sources do not appear explicitly in the corresponding gasdynamic equations; instead, they come up in expressions for the tensor of viscous stress π_{ij} and the vector of thermal flux q. Without sources, these quantities are defined by (11.3.6), (11.3.8); their fluctuations with appropriate Langevin sources are given by

$$\delta\pi_{ij} = -\eta\left[\frac{\partial\delta u_i}{\partial r_j} + \frac{\partial\delta u_j}{\partial r_i} - \frac{2}{3}\delta_{ij}\frac{\partial\delta u_k}{\partial r_k}\right] + \pi_{ij}^{source} ,$$

(19.3.8)

$$\delta q_i = -\lambda\frac{\partial\delta T}{\partial r_i} + q_i^{source} .$$

(19.3.9)

The mean values of these Langevin sources are zero. Correlators of Langevin sources for the states of complete or local equilibrium can be obtained by methods of thermodynamics of nonequilibrium processes (Landau – Lifshitz 1957). For the more general case, when the distribution function is defined by the solution of Boltzmann equation, one should use the method employed in Ch. 11 for deriving the equations of gas dynamics. Application of this method for approximate solution of (19.3.6) yields the following expressions:

$$\left\langle \pi_{ij}\pi_{kl}\right\rangle_{r,t,r',t'}^{source} = 2\eta\kappa T\left[\delta_{ik}\delta_{jl} + \delta_{il}\delta_{jk} - \frac{2}{3}\delta_{ij}\delta_{kl}\right]\delta(r-r')\delta(t-t') ,$$

(19.3.10)

$$\left\langle q_i q_j\right\rangle_{r,t,r',t'}^{source} = 2\lambda\kappa T^2\delta_{ij}\delta(r-r')\delta(t-t') .$$

(19.3.11)

Like in (19.2.8), along with such sources we may also introduce the relevant random "forces"

$$f_i^{(\eta)} = -\frac{\partial \pi_{ij}^{\text{source}}}{\partial r_j} \ , \quad f^{(\lambda)} = -\frac{\partial q^{\text{source}}}{\partial r} \ . \tag{19.3.12}$$

Correlators of these forces are represented in the following form:

$$\left\langle f^{(\eta)} f^{(\eta)} \right\rangle_{r,t,r',t'} = 2A^{(\eta)}(r,r',t)\delta(t-t') \ , \tag{19.3.13}$$

$$\left\langle f^{(\lambda)} f^{(\lambda)} \right\rangle_{r,t,r',t'} = 2A^{(\lambda)}(r,r',t)\delta(t-t') \ , \tag{19.3.14}$$

where the corresponding intensities are

$$A^{(\eta)}(r,r',t) = \frac{10}{3}\eta \frac{\partial^2}{\partial r \partial r'}\left[\delta(r-r')\kappa T(r,t)\right] \ , \tag{19.3.15}$$

$$A^{(\lambda)}(r,r',t) = \lambda \frac{\partial^2}{\partial r \partial r'}\left[\delta(r-r')\kappa T^2(r,t)\right] \ . \tag{19.3.16}$$

These results are examples of fluctuation-dissipation relations (FDR) in hydrodynamics. They can, of course, be written in spectral form; for instance, from (19.3.10) we get

$$\left(\pi_{ij}\pi_{kl}\right)_{\omega,k}^{\text{source}} = 2\eta\kappa T\left[\delta_{ik}\delta_{jl} + \delta_{il}\delta_{jk} - \frac{2}{3}\delta_{ij}\delta_{kl}\right] \ , \tag{19.3.17}$$

which is a space-time white noise. At the same time, the spectral functions of "forces" are proportional to k^2, which is typical of the diffusion processes (cf. (19.2.15)).

Let us once again repeat that these random sources reflect the existence of atomic structure of "continuous medium"; for this reason they are nonzero even at equilibrium. Additional sources of noise may exist in the states far from equilibrium. Such noises can be referred to as 'turbulent". They are characteristic of open systems, and will be discussed later.

19.4. Traditional Calculation of Fluctuations of Gasdynamic Functions. Langevin Method

19.4.1. *Langevin Equations for Gasdynamic Functions*

Let us return to equations of gas dynamics (11.1.6), (11.3.9), (11.3.11) for the density ρ, velocity u, and pressure p. We introduce Langevin sources and consider fluctuations with respect to the state of equilibrium with $\rho = $ const, $T = $ const, $u = 0$. In linear approximation with respect to fluctuations we obtain the following set of Langevin equations:

$$\frac{\partial \delta \rho}{\partial t} + \rho \operatorname{div} \delta u = 0 \ , \tag{19.4.1}$$

$$\rho \frac{\partial \delta u_i}{\partial t} = -\frac{\partial}{\partial r_i} \delta p + \eta \left(\Delta \delta u_i + \frac{1}{3} \frac{\partial}{\partial r_i} \operatorname{div} \delta u \right) - \frac{\partial \pi_{ij}^{\text{source}}}{\partial r_j} \ , \tag{19.4.2}$$

$$\frac{\partial \delta p}{\partial t} + \frac{5}{3} p \operatorname{div} \delta u = \frac{2}{3} \lambda \Delta \delta T - \frac{2}{3} \operatorname{div} q^{\text{source}} \ , \tag{19.4.3}$$

$$p = \frac{\rho}{m} \kappa T \ , \quad \delta p = \frac{\kappa T}{m} \delta \rho + \frac{\rho \kappa}{m} \delta T \ . \tag{19.4.4}$$

The moments of Langevin source are given by (19.3.10), (19.3.11). The corresponding equations for the Fourier components are:

$$-i\omega \delta \rho + i(k \delta u) = 0 \ ,$$

$$-i\omega \delta u_i = -\frac{i}{\rho} k_i \delta p - v \left(k^2 \delta u_i + \frac{1}{3} k_i (k \delta u) \right) - \frac{i}{\rho} k_k \pi_{ik}^{\text{source}} \ ,$$

$$\tag{19.4.5}$$

$$-i\omega \delta p + i \frac{5}{3} p(k \delta u) = -\frac{2}{3} \left(k q^{\text{source}} \right) \ ,$$

$$\delta p = \frac{\kappa T}{m} \delta \rho + \frac{\rho \kappa}{m} \delta T \ .$$

These equations must be supplemented by expressions for spectral densities of Langevin source, which, with due account for (19.3.17), are:

$$\left(k_k \pi_{ik} k_l \pi_{jl}\right)^{\text{source}}_{\omega,k} = 2\eta\kappa T\left(\delta_{ij}k^2 + \frac{1}{3}k_ik_j\right),$$ (19.4.6)

$$\left(k_i q_i k_j q_j\right)^{\text{source}}_{\omega,k} = 2\lambda\kappa T^2 k^2, \quad \eta = \rho v, \quad \lambda = \rho c_p \chi.$$ (19.4.7)

So, we have a set of linear algebraic equations with random sources in three scalar ($\delta\rho$, δp, δT) and one vector (δu) variables. The velocity vector δu may be represented as a sum of longitudinal and transverse (with respect to the wave vector) components:

$$\delta u = \delta u^\| + \delta u^\perp = \frac{k(ku)}{k^2} + \frac{[k[ku]]}{k^2}.$$ (19.4.8)

Now let us find the spectral density of fluctuations $\delta u^\perp$. From the second equation in (19.4.5) it follows that

$$\left(-i\omega + vk^2\right)\delta u_i^\perp = -\frac{i}{\rho}\left(k_j \delta\pi_{ij}^{\text{source}}\right)^\perp.$$ (19.4.9)

After some straightforward manipulations we find the desired expression for the tensor of spectral density:

$$\left(\delta u_i^\perp \delta u_j^\perp\right)_{\omega,k} = \frac{2vk^2}{\omega^2 + (vk)^2}\left(\delta_{ij} - \frac{k_ik_j}{k^2}\right)\frac{\kappa T}{\rho}.$$ (19.4.10)

Carrying out integration with respect to frequency, we find the expression for the tensor of spatial spectral density,

$$\left(\delta u_i^\perp \delta u_j^\perp\right)_k = \left(\delta_{ij} - \frac{k_ik_j}{k^2}\right)\frac{\kappa T}{\rho}.$$ (19.4.11)

The sum of diagonal elements is

$$\left(\delta u^\perp \delta u^\perp\right)_k = 2\frac{\kappa T}{\rho}, \quad \text{and hence} \quad \left(\delta u \delta u\right)_r = 2\frac{\kappa T}{\rho}\delta(r).$$ (19.4.12)

Factor two indicates that the transverse velocity $\delta u^\perp$ has two independent components (two degrees of freedom). The remaining degree of freedom pertains to the longitudinal velocity $\delta u^\|$.

Here, like in the case of diffusion equation with Langevin source, the dispersion of velocity is infinite. As we know, this difficulty can only be overcome by defining what we mean by the "point" of "continuous medium".

Now we shall calculate the fluctuations of velocity, pressure, temperature and longitudinal velocity. Observe that equations (19.4.1) – (19.4.4) for fluctuations are similar to equations (14.2.2) – (14.2.4) for deviations of gasdynamic functions from their equilibrium values. Random sources in these equations reflect the existence of atomic structure of continuous equation. A similar correspondence exists between equations (19.4.5) and (14.3.2) for Fourier components.

19.4.2. High-Frequency Fluctuations $\delta\rho$, δp, δT, $\delta u^{\|}$

Since the longitudinal velocity component is proportional to the scalar product $(k\delta u)$, these fluctuations may be calculated from equations for ρ, δp, δT, $(k\delta u)$, which follow from (19.4.5):

$$-i\omega\delta\rho + i(k\delta u) = 0 \ ,$$

$$-i\omega(k\delta u) = -\frac{i}{\rho}k^2\delta p - \frac{4}{3}vk^2(\delta u) - \frac{i}{\rho}k_i k_k \pi_{ik}^{\text{source}} \ ,$$

$$\tag{19.4.13}$$

$$-i\omega\delta p + i\frac{5}{3}p(k\delta u) = -\frac{2}{3}(kq^{\text{source}}) \ ,$$

$$\delta p = \frac{\kappa T}{m}\delta\rho + \frac{\rho\kappa}{m}\delta T \ .$$

These must be supplemented by appropriate expressions for the spectra of intensities of random sources:

$$\left(k_i \pi_{ik} k_j \pi_{jl} k_l\right)_{\omega,k}^{\text{source}} = 2\eta\kappa T\frac{4}{3}k^4 \ ,$$

$$\tag{19.4.14}$$

$$\left(k_i q_i k_j q_j\right)_{\omega,k}^{\text{source}} = 2\lambda\kappa T^2 k^2 \ .$$

The general portrait of fluctuations is obtained by solving the set (19.4.13). This task, however, is quite complicated. For the sake of simplicity we shall once again single out the main fragments of this picture, like we did in Sect. 14.2 and 14.3.

First we consider the fast processes as defined by (14.2.11). The main contribution to the spectra comes from the narrow region of sonic resonance, $\omega \sim kv_s$, where $\omega \gg vk^2$ (or χk^2) which implies that the corresponding diffusion time is much greater than the period of oscillations. Entropy variations are then negligibly small ($\delta s = 0$), and so the fast processes are close to being adiabatic. Because of this, the solution of the problem becomes much simpler.

We return to equations (19.4.13) and eliminate variable $(k\delta u)$, getting a set of two equations in $\delta\rho$ and δp:

$$\omega^2 \delta\rho - k^2 \delta p + i\frac{4}{3}\omega vk^2 \delta\rho = k_i \pi_{ij}^{\text{source}} k_j ,$$

$$\delta p - \frac{5}{3}p\frac{\delta\rho}{\rho} + i\frac{2}{3}\frac{\lambda}{\omega}k^2 \delta T = \frac{2}{3\omega}\left(kq^{\text{source}}\right) .$$

(19.4.15)

For the region of fast processes near the resonance $\omega - kv_s \sim vk^2$ the nondissipative and the dissipative terms on the left-hand side of the first equation are of the same order of magnitude. In the second equation, however, the first two terms cancel out for the adiabatic process, since

$$p = C\rho^{5/3} , \quad \text{and} \quad \delta p = \frac{5}{3}p\frac{\delta\rho}{\rho} , \quad \left(\frac{c_p}{c_v} = \frac{5}{3}\right).$$

(19.4.16)

Then the two equations are reduced to one:

$$\left(\omega^2 - v_s^2 k^2 + i2\omega\gamma\right)\delta\rho = k_i k_j \pi_{ij}^{\text{source}} + \frac{2}{3}\frac{k^2}{\omega}\left(kq^{\text{source}}\right) ,$$

(19.4.17)

where the dissipative coefficient and the speed of sound are, respectively,

$$\gamma = \frac{1}{2}\left[\frac{4}{3}v + \left(\frac{c_p}{c_v} - 1\right)\chi\right]k^2 ; \quad v_s^2 = \frac{c_p}{c_v} .$$

(19.4.18)

The dissipation is due to both viscous friction and heat conduction (cf. (14.3.7)).

Now we use (19.4.17) to find the desired expression for spectral density of fast (high-frequency) fluctuations of density, and separate the resonant contribution:

$$\left(\delta\rho\delta\rho\right)^{\text{h.f.}}_{\omega,k} = \frac{\gamma}{\left(\omega - kv_{\text{s}}\right)^2 + \gamma^2}\frac{c_p}{c_v}m\rho \ .$$

(19.4.19)

Now it is easy to obtain expressions for spectral density of fluctuations of pressure and temperature:

$$\left(\delta p\delta p\right)^{\text{h.f.}}_{\omega,k} = \frac{\gamma}{\left(\omega - kv_{\text{s}}\right)^2 + \gamma^2}\rho v_{\text{s}}^2\kappa T \ ,$$

(19.4.20)

$$\left(\delta T\delta T\right)^{\text{h.f.}}_{\omega,k} = \frac{\gamma}{\left(\omega - kv_{\text{s}}\right)^2 + \gamma^2}\frac{2}{3}\frac{1}{c_p}\frac{\kappa T^2}{\rho} \ .$$

(19.4.21)

The one-time correlators are found by carrying out integration with respect to ω. The spectral density of fluctuations of velocity $\delta u^{\parallel}$ is

$$\left(\delta u^{\parallel}\delta u^{\parallel}\right)^{\text{h.f.}}_{\omega,k} = \frac{\gamma}{\left(\omega - kv_{\text{s}}\right)^2 + \gamma^2}\frac{\kappa T}{\rho} \ .$$

(19.4.22)

From (19.4.20), (19.4.22) we find the one-time correlators of fluctuations of pressure and velocity:

$$\left(\delta p\delta p\right)^{\text{h.f.}}_{k} = \rho v_{\text{s}}^2\kappa T \ , \quad \left(\delta u^{\parallel}\delta u^{\parallel}\right)^{\text{h.f.}}_{k} = \frac{\kappa T}{\rho} \ .$$

(19.4.23)

Let us consider important implications of these formulas. First, the combined contribution of correlators (19.4.12), (19.4.23) of fluctuations $\delta u^{\perp}$ and $\delta u^{\parallel}$ is $3\kappa T/\rho$. This exhausts the limit of kinetic energy per particle available for all three degrees of freedom. Accordingly, in this approximation the correlator of slow (low-frequency) fluctuations of longitudinal velocity is zero,

$$\left(\delta u^{\parallel}\delta u^{\parallel}\right)^{\text{l.f.}}_{k} = 0 \ , \ \text{and hence} \ \left(\delta u^{\parallel}\right)^{\text{l.f.}} = 0 \ .$$

(19.4.24)

A similar result holds for the fluctuations of pressure. Juxtaposition of expressions (19.4.23) for the one-time correlator and (4.9.5) for the dispersion of pressure reveals that the entire contribution comes from the high-frequency fluctuations of pressure, so that for the region of low-frequency fluctuations we have

$$(\delta p \delta p)_k^{\text{l.f.}} = 0 \text{ , and hence } (\delta p)^{\text{l.f.}} = 0 \text{ .} \tag{19.4.25}$$

To carry out the comparison, one must replace $V \Rightarrow V_{\text{ph}}$ in (4.9.5), and use the definition of "point" $\delta(r - r')\big|_{r=r'} = 1/V_{\text{ph}}$ in the expression for spatial correlator which follows from (19.4.23). The last two results will be used in the next section, and here we shall just make one final remark.

All spectral lines (19.4.19) – (19.4.22) of high-frequency oscillations have the same shape. The structure of the corresponding one-time correlators is such that the relative fluctuations of all variables are of the same order of magnitude,

$$\left(\frac{\delta p}{p}\right)^{\text{h.f.}} \sim \left(\frac{\delta u^{\parallel}}{v_T}\right)^{\text{h.f.}} \sim \left(\frac{\delta \rho}{\rho}\right)^{\text{h.f.}} \sim \left(\frac{\delta T}{T}\right)^{\text{h.f.}} . \tag{19.4.26}$$

19.4.3. Low-Frequency Fluctuations $\delta\rho$, δT, δs. Inconsistencies of Traditional Calculation of Gasdynamic Fluctuations

Return to equations (19.4.13). The results (19.4.24), (19.4.25) indicate that in this approximation the low-frequency fluctuations of velocity $\delta u^{\parallel}$ and pressure δp are zero, $\left(\delta u^{\parallel}\right)^{\text{l.f.}} = 0$ and $(\delta p)^{\text{l.f.}} = 0$. Then, from (19.4.13) it follows that negligibly small are also the fluctuations of density, and hence of temperature.

How can one derive the known expressions for spectral densities of fluctuations of density and temperature (Lifshitz – Pitayevsky 1979; Klimontovich 1982):

$$(\delta T \delta T)_{\omega,k}^{\text{l.f.}} = \frac{T^2}{\rho^2}(\delta\rho\delta\rho)_{\omega,k}^{\text{l.f.}} = \frac{2\chi k^2}{\omega^2 + \left(\chi k^2\right)^2}\frac{\kappa T^2}{c_p \rho} . \tag{19.4.27}$$

The coefficient of temperature conductivity χ here defines not only the width of spectral function of fluctuations of temperature, but also the width of spectral function of fluctuations of density, which has no clear-cut physical interpretation.

The fact is that expression (19.4.27) exceeds the accuracy of the approximation in question. It is derived under assumption that $\rho(\delta u)^{\text{l.f.}} = \omega(\delta\rho)^{\text{l.f.}} \sim vk^2(\delta\rho)^{\text{l.f.}}$. Then from (19.4.27) we obtain the following estimate for the fluctuations of density:

$$\rho^2\left(\delta u^{\parallel}\delta u^{\parallel}\right)_{\omega,k}^{\text{l.f.}} \sim (vk)^2(\delta\rho\delta\rho)_{\omega,k}^{\text{l.f.}} \sim v_T^2(lk)^2(\delta\rho\delta\rho)_{\omega,k}^{\text{l.f.}} . \tag{19.4.28}$$

Hence it follows that the relative fluctuations of velocity are much less than the relative fluctuations of density, since Knudsen number is small, $Kn = l/\lambda \sim lk \ll 1$. All this implies that equation (19.4.27) is not quite consistent. Let us demonstrate that fluctuations of gasdynamic functions can be calculated in a more rational way when self-diffusion is taken into account.

19.5. Calculation of Hydrodynamic Functions with Self-Diffusion

We start with equations of gas dynamics (14.1.2), (14.1.7), (14.1.8), based on the generalized kinetic equation (13.4.1). Dissipative processes are characterized in these equations by three independent kinetic coefficients D, v, χ, rather than by the usual two.

Let us introduce Langevin sources into these equations. This can be done by using the general expression for the intensity of Langevin source (9.7.7). The linearized collision operators are given by (19.1.1), (19.1.2), (19.1.5). By virtue of the properties (19.1.4) the terms with operators $\delta \hat{I}_p$ do not enter explicitly the equations of gas dynamics.

This allows us to drop the terms with $\delta \hat{I}_p$ straight from the outset, and get the following expression for the correlator of the source:

$$\langle yy \rangle_{r,p,t,r',p',t'} = 2 \frac{D}{n} \frac{\partial^2}{\partial r \partial r'} \delta(r - r') f(r,p,t) \delta(t - t') . \tag{19.5.1}$$

Once again we consider small fluctuations with respect to the state with $\rho = \text{const}$, $T = \text{const}$, $u = 0$. In the linear equations for fluctuations we carry out expansion in Fourier series, and arrive at the set of algebraic equations with Langevin sources. Again we separate the equation for the transverse velocity:

$$\left(-i\omega + vk^2\right)\delta u_i^\perp = \frac{1}{\rho}\left(y_i^{(u)}(\omega,k)\right)^\perp . \tag{19.5.2}$$

For convenience, we mark off Langevin sources in equations for density, velocity and temperature with superscripts "ρ", "u", "T".

With the aid of general expression (19.5.1) we find the Fourier components of the source:

$$\left(\frac{1}{\rho}\right)^2 \left(y_i^{(u)} y_j^{(u)}\right)_{\omega,k}^\perp = 2vk^2\left(\delta_{ij} - \frac{k_i k_j}{k^2}\right)\frac{\kappa T}{\rho} . \tag{19.5.3}$$

As a result, for the spectral density of $\delta u^{\perp}$ we again obtain expression (19.4.10).

The generalized kinetic equation (13.4.1) was derived under assumption that the three kinetic coefficients are equal, $D = v = \chi$. In gas dynamics, however, it is more convenient to distinguish between these coefficients in expressions like (19.5.3). Then the results will hold good even if the values of kinetic coefficients are not the same.

Let us now write the set of equations for the Fourier components of fluctuations of density, temperature, and longitudinal velocity:

$$\left(-i\omega + Dk^2\right)\delta\rho + i\rho(k\delta u) = y_{(\rho)}(\omega,k) , \tag{19.5.4}$$

$$\rho\left(-i\omega + vk^2\right)(k\delta u) = k^2\delta p + \rho\left(ky_{(u)}(\omega,k)\right) , \tag{19.5.5}$$

$$-i\omega\delta p + \frac{p}{\rho}Dk^2\delta\rho + \frac{p}{T}\chi k^2\delta T + \frac{c_p}{c_v}i\rho(k\delta u) = y_{(p)}(\omega,k) , \tag{19.5.6}$$

$$\delta p(\omega,k) = \frac{\kappa T}{m}\delta\rho(\omega,k) + \frac{\kappa\rho}{m}\delta T(\omega,k) . $$

These are similar to equations (19.4.13) obtained earlier without taking self-diffusion into account. The new sources introduced here are linked with the Langevin source in the kinetic equation by

$$y_{(\rho)}(\omega,k) = \rho\int y(\omega,k,p)dp , \quad y_{(u)}(\omega,k) = \int vy(\omega,k,p)dp ,$$

$$y_{(p)}(\omega,k) = \frac{\kappa T}{m}y_{(\rho)} + \frac{\kappa\rho}{m}y_{(T)} = \rho\int\frac{1}{3}v^2 y(\omega,k,p)dp . \tag{19.5.7}$$

Given that the kinetic coefficients are equal ($D = v = \chi$), we use the general formula (19.5.1) to find the intensities of Langevin sources:

$$\left(y_{(\rho)}y_{(\rho)}\right)_{\omega,k} = 2Dk^2 m\rho , \quad \left(y_{(u)}y_{(u)}\right)_{\omega,k} = 2vk^2\frac{\kappa T}{\rho} ,$$

$$\left(y_{(p)}y_{(p)}\right)_{\omega,k} = 2Dk^2\rho v_s\kappa T . \tag{19.5.8}$$

As before, we give separate treatment to the high-frequency and low-frequency fluctuations. The process at high frequencies may again be considered adiabatic; there is no need to repeat the calculations since the result is the same. The only difference is that the dissipative coefficient is now a combination of three (rather than two, as in (19.4.18)) kinetic coefficients,

$$\gamma = \frac{1}{2}\left[v + \frac{c_p}{c_v}D + \frac{\kappa}{mc_p}\chi \right]k^2 . \tag{19.5.9}$$

If the values of the kinetic coefficients are the same, however, both expressions reduce to $\gamma = Dk^2$.

More interesting are the results for the low-frequency fluctuations. As follows from (19.4.24), (19.4.25), in the low-frequency region the fluctuations of pressure and longitudinal velocity are negligibly small. Let us analyze equations (19.5.4) – (19.5.6) in this approximation. The first of these then takes on the form

$$\left(-i\omega + Dk^2\right)\delta\rho = y_{(\rho)}(\omega,k)^{\text{l.f.}} , \tag{19.5.10}$$

All terms in the second equation in this approximation are negligibly small, and so the equation falls out of consideration. The last two equations, by virtue of (19.5.4) and (19.5.7), are reduced to one:

$$\left(-i\omega + \chi k^2\right)\delta T = y_{(T)}(\omega,k)^{\text{l.f.}} . \tag{19.5.11}$$

In this way, calculation of low-frequency fluctuations of density and temperature amounts to solving two independent diffusion equations with appropriate Langevin sources. Together with (19.5.2), we thus have a complete set of diffusion equations for describing the equilibrium hydrodynamic fluctuations in the low-frequency region.

In Sect. 14.2 we considered the diffusion stage of time relaxation towards equilibrium, which is also described by the system of three diffusion equations — (14.2.8) for the transverse velocity, (14.2.13) for the density, and (14.2.14) for the temperature. Now we see that the diffusion processes play a dual role: they govern the last stage of relaxation towards equilibrium, and are definitive for the low-frequency fluctuations.

Now we have to find the structure of moments of Langevin sources in (19.5.10), (19.5.11). Note that the Fourier component of the source intensity $y_{(\rho)}$ can be obtained with the general expression (19.5.1); it has the form of (19.5.8). The following, however, must be borne in mind.

As established in the preceding section, the fluctuation $\delta u^{\parallel}$ is zero only in the low-frequency region. On the contrary, the fluctuations of velocity in the high-frequency region are important. Because of this, the intensity of the source must be split into two parts, corresponding to the low-frequency and the high-frequency regions. The high-frequency contribution is already included in (19.4.19), whereas the remaining part is

$$\left(y_{(\rho)}y_{(\rho)}\right)^{\text{l.f.}}_{\omega,k} = 2Dk^2 m\rho\,\frac{1}{c_p}\,.$$
(19.5.12)

This brings us to the desired expression for the spectrum of low-frequency fluctuations of density $\delta\rho$:

$$\left(\delta\rho\delta\rho\right)^{\text{l.f.}}_{\omega,k} = \frac{2Dk^2}{\omega^2 + \left(Dk^2\right)^2}\,\frac{\kappa\rho}{c_p}\,.$$
(19.5.13)

Now let us calculate the spectral density of fluctuations of temperature. The high-frequency part of the corresponding source of fluctuations δT in included in (19.4.21); the "total" follows from the thermodynamic equality (4.2.6) with due account for the fact that $\delta E = c_v \delta T$. Then the "remainder" pertains to the low-frequency region:

$$\left(y_{(T)}y_{(T)}\right)^{\text{l.f.}}_{\omega,k} = 2\chi k^2\,\frac{\kappa T^2}{\rho c_p}\,,$$
(19.5.14)

and the spectral density of fluctuations of temperature is

$$\left(\delta T\delta T\right)^{\text{l.f.}}_{\omega,k} = \frac{2\chi k^2}{\omega^2 + \left(\chi k^2\right)^2}\,\frac{\kappa T^2}{\rho c_p}\,.$$
(19.5.15)

So, the equations of gas dynamics with self-diffusion allow one to carry out a complete and consistent description of equilibrium fluctuations of gasdynamic functions. Since these fluctuations are definitive, for instance, for the spectrum of molecular light scattering by gases, we may speak of experimental proof of existence of self-diffusion. This in its turn serves as indirect evidence of efficiency of the generalized kinetic equation (13.4.1), which has been used for deriving the equations of gas dynamics with self-diffusion.

19.6. Kinetic Fluctuations in Active Medium

Now at last we may address the main theme of the present chapter — the calculation of nonequilibrium fluctuations in active media.

Such calculations are necessary, for instance, in the analysis of thermal convection at large temperature gradients, when the laminar flow becomes unstable. The problem of instability of convective flow was discussed in Sect. 14.6, 14.7. The nonlinear theory of fluctuations at convective motion can be found, for instance, in (Haken 1978); presently we shall be concerned with a somewhat different scope of problems.

19.6.1. *Langevin Source in Kinetic Equation*

We saw that the transition from the kinetic Boltzmann equation to the generalized kinetic equation (13.4.1) opens up new possibilities in the kinetic and hydrodynamic description of nonequilibrium processes. The generalization consists not only in the inclusion of collisions between the gas particles, but also in the inclusion of additional dissipation caused by dynamic instability of motion of the same particles.

In Ch. 18 this approach has been extended to the case of active media. To wit, we considered some examples of the appropriate generalized kinetic equations. Now the task is to calculate the corresponding kinetic and hydrodynamic fluctuations.

So we return to Sect. 18.3, devoted to the generalized kinetic equations for an active medium consisting of active bistable elements (like Duffing oscillators). In this case, equation (18.3.5) may be replaced (under condition $\gamma \gg \omega_{tr}$) by the simpler equation (18.3.6) for the distribution function $f(x,R,t)$. Let us write down the relevant Langevin equation,

$$\frac{\partial f}{\partial t} = \frac{\partial}{\partial x}\left[D_{(x)}(x)\frac{\partial f}{\partial x} - \frac{F(x)}{m\gamma}f\right] + D\frac{\partial^2 f}{\partial R^2} + y(x,R,t) ,$$

$$\int f dx \frac{dR}{V} = 1 ,$$

(19.6.1)

and recall the expressions for the nonlinear force and the coefficient of diffusion,

$$F(x) = -m\omega_0^2 x\left(1 - a_f + bx^2\right) , \quad D_{(x)}(x) = \frac{\kappa T}{m\gamma}\left(1 + bx^2\right) .$$

(19.6.2)

When the distribution with respect to R is spatially homogeneous, the stationary solution is given by (18.4.1). If the coefficient of effective field a_f is zero, the stationary solution coincides with the Boltzmann distribution for linear harmonic oscillator. Finally, if the coefficient of diffusion is linear, $D_{(x)} = \text{const}$, the solution coincides with the Boltzmann distribution with bistable potential,

$$f_0(x) = C \exp\left\{-\frac{m\omega_0^2}{\kappa T}\left[(1 - a_f)\frac{x^2}{2} + b\frac{x^4}{4}\right]\right\} . \tag{19.6.3}$$

The moments of Langevin source in the kinetic equation (19.6.1) can be found with the general formula (19.1.6), with due account for the new set of variables (x, R), and for the structure of the corresponding "linearized collision operator":

$$\langle y(x,R,t)\rangle = 0 ,$$

$$\langle y(x,R,t)y(x',R',t')\rangle = 2\left[\frac{\partial^2}{\partial x \partial x'}D_{(x)}(x) + D\frac{\partial^2}{\partial R \partial R'}\right]$$

$$\times \frac{1}{n}\delta(x - x')\delta(R - R')f(x,R,t)\delta(t - t') . \tag{19.6.4}$$

As before, distribution $f(x,R,t)$ satisfies the kinetic equation without Langevin source. The intensity of the source is determined by the sum of two diffusion terms. Let us discuss some implications of this general result.

19.6.2. Langevin Source in Reaction-Diffusion Equation

When Brownian particles are represented by Duffing oscillators, the kinetic equation (19.6.1) can be used for obtaining the diffusion equation for fluctuations of density of oscillators $\delta n(R,t) = n\delta f(R,t) = n\int \delta f(x,R,t)dx$ (the zero moment of distribution δf):

$$\frac{\partial}{\partial t}\delta n = D\frac{\partial^2}{\partial R^2}\delta n + y(R,t) ,$$

$$\langle y(R,t)\rangle = 0 , \tag{19.6.5}$$

$$\langle y(R,t)y(R',t')\rangle = 2D\frac{\partial^2}{\partial R \partial R'}\delta(R - R')n(R,t)\delta(t - t') .$$

This equation coincides with equation (19.2.10) for diffusion of Brownian particles in thermostat, with all pursuant consequences.

Assume now that the system of oscillators is incompressible. Then the zero moment of the distribution δf is zero, whereas possible is the evolution of the first moment of $\delta f(x,R,t)$,

$$\delta x(R,t) = \int x \delta f(x,R,t) dx \ . \tag{19.6.6}$$

To demonstrate the efficiency of calculation of these fluctuations on the basis of kinetic equation (19.6.1), consider fluctuations with respect to the stationary solution of (18.3.6) which is characterized by distribution (18.4.1). The mean value of $\langle x \rangle$ is zero, and the second moment (for arbitrary a_f) is given by

$$\langle x^2 \rangle = C \int x^2 \exp\left[-\frac{m\omega_0^2}{2\kappa T}\left(x^2 - \frac{a_f}{b}\ln(1 + bx^2)\right)\right] dx \ . \tag{19.6.7}$$

If $D_{(x)} = $ const, we may use the simpler distribution (19.6.3). Then the expression for the second moment is

$$\langle x^2 \rangle = C \int x^2 \exp\left\{-\frac{m\omega_0^2}{\kappa T}\left[(1 - a_f)\frac{x^2}{2} + b\frac{x^4}{4}\right]\right\} dx \tag{19.6.8}$$

(feasibility of transition from (18.4.1) to (19.6.3) has been discussed in Sect. 18.6.2).

Now we return to equation (19.6.1) and use it for obtaining a simpler equation for fluctuation $\delta x(R,t) = \int x \delta f(x,R,t) dx$. This equation is not closed, since it involves the higher moments of distribution δf. Once again we make use of the approximation (18.3.9), this time for the fluctuations

$$\int x^3 \delta f(x,R,t) dx \equiv \int Ex \delta f(x,R,t) dx = \langle x^2 \rangle \delta x \ , \quad \langle x \rangle = 0 \ , \tag{19.6.9}$$

and arrive at the following equation:

$$\frac{\partial}{\partial t}\delta x(R,t) + \gamma_{(x)}\delta x(R,t) - D\frac{\partial^2}{\partial R^2}\delta x(R,t) = y(R,t) \ , \tag{19.6.10}$$

where

$$\gamma_{(x)} = \frac{\omega_0^2}{\gamma}\left(1 - a_f + b\langle x^2 \rangle\right) \tag{19.6.11}$$

is the effective coefficient of friction.

There is also another useful form of this expression. Observe that, in the approximation (19.6.9), from the second equation in (18.3.10) for the stationary homogeneous state follows an algebraic equation in $\langle x^2 \rangle \equiv \langle E \rangle$:

$$\frac{\omega_0^2}{\gamma}\left(1 - a_f + b\langle x^2 \rangle\right)\langle x^2 \rangle = \frac{\kappa T}{m\gamma}\left(1 + b\langle x^2 \rangle\right) . \tag{19.6.12}$$

This can be used to rewrite (19.6.11) as

$$\gamma_{(x)} = \frac{\kappa T}{m\gamma} \frac{1 + b\langle x^2 \rangle}{\langle x^2 \rangle} . \tag{19.6.13}$$

The moments of Langevin source in (19.6.10) are found from the general expressions (19.6.4):

$$\langle y_{(x)}(R,t) \rangle = 0 , \tag{19.6.14}$$

$$\langle y_{(x)}(R,t) y_{(x)}(R',t') \rangle = 2\left[\frac{\kappa T}{m\gamma}\left(1 + b\langle x^2 \rangle\right) + D\frac{\partial^2}{\partial R \partial R'}\langle x^2 \rangle\right]\frac{1}{n}\delta(R - R')\delta(t - t') .$$

Now by solving (19.6.10) we find the following expression for the space-time spectral density of fluctuations δx:

$$(\delta x \delta x)_{\omega,k} = \frac{2\left(\gamma_{(x)} + Dk^2\right)}{\omega^2 + \left(\gamma_{(x)} + Dk^2\right)^2} \frac{\langle x^2 \rangle}{n} . \tag{19.6.15}$$

Expressions for the spatial spectrum and spatial correlations of fluctuations δx follow after integration with respect to ω and ω, k:

$$(\delta x \delta x)_k = \frac{\langle x^2 \rangle}{n} , \quad (\delta x \delta x)_{R,R'} = \frac{\langle x^2 \rangle}{n}\delta(R - R') . \tag{19.6.16}$$

The second of these expressions implies that the dispersion $(R = R')$ is infinite. To avoid this nonphysicality, we again define physically infinitesimal volume and make use of equation $\delta(R - R) = \delta(0) = V_{ph}$. The expression for dispersion then becomes

$$\langle \delta x \delta x \rangle = \frac{1}{N_{ph}}\langle x^2 \rangle = \frac{C}{N_{ph}}\int x^2 \exp\left\{-\frac{m\omega_0^2}{\kappa T}\left[(1 - a_f)\frac{x^2}{2} + b\frac{x^4}{4}\right]\right\}dx \tag{19.6.17}$$

(for illustration we have used here expression (19.6.8) for the dispersion $\langle x^2 \rangle$ of individual linear oscillator).

These formulas carry complete information about equilibrium fluctuations of active medium under consideration (per one particle!) at any admissible value of controlling parameter a_f. The factor of $1/N_{ph}$ implies that the dispersion in the medium is much less than that of an individual oscillator. If the total number of oscillators in the medium is N, then from (19.6.17) we find the total fluctuation energy of the medium:

$$N \frac{m\omega_0^2}{2} \langle \delta x \delta x \rangle = \frac{N}{N_{ph}} \langle x^2 \rangle = N_{freedom} \langle x^2 \rangle , \qquad (19.6.18)$$

where $N_{freedom}$ is the effective number of degrees of freedom in continuous medium.

In Ch. 13 we have given a consistent definition of kinetic and hydrodynamic physically infinitesimal scales for Boltzmann gas in connection with construction of the generalized kinetic equation (13.4.1). Applying formally these results to condensed media we find that the number of particles within physically infinitesimal volume of condensed continuous medium is $N_{ph} \geq 1$. Accordingly, for condensed media the effective number of degrees of freedom differs little from the total number of particles N.

19.6.3. *Kinetic Fluctuations at Heat Transfer in Active Medium*

We have already discussed several problems of the kinetic theory of heat transfer. In Sect. 12.5, for example, we introduced kinetic equation (12.5.10) with heat source; the stationary distribution with respect to velocities was then given by (12.5.12). In the presence of heat source it may differ considerably from Maxwell distribution. In Sect. 12.6 we considered the distribution of values of kinetic energy ("current temperature").

Kinetic and gasdynamic description of heat transfer in Sect. 14.8 was based on the generalized kinetic equation (13.4.1). In particular, we saw that equation (14.8.7) for the dispersion of kinetic energy has the structure of a reaction-diffusion equation.

Then, in Sect. 18.1 we returned to the theory of heat transfer in the presence of heat source, this time on the basis of the generalized kinetic equation (18.1.5) containing two dissipative terms. In the approximation of first moment we obtained equation (18.1.8) of reaction-diffusion type for the mean kinetic energy (or temperature).

Let us go back to equation (18.1.5) and use it for obtaining the equation for distribution function of kinetic energy $f(E,R,t)$:

$$\frac{\partial f}{\partial t} = 2\frac{\partial}{\partial E}\left[D(E)\left(E\frac{\partial f}{\partial E} - \frac{1}{2}f\right)\right] + 2\frac{\partial}{\partial E}[\gamma(E)Ef] + \chi\frac{\partial^2 f}{\partial R^2} \ . \tag{19.6.19}$$

The same scheme can also be used for constructing the kinetic theory of fluctuations at heat transfer. We assume that fluctuations of density and velocity are zero. Introducing the appropriate Langevin source, we rewrite this equation for fluctuations of distribution function of kinetic energy $f(E,R,t)$:

$$\frac{\partial \delta f}{\partial t} = 2\frac{\partial}{\partial E}\left[D(E)\left(E\frac{\partial \delta f}{\partial E} - \frac{1}{2}\delta f\right)\right] + 2\frac{\partial}{\partial E}[\gamma(E)E\delta f] + \chi\frac{\partial^2 \delta f}{\partial R^2} + y(E,R,t) \ . \tag{19.6.20}$$

The moments of Langevin source now are

$$\langle y(E,R,t)\rangle = 0 \ , \quad \langle y(E,R,t)y(E',R',t')\rangle = A(E,E',R,R',t,t')\delta(t-t') \ . \tag{19.6.21}$$

Once again we assume that without fluctuations the state of the system is spatially homogeneous and stationary. Then the intensity of the source is

$$A = 2\left\{2\frac{\partial^2}{\partial E\partial E'}D(E)E\delta(E-E')f(E)\right.$$

$$\left. + \ \chi\frac{\partial^2}{\partial R\partial R'}[\delta(E-E')f(E) - f(E)f(E')]\right\}\frac{1}{n}\delta(R-R') \ . \tag{19.6.22}$$

To eliminate the fluctuations of density, which are disregarded in this approximation, we have put a second term into square brackets, so as to make sure that the intensity of the source of fluctuations $A(R,R')$ in the expression for fluctuations of density $\delta\rho = n\int \delta f(E,R,t)dE$ is zero. The stationary solution $f(E)$ of (19.6.19) is

$$f(E) = C\exp\left[-\frac{E}{\kappa T} + \frac{\alpha_f}{\kappa T\beta}\ln\left(1 + \frac{\beta}{\gamma}E\right) + \frac{1}{2}\ln E\right] , \tag{19.6.23a}$$

where T is the temperature of thermostat, and $\gamma = 1/\tau$. In the absence of heat source ($\alpha_f = 0$), $f(E)$ is the Maxwell distribution, and the most probable energy value is defined by the temperature, $E_{m.p.} = \kappa T/2$. On the contrary, when the heat source is "strong", the effective distribution is Gaussian,

$$f(E) = C \exp\left[-\frac{(E - \langle E \rangle)^2}{2} \langle (\delta E)^2 \rangle\right], \quad \int\limits_0^\infty f dE = 1 \ . \tag{19.6.23b}$$

The mean value and the dispersion at $D\beta/\alpha^2 \ll 1$, and $\alpha = \alpha_f - \gamma \ll \gamma$, are given by

$$\langle E \rangle = \frac{\alpha}{\beta} = \frac{\alpha_f - \gamma}{\beta} \ , \quad \langle (\delta E)^2 \rangle = \frac{\gamma \kappa T}{\beta} \ . \tag{19.6.24}$$

Equation (19.6.20) with formulas (19.6.21) – (19.6.24) is suitable for calculating the fluctuations of distribution function $\delta f(E, R, t)$ with respect to the stationary state with distribution (19.6.23a) at all values of controlling parameter α_f.

Now we can use the kinetic equation (19.6.20) for obtaining the Langevin equation, for instance, for fluctuations of local temperature

$$\delta E(R, t) = \frac{3}{2} \kappa \delta T(R, t) = \int E \delta f(E, R, t) dE \ . \tag{19.6.25}$$

We start with defining the intensity of Langevin source in the equation for fluctuation of temperature at the state of equilibrium, when $\alpha_f = 0$. Then the first "collision integral" in the kinetic equation vanishes, and the expression for correlator of the source is

$$\langle y_{(T)}(R, t) y_{(T)}(R', t') \rangle = 2\chi \langle (\delta T)^2 \rangle \frac{\partial^2}{\partial R \partial R'} \frac{1}{n} \delta(R - R') \delta(t - t') \ . \tag{19.6.26}$$

Fluctuations of temperature satisfy the equation of heat conduction,

$$\frac{\partial \delta T}{\partial T} = \chi \frac{\partial^2}{\partial R^2} \delta T + y_{(T)}(R, t) \ , \quad \delta E = \frac{3}{2} \kappa \delta T \ . \tag{19.6.27}$$

The intensity of the source is proportional to the dispersion of temperature. At equilibrium, $\langle (\delta T^2) \rangle = 2T^2/3$. Given all this, we get the following expression for the spectral density of fluctuations δT:

$$(\delta T \delta T)_{\omega, k} = \frac{2\chi k}{\omega^2 + \left(\chi k^2\right)^2} \frac{\kappa T}{\rho c_v} \ . \tag{19.6.28}$$

This result corresponds to (19.5.15), with the only difference that (19.6.28) involves the total intensity, since the fluctuations of density and velocity are not taken into account.

Now let us carry out our calculation for the case when the system is far from equilibrium. We assume that the state of the system is described by Gaussian distribution (19.6.23b), and the relative dispersion is small (which corresponds to large values of α). The coefficient of diffusion is constant, $D = \gamma \kappa T$. The correlator of the source is given by

$$\left\langle y_{(T)}(R,t) y_{(T)}(R',t') \right\rangle = 2 \left[2D\langle E \rangle + \chi \left\langle (\delta E)^2 \right\rangle \frac{\partial^2}{\partial R \partial R'} \right] \frac{1}{n} \delta(R - R') \delta(t - t') , \quad (19.6.29)$$

and is defined by two moments of the stationary distribution, the mean energy and the dispersion of energy. The relationship between these two quantities follows from (19.6.24):

$$D\langle E \rangle = \left\langle (\delta E)^2 \right\rangle \alpha . \quad (19.6.30)$$

Recall also that $D = \gamma \kappa T$, and $\alpha = \alpha_f - \gamma > 0$. Then (19.6.29) can be rewritten as

$$\left\langle y_{(T)}(R,t) y_{(T)}(R',t') \right\rangle = 2 \left\langle (\delta E)^2 \right\rangle \left[2\alpha + \chi \frac{\partial^2}{\partial R \partial R'} \right] \frac{1}{n} \delta(R - R') \delta(t - t') . \quad (19.6.31)$$

Using Langevin equation (19.6.20) for δf, we find the appropriate equation for fluctuation of temperature (19.6.25). In this approximation the contribution of diffusion $D_{(E)}$ is negligible; with due account for the definition of $\langle E \rangle$ we get the following equation:

$$\frac{\partial}{\partial t} \delta T = -2\alpha \delta T + \chi \frac{\partial^2}{\partial R^2} \delta E + y_{(E)}(R,t) , \quad \alpha = \alpha_f - \gamma > 0 . \quad (19.6.32)$$

Hence follows the expression for spectral density of fluctuations,

$$(\delta T \delta T)_{\omega,k} = \frac{2(2\alpha + \chi k^2)}{\omega^2 + (2\alpha + \chi k^2)^2} \left\langle (\delta T)^2 \right\rangle . \quad (19.6.33)$$

The dispersion of temperature follows from (19.6.24),

$$\left\langle (\delta T)^2 \right\rangle = \frac{4}{9} \left\langle (\delta E)^2 \right\rangle = \frac{4}{9} \frac{\gamma}{\beta} \kappa T . \quad (19.6.34)$$

The ratio of this quantity to the dispersion of temperature at equilibrium is defined by dimensionless parameter $\gamma/\beta\kappa T$.

So, we have calculated the fluctuations for two nonlinear open systems of different nature. The first was represented by the medium of nondissipative bistable elements. The controlling parameter (the effective field) controls the height of the barrier. In the second example the active dissipative medium was Boltzmann gas in the presence of heat source.

Despite the considerable difference between these systems, the intensities of Langevin sources in equations for fluctuations of distribution functions (19.6.4), (19.6.22), and the corresponding spectral densities of fluctuations (19.6.15), (19.6.33), have similar structure. There are, of course, some important distinctions.

For example, the width of spectral line (19.6.15) for large (but not too large, $\alpha_f - 1 \ll 1$) deviations from the critical point decreases as $1/\langle x^2 \rangle$. By contrast, the width of spectral line (19.6.33) (far above the critical point) decreases when the critical point is approached. It is important that in both cases the width of spectral line depends on the controlling parameter. As we shall see in the next chapter, it is because of this that flicker noise ($1/f$ -noise) may exist in nonequilibrium systems in a certain range of frequencies.

The list of examples could, of course, be continued. Of great interest, for instance, are active media comprised of oscillators of varying degree of sophistication. For some characteristics of such systems (for instance, for positions and velocities) the width of spectral lines will decrease upon transition into the region of well-developed generation. On the contrary, the line structure for fluctuations of energy is similar to (19.6.33).

In all our examples we considered diffusion motion in boundless media. This restricts the possibility of studying the time fluctuations at times greater than the diffusion time, and the time spectra at low frequencies. In the next chapter we shall be dealing with diffusion in finite systems. This generalization will allow us to gain useful information about large-scale correlations and low-frequency spectra. We shall see that the diffusion in space with fractal dimensions plays an important role; this kind of diffusion may be referred to as anomalous.

CHAPTER 20

ANOMALOUS BROWNIAN MOTION. EQUILIBRIUM AND NONEQUILIBRIUM NATURAL FLICKER NOISE AND RESIDUAL TIME CORRELATIONS

The present chapter is devoted to the statistical theory of diffusive motion in finite systems. The main problems we are going to be concerned with are enumerated in the heading; the chapter begins with the "diffusion theory of flicker noise" (Klimontovich 1983; IV), one of the most fascinating and intriguing phenomena in the physics of fluctuative processes.

Flicker noise was first discovered some seventy years ago by Johnson as slow fluctuations (flicker) of cathode emission in electron tubes. It was found that, as the frequency f decreases, the intensity of noise more or less follows the $1/f$ law. Numerous studies indicate that this effect is seemingly universal and is observed in the most diverse systems. In defiance of the earnest efforts of numerous students, however, there is still no generally accepted view on the nature of this phenomenon (Kogan 1985; Weissmann 1988). In the words of one of the leading specialists in the physics of fluctuative processes (Voss 1992), "these $1/f$ noises exhibit $S(f) = 1/f^{\beta}$ with $\beta \approx 1$ over many decades. In most cases, the physical reason for this long-range power law behavior remains a mystery".

Our presentation of the theory of flicker noise will be based on (Klimontovich 1983b, IV, 1990b). In the course of this work the author discussed the main ideas with professor Hermann Haken; the first discussion took place in 1982 at a conference in Tallinn dedicated, among other issues, to the problems of synergetics. Then Haken, with tongue in cheek, observed that "each theoretical physicist may have his own theory of flicker noise". So the author takes advantage of this kind permission to expose his own views in a way as clear and convincing as possible. One of the tasks of the present chapter is to demonstrate how this theory fits in with the structure of the general theory of statistical processes in open systems.

Recall that the important concepts in the statistical theory are the limiting thermodynamic transition ($N \to \infty$, $V \to \infty$, but $N/V = \text{const}$), and the ergodicity. The situation is much different in the theory of flicker noise: namely, in the domain of flicker noise any sample of arbitrary dimensionality d may be regarded as a point — that is, as a zero-dimensional item.

In this way, here we are dealing with a limit which is opposite to the thermodynamic one.

What is more, the spectrum of flicker noise is not limited from the side of low frequencies: it is restricted from below only by the observation time, $\omega_{min} \sim 1/\tau_{obs}$. The time of observation also defines the half-width of the spectrum. Because of this, the traditional concept of ergodicity cannot be used in the construction of the theory of flicker noise.

In this chapter we shall also discuss the linkage between flicker noise and superconductivity. There are good reasons to believe that it is the existence of flicker noise that causes disappearance of electric resistance in a dissipative medium at zero frequency. In this way superconduction (undamped flow of electric current) is made possible in a dissipative medium.

The situation in which noise is a crucial factor for the existence of a coherent state of superconduction may seem paradoxical. We shall see, however, that flicker noise is a coherent state which may be assimilated to the condensation of Bose gas. In this case, however, the condensation takes place in the space of wave numbers (moments) in a dissipative system.

Flicker noise will be treated as an example of anomalous diffusion in a finite medium — an example of anomalous Brownian motion.

This is just one example of such motion; anomalous is also Brownian motion in systems with fractional or fractal dimensionality. We shall discuss the reasons which allow us to consider the motion of this kind as something in between the "normal" diffusion process and flicker noise. The description of anomalous Brownian motion requires generalizing the appropriate diffusion equation, in which the conventional time derivative will be replaced by fractional derivative with respect to time.

The theory of anomalous Brownian motion is currently based on a rather limited number of experimental facts. Because of this, at this time we find ourselves at the cradle of this new fundamental direction of scientific research.

The term "natural flicker noise" is used to emphasize that this noise is due to the "atomic structure" of the medium under consideration. Natural flicker noise may be equilibrium and nonequilibrium; universal is the equilibrium flicker noise, whereas the nonequilibrium flicker noise can only be observed under certain special conditions. Both can be described on the basis of reaction diffusion equations.

20.1. Equilibrium Natural Flicker Noise

As indicated above, the epithet "natural" emphasizes the fact that the noise we are dealing with is due to the "atomic structure" of the system which is at statistical equilibrium. As we shall see, however, the concept of statistical equilibrium itself in the theory of flicker noise needs to be revised, since the lower limit of the spectrum depends on the time of observation, and the process is therefore not ergodic.

The second part of the term ($1/f$ noise) indicates that in the region of low frequencies the noise increases as $1/f$ (or $1/\omega$). Of special importance are the following four properties of flicker noise.

1. The region of flicker noise is restricted from above by a certain frequency ω_{max} which depends on the properties of the system. The properties of the system, however, have no effect on the lower frequency limit ω_{min}, which becomes lower and lower as the time of observation increases. Accordingly, the spectrum of flicker noise lies between

$$\omega_{max} > \omega > \omega_{min} = \frac{1}{\tau_{obs}} \ . \tag{20.1.1}$$

At nonequilibrium conditions, the lower limit of the spectrum of flicker noise does depend on the properties of the system in question (see Sect. 20.4).

2. The intensity of natural flicker noise is inversely proportional to the number of structural "particles" of the system N, which may be microscopic or macroscopic.

3. For sufficiently long observation times the time correlations in the region of natural flicker noise exhibit logarithmic decrease for large arguments. This implies that there are "residual" time correlations, which persist over the entire period of observation.

4. In the region of flicker noise the nature of time correlations and the spectrum are independent of the dimensionality d of the system. This is one of the peculiar features of anomalous Brownian motion.

20.2. Theory of Equilibrium Natural Flicker Noise

Let us show that the equilibrium natural flicker noise may be regarded as a diffusion process in a finite medium.

Recall some results of our calculations of fluctuations for the process of diffusion in unlimited medium. We started with the diffusion equation (19.2.10) with the appropriate Langevin source (19.2.12). Function $\rho(R,t)$ satisfies the diffusion equation without source.

Now in place of $\rho(R,t) = mn(R,t)$ it is more convenient to use function $n(R,t)$, by which we refer to an arbitrary physical characteristic whose space-time evolution is a process of diffusion. Then, in accordance with (19.2.16), the space-time spectral density of fluctuations $\delta n(R,t)$ at equilibrium is defined by

$$(\delta n \delta n)_{\omega,k} = \frac{2Dk^2}{\omega^2 + \left(Dk^2\right)^2} (\delta n \delta n)_k \, , \qquad (20.2.1)$$

where δn is the one-point correlator of fluctuations. For ideal gas $(\delta n \delta n)_k = n$, and (20.2.1) coincides with (19.2.16).

Observe that the validity of this result is limited from the side of both small and large scales: we disregard both the structure of "continuous medium", and the finiteness of the system. Let us remove these constraints, starting with the "atomic structure" of continuous medium.

The one-time spectral density $(\delta n \delta n)_k$ in this approximation does not depend on k, and so we are dealing with "white noise" in the space of wave numbers. The corresponding correlator is proportional to delta-function $\delta(R - R')$. As before, we carry out smoothing with respect to physically infinitesimal volume V_{ph} — in other words, with respect to the volume of the "point" of continuous medium. Then the one-time correlator is given by

$$\left\langle \delta n_{V_{ph}} \, \delta n_{V_{ph}} \right\rangle_r = \left\langle \delta n_{V_{ph}} \, \delta n_{V_{ph}} \right\rangle \exp\left(-\frac{r^2}{2l_{ph}^2} \right) , \quad r = R - R' \, . \qquad (20.2.2)$$

Expression (20.2.1) is changed accordingly:

$$\left\langle \delta n_{V_{ph}} \, \delta n_{V_{ph}} \right\rangle_{\omega,k} = \frac{2Dk^2}{\omega^2 + \left(Dk^2\right)^2} N_{ph} \left\langle \delta n_{V_{ph}} \, \delta n_{V_{ph}} \right\rangle \exp\left(-\frac{l_{ph}^2 k^2}{2} \right) . \qquad (20.2.3)$$

The properties of the medium are taken into account via the structure of the one-time and one-point correlator $\left\langle \delta n_{V_{ph}} \, \delta n_{V_{ph}} \right\rangle$ — that is, via the dispersion of fluctuations $\delta n_{V_{ph}}$. Recall that for ideal gas this correlator is n/V_{ph} ; for a real gas or liquid it is given by

$$\left\langle \delta n_{V_{ph}} \, \delta n_{V_{ph}} \right\rangle = \frac{n}{V_{ph}} \left(1 + n \int g_2(r)dr \right) = \frac{n}{V} \beta_T \, , \quad \beta_T = \kappa T \left(\frac{\partial n}{\partial p} \right)_T \, , \qquad (20.2.4)$$

where g_2 is the two-particle correlation function, β_T is the coefficient of isothermal compressibility, and which follows from (8.2.8) with the replacement $\Delta V \Rightarrow V_{ph}$.

From (20.2.3) follows the expression for the spectral density of Langevin source,

$$(yy)_{\omega,k} = 2Dk^2 V_{ph} \left\langle \delta n_{V_{ph}} \, \delta n_{V_{ph}} \right\rangle \exp\left(-\frac{l_{ph}^2 k^2}{2}\right).$$

(20.2.5)

By contrast to the corresponding expression for the source of noise, here we have eliminated the scales which fall within the "point" of the medium. Because of this, in place of "white noise" in the space of wave numbers we now have a distribution with the maximum at $k = 0$. The dispersion of distribution depends on the size of the "point". Let us show that this distribution is further narrowed down when we go over into the domain of flicker noise; the structure of Langevin source is changed accordingly.

The above expressions do not take the finite size of the system into account, and therefore are only valid for those values of τ which are much smaller than the time of diffusion τ_D, in the corresponding range of frequencies:

$$\tau \ll \tau_D = \frac{L^2}{D}, \quad \sqrt{D\tau} \ll L \; ; \quad \frac{1}{\omega_D} \gg \tau \, , \quad \sqrt{\frac{D}{\omega}} \ll L \, .$$

(20.2.6)

Now let us consider those changes which arise because of the finite size of the system in question.

As we know, the time of diffusion $\tau_D = L^2/D$ is the time over which the homogeneous distribution of particles with the density $n = N/V$ is established; the fluctuations $\delta n(R,t)$ take place against the background of this distribution.

Depending on the relative values of τ and τ_D (or ω and $1/\tau_D$), the fluctuations may be divided into two classes. The first is defined by conditions (20.2.6); the boundary conditions are then not important, and the description of fluctuations may be based on the above formulas. The situation is essentially different for the other class, when the values of τ and ω are such that

$$\sqrt{D\tau} \gg L \, , \quad \sqrt{\frac{D}{\omega}} \gg L \, .$$

(20.2.7)

In this region τ is much greater than the diffusion time. Because of this, the particle manages to diffuse many times throughout the volume of the sample V. The effective volume covered by the moving particle is

$$V_{eff} = (D\tau)^{3/2} \sim \left(\frac{D}{\omega}\right)^{3/2} \gg V \, .$$

(20.2.8)

We see that, when τ is large (or ω is small), the effective volume of diffusion V_{eff} is much larger than the volume of the system V. This allows us to assume that the dimensionality of the sample under consideration is zero.

Now let us return to expression (20.2.5) and show that this distribution becomes much narrower when we go into the domain of flicker noise. To elucidate the corresponding changes in the structure of the Langevin source in equation for fluctuations δn, let us refine the above properties of equilibrium natural flicker noise for the case of diffusion model of random process.

The maximum frequency of the spectrum is determined by the diffusion time,

$$\frac{1}{\tau_{obs}} << \omega << \omega_{max} = \frac{1}{\tau_D} = \frac{D}{L^2} . \qquad (20.2.9)$$

In addition to the characteristic size of the sample, we now have a new length scale

$$L_\omega = L + \sqrt{\frac{2D}{\omega}} . \qquad (20.2.10)$$

In the domain of flicker noise the effective volume of multiple diffusion is much greater than the volume of the sample:

$$V_\omega \equiv L_\omega^3 = \left(\frac{D}{\omega}\right)^{3/2} >> V , \quad L >> \sqrt{\frac{2D}{\omega}} . \qquad (20.2.11)$$

Under these conditions the dimensionality of the sample d is not important, and may be assumed zero in the limit of very large observation times.

Since the lowest frequency in the diffusion process is limited only by the time of observation, the important feature of the system in question is its lifetime τ_{life}. For instance, in the study of superconductivity this is the time during which the sufficiently low temperature is maintained. It is obvious that the time of observation cannot exceed the lifetime, $(\tau_{obs})_{max} \leq \tau_{life}$. Assume that the observation time is much smaller than the lifetime; then conditions (20.2.9) can be supplemented by yet another inequality, so that

$$\frac{1}{\tau_{life}} << \frac{1}{\tau_{obs}} << \omega << \omega_{max} = \frac{1}{\tau_D} = \frac{D}{L^2} , \qquad (20.2.12)$$

whence follows the corresponding chain of inequalities for the characteristic volumes:

$$V = V_{max} \ll V_\omega \ll V_{obs} \ll V_{life} \,. \tag{20.2.13}$$

Now we may return to the question of the structure of Langevin source. It should be borne in mind that a consistent definition of statistical characteristics must be based on the appropriate kinetic equation. This is how we proceeded with the construction of the theory of kinetic fluctuations in Ch. 9, 10. Since no kinetic equation is available yet for the construction of the theory of flicker noise, let us try to "design" a Langevin source from physical premises. With this purpose we turn to expression (20.2.4).

Recall that $\left\langle \delta n_{V_{ph}} \delta n_{V_{ph}} \right\rangle$ is a characteristic of matter of which the system under consideration is composed. Let us illustrate this with the example of nonideal liquid or gas. Looking at formula (20.2.4) for the dispersion $\delta n_{V_{ph}}$, we see that the dispersion decreases as the procedure of smoothing is carried on, and arrives at its minimum when $V_{ph} \to V$. Then it can be written as

$$\left\langle \delta n_V \delta n_V \right\rangle = \frac{n}{V} \left(1 + n \int g_2(r) dr \right). \tag{20.2.14}$$

So, the dispersion δn_V depends on the properties of the substance of the system under consideration, and is the source of fluctuations. In the thermodynamic limit, when $N \to \infty$, the dispersion is zero. We see that the only "free" parameter in (20.2.5) is then V_{ph} (or l_{ph}). By definition of physically infinitesimal volume in hydrodynamics, V_{ph} is the smallest volume in which the hydrodynamic motion (or the diffusion motion in our present context) is still preserved.

In the theory of flicker noise the value of V_ω at a certain frequency ω may be taken for the diffusion volume. Given that the observation time is known, the largest volume of diffusion is V_{obs}. This allows us to extend the region of smoothing in constructing the Langevin source in the diffusion equation for the Fourier component $\delta n(\omega,k)$ by replacing

$$V_{ph} \Rightarrow V_\omega \,, \quad V \Rightarrow V_{\omega_{min}} \equiv V_{obs} \,. \tag{20.2.15}$$

Then from (20.2.5) we get the desired expression for the intensity of Langevin source:

$$(y,y)_{\omega,k} = 2Dk^2 A V_\omega \left\langle \delta n_V \delta n_V \right\rangle \exp\left(-\frac{L_\omega^2 k^2}{2} \right). \tag{20.2.16}$$

The normalization constant A will be defined shortly.

We see that the dispersion of distribution with respect to wave numbers is determined by L_ω^{-2}, and therefore depends on the frequency. As follows from (20.2.10), at low frequencies it is proportional to ω, and thus tends to zero when $\omega \to 0$. Of course, this limit is unattainable because the time of observation is finite. There is, however, a tendency towards "condensation" in the space of wave numbers — in other words, towards formation of a coherent state.

Now let us write the appropriate expression for the space-time spectral density,

$$(\delta n \delta n)_{\omega,k} = \frac{2Dk^2}{\omega^2 + (Dk^2)^2} A V_\omega \langle \delta n_V \delta n_V \rangle \exp\left(-\frac{L_\omega^2 k^2}{2}\right),$$
(20.2.17)

and separate the region of low frequencies for which we have

$$\langle (\delta k)^2 \rangle = \frac{\omega}{D} \, , \quad V_\omega = \left(\frac{D}{\omega}\right)^{3/2} .$$
(20.2.18)

This is the analog of Einstein's formula for the domain of flicker noise, which allows us to carry out integration with respect to k and obtain the expression for the temporal spectral density in the low-frequency range:

$$(\delta n \delta n)_\omega = \frac{\pi}{\ln \dfrac{\tau_{obs}}{\tau_D}} \frac{\langle \delta n_V \delta n_V \rangle}{\omega} \, , \quad \frac{1}{\tau_{obs}} \ll \omega \ll \frac{1}{\tau_D} .$$
(20.2.19)

This spectrum exhibits the frequency dependence of flicker noise. The normalizing constant A has been found from the approximate condition

$$\int_{1/\tau_{obs}}^{1/\tau_D} (\delta n \delta n)_\omega \frac{d\omega}{\pi} = \langle \delta n_V \delta n_V \rangle ,$$
(20.2.20)

which amounts to assuming that the main contribution comes from the region of low frequencies (the domain of flicker noise).

Expression (20.2.19) can be represented in the canonical form,

$$\frac{\langle \delta n \delta n \rangle_\omega}{n^2} = \frac{2\pi a}{N\omega} ,$$
(20.2.21)

where

$$a = \frac{1}{2\ln\frac{\tau_{obs}}{\tau_D}} \frac{\langle \delta n_V \delta n_V \rangle}{\frac{n}{V}}$$

is the so-called Hoog's constant (Kogan 1985; Klimontovich V). For ideal gas $\langle \delta n_V \delta n_V \rangle = n/V$, and the Hoog's constant is

$$a = \frac{1}{2\ln\frac{\tau_{obs}}{\tau_D}} . \qquad (20.2.22)$$

Hoog's constant decreases when the correlations are taken into account; it can also be defined for the case when a normalization condition is used which is more general than (20.2.20).

The arguments developed above support the view on flicker noise as a diffusion process in a finite system. Since the diffusion processes are most typical for the last stage of relaxation towards equilibrium, the universality of this phenomenon becomes quite understandable. Let us now consider the corresponding time correlations.

20.3. Residual Time Correlations

The time correlator of fluctuations δn in the domain of natural flicker noise is defined by the integral

$$\langle \delta n \delta n \rangle_\tau = \int_{1/\tau_{obs}}^{1/\tau_D} (\delta n \delta n)_\omega \cos \omega\tau \, \frac{d\omega}{\pi} . \qquad (20.3.1)$$

Carrying out integration with respect to ω, we come to the desired expression:

$$(\delta n \delta n)_\tau = \left[C - \frac{\ln\frac{\tau}{\tau_D}}{\ln\frac{\tau_{obs}}{\tau_D}} \right] \langle \delta n_V \delta n_V \rangle ; \quad \tau_D \ll \tau \ll \tau_{obs} ;$$

$$(20.3.2)$$

$$C = 1 - \frac{\gamma}{\ln\frac{\tau_{obs}}{\tau_D}} , \quad \gamma = 0.577 .$$

We see that the dependence on τ in the domain of flicker noise is very weak (logarithmic with very large argument). This allows us to speak of residual time correlations.

Observe that the magnitude of residual correlations depends via the one-time correlator $\langle \delta n_V \delta n_V \rangle$ on the properties of matter of which the system in question is composed.

20.4. Langevin Equation for Domain of Flicker Noise

As we know, the Langevin equation for the diffusion process may be written as

$$\left(-i\omega + Dk^2\right)\delta n(\omega,k) = y(\omega,k) \ . \tag{20.4.1}$$

The space-time spectral density of Langevin source is now given by (20.2.16).

Since the dispersion of distribution with respect to wave numbers in the domain of flicker noise is very small (see (20.2.18)), we may carry out integration over k in the Langevin equation. Taking advantage of the fact that the distribution is narrow, we make the replacement

$$k^2 \Rightarrow \overline{k^2}^k = \frac{1}{L_\omega^2} = \frac{\omega}{D} \tag{20.4.2}$$

and arrive at the Langevin equation for a simpler distribution function $\delta n(\omega)$:

$$\left(-i\omega + \gamma_\omega\right)\delta n(\omega) = y(\omega) \ ; \quad \gamma(\omega) = |\omega| \ . \tag{20.4.3}$$

This replacement actually implies that we take into account the nonstationarity of the process — that is, the dependence of dispersion on the frequency ω. The corresponding expression for the spectral density of the intensity of Langevin source is

$$(yy)_\omega = 2\gamma_\omega \frac{\pi}{\ln \dfrac{\tau_{\text{obs}}}{\tau_D}} \langle \delta n_V \delta n_V \rangle \ ; \quad \gamma(\omega) = |\omega| \ . \tag{20.4.4}$$

We see that for the domain of flicker noise the dissipative coefficient is $\gamma(\omega) = |\omega|$, and tends to zero when $\omega \to 0$.

This expression links the spectral density of fluctuations of the source to the appropriate dissipative coefficient, and is therefore a fluctuation-dissipation relation (FDR) for the domain of flicker noise. The intensity of the noise is proportional to the one-time correlator

$\langle \delta n_V \delta n_V \rangle$. For the spectral density of fluctuations $\delta n(\omega)$ we once again come to expression (20.2.19).

20.5. Flicker Noise in Space of Wave Numbers

Expression (20.2.19) for the temporal spectral density has been obtained from the more general expression (22.2.17) by integration with respect to k. Let us now demonstrate that flicker noise for the distribution of mean "energy" $E(k)$ with respect to wave numbers can be obtained by integration over ω (not over k!). The desired function is found by integrating the spectral density (22.2.17) with respect to ω:

$$E(k) = 4\pi k^2 \int\limits_{1/\tau_{\text{obs}}}^{1/\tau_D} (\delta n \delta n)_{\omega,k} \frac{d\omega}{2\pi} = B \frac{1}{|k|} \langle \delta n_V \delta n_V \rangle . \tag{20.5.1}$$

The new constant B is found from the normalization condition with respect to values of wave numbers. If the main contribution corresponds to the region of small wave numbers, then

$$\int\limits_{k_{\min}}^{k_{\max}} E(k) dk = \langle \delta n_V \delta n_V \rangle , \quad k_{\max} = \sqrt{\frac{1}{D\tau_D}} , \quad k_{\min} = \sqrt{\frac{1}{D\tau_{\text{obs}}}} . \tag{20.5.2}$$

So, when a random diffusion process takes place in a finite system, there exists not only temporal, but also spatial flicker noise ($1/k$–noise).

We have considered the main results of the theory of natural flicker noise. This theory is based on the analysis of random diffusion process in a finite system. The appearance of $1/f$-type noise and the residual time correlations is a consequence of rearrangement of motion due to multiple diffusion.

The real structure of the medium enters via two parameters. The first of these is statical; it is the dispersion of δn_V. The second is dynamic, and is represented by the diffusion time τ_D. Recall that the value of δn is not necessarily linked with fluctuations of the number of particles: it may be any characteristic controlled by the diffusion equation. Later we shall discuss some applications of the theory; first, however, let us consider the problem of flicker noise in nonequilibrium active systems.

20.6. Natural Flicker Noise in Finite Active Systems at Reaction-Diffusion Processes

20.6.1. *Active Medium of Bistable Elements*

Let us recall some results of Ch. 18. We return to the kinetic equation (18.3.5) for distribution function $f(x,v,R,t)$ of states of the medium comprised of bistable elements. This equation involves two diffusion coefficients in the space of internal variables v, x,

$$D_{(v)} = \gamma \frac{\kappa T}{m}, \quad D_{(x)} = D_{(0)}\left(1 + bx^2\right), \quad \text{where} \quad D_{(0)} = \frac{\kappa T}{m\gamma}, \tag{20.6.1}$$

and the coefficient of diffusion in space R. The effective force F_{eff} is defined by (17.5.4), and the equilibrium solution is the Maxwell – Boltzmann distribution.

For the class of distribution functions

$$f(x,v,R,t) = f\left(x,|v|,R,t\right) \tag{20.6.2}$$

it is possible to carry out integration with respect to v in (18.3.5). This brings us to the Einstein – Smoluchowski equation (18.3.6) for a simpler distribution function $f(x,R,t)$.

Recall also that the transition from the generalized kinetic equation (18.3.5) to the Einstein – Smoluchowski equation for the class of distribution functions (20.6.2) is accomplished without resorting to perturbation theory in Knudsen number.

In Sect. 19.6 we introduced the Langevin source into the Einstein – Smoluchowski equation. The moments of the source are given by (19.6.4). We have considered two particular cases.

For a one-component system of free nonlinear oscillators, it is possible to go over from the Langevin equation (19.6.1) to the appropriate diffusion equation (19.6.5) for function $\delta n(R,t)$ with Langevin source. Then we come back to the problem of fluctuations at a diffusion process. Calculation of fluctuations was carried out for infinite medium in Sect. 19.2, and in the present chapter for finite medium.

Now let us attack the main task of this section, the calculation of low-frequency fluctuations on the basis of reaction-diffusion equation. We return to the Langevin equation (19.6.1), where the moments of the source are given by (19.6.4), and assume again that the stationary distribution without fluctuations is defined by (18.4.1), or by a simpler expression (19.6.3). To find the equation for fluctuations $\delta x(R,t) = \int x \delta f(x,R,t) dx$ we use equation (19.6.1) and take advantage of the approximation (19.6.9). As a result, we come to the reaction-diffusion equation (19.6.10) with the Langevin source. Let us reproduce it once again:

$$\frac{\partial \delta x(R,t)}{\partial t} + \gamma_{(x)}\delta x(R,t) - D\frac{\partial^2 \delta x(R,t)}{\partial R^2} = y(R,t) \ . \tag{20.6.3}$$

The moments of the source are, as before, given by (19.6.14):

$$\left\langle y_{(x)}(R,t)\right\rangle = 0 \ ,$$

$$\left\langle y_{(x)}(R,t)y_{(x)}(R',t')\right\rangle = 2\left[D_{(0)}\left(1+b\left\langle x^2\right\rangle_{R,t}\right) + D\frac{\partial^2}{\partial R\partial R'}\left\langle x^2\right\rangle_{R,t}\right]\frac{1}{n}\delta(R-R')\delta(t-t') \ . \tag{20.6.4}$$

We see that the intensity of the source is determined by the dispersion of variable x. In the steady homogeneous state the dispersion $\left\langle x^2\right\rangle$ is independent of R, t, and is defined, depending on the accepted conditions, by either (19.6.7) or (19.6.8). The effective coefficient of damping in (20.6.3) is given by (19.6.13),

$$\gamma_{(x)} = \frac{D_{(0)}}{\left\langle x^2\right\rangle}\left(1+b\left\langle x^2\right\rangle\right) \ , \quad D_{(0)} = \frac{\kappa T}{m\gamma} \ . \tag{20.6.5}$$

The solution of (20.6.3) leads to expression (19.6.15) for the space-time spectral density of fluctuations $\delta x(R,t)$:

$$(\delta x\delta x)_{\omega,k} = \frac{2\left(\gamma_{(x)}+Dk^2\right)}{\omega^2+\left(\gamma_{(x)}+Dk^2\right)^2}\frac{\left\langle x^2\right\rangle}{n} \ . \tag{20.6.6}$$

This expression only holds for an infinite system. Because of this, we are once again facing the problem of the spectrum at low frequencies, when multiple diffusion is possible.

Feasibility of the domain of flicker noise ($1/\omega$-spectrum) depends on the relationship between the two dissipative contributions in the expression for spectral density. The largest coefficient of damping due to diffusion is determined by the diffusion time $1/\tau_D = D/L^2$. If $\gamma_{(x)} > 1/\tau_D$ — that is, if the process of diffusion is relatively slow, and multiple diffusion is of little importance — the dissipation is described by coefficient $\gamma_{(x)}$ which characterizes the internal dissipation in bistable element. This dissipative coefficient exhibits strong dependence on the controlling parameter a_f. Depending on the value of the latter, two extreme cases are possible:

$$\gamma_{(x)}(a_f = 0) >> \frac{1}{\tau_D} , \quad \text{and} \quad \gamma_{(x)}(a_f >> 1) << \frac{1}{\tau_D} . \tag{20.6.7}$$

The former corresponds to the state of equilibrium with the Boltzmann distribution for linear oscillator, and the latter to the bistable state, when there are two minima of potential energy. If the potential barrier is high enough, a slow relaxation process of crossing the barrier takes place. As before, by τ_{tr} we denote the transition time (the time of crossing the barrier). If $\tau_{tr} >> \tau_D$, we can single out a region of time spectrum such that

$$\frac{1}{\tau_{tr}} = \omega_{min} << \omega << \omega_{max} = \frac{1}{\tau_D} . \tag{20.6.8}$$

The left-hand limit can also be defined by the coefficient of friction $\gamma_{(x)}(a_f >> 1)$ (see (20.6.7)).

If the right-hand inequality is satisfied, the process of multiple diffusion is possible, and therefore possible is flicker noise. Now, however, the lower limit is in general defined by the internal process of relaxation within bistable element rather than by the time of observation τ_{obs}.

The intensity of Langevin source for the domain of flicker noise is defined by expression similar to (20.2.16). Again we carry out averaging with respect to wave numbers, and arrive at the following expression for the temporal spectral density:

$$(\delta x \delta x)_\omega = \frac{\dfrac{1}{\tau_{tr}} + |\omega|}{\omega^2 + \left(\dfrac{1}{\tau_{tr}} + |\omega|\right)^2} A\langle \delta x_V \delta x_V \rangle , \quad \omega >> \frac{1}{\tau_D} . \tag{20.6.9}$$

Constant A is found approximately from the normalization condition for the region of flicker noise:

$$A = \frac{4}{\pi} \int\limits_{1/\tau_{tr}}^{1/\tau_D} \frac{\dfrac{1}{\tau_{tr}} + |\omega|}{\omega^2 + \left(\dfrac{1}{\tau_{tr}} + |\omega|\right)^2} d\omega , \quad \tau_{obs} >> \tau_{tr} . \tag{20.6.10}$$

In this way, now the domain of flicker noise is delimited by inequalities (20.6.8). At $\omega < 1/\tau_{tr}$ the spectral density ceases to grow, and the spectrum comes to a plateau.

Let us discuss some other examples of nonequilibrium flicker noise in open systems.

20.6.2. *Flicker Noise at Heat Transfer in Medium with Heat Source*

Let us return to Sect. 19.6.3, where we calculated nonequilibrium fluctuations at heat transfer in active medium. The space-time spectral density of fluctuations for infinite medium far from equilibrium is given by (19.6.35). For a given value of wave vector, the width of spectral line is, like in (20.6.6), a sum of two dissipative terms, of which the second is determined by thermal conductivity, and the first by the overrun of coefficient of feedback α above the threshold: $\alpha = \alpha_f - \gamma > 0$.

Hence it follows that one may vary the ratio of dissipation times due to heat conduction and to internal friction in active elements of the medium. In the neighborhood of instability threshold the conditions of feasibility of flicker noise may be satisfied,

$$\alpha = \omega_{min} << \omega << \omega_{max} = \frac{1}{\tau_\chi}, \quad \tau_\chi = \frac{L^2}{\chi}. \tag{20.6.11}$$

Once again we carry out averaging with respect to the distribution of wave numbers, and find the spectral density of time spectrum:

$$(\delta T \delta T)_\omega = \frac{\alpha + |\omega|}{\omega^2 + (\alpha + |\omega|)^2} A \langle \delta T_V \delta T_V \rangle, \quad \omega >> \frac{1}{\tau_\chi}. \tag{20.6.12}$$

Constant A is, as before, found from the normalization condition.

These examples demonstrate two different possibilities of emergence of the domain of flicker noise at nonequilibrium states. In the first case the coefficient of internal dissipation decreases as the "energy" $\langle x^2 \rangle$ increases, whereas in the second case the same happens when the coefficient of feedback α decreases (the overshoot of stationary distribution of the mean temperature above the instability threshold becomes smaller). Interestingly enough, both these possibilities may be realized in active media composed of various oscillators. Let us take a brief look at such systems.

20.7. Flicker Noise in Media Composed of Elements with Complex Behavior

In Sect. 18.7 we considered an example of active medium whose elements were represented by Van der Pol oscillators. The internal fluctuations in oscillators may be divided into the slow and the fast (Stratonovich 1967; Malakhov 1968; Rytov 1976; Klimontovich IV, Ch. 12). Fast are the fluctuations of amplitude or energy of oscillations, whereas the fluctuations of phase drift, for instance, are slow.

In the first case, like that of the temperature fluctuations, the corresponding dissipative coefficient is proportional to the overrun above the generation threshold. Flicker noise is then feasible under conditions similar to (20.6.11), and α is interpreted as the deviation from the generation threshold.

In the second case (fluctuations of phase drift in the oscillator) the conditions of feasibility of flicker noise are similar to (20.6.7), (20.6.8). The time of diffusion of phase plays the role of $1/\gamma_{(x)}$ or τ_{tr}. Thus, the region of flicker noise is defined by the following conditions:

$$\frac{1}{\tau_{phase}} \equiv D_{phase} = \omega_{min} \ll \omega \ll \omega_{max} = \frac{1}{\tau_D} . \qquad (20.7.1)$$

The coefficient of diffusion of phase D_{phase} is inversely proportional to the energy of oscillations; because of this, the domain of flicker noise expands as the overrun above the threshold increases and the system passes into the region of well-developed generation.

Of both academic and practical interest are active media consisting of elements with complex behavior. Such elements may be represented, for instance, by oscillators with inertial nonlinearity (Anishchenko – Astakhov oscillators; see Sect. 16.7), by Van der Pol oscillators with multistable stationary states of energy of oscillations, Chua's circuits (Chua 1993), and oscillators in discrete time, where bifurcations of both energy and period of oscillations are possible (Sect. 16.9). All these systems display both plain and strange attractors.

Of course, analytical methods are of limited use in view of the complexity of such systems. For this reason most of the studies are based on numerical simulations.

The results of numerical experiments for various mappings which may be interpreted as elements of active medium are reproduced in (Schuster 1984). However, the question remains open whether and how the results of such experiments can be useful for explaining the phenomenon of flicker noise.

Low-frequency noises in the so-called Chua's circuits, which are similar to Duffing oscillator but exhibit lagging in the nonlinear term, are studied in (Anishchenko – Neiman – Chua 1993). Calculations indicate that in the critical region of chaos-chaos transition, at small elevations above the threshold $(a - a_{cr} \ll a_{cr})$, the spectrum of power in a limited range of low frequencies is flicker noise $1/\omega^\delta$ with $\delta \approx 1.1$. At low frequencies the spectrum comes to a plateau, whose width is $\Delta\omega = (a - a_{cr})^\gamma$, with $\gamma = 0.62$, and depends on the deviation from the threshold. At large deviations above the threshold, when $a - a_{cr} \gg a_{cr}$, the spectrum transforms into a Lorentz line.

When, however, a tends to the critical value, the plateau does not disappear, and its width is determined by the accuracy of computation σ. The experimentally found dependence is $\Delta\omega \propto \sigma^{\mu}$ with $\mu = 0.32$.

When interpreting these results in the context of our theory of flicker noise, the following must be borne in mind.

Although the numerical experiment is carried out with just one element, the computational error, associated, for one thing, with the rounding-off procedure, may be regarded as an additional dissipative factor. This dissipation is similar to spatial diffusion in that it defines the upper limit ω_{max} of the domain of flicker noise. A more detailed comparison between the results of numerical experiment and the theory of flicker noise calls for a special study.

20.8. Natural Flicker Noise in Music

Experimental investigation of music written by different composers from Bach to Stockhausen (Voos –Clarke 1978) indicates that in the region of low frequencies $\omega < \omega_{max}$ the spectrum of fluctuations of intensity of musical texts is a flicker noise $(1/f$ noise). Frequency f_{max} is of the order of 1 Hz.

This discovery stimulated a series of studies of musical texts which occupy an intermediate position between completely chaotic and entirely predictable time processes. Musical texts also became important objects in the studies of various characteristics of complex signals, in particular, those with fractal dimensionality (Mandelbrot 1982). Theoreticians were once again challenged by the omnipresent and enigmatic phenomenon of flicker noise.

In this connection an attempt has been made to apply the theory of flicker noise as developed above to flicker noise in music (Klimontovich – Boon 1987). Of course, the analysis could have been concerned with recorded speech, radio broadcasts, or, for that matter, with any acoustic signals. Music, however, adds a romantic touch to this enterprise.

For a sample taken from Bach, Voos and Clarke (1978) obtained both the spectrum of amplitude of musical signal $S_v(f)$ and the corresponding spectrum of intensities $S(f)$. The quantity $I(t)$ characterizes the volume (loudness) of music. The spectrum $S_v(f)$ consists of a series of sharp peaks in the range from 100 Hz to 2 kHz, which reflect the spectral structure of the music.

To obtain the function $I(t)$, the sound signal $V(t)$ was amplified and passed through a filter with the passband from 100 Hz to 10 kHz. The output voltage was then squared and passed through a low-frequency filter with the bandwidth of 20 Hz. The spectrum $S_I(f)$ below 1 Hz exhibits $1/f$ behavior, whereas at higher frequencies it is constant in magnitude and close to zero. The lower limit of the domain of flicker noise is determined by the length of musical piece ($\omega_{min} \approx 10^{-2}$ Hz). To obtain the function $S_I(f)$ at lower frequencies, a lengthier sound signal was used. As a result, the domain of flicker noise expanded into the lower frequencies.

We see that the spectrum of the sound signal, which carries information about the original musical text, and the spectrum of the intensity fluctuations fall into different frequency ranges. The low frequency spectrum $S_I(f)$ is generated by the musical instrument or by other source of sound signal.

Observe that musical instruments are distributed systems, in which the processes are described by dissipative wave equations with appropriate Langevin sources. Let $V(R,t)$ be a pulsating (random) function which satisfies such equation. To go over to the equation for intensity one must carry out averaging with respect to the fast variables. The characteristic frequencies of such processes are determined by the locations of the peaks of $S_v(f)$. Then we arrive at the relaxation equation for function $I(R,t)$ which describes the slow processes in the instrument.

Since the coefficient of resonator of the instrument, considered as a distributed medium, is proportional to squared wave vector k^2, the function $I(R,t)$ satisfies the diffusion equation with the Langevin source:

$$\frac{\partial I}{\partial t} = D\frac{\partial^2 I}{\partial R^2} + y(R,t) .$$

(20.8.1)

Let L be the characteristic size of the resonator. Then we once again have to deal with the problem of fluctuation diffusion process in a finite system. Under conditions (20.2.12) we obtain the following expression for the time spectrum of intensity of fluctuations of sound signal:

$$\left(\delta I \delta I\right)_\omega = \frac{\pi}{\ln\dfrac{\tau_{obs}}{\tau_D}}\frac{\langle\delta I_v \delta I_v\rangle}{\omega} , \quad \frac{1}{\tau_{obs}} << \omega << \frac{1}{\tau_D} ,$$

(20.8.2)

which is similar to (20.2.19), but now relates to the flicker noise in the musical instrument or other source of sound. The properties of musical instrument are again described by two characteristics: the diffusion time, which defines the upper limit of the spectrum, and the dispersion of fluctuations δI_v, which reflects the existence of "atomic structure" of the resonator medium. Finally, the observation time is delimited by the duration of musical piece.

These reflections on the nature of flicker noise in music allow us to conclude that flicker noise is part of music to the extent that music is inseparable from musical instrument. In any case, the composer does not compose flicker noise. In this connection it would be interesting to analyze flicker noise in one and the same piece performed on essentially different instruments — for example, the organ and the piano.

20.9. Flicker Noise and Superconductivity

In the preamble to this chapter we have already spoken of the possible connection between two seemingly antagonistic phenomena, flicker noise and superconductivity. Now we have even better reasons to anticipate the existence of such relationship (Klimontovich 1990, V).

First, both flicker noise and superconductivity may be regarded as coherent phenomena. This follows from the fact that the distribution with respect to wave numbers in the domain of flicker noise displays a very sharp maximum near the zero value of wave number. Moreover, dispersion is proportional to the frequency, and therefore tends to zero together with ω.

Secondly, in the domain of flicker noise the coefficient of diffusion friction Dk^2 becomes equal to ω, and therefore vanishes for the steady component of the fluctuation process. As we shall see, this is exactly what is responsible for the disappearance of electric resistance in superconductors at zero frequency. In this way, the possibility is opened for the existence of undamped electric current in a dissipative medium.

Before the discovery of high-temperature superconductivity (Bednorz – Müller 1988), the theory of superconductivity seemed to have been completed, at least in general (Bardeen – Cooper – Schrieffer 1957; Ginzburg – Landau 1950; Bogolyubov 1959). Now, however, new problems emerged. In particular, open remains one of the main questions: how can permanent electric current exist in a dissipative system — the system of many Bose particles like Cooper pairs.

The transition from normal to superconducting state is an example of phase transition of the second kind. The role of parameter of order is played by the coherent wave function of pairs of bound electrons (Cooper pairs), in terms of which the superconduction current is expressed.

At the critical point of phase transition from the system of electrons to the system of Cooper pairs the number of pairs is zero. This number starts to grow as the temperature decreases. Let us consider the state of the system at temperatures far below critical, when practically all electrons are bound in pairs and make up a system of Bose particles.

Observe that the concept of "bound Cooper pairs" does not reflect the state of things quite accurately. It is more consistent to speak of paired correlations of electrons with opposite spins. This correlation radius is usually denoted by ξ; it depends on the temperature and is the largest at T_0. The value of ξ_0 is commonly referred to as the *coherence length*.

Another characteristic length is δ_L (for Heinz and Fritz London), which is defined by

$$\delta_L^2 = \frac{mc^2}{4\pi e^2 n_s} \,, \tag{20.9.1}$$

and determines the depth of penetration of magnetic field into the bulk of superconductor (this will be discussed later). Here m and e are the mass and the charge of electron, and n_s is the density of electrons which create the superconduction current. At $T = 0$ the value of n_s coincides with the total density of free electrons in the system n.

For subsequent discussion we shall require the relationship between these two characteristic length scales. Superconductors are divided into two classes. In superconductors of the first class the coherence length is much greater than the depth of penetration, $\xi_0 \gg \delta_L$. By contrast, for superconductors of the second class (which includes also the high-temperature superconductors) we have

$$\xi_0 \ll \delta_L \,, \text{ but } n\xi_0^3 \gg 1 \,. \tag{20.9.2}$$

The second of these inequalities indicates that, although ξ_0 is small, the volume ξ_0^3 contains "many" (about a hundred) electrons.

Assume that conditions (20.9.2) are satisfied. The second of these justifies the transition from the system of particles to the approximation of continuous medium, whereas the first allows us to describe superconductivity on the basis of the macroscopic Maxwell equations for the field and the material equations of hydrodynamics.

Now our aim is to demonstrate the feasibility of practically undamped electric current and the Meissner – Ochsenfeld effect of expulsion of permanent magnetic field from the bulk of semiconductor in a dissipative macroscopic system of charged Bose particles. We shall show that this is possible exactly because of the existence of natural flicker noise, which is a consequence of the "atomic structure" of superconductor, the "atoms" being represented this time by Cooper pairs of electrons. It would be natural to illustrate this with the simplest examples of dissipative systems.

1. *Dissipation due to electric conductivity.* Assume that the mean velocity of charged Bose particles is small, and the dissipative term in the equation for electric current may be written as vj, where v is the effective frequency of collisions. To eliminate the gradient of pressure in the equation for electric current, we write the equation for $\Omega = \text{curl } j$:

$$\frac{\partial \Omega}{\partial t} + v\Omega = \frac{e^2 n_s}{m} \text{curl } E \,, \quad \Omega = \text{curl } j \,, \quad j = e n_s u \,, \tag{20.9.3}$$

where E is the electric field.

This equation is supplemented by the Maxwell equations:

$$\text{curl } B = \frac{4\pi}{c} j \,, \quad \text{curl } E = -\frac{1}{c}\frac{\partial B}{\partial t} \,, \quad \text{div } B = 0 \,. \tag{20.9.4}$$

Here we have neglected the "displacement current", which at low frequencies is small, and assumed that $B = H$. Eliminating curl E from these equations, we get the set of equations in Ω and B:

$$\frac{\partial \Omega}{\partial t} + v\Omega = -\frac{e^2 n_s}{mc}\frac{\partial B}{\partial t} \,, \quad \Omega = -\frac{c}{4\pi}\frac{\partial^2 B}{\partial R^2} \,, \tag{20.9.5}$$

whence follows the dissipative equation for B:

$$\frac{\partial}{\partial t}\left(B - \delta_L^2 \frac{\partial^2 B}{\partial R^2} \right) = D\frac{\partial^2 B}{\partial R^2} \,, \tag{20.9.6}$$

where the coefficient of diffusion is

$$D = \delta_L^2 v \equiv \frac{c^2}{4\pi\sigma} \,, \quad \text{and} \quad \sigma = \frac{e^2 n_s}{mv} \tag{20.9.7}$$

is the electric conductivity.

In this way we have obtained an equation with the dissipative term of diffusion type. If $D \equiv 0$, which in this context implies that the conductivity is infinite, equation (20.9.6) reduces to the London equation for the magnetic field:

$$B - \delta_L^2 \frac{\partial^2 B}{\partial R^2} = 0 \,. \tag{20.9.8}$$

Its solution describes the Meissner – Ochsenfeld effect, the expulsion of external magnetic field from the bulk of superconductor.

Because of the existence of "atomic structure" of continuous medium, which is comprised of Cooper pairs, fluctuations are inevitable, and therefore condition $D \equiv 0$ is incompatible with the fluctuation dissipation relation.

This example is just a useful illustration. Dissipation by viscous friction is more pertinent to the description of slow processes.

2. *Dissipation due to viscous friction.* Superconduction current flows in a thin but macroscopic skin layer, whose thickness is $\delta_L \approx 10^{-5}$ cm. There exists a velocity gradient, since velocity is the largest on the outside surface of superconductor, and zero on the inner surface of the skin layer. This allows us to introduce viscous friction into equation for the current, so that in place of (20.9.5) we get

$$\frac{\partial}{\partial t}\left(\Omega + \frac{e^2 n_s}{mc}B\right) = D\frac{\partial^2 \Omega}{\partial R^2} , \quad \Omega = \text{curl}\, j , \quad D = \frac{\eta}{\rho} = v . \tag{20.9.9}$$

Now the coefficient of diffusion is expressed in terms of the coefficient of viscosity.

Thus, we come again to the diffusion-type equation. Assuming that $D \equiv 0$, equation (20.9.9) reduces to the London equation which links the current and the magnetic field in a superconductor:

$$\text{curl}\, j = -\frac{e^2 n_s}{mc}B . \tag{20.9.10}$$

This assumption, however, is not justified for the same reasons as before. After all, our aim is not to do away with the dissipation, but rather to explain the phenomenon of superconduction in a dissipative system.

If the system displays some kind of "atomic structure", fluctuations inevitably exist against the background of averaged motion. Then, in accordance with the fluctuation dissipation relation, the dissipation must exist.

This statement can also be read the other way round: if there is dissipation, then fluctuations must be taken into account. For us this is an "old truth", since we have more than once dealt with the calculation of fluctuations in dissipative systems.

In other words, superconductivity must exist in good agreement with the fluctuation dissipation relation.

The important question now is, what is "permanent current". After all, the time of its existence is limited by the lifetime of the installation τ_{life}, which in reality cannot be infinite. It can only be much greater than the time of observation τ_{obs} — that is,

$$\tau_{\text{life}} \gg \tau_{\text{obs}} . \tag{20.9.11}$$

This complication is removed if the fluctuations of current and magnetic field at low frequencies behave as flicker noise.

Indeed, then the intensity of Langevin source is proportional to the frequency (see (20.4.4)). Because of this, the dissipative coefficient Dk^2 is replaced by ω, and "permanent current" is then defined as Fourier component of the current at the lowest possible frequency,

$$j_0 = j\left(\omega = \frac{1}{\tau_{\text{life}}}\right).$$ (20.9.12)

Hence it follows that the linewidth of time spectrum of "permanent current" is of the order of $1/\tau_{\text{life}}$. Under condition (20.9.11), flicker noise occurs against the background of this "steady component" of the current.

Let us illustrate this with the example of equation for Ω.

Since dissipation is determined by diffusion, for the domain of flicker noise we may introduce, in accordance with (20.2.16), the following distribution function of wave numbers:

$$f(k) = \left(\frac{L_\omega^2}{2\pi}\right)^{3/2} \exp\left(-\frac{L_\omega^2 k^2}{2}\right), \quad \int f(k)dk = 1, \quad L_\omega^2 = \frac{D}{\omega}.$$ (20.9.13)

Then from (20.9.9) follows the Langevin equation

$$\left[\frac{\partial}{\partial t}\left(\Omega + \frac{e^2 n_s}{mc}B\right)\right]_{\omega,k} + Dk^2\Omega(\omega,k) = y_\Omega(\omega,k).$$ (20.9.14)

After averaging over the distribution (20.9.13) with respect to k, we get the following equation for the time Fourier component of fluctuations of the current:

$$\left[\frac{\partial}{\partial t}\left(\Omega + \frac{e^2 n_s}{mc}B\right)\right]_\omega + |\omega|\Omega(\omega) = y_\Omega(\omega).$$ (20.9.15)

The spectral density of the source intensity is

$$(y_\Omega y_\Omega)_\omega = 2|\omega|\frac{\pi}{\ln\dfrac{\tau_{\text{obs}}}{\tau_D}}\langle\delta\Omega_V\delta\Omega_V\rangle.$$ (20.9.16)

Given the definition of the "steady component" (20.9.12), the source of noise and the dissipative term go to zero, and we come to the London equation (20.9.10), whence, with the aid of Maxwell equations, follows the closed equation for the magnetic field (20.9.8). The latter describes the effect of Meissner – Ochsenfeld of expulsion of magnetic field.

In this way, the existence of flicker noise opens up the feasibility of "permanent current" in the sense of (20.9.12), and explains the Meissner – Ochsenfeld effect.

In the frequency range

$$\frac{1}{\tau_D} \gg \omega \gg \frac{1}{\tau_{obs}} , \quad (\tau_{obs} \ll \tau_{life}) , \tag{20.9.17}$$

the spectrum of fluctuations of the curl of current against the background of "permanent current" is given by

$$(\delta\Omega\delta\Omega)_\omega = \frac{\pi}{\ln\dfrac{\tau_{obs}}{\tau_D}} \frac{\langle\delta\Omega_V\delta\Omega_V\rangle}{\omega} . \tag{20.9.18}$$

The same equations can be used for finding the spectrum of induced current fluctuations, proportional to δB.

Observe finally that we have considered the problem of flicker noise in the system of charged Bose particles. In the system of neutral Bose particles (He^4) the existence of flicker noise explains the effect of superfluidity of helium, the zero-viscosity flow in capillaries or narrow channels without pressure difference. In this case the coefficient of diffusion is of the order of Planck's constant $\hbar$.

It should be noted that "technical" low-frequency noises, caused by imperfections in the experimental setup, may exist in superconductors along with the low-frequency flicker noise.

20.10. Spectrum of Thermal Radiation in Superconductors: Violation of Rayleigh – Jeans Law in Domain of Flicker Noise

Let us consider one of the consequences of flicker noise in superconductors, the low-frequency fluctuations of the current.

It is well known that the spectrum of fluctuations of thermal electromagnetic field in the classical limit ($\hbar\omega \ll \kappa T$) is given by Rayleigh – Jeans formula,

$$(\delta E\delta E)_\omega = \frac{4\omega^2}{c^3} n(\omega)\kappa T \propto \omega^2 , \tag{20.10.1}$$

where n is the index of refraction at frequency ω. This formula does not take into account explicitly the finite size of the region filled by radiation. The frequency dependence of fluctuations of electric δE and magnetic δB fields, and that of the vector potential δA, is the same.

In the domain of flicker noise the spectral distributions of these functions become different. For instance, the spectral density of fluctuations of the vector potential δA is

$$(\delta A \delta A)_\omega = \left(\frac{mc}{e^2 n}\right)^2 (\delta j \delta j)_\omega \propto \frac{1}{\omega} . \tag{20.10.2}$$

At the same time, the spectral functions of the field exhibit different dependence on ω.

$$(\delta E \delta E)_\omega \propto \omega , \quad (\delta B \delta B)_\omega \propto \omega^0 . \tag{20.10.3}$$

Let us evaluate the ratio of spectral functions of electric field for the domain of flicker noise and for the Rayleigh – Jeans law:

$$\frac{(\delta E \delta E)_\omega}{(\delta E \delta E)_\omega^{R-J}} \sim \frac{\omega_{max}}{\omega} , \quad \omega_{max} \sim \frac{c}{L} \frac{\delta_L^2}{L^2} , \tag{20.10.4}$$

where L is the characteristic size of the sample.

We see that the intensity of spectrum of field fluctuations for the region of flicker noise is greater than the corresponding Rayleigh – Jeans value.

We have analyzed the results of calculations of natural flicker noise for different systems. Flicker noise has been regarded as a random diffusion process in a finite system. It is caused by the restructuring of the noise spectrum because of multiple diffusion over the time which is much larger than the diffusion time. This approach has allowed us to explain the main features of flicker noise, established in numerous experimental studies.

Since, however, there is no consensus of opinion as to the nature of flicker noise, it will be worthwhile to give a brief outline of a different view on the nature of this mysterious phenomenon.

20.11. Flicker Noise of System of Independent Sources with Exponential Distribution of Relaxation Times

Let the motion of individual element be described by the Langevin equation

$$\frac{dx}{dt} + \frac{1}{\tau} x = y(t) , \quad \langle y \rangle = 0 , \quad \langle y(t) y(t') \rangle = 2\frac{1}{\tau} \langle x^2 \rangle \delta(t - t') . \tag{20.11.1}$$

The spectral density is given by

$$(xx)_\omega = \frac{\frac{2}{\tau}}{\omega^2 + \left(\frac{1}{\tau}\right)^2}\langle x^2\rangle , \quad \int(xx)_\omega \frac{d\omega}{2\pi} = \langle x^2\rangle . \tag{20.11.2}$$

Assume now that separate elements have different relaxation times τ, and introduce distribution function $f(\tau)$ of the values of relaxation times. Then we may carry out averaging of spectrum (20.11.2) with respect to $f(\tau)$:

$$\overline{(xx)_\omega} = \int \frac{\frac{2}{\tau}}{\omega^2 + \left(\frac{1}{\tau}\right)^2}f(\tau)d\tau\langle x^2\rangle , \quad \int\overline{(xx)_\omega} \frac{d\omega}{2\pi} = \langle x^2\rangle , \quad \int f(\tau)d\tau = 1 . \tag{20.11.3}$$

Next, an important assumption concerning the form of distribution function $f(\tau)$ within a certain range of τ:

$$f(\tau) = A\frac{1}{\tau} , \quad \tau_1 << \tau << \tau_2 . \tag{20.11.4}$$

Substitution of this distribution into (20.11.3) leads to

$$\overline{(xx)_\omega} \propto \frac{1}{\omega} \quad \text{if} \quad \frac{1}{\tau_2} << \omega << \frac{1}{\tau_1} . \tag{20.11.5}$$

So, there is a region of $1/\omega$-noise for a specially selected distribution function with respect to the relaxation times. Let the reader decide whether this explanation of flicker noise is convincing enough. Some cases when the distribution (20.11.4) is physically justified are discussed in (Kogan 1985). It is hard to explain in this way, however, the universal presence of flicker noise.

20.12. Diffusion in Space of Fractal (Fractional) Dimensionality. Anomalous Brownian Motion

Drawing comparison with a diffusion process in an infinite system, flicker noise, as indicated above, may be regarded as an example of anomalous Brownian motion. In particular, this is manifested through the change in the structure of the classical Einstein formula. Now the dispersion of values of wave number is limited not by the size of the system, but rather by the observation time:

$$\left\langle (\delta k)^2 \right\rangle_{\min} \sim \frac{1}{D\tau_{\text{obs}}} .$$

Because of this, the nature of time spectrum also becomes much different.

There are, however, other anomalies, even when the finite size of the system may be not taken into account. This concerns diffusion in the space of fractal (fractional) dimensionality, where the anomaly is detected through the violation of the Einstein relation which states that mean square displacement is proportional to the time t, with the proportionality coefficient equal to the coefficient of diffusion D.

This law was first established for Brownian particles, but it applies also to certain generalized characteristics. Taking advantage of this fact, we may express the coefficient of diffusion in terms of the "free path" effective length l and time τ, and write Einstein's relation as

$$\left\langle (\delta r)^2 \right\rangle = 6Dt = 6l^2 \frac{t}{\tau} . \tag{20.12.1}$$

Analytical and numerical analysis of most diverse phenomena (such as turbulence, biopolymer structure, cardiograms, texts, music) reveals that the characteristics of these processes (for instance, occurrences of characters or words in texts) resemble Brownian motion. However, the Einstein relation for one-dimensional motion must then be replaced by the more general relation,

$$\left\langle (\delta r)^2 \right\rangle = l^2 \left(\frac{t}{\tau} \right)^\alpha . \tag{20.12.2}$$

In other words, the mean square displacement of a given characteristic is proportional not to the time, but rather to a certain power of the time. The exponent is different for different processes (some authors use 2α instead of α).

Interestingly, the value of α may deviate from Einstein's $\alpha = 1$ both ways.

"Brownian motion" of alphabetical characters, and of some words and phrases, was studied in (Ebeling – Klimontovich – Neiman 1994) for a number of long texts: Luther's *Bible* ($L \approx 4,423,000$ characters), Grimm's *Tales* ($L \approx 1,436,000$ words), Melville's *Moby Dick* ($L \approx 1,170,000$ words). In all cases the value of α was greater than one: $\alpha = 1.67$ for *The Bible*, $\alpha = 1.24$ for Grimm's *Tales*, and $\alpha = 1.32$ for *Moby Dick*.

It was possible to change the value of α by shuffling characters and words in the text — that is, by destroying the internal correlations. Then α comes close to unity, $\alpha \approx 1$. This indicates that the classical Brownian motion corresponds to the most chaotic arrangement of characters in the body of text.

Cardiograms corresponding to different heart conditions were analyzed in (Peng – Mietus – Hausdorff – Havlin – Stanley – Goldberger 1993). It was found that for a healthy organism the value of α is close to zero, whereas for patients with heart trouble α is close to one (in this paper $\alpha \Rightarrow 2\alpha$, and so $\alpha \approx 0.5$). In other words, the nature of "Brownian motion" is close to classical for a sick person.

Now does this imply that the disease destroys internal correlations, and the state of the system becomes more chaotic? To answer this question we need a criterion of the relative degree of order of different states of open systems. Such criterion will be considered in the next chapter.

Anomalous diffusion may be described by a diffusion equation, in which the conventional time derivative must be replaced by a corresponding fractional derivative:

$$\frac{\partial^\alpha f}{\partial t^\alpha} = D_\alpha \frac{\partial^2 f}{\partial r^2} \ , \quad D_\alpha = \frac{l^2}{\tau^\alpha} \ , \quad \int f(r,t)dr = 1 \ , \tag{20.12.3}$$

where $f(r,t)$ is the distribution function of "Brownian particles".

The solution of this equation at the given initial position of "Brownian particle" is

$$f(x,t;x_0,t_0) = \frac{1}{\sqrt{4\pi l^2 \left(\dfrac{t}{\tau}\right)^\alpha}} \exp\left[-\frac{(x-x_0)^2}{4l^2 \left(\dfrac{t}{\tau}\right)^\alpha}\right] . \tag{20.12.4}$$

At the initial time the distribution function is $f = \delta(x - x_0)$.

Of course, all those questions we had to deal with in connection with "normal" Brownian motion arise also in the theory of anomalous Brownian motion. In particular, there is the problem of structure of Langevin source in the equation of anomalous diffusion, and anomalous diffusion in a finite system. Then there is the problem of *anomalous flicker noise*. Many new problems arise in the study of anomalous Brownian motion in nonequilibrium systems.

As indicated above, the theory of anomalous Brownian motion is still in its early stage. We may only hope that these results will stimulate further developments on this new and promising avenue of scientific research.

By now we have compiled a list of questions to answer which we need a criterion of the relative degree of order of different states of open systems. One of such problems is whether the state of "normal" Brownian motion is more chaotic than the state of anomalous Brownian motion. When we shuffle characters or words in a text, the degree of mixing acts as the controlling parameter. When Brownian motion in cardiograms is analyzed, the "controlling" parameters may be various stresses which give a shake to the organism. According to the paper quoted above, the nature of Brownian motion will then become more "classical". So, by medication or by some other means one may try to bring the patient into a better state of health, when Brownian motion of his cardiogram will become anomalous.

Of course, these are just particular examples of the general problem of which we have already spoken in the Introduction. In Sect. 1.6.1 we discussed the two opposite directions of evolution: degradation and self-organization. An example of degradation is the path towards equilibrium in a closed system. In general, however, it may be quite difficult to distinguish between these two processes: one needs reliable criteria of the relative degree of order of two arbitrary states of the open system under consideration. This is what the next chapter is about.

CHAPTER 21

CRITERIA OF SELF-ORGANIZATION

As we have already said in the Introduction, one of the main tasks of statistical theory of open systems consists in establishing the criteria of self-organization. We touched upon this problem now and then in the course of our discussion. Now is the time to draw conclusions, to compare various criteria of self-organization, to illustrate their efficiency with concrete examples.

In this chapter we shall discuss various criteria of relative degree of order of different states in systems of most diverse nature. Our attention will be focused mainly on the criterion of S–theorem, which has already been used in the preceding chapters. We shall establish connection with other criteria widely used in the analysis of complex motion: Krylov – Kolmogorov – Sinai entropy (K–entropy), Lyapunov indices, Kullback entropy, Renyi entropy, nonlinear K–entropy (K_{nl}), statistical K–entropy (K_{st}).

Processes of self-organization may be regarded as chains of nonequilibrium phase transitions, resulting in new dissipative structures. Because of this, it is important to be able to compare the relative degree of order of different states of open systems. In case of physical systems the state of equilibrium can be taken for the reference state. This possibility is based on the Gibbs theorem; let us briefly review this important result (see Sect. 1.6.4).

The state of equilibrium is described by the canonical Gibbs distribution $f_N^{(0)}(X)$, the particular form of which depends on the Hamilton function $H(X)$ of the system in question. Let $f_N(X)$ be an arbitrary normalized distribution with one (important!) constraint: the mean value of the Hamilton function must be the same as that for the equilibrium distribution, as given by (1.6.5). Then, according to (1.6.6), the state of equilibrium corresponds to the maximum entropy value.

In the theory of Brownian motion (Ch. 15, 16) we considered a number of examples when the "arbitrary" function $f_N(X)$ was represented by the nonequilibrium distribution for the stationary state of the system under consideration. The degree of nonequilibrium in Van der Pol oscillator, for instance, depends on the magnitude of the feedback parameter α_f. The structure of the coefficient of diffusion for nonlinear process is determined from condition that the system should be at equilibrium when the feedback parameter is zero; in other words, the state of equilibrium us taken for the point of reference in comparing the order of different states. At the same time, the situation now is not the same as in case of Gibbs formulation. The stationary distribution is not arbitrary; it is given by the solution of the appropriate equation. In general, the mean value of the energy will not be the same. Because of this, Gibbs theorem (1.6.6) cannot be immediately applied to open systems.

To construct a criterion of self-organization one requires not only the reference point, but also the quantitative measure of order. Therefore, the entropy values must be renormalized to the given values of the mean energy. This is the essence of S–theorem (Sect. 1.6.6). As a rule, the *effective* Hamilton function H_{eff} is used in place of the conventional Hamilton function in the "mechanical" sense.

The process of self-organization may also go via a sequence of equilibrium states. Such process was considered in Sect. 18.6, devoted to the theory of equilibrium phase transitions of the second kind. The role of the controlling parameter is played by the temperature. "The most equilibrium" is the state with the highest symmetry, which is realized when the parameter of effective field a_f is zero. We used S–theorem to demonstrate that the degree of order increases when the system undergoes a phase transition to the state of lower symmetry (from monostable state to bistable one). This is an example of the process of self-organization formed by a sequence of equilibrium states.

This fact does not contradict the general assumption that processes of self-organization are only possible in open systems, since the system in a thermostat is not closed. As a matter of fact, the system exchanges energy with the thermostat when the temperature is varied.

There is another issue associated with the choice of the reference point of chaoticity. In open systems we have to compare the relative degree of order of essentially nonequilibrium states. This situation is typical, for instance, in the study of turbulence, or in the analysis of biological objects.

In such situations, one of the states is a priory assumed to be the most chaotic one. This assumption is later proved (or disproved) by calculations based on the S–theorem. This is the only option available when all information about the system is drawn from experiment. In addition to those considered in the preceding chapters, we shall presently discuss a few more examples of this kind. For the sake of integrity we shall repeat here some of our previous results (especially those of Sect. 1.6, 12.7 – 12.9, 16.5, 16.6), adding some important details which could not have been introduced earlier.

21.1. Evolution in Time. Degradation and Self-Organization. H–Theorem

Two important classes may be distinguished among the various processes of evolution. First, there is evolution in time when the values of controlling parameters are fixed.

Secondly, there are sequences of stationary states corresponding to different values of controlling parameters. Of course, a combination of these two cases is also possible.

Evolution may take the path of degradation (transition to a more chaotic state), or self-organization (transition to a more ordered state). An physical example of degradation is given by the transition to the state of equilibrium at a fixed value of the mean energy. Such is the case when the time evolution is described by the kinetic Boltzmann equation. Then Boltzmann's H–theorem holds, as expressed by the two inequalities (1.6.3), (1.6.4):

$$\Lambda_S = S_0 - S(t) = \kappa n \int \ln \frac{f(r,p,t)}{f_0(p)} f(r,p,t) drdp \ge 0 \ , \tag{21.1.1}$$

$$\frac{d\Lambda_S}{dt} = \frac{d}{dt}\left(S_0 - S(t)\right) \le 0 \ , \quad \left\langle \frac{p^2}{2m} \right\rangle = \text{const} \ . \tag{21.1.2}$$

These inequalities indicate that in the course of time evolution at fixed value of mean energy the entropy of the system increases and reaches its maximum at the state of equilibrium. It is the condition of constancy of the mean energy that allows us to define Lyapunov functional as the difference in entropies.

In open systems, however, the mean energy does not remain the same in the course of time evolution, and the appropriate Lyapunov functionals cannot be defined. To overcome this difficulty, one has to extend the concept of thermodynamic free energy to nonequilibrium states (Sect. 12.4). Then the Lyapunov functional Λ_F can be defined by (12.4.9), (12.4.10), and (15.11.5) as the difference in free energies of the current and the stationary states.

These inequalities imply that the free energy is at minimum in the stationary state. The question now is whether the decrease is monotonous. If the answer is positive, Λ_F is a Lyapunov functional.

In Sect. 15.11 we have proved the relevant inequality (15.11.9) for the master equation. This result was particularized in Sect. 16.5 for Brownian motion in Van der Pol oscillator. This brought us to the counterpart of Boltzmann's H–theorem:

$$\Lambda_F(t) = F(t) - F_0 = \kappa T \int_0^\infty \ln \frac{f(x,t)}{f_0(x,\alpha_f)} f(x,t) dx \ge 0 \ , \tag{21.1.3}$$

$$\frac{d\Lambda_F}{dt} = -\sigma_F \le 0 \ , \tag{21.1.4}$$

where σ_F is a function similar to the entropy production. Let us recall some results of Sect. 16.5 for the Van der Pol oscillator.

The Fokker – Planck equation for the distribution function of energy of oscillations is (see (16.2.8))

$$\frac{\partial f(E,t)}{\partial t} = \frac{\partial}{\partial E}\left[D(E)E\frac{\partial f}{\partial E}\right] + \frac{\partial}{\partial E}[(-\alpha + \beta E)Ef] \, . \qquad (21.1.5)$$

Here, as before, we have $\alpha = \alpha_f - \gamma$, and the feedback parameter acts as the controlling parameter. Function $D(E)$ is found from condition that the stationary solution coincides with the equilibrium solution when $\alpha_f = 0$. In this way, the state of equilibrium is taken for the reference point of chaoticity. At $\alpha_f \neq 0$ the stationary solution is represented as the canonical Gibbs distribution,

$$f(E) = \exp\frac{F(T,\alpha_f) - H_{eff}(E,\alpha_f)}{\kappa T} \, , \quad H_{eff} = E - \frac{\alpha_f}{\beta}\ln\left(1 + \frac{\beta}{\gamma}E\right) \, , \qquad (21.1.6)$$

where $F(T,\alpha_f)$ is the free energy of the stationary nonequilibrium ($\alpha_f \neq 0$) state. The current value of the free energy is, as before, defined as

$$F(t) = \langle H_{eff}(E)\rangle - TS(t) \, , \quad S(t) = -\kappa\int_0^\infty \ln f \, fdE \, . \qquad (21.1.7)$$

The difference in the free energies reduces to

$$\Lambda_F = F(t) - F(E,\alpha_f) = \kappa T\int_0^\infty \ln\frac{f(E,t)}{f(E,\alpha_f)} f(E,t)dE \geq 0 \, , \qquad (21.1.8)$$

and is a decreasing function:

$$\frac{d(F(t) - F(E,\alpha_f))}{dt} = \kappa T\int D(E)Ef(E,t)\left(\frac{\partial}{\partial E}\ln\frac{f(E,t)}{f(E,\alpha_f)}\right)^2 dE \leq 0 \, . \qquad (21.1.9)$$

These two inequalities express the H–theorem for the functional Λ_F. In Sect. 16.5 this result was also reformulated for the Lyapunov functional Λ_S, defined as the difference in the entropies. The value of entropy renormalized to the given value of the mean effective Hamilton function is found to increase in the course of evolution towards the stationary state.

Now what is the implication of the increase in the renormalized entropy (or the decrease in the nonequilibrium free energy $F(t)$) in the course of time evolution in an open system?

The value of the controlling parameter α_f and the function $D(E)$ define uniquely only the stationary state. The current distribution can always be represented as $f(E,t) = f(E, \alpha_f) F(E,t)$, where $F(E,t)$ is some function which takes on the value of $F(E,t) = 1$ at the stationary state. Hence it follows that one needs additional information so as to define unambiguously the state at any given instant. The lack of such information is reflected by the growth of renormalized entropy (or the fall of free energy) in the course of time evolution towards the stationary state.

Let us once again point to the advantages of the Lyapunov functional Λ_S as defined by the difference in the entropies rather than in any other nonequilibrium thermodynamic potentials.

First, the entropy can be defined in terms of the distribution function for arbitrary nonequilibrium states. This allows one to define the entropy directly from experimental data, without using mathematical models. This is an important gain, since adequate mathematical simulation of complex systems is often very difficult.

Secondly, it is only the functional Λ_S that can be used as measure of the relative degree of order of various states of open systems.

21.2. Evolution in Space of Controlling Parameters. S–Theorem

In recent years the concept of "dynamic chaos" has emerged in connection with the discovery and study of complex motion in relatively simple systems. Dynamic chaos results from dynamic instability of motion: the divergence of trajectories which is extremely sensitive to small changes in the initial conditions.

In the Introduction we raised the question of the relationship between dynamic and physical chaos. We can put this question in a different way: can dynamic instability lead to a more ordered, although perhaps more complicated, type of motion? Can dynamic instability play a constructive role in the statistical theory of open systems? The answers to these questions were positive. In particular, we have shown that it is possible to construct the equations in macroscopic functions (kinetic, hydrodynamic, reaction-diffusion equations) exactly because of the complexity of motion on microscopic level, which ensures mixing in the phase space due to the dynamic instability of motion of "atoms" of the system under consideration.

This conclusion, however, ought to be thoroughly checked, and for this we need reliable criteria of the relative degree of order of nonequilibrium states of the system. Such criteria will provide a quantitative measure of order, and will enable one to tell chaos from order. For a sequence of states such analysis allows us to distinguish between degradation from self-organization, with checking simultaneously the correct choice of controlling parameters.

Of course, the criteria of self-organization are useful not only for studying physical or chemical systems, but also for medico-biological diagnostic purposes. We shall discuss some examples of such applications. First, however, we must consider various possible criteria of order, so as to gain a better overview of the problem.

In this section we return to the criterion of S–theorem (Klimontovich 1983c, 1984a, V). This criterion has been used more than once in the preceding chapters for analyzing the relative degree of order of states of open systems. Now we shall repeat the formulation of S–theorem only to the extent required for drawing the relevant conclusions directly from the experimental data (Klimontovich 1988, V). This recapitulation will also be useful for those specialists who intend to use criteria of self-organization as a practical tool.

Assume that information about the system is contained in the time realizations obtained for different values of controlling parameters. Like we did in Sect. 1.6.1, 12.8, we select two stationary states corresponding to two different values of the controlling parameter, a_0 and $a_0 + \Delta a$. The corresponding time realizations are

$$X(t,a_0) , \quad X(t,a_0 + \Delta a) . \tag{21.2.1}$$

The external conditions being stationary, the realizations can be made long enough to ensure that the distribution functions

$$f_0(X,a_0) , \quad f(X,a_0 + \Delta a) , \quad \int f_0 dX = \int f dX = 1 \tag{21.2.2}$$

are practically time-independent.

Obviously, the assumption of "stationarity" is an idealization. In practice, we may only speak of separating the fast processes from the "slow" background. For example, a cardiogram suitable for constructing the distributions (21.2.2) can be taken within the time which is much shorter than the time of the patient's recovery.

Recall that the concept of "physical chaos" has been used in constructing the criterion of S–theorem, since the layman's notion of chaos is too vague. Indeed, according to Maxwell, Boltzmann and Gibbs, "chaotic" is the motion of particles at the state of equilibrium. At the same time, the states far from equilibrium, like turbulent motion or dynamic chaos, are often commonly regarded as chaotic.

It is hardly possible to give a precise definition of "chaos" and "order". As a matter of fact, such a definition is not much needed. What is really necessary is the ability to determine the relative degree of order of states of open systems. As indicated above, this would allow one to distinguish chaos from order, and degradation from self-organization.

Let us return to the distributions (21.2.2) based on the experimental data for the states corresponding to different values of the controlling parameter a. Which of the two corresponds to the more chaotic state? The answer to this question might seem quite easy. We know that entropy is the measure of uncertainty in the statistical description; it is the entropy (see Sect. 3.5) that possesses all the necessary properties. Then from distributions (21.2.2) we simply find the values of Shannon entropy,

$$S_0(a_0) = -\int \ln f_0(X, a_0) f_0(X, a_0) dX ,$$

$$(21.2.3)$$

$$S(a_0 + \Delta a) = -\int \ln f(X, a_0 + \Delta a) f(X, a_0 + \Delta a) dX ,$$

and use the difference between these to assess the relative degree of order of the states in question.

This "straightforward" approach, however, misses the target. We have seen this in Sect. 16.6 in connection with the analysis of evolution of Van der Pol oscillator as the generation develops. To wit, the result obtained in this way told us that the chaoticity increases, which does not agree with the physicist's view, according to which the order is higher in the state of well-developed generation.

The wrong conclusion resulted from the neglect of the fact that the mean energy of oscillations changes as the generation develops. Everything "clicked into place" after renormalization to the given value of the mean energy. This was possible because the stationary solution of Fokker – Planck equation was available for all values of the feedback parameter. The zero value of this parameter corresponds to the state of equilibrium; thus, we had the reference point for counting the degree of chaoticity.

What then do we do if the relative degree of order has to be assessed on the basis of experimental data which are available only for nonequilibrium states? This is when it becomes necessary to use the concept of physical chaos.

Let us return to the distribution (21.2.2) for the two states at two different values of the controlling parameter. In general, both states are nonequilibrium. So far we have no grounds for selecting one of these as the more chaotic. Straightforward use of the entropy values is prevented by the fact that the mean energy may be not constant. Moreover, it would be hard to formulate such a condition for, say, nonequilibrium biological systems. We may only assume that one of these (not known which) is more chaotic than the other.

So, let us assume (just assume!) that the more chaotic state corresponds to $a = a_0$. The check of this assumption must be incorporated in the criterion of the relative degree of order. The more chaotic state is taken for the state of "physical chaos". The adjective "physical" indicates that for "chaos" we take a nonequilibrium state which is just relatively more chaotic than the other, more ordered, state.

By this assumption, the distribution f_0 characterizes the state of physical chaos. We use this to define the effective Hamilton function

$$H_{eff}(X,a_0) = -\ln f_0, \text{ or } f_0 = \exp(-H_{eff}) . \tag{21.2.4}$$

In general, the mean value of the effective energy introduced in this way is not the same for the two selected states. To equalize the two values of the effective energy, we replace f_0 by the renormalized distribution $\tilde{f}_0$, which is represented as the canonical Gibbs distribution:

$$\tilde{f}_0(X,a_0,\Delta a) = \exp\frac{F_{eff} - H_{eff}(X,a_0)}{T_{eff}(\Delta a)} , \quad \int \tilde{f}_0 dX = 1 , \tag{21.2.5}$$

where F_{eff} and T_{eff} are the effective free energy and the effective temperature. The effective Hamilton function H_{eff} has, in general, nothing to do with the conventional concept of energy.

Owing to the normalization condition, the effective free energy is expressed via the effective temperature by

$$F_{eff} = -T_{eff} \ln \int \exp\left(-\frac{H_{eff}(X,a_0)}{T_{eff}}\right) dX . \tag{21.2.6}$$

Given this relationship, expression (22.2.5) contains only one unknown, the effective temperature. We define it from condition that the values of mean effective energy should be the same for the two selected states:

$$\int H_{eff}(X,a_0)\tilde{f}_0(X,a_0,\Delta a)dX = \int H_{eff}f(X,a_0 + \Delta a)dX . \tag{21.2.7}$$

This condition may be regarded as an equation in the effective temperature, since its right-hand side contains only the known functions. This allows us to find the only unknown function in the renormalized distribution:

$$T_{eff} = T_{eff}(\Delta a) ; \quad T_{eff}(\Delta a)\big|_{\Delta a=0} = 1 . \tag{21.2.8}$$

The second of these is the "initial condition": when the change in the controlling parameter is zero, there is no need to carry out renormalization, and the effective temperature is equal to the initial temperature (that is, to unity).

So we have two distributions, f and $\tilde{f}_0$, for which the mean values of the effective Hamilton function are the same. In place of (21.2.3) we use the expression for the entropy defined in terms of $\tilde{f}_0$; then

$$\tilde{S}_0 - S = \int \ln \frac{f}{\tilde{f}_0} f dx \geq 0 \quad \text{at} \quad \langle H_{\text{eff}}(X, a_0) \rangle = \text{const} . \tag{21.2.9}$$

This inequality indicates that the renormalized entropy for state "0" of physical chaos is greater than for the second selected state. To adopt this difference for the measure of the relative degree of order of the two nonequilibrium states in question, we must verify the correct choice of the state of physical chaos. For this purpose we turn to the solution of equation (21.2.7).

Apart from the trivial solution $\Delta a = 0$, we have just two options: $\Delta a > 1$ and $\Delta a < 1$. Let us consider them separately. Assume that the solution is

$$T_{\text{eff}}(\Delta a) > 1 ; \; T_{\text{eff}}(\Delta a)\big|_{\Delta a=0} = 1 . \tag{21.2.10}$$

This implies that in order to satisfy condition (21.2.7) we must raise the effective temperature and add heat to the system at state "0", which will transform into the more organized motion at state "1". Accordingly, the choice of state "0" as the more chaotic state is justified, and expression (21.2.9) gives the relative measure of chaoticity of the states of open system.

Obviously, the representation of renormalized distribution in the canonical form (21.2.5), justified by the specific properties of exponential distribution, is still somewhat arbitrary. For this reason it would be useful to get additional evidence that our choice of the more chaotic state has been correct. By the rule of contraries, we may assume that the more chaotic state is state "1"; the temperatures for the "forward" and "back" processes we denote, respectively, by $\vec{T}_{\text{eff}}$ and $\overleftarrow{T}_{\text{eff}}$. If we find that $\vec{T}_{\text{eff}} < 1$, we get another argument in favor of our choice of state "0" as the more chaotic state.

It may happen, however, that both effective temperatures are greater than unity:

$$\vec{T}_{\text{eff}}(\Delta a) > 1 , \; \overleftarrow{T}_{\text{eff}}(\Delta a) > 1 . \tag{21.2.11}$$

In this "arguable" case it would be natural to assume that the more chaotic state is that for which inequality (21.2.11) is "stronger".

In connection with the above discussion, let us recall the following. Time evolution at constant mean energy (for example, in Boltzmann gas) is associated with the increase in entropy. The entropy ceases to grow only when equilibrium (the most chaotic state) is established. This is the second law of thermodynamics.

The situation is different in case of evolution of stationary states in open systems as the controlling parameters are varied: the entropy may either decrease or increase. The decrease in entropy corresponds to self-organization, the increase to degradation. The constraints of the second law of thermodynamics are thus loosened.

Of course, these two types of evolution represent the two extremes of the more general process of development, when variation of external parameters causes the entropy to exhibit nonmonotonous changes.

21.3. Assessment of Relative Degree of Order from Spectra

In place of time realizations (21.2.1) one may use the relevant time spectra, corresponding to the same values of controlling parameters,

$$I(\omega, a_0) , \quad I(\omega, a_0 + \Delta a) . \tag{21.3.1}$$

We introduce the appropriate distribution functions

$$f_0(I, a_0) , \quad f(I, a_0 + \Delta a) , \tag{21.3.2}$$

and assume again that the more chaotic state is state "0". The effective Hamilton function is defined by

$$H_{\text{eff}}(I, a_0) = -\ln f_0(I, a_0) . \tag{21.3.3}$$

Following the guidelines developed above, we carry out renormalization to the given value of effective energy $H_{\text{eff}}(I, a_0)$, solve the corresponding equation for the effective temperature, and calculate the difference in the temperatures.

It may be more convenient to use, in place of (21.3.2), the distribution function of frequency values in the most interesting range $\omega_1 < \omega < \omega_2$:

$$f_0(\omega, a_0) = \frac{I(\omega, a_0)}{\displaystyle\int_{\omega_2}^{\omega_2} I(\omega', a_0) d\omega'} , \quad \int_{\omega_2}^{\omega_2} f(\omega', a_0) d\omega' = 1 . \tag{21.3.4}$$

If, by assumption again, we take state "0" for the state of physical chaos, the effective Hamilton function is defined by

$$H_{\text{eff}}(\omega, a_0) = -\ln f_0(\omega, a_0) \tag{21.3.5}$$

and depends on the frequency. Now the above-described method can be applied.

Thus, we now have a criterion which allows us to compare the relative order of different states of open systems on the basis of straightforward experimental data. The difference in entropies reduced to the preassigned value of the mean effective energy serves as the quantitative measure of chaoticity, and furnishes us with a tool for distinguishing between self-organization and degradation.

Moreover, this criterion allows one to verify the correct choice of controlling parameters, by identifying the range where these parameters "stand up to the expectations". Outside of this range other controlling parameters have to be employed.

Observe finally that the availability of several controlling parameters makes it possible to optimize the path of self-organization in the space of controlling parameters. Examples of this kind can be found in (Klimontovich – Bonitz 1988; Klimontovich V). They, however, represent just first steps towards solution of this complicated problem.

The criterion of S–theorem (where "S" stands for *self-organization*) has been first formulated for two concrete examples in (Klimontovich 1983c, 1984a). The first was concerned with self-organization in the system of Van der Pol oscillators, whereas the second dealt with the comparison of the relative degree of order of laminar and steady turbulent flows in channels. It was the first proof that, contrary to the predominant opinion, the steady turbulent flow is more ordered. The physical background of this will be explained in the next chapter. The results of first investigations are summarized in (Ebeling – Klimontovich 1984).

Now it is time to consider some other characteristics of complex motion.

21.4. Distribution Function of Separation of Trajectories

As we already know, dynamic instability of motion (exponential divergence of the initially close trajectories) may give rise to "dynamic chaos", complicated and unpredictable motion in dynamic systems.

Obviously, there is the need for a quantitative characteristic of dynamic instability. Commonly used for this purpose are such characteristics of dynamic systems as Krylov – Kolmogorov – Sinai entropy and Lyapunov indices. These characteristics allow one to detect the transition to the state of dynamic chaos. It is important, however, to be aware of the following.

Dynamic description of dissipative processes in nonlinear systems is based on equations in deterministic (non-random) macroscopic characteristics of continuous medium. Such are, for example, equations of chemical kinetics: the set of equations in concentrations of chemically reacting components of the continuous medium under consideration. In this case the problem reduces to solving a set of ordinary differential equations. Dynamic description on a higher level may be based on kinetic, hydrodynamic, or reaction-diffusion equations.

In all cases, however, this description is not complete. As a matter of fact, in accordance with the general fluctuation-dissipation relations of the statistical theory, fluctuations in dissipative systems are inevitable; they reflect the existence of "atomic structure" of continuous medium. Due account for fluctuations brings us closer to the complete description, which in principle is feasible on the basis of the reversible (!) microscopic dynamic equations for the elements of the medium. We know that this approach opens new possibilities in the description of, for instance, Brownian motion.

For these reasons it is desirable to find such characteristics which would describe not only the dynamic behavior, but also the statistical trends of the system. Since the complexity of dynamic motion is mostly due to the dynamic instability of motion (exponential divergence of trajectories), it would be natural to characterize the state of the system by the distribution function for the distance between the trajectories. This function is constructed on the basis of experimental time realizations. Owing to this fact, the structure of the distribution will reflect not only the dynamic instability of motion, but also the statistical properties of the system (Klimontovich V).

Let us consider two time realizations: two "trajectories"

$$X_1(t,a) , \quad X_2(t,a) , \tag{21.4.1}$$

which at the given value of controlling parameter a correspond to two different initial conditions $X_1(0,a)$, $X_2(0,a)$. Assume that the realizations are long enough to allow constructing the stationary distribution of the two variables X_1, X_2:

$$f(X_1,X_2,a) , \quad \int f(X_1,X_2,a)dX_1dX_2 = 1 . \tag{21.4.2}$$

In this way, we find the stationary distribution of coordinates of the two trajectories for the given value of controlling parameter a and two different initial conditions. This distribution includes information about the internal noise, which reflects the existence of "atomic structure" of the system in question, and can be used for finding the distribution function of the distance between the trajectories:

$$f(D,a) = \int \delta\left(D - \sqrt{(X_1 - X_2)^2}\right)f(X_1,X_2,a)dX_1dX_2 , \tag{21.4.3}$$

and the corresponding expression for Shannon entropy

$$S(a) = -\int \ln f(D,a)f(D,a)dD . \tag{21.4.4}$$

To use this entropy for assessing the relative degree of order of the system at different values of controlling parameter a, we employ once again the criterion of S–theorem. Again we select two states at different values of a, take one of them for the state of "physical chaos", etc. It is possible, however, to obtain useful information about the system directly from the distribution (21.4.3).

Making use of the logistic equation (16.8.3) with $E_n \Rightarrow X_n$,

$$X_{n+1} = (a+1-X_n)X_n \equiv F(X_n) , \quad 0 \le X \le 4 , \quad 0 \le a+1 \le 4 , \tag{21.4.5}$$

at two different initial conditions, we construct the distribution function $f(D,a)$ from solutions of the two resulting equations. At those values of a which correspond to simple attractors (below the critical Feigenbaum point), this distribution is represented by a set of lines. In the domain of strange attractor (above the critical Feigenbaum point) the distribution function exhibits complicated structure, and at the maximum value of $a = 3$ is a continuous distribution, similar to (16.7.2).

If the motion takes place in n-dimensional space, one of the most informative is the distribution function for the distance between the trajectories

$$D = \sqrt{\sum_{1 \le i \le n} (x_{1i} - x_{2i})^2} . \tag{21.4.6}$$

For assessing the relative degree of order with the aid of distributions $f_0(D,a_0)$, $f(D,a_0 + \Delta a)$, we again assume a priori that state "0" is the more chaotic one, define the effective Hamilton function

$$H_{\text{eff}}(D,a_0) = -\ln f_0(D,a_0) , \tag{21.4.7}$$

and apply the criterion of S–theorem.

21.5. K–Entropy and Lyapunov Indices at Dynamic and Statistical Description of Complex Motion

Let us start with the definitions. We distinguish dynamic and stochastic (statistical) motion. This distinction is, of course, a matter of convention. In numerical experiment, however, it may be based on the reproducibility of motion at the given initial conditions.

If the motion in a dissipative system is reproducible, it is referred to as *dynamic motion*. Otherwise we are dealing with *stochastic (statistical) motion*.

Obviously, in real systems, where natural noise caused by the "atomic structure" of continuous medium cannot be eliminated, the motion is stochastic. But in numerical experiment with dynamic systems the initial conditions may be reproduced within the accepted accuracy. If the equations do not contain random sources, the motion will also be reproducible. At the same time, it may be unpredictable because of the dynamic instability making it extremely sophisticated. When there are random sources, the motion will be both unpredictable and irreproducible, which is characteristic of the stochastic processes.

In numerical simulations of stochastic processes the following is important.

Generators of random numbers in computers work according to a certain algorithm, and are therefore essentially deterministic. They may be considered as random sources, however, if the characteristic sequence repetition time is much greater than the characteristic time of relaxation in the dynamic dissipative system under consideration.

Complex dynamic motion was first discovered in Hamiltonian systems, and it was then that the concept of dynamic chaos was introduced. Currently this term is also commonly used for complex motion in dissipative dynamic systems.

Let us recall the main characteristic features of dynamic chaos.

K–entropy. Lyapunov indices. The main distinctive feature of dynamic chaos is the dynamic instability of motion (exponential divergence of the initially close trajectories). In other words, the motion is extremely sensitive to small changes in the initial conditions.

The measure of exponential divergence is furnished by K–entropy (Krylov – Kolmogorov – Sinai entropy). K–entropy is in turn linked with the Lyapunov indices, which characterize divergence of the initially close trajectories in case of unstable motion, or their convergence in case of stable motion.

For a broad class of systems, K–entropy may be expressed via the positive Lyapunov indices as

$$K = \sum_i \lambda_i \ , \ \lambda_i > 0 \ . \tag{21.5.1}$$

If positive Lyapunov indices are absent, and the motion is therefore stable, then the value of K–entropy is zero. The total number of Lyapunov indices coincides with the number of degrees of freedom of the dynamic system.

Let us, by way of example, consider one-dimensional motion whose stability is characterized by one Lyapunov index. If the motion is unstable, then $K = \lambda > 0$. For a continuous process these quantities are expressed by (see, for example, Schuster 1984)

$$K = \lambda = \lim_{t \to \infty} \lim_{D(t=0) \to 0} \frac{1}{t} \ln \frac{D(t)}{D(t=0)} \ , \tag{21.5.2}$$

where

$$D(t) = \sqrt{(x_1(t) - x_2(t))^2} \ , \ D(t = 0) \tag{21.5.3}$$

are the distances between the trajectories at the current time t and the initial instant $t = 0$, respectively. From this definition it follows that K–entropy is positive only in case of exponential divergence of trajectories.

For a nonlinear dissipative system the values of K and λ can only be found from numerical experiment, after the transition to discrete time. The limiting transition $D \to 0$, however, cannot be accomplished, since the value of D_{min} is finite. For this reason, the quantities K and λ for the purposes of numerical simulation are defined in a different way.

K–entropy and Lyapunov index for logistic equation. Let us return to logistic equation (21.4.5). In this case the Lyapunov index is defined by

$$K = \lambda = \lim_{n \to \infty} \frac{1}{n} \sum_{k=0}^{n-1} \ln \left| \frac{\partial F}{\partial x} \right|_{x_k} \tag{21.5.4}$$

(see Eq. (2.9) in Schuster 1984). In fact, this is an independent (of (21.5.2)) definition of Lyapunov index for a process in discrete time with unit step. It characterizes (on the average over the trajectory) the linear deviation for each separate step of the process in question with zero initial deviation at each step. The result (21.5.4) is finite when the divergence of trajectories is exponential.

Typical of the linear index of instability is the fact that the Lyapunov index vanishes at points of bifurcation (curve 1 in Fig. 6), which corresponds to infinitely large relaxation times. This is a limitation of the instability criterion in linear approximation: the details of transition across the bifurcation point in this approximation cannot be described.

This limitation can be removed by using the relevant nonlinear characteristic of divergence of trajectories.

21.6. Nonlinear Characteristic of Divergence of Trajectories. K_{nl}–Entropy

To elucidate the character of divergence of trajectories with due account for the nonlinearity, in place of (21.5.4) one may use the time-average logarithm of the ratio of distances between trajectories at the current time and the initial time:

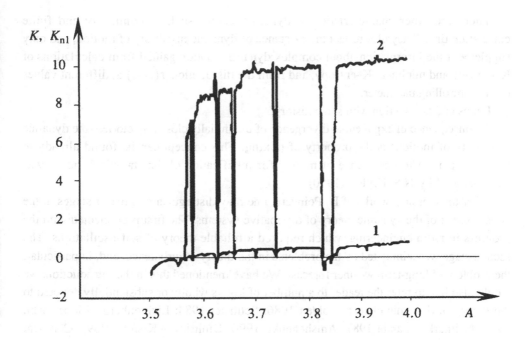

Figure 6: K–entropy (Lyapunov index $\lambda \equiv K$) (curve 1) as function of the controlling parameter $A = \alpha + 1$. Calculated according to (21.5.4) for the logistic equation (21.4.5) for the range $3.5 \le \alpha \le 4$.
$K_{nl} \equiv L_{nl}$ (curve 2), calculated according to (21.6.1) for the range $3.5 \le \alpha \le 4$. The neighborhood of $A = 3.83$ corresponds to the broadest "window of order".

$$K_{nl} = \lim_{n \to \infty} \frac{1}{n} \sum_{k=1}^{n} \ln \frac{D(k)}{D(0)} \,, \quad D(k) = \sqrt{(x_1(k) - x_2(k))^2} \ge D(0) \,. \tag{21.6.1}$$

By contrast to (21.5.4), the result now depends on the initial separation $D(0)$. To eliminate the negative contributions so as to characterize only the divergence of trajectories of the dynamic system, retained in the summation are only those contributions which characterize the divergence, $D(k) \ge D(0)$. Note that a similar condition is actually present in the definition (21.5.2) of K–entropy, as ensured by the first transition to the limit $D(0) \to 0$. Because of this, only the contributions with $D(t) \ge D(t = 0)$ are taken into consideration, and K–entropy is nonzero only in case of exponential divergence of trajectories.

Curve 2 in Fig. 6 shows the dependence of K_{nl} on the controlling parameter $a + 1$ in the range $3.5 \le a + 1 \le 4$. For comparison, a corresponding graph for Lyapunov index is shown in the same diagram. We see that function $K_{nl}(a + 1)$ portrays the domain of dynamic chaos in greater detail than does the Lyapunov index. Moreover, owing to the fact that the nonlinearity is taken into account, it also gives a notion about the size of the domain of strange attractor.

There are other characteristics of dynamic chaos, such as continuity and finite correlation time. They also reflect the existence of dynamic instability of motion, and only supplement the information about complex dynamic motion gained from calculations of K–entropy and nonlinear K–entropy, and from the distribution $f(D,a)$ at different values of the controlling parameter.

Let us end this section with some history.

A consequence of exponential divergence of the initially close trajectories (the dynamic instability of motion) is the property of mixing. This concept can be found already in Gibbs' papers. The importance of mixing for justification of the statistical theory was demonstrated by N.S. Krylov (1950).

After the classical works of H. Poincaré, one may distinguish two major stages in the development of the dynamic theory of dissipative systems. The first is connected with the progress in radio engineering, which required a reliable theory of self-oscillations. The second stage was stimulated by the problems in the theory of turbulence, and, in particular, the problem of long-term weather forecast. We have mentioned this in the Introduction; we would also like to refer the reader to a number of books wholly or substantially devoted to the theory of dynamic chaos: (Holden 1986; Schuster 1984; Lichtenberg –Liebermann 1982; Neimark – Landa 1987; Anishchenko 1990; Dmitriev – Kislov 1989). Classical papers dealing with the dynamic chaos are collected in (Hao Bai-lin 1985).

Now let us establish the connection between K–entropy and nonlinear K–entropy with the corresponding more general characteristics of the statistical theory of open systems.

21.7. K–Entropy and Entropy Production. Statistical Analog of K–entropy

Time of evolution. Consider again the two trajectories of (24.4.1) at two different initial conditions. For this we introduce the appropriate dynamic distribution function of variables X_1, X_2,

$$f^{(\mathrm{d})}(X_1,X_2,t,a) = \delta(X_1 - X_1(t,a))\delta(X_2 - X_2(t,a)) , \qquad (21.7.1)$$

and carry out averaging over the time interval T:

$$\tilde{f}^{(T)}(X_1,X_2,t) = \int_t^{t+T} \delta(X_1 - X_1(t',a))\delta(X_2 - X_2(t',a))\frac{dt'}{T} . \qquad (21.7.2)$$

The condition of local ergodicity is assumed to hold. Then we may introduce the corresponding averaging over the ensemble. If the time of averaging T is much smaller than the time of relaxation towards the stationary state, then the distribution function

$f(X_1, X_2, a, t)$ will be time-dependent. We use it to define the distribution function of distances between the trajectories as

$$f(D, t, a) = \int \delta \left(D - \sqrt{(X_1 - X_2)^2} \right) f(X_1, X_2, t, a) dX_1 dX_2 \ . \tag{21.7.3}$$

Using this distribution, we define two statistical characteristics: the mean distance between the trajectories

$$\bar{D}(t) = \int D f(D, t) dD \ , \tag{21.7.4a}$$

and the effective distance between the trajectories

$$\Delta D(t) = \frac{1}{f(\bar{D}, t)} \ . \tag{21.7.4b}$$

The second of these characteristics has been obtained from the normalization condition by substituting there $f(D, t)$ with $D = \bar{D}$.

Return now to the expression for the entropy

$$S(t) = -\int \ln f(D, t) f(D, t) dD \tag{21.7.5}$$

and expand the integrand with respect to small deviation $\bar{D} - D$. As a result, we get the analog of "Boltzmann's formula"

$$S(t) = \ln \Delta D(t) \ , \tag{21.7.6}$$

which can be used to find the difference in entropies at the initial and the current time:

$$S(t) - S(t_0) = \ln \frac{\Delta D(t)}{\Delta D(t_0)} \ . \tag{21.7.7}$$

We see that, as the entropy increases, the effective distance between the trajectories increases too. The last expression can be used for finding the average entropy change in unit time:

$$K_{\text{stat}} = \frac{S(t) - S(t_0)}{t - t_0} = \frac{1}{t - t_0} \ln \frac{\Delta D(t)}{\Delta D(t_0)} \ , \tag{21.7.8}$$

where K_{stat} is the statistical analog of K–entropy. To elucidate this analogy, let us look at this expression in the limit of small variations:

$$K_{stat} = \lambda_{stat} = \lim_{\Delta t \to \infty} \lim_{\Delta D(t_0) \to 0} \frac{1}{\Delta t} \ln \frac{\Delta D(t)}{\Delta D(t_0)} . \tag{21.7.9}$$

The structural analogy between (21.5.2) and (21.7.9) is obvious. However, there also is a fundamental difference.

As a matter of fact, the statistical K–entropy is found not by solving dynamic equations (for instance, the logistic equation), but directly from the experiment-based time realizations. In this way, into account is taken not only the complexity of motion (dynamic chaos), but also the fluctuations due to the "atomic structure" of the macroscopic open system under consideration. In this respect K_{stat} is a more universal characteristic than K–entropy. At the same time, there are some snags: while K–entropy is expressed in terms of exact dynamic characteristics (distances between the trajectories), K_{stat} is defined via the effective distances between the trajectories. This may be a blessing in disguise, because the effective distances can be determined in the most general case without the assumption of small deviations.

From arguments developed above it also follows that Krylov – Kolmogorov – Sinai entropy characterizes not the entropy itself, but rather the production of entropy — that is, the change in entropy per unit time. The sign of entropy production depends on the nature of the process. If, for example, the effective distance between the trajectories increases in the course of evolution ($\Delta D(t) > \Delta D(t_0)$ — the analog of dynamic instability), the entropy also increases. By contrast, when the distributions converge (that is, $\Delta D(t) < \Delta D(t_0)$), the entropy decreases.

This, however, does not necessarily imply that the relative degree of chaoticity increases, since the condition of constancy of the mean effective energy may be not satisfied in the course of time evolution. We shall discuss this point in the next section.

21.8. Evolution in Space of Controlling Parameters. S–Theorem

Now let us change the statement of the problem. Namely, instead of time evolution at the given value of a, we shall once again consider the evolution of stationary distributions of distances between trajectories in the space of controlling parameters. Then, in place of (21.7.8), the following expression must be used for the statistical K–entropy:

$$K_{\text{stat}} = \frac{S(a) - S_0(a_0)}{a - a_0} = \frac{1}{a - a_0} \ln \frac{\Delta D(a)}{\Delta D(a_0)} \ . \tag{21.8.1}$$

Expression (21.7.9) is modified in a similar way:

$$K_{\text{stat}} = \lim_{\Delta a \to \infty} \lim_{\Delta D(a_0) \to 0} \frac{1}{a - a_0} \ln \frac{\Delta D(a)}{\Delta D(a_0)} \ . \tag{21.8.2}$$

While the analysis of dynamic systems was concerned with the divergence of trajectories, now we may speak of the "divergence" of statistical characteristics. With this purpose we single out those contributions which determine the increase in the effective distance between the trajectories, $\Delta D(a) > \Delta D(a_0)$. Then the nonzero positive contributions to K_{stat} will only occur on the segments where $\Delta D(a)$ exponentially increases.

To gain more information about the statistical behavior of the system, we must again use the distribution function $f(D, a)$ in place of $\Delta D(a)$. This allows us to apply the criterion of S–theorem.

From definitions (21.7.4) it follows that the effective distance between the trajectories $\Delta D(a)$ is defined by the distribution function $f(D, a)$ at point $D = \overline{D}$. For a more comprehensive description we may use the appropriate local relation in place of (21.8.1):

$$S_0(D, a_0) - S(D, a) = \ln \frac{f(D, a)}{f_0(D, a_0)} \ , \tag{21.8.3}$$

whence, carrying out averaging and using distribution $f(D, a)$, we find:

$$\int S_0(D, a_0) f(D, a) dD - S(a) = \int \ln \frac{f(D, a)}{f_0(D, a_0)} f(D, a) dD \geq 0 \ . \tag{21.8.4}$$

Here we again have taken advantage of the fact that $\ln A \geq 1 - 1/A$.

To represent the left-hand side as the difference in entropies and obtain the criterion of the relative degree of order of states at different values of a, we must carry out renormalization to the specified value of the mean effective Hamilton function for the state of "physical chaos" at $a = a_0$.

As before, we represent the renormalized distribution function in the form of canonical Gibbs distribution and use the condition of constancy of $\langle H_{\text{eff}}(D, a_0) \rangle$:

$$\int H_{\text{eff}}(D, a_0) \tilde{f}_0(D, a_0, \Delta a) dD = \int H_{\text{eff}}(D, a_0) f(D, a_0 + \Delta a) dD \ . \tag{21.8.5}$$

By solving this equation we find the effective temperature T_{eff} as a function of Δa. If the solution is such that $T_{\text{eff}} > 1$, then the state with $a = a_0$ is more chaotic, and the state with $a = a_0 + \Delta a$ is more ordered.

Then (21.8.4) can be represented in the form

$$\tilde{S}_0(a_0, \Delta a) - S(a_0 + \Delta a) = \int \ln \frac{f(D, a_0 + \Delta a)}{\tilde{f}_0(D, a_0, \Delta a)} f(D, a_0 + \Delta a) dD \geq 0 \ . \tag{21.8.6}$$

This expression defines the difference in entropies at constant mean value of the effective Hamilton function, which serves as quantitative measure of the relative degree of order of the selected states of the open system under consideration. The statistical information about the system is provided by the distribution functions of distances between the trajectories.

Shortly we shall consider a number of concrete applications of S–theorem. First, however, let us establish the connection between this criterion and the efficient method of analysis of complex motion based on the so-called Renyi entropy (Beck – Schlögl 1993).

21.9. Renyi Entropy in Statistical Theory of Open Systems

In this section we are going to establish the linkage between the criterion of S–theorem and the Renyi entropy, as well as between the above distribution $\tilde{f}_0$ for the state of physical chaos and the so-called *escort distribution* (Beck – Schlögl 1993). This will help us to explain the popularity of Renyi entropy, and to employ the corresponding technique in the statistical theory of open systems.

To begin with, let us recall some of our earlier results and definitions.

In numerical simulation, the continuous variable X is replaced by the discrete variable X_i. In place of (21.2.2) we shall use the corresponding distribution functions

$$f_0(X_i, a_0) , \quad f(X_i, a_0 + \Delta a) , \quad \sum_i f_0(X_i, a_0) = \sum_i f(X_i, a_0 + \Delta a) = 1 \ . \tag{21.9.1}$$

Shannon entropy and the corresponding information are defined by

$$I_0[f_0] = -S[f_0] = \sum_i f_0(X_i, a_0) \ln f_0(X_i, a_0) , \quad \sum_i f_0 = 1 \ . \tag{21.9.2}$$

Summation is carried out over all states for which the distribution f_i is nonzero. The effective Hamilton function is again defined via the distribution function of the state of "physical chaos":

$$H_{eff}(X_i, a_0) = -\ln f_0(X_i, a_0) , \quad \sum_i \exp(-H_{eff}) = 1 .$$ (21.9.3)

We may also write the renormalized distribution function as

$$\tilde{f}_0(X_i, a_0) = \exp\frac{F_{eff} - H_{eff}(X_i, a_0)}{T_{eff}} , \quad \sum_i \tilde{f}_0 = 1 .$$ (21.9.4)

The effective functions are linked by the appropriate thermodynamic relation

$$F_{eff}(T_{eff}, a_0) = \langle H_{eff}\rangle_0 - T_{eff}\tilde{S}_0 ,$$ (21.9.5a)

where

$$\langle H_{eff}\rangle_0 = \sum_i H_{eff}(X_i, a_0)\tilde{f}_0(X_i, a_0) ,$$ (21.9.5b)

$$\tilde{S}_0 = -\sum_i \tilde{f}_0(X_i, a_0)\ln \tilde{f}_0(X_i, a_0) .$$ (21.9.5c)

As before, the mean value of the effective Hamilton function depends on the effective temperature T_{eff} and the parameter a_0.

Observe that the effective temperature in (21.9.4), (21.9.5) may be regarded as a free parameter, which will be shown to be related to the characteristic free parameter β in Renyi's theory. In the formulation of the criterion of S–theorem the choice of T_{eff} as function of Δa, as well as the choice of parameter β in Renyi's theory, is based on the assumption that the value of $\langle H_{eff}\rangle$ remains the same when we go over from a_0 to $a_0 + \Delta a$. The solution of the relevant equation yields

$$T_{eff} = T_{eff}(\Delta a) , \quad T_{eff}\big|_{\Delta a=0} = 1 .$$ (21.9.6)

Observe finally that distribution (21.9.4) allows one to obtain two useful formulas: the mean effective Hamilton function

$$\langle H_{eff}\rangle_{\Delta a} = T_{eff}^2(\Delta a)\frac{\partial}{\partial T_{eff}}\ln\sum_i \exp\left[-\frac{H_{eff}}{\kappa T_{eff}(\Delta a)}\right] ,$$ (21.9.7)

and the dispersion of the effective Hamilton function

$$\left\langle \left(\delta H_{\text{eff}} \right)^2 \right\rangle_{\Delta a} = T_{\text{eff}}^2 \frac{\partial \langle H_{\text{eff}} \rangle}{\partial T_{\text{eff}}} \geq 0 \ . \tag{21.9.8}$$

These expressions are similar to (4.2.3), (4.2.4) for the mean value and the dispersion of the internal energy, obtained on the basis of the canonical Gibbs distribution. This similarity is of course not accidental, since the renormalized distribution (21.9.4) also has the form of the canonical Gibbs distribution.

In a sense, however, (21.9.7) and (21.9.8) are more general, because the effective Hamilton function is derived directly from experimental data for any open system.

Useful is also another implication of distribution (21.9.4). To wit, by analogy with (4.4.2) one may introduce the distribution function of the values of local entropy

$$\tilde{S}_0(X_i, a_0) = -\ln \tilde{f}_0(X_i, a_0) \ , \tag{21.9.9}$$

which is defined as

$$f\left(\tilde{S}_0, T_{\text{eff}}, a_0 \right) = \sum_i \delta\left(\tilde{S}_0 + \ln \tilde{f}_0(X_i, T_{\text{eff}}, a_0) \right) \tilde{f}_0(X_i, T_{\text{eff}}, a_0) \ . \tag{21.9.10}$$

Hence follows the expression (21.9.5c) for the mean entropy, and the relevant expression for the dispersion:

$$\left\langle \left(\delta \tilde{S}_0 \right)^2 \right\rangle = \frac{\left\langle \left(\delta H_{\text{eff}} \right)^2 \right\rangle_{\Delta a}}{T_{\text{eff}}^2} = \frac{\partial \langle H_{\text{eff}} \rangle}{\partial T_{\text{eff}}} \geq 0 \ . \tag{21.9.11}$$

In other words, the dispersion of entropy is determined by the derivative of the mean effective energy with respect to the effective temperature. The derivative is taken at a fixed value of Δa.

Renyi entropy. Let us now give definition of Renyi entropy. We shall stick to the notation used in (Beck – Schlögl 1993), and then establish correspondence with the conventions accepted herein.

Renyi entropy and information are given by

$$I_\beta[p] = -S_\beta[p] = \frac{1}{1-\beta} \ln \sum_i (p_i)^\beta \ , \tag{21.9.12}$$

where p_i is the distribution function of states of system i, and β is the free parameter of the distribution. At $\beta = 1$ Renyi entropy and information coincide with the corresponding Shannon characteristics:

$$I_{\beta=1}[p] = I[p] = -S[p] = \sum_i p_i \ln p_i \ .$$

(21.9.13)

Escort distribution. We begin by representing the distribution p_i in the form

$$p_i = \exp(-b_i) \ , \quad \sum_i p_i = \sum_i \exp(-b_i) = 1 \ .$$

(21.9.14)

Then the escort distribution is defined as

$$P_i = \frac{(p_i)^\beta}{\sum_i (p_i)^\beta} \ , \quad \sum_i P_i = 1 \ .$$

(21.9.15)

Making use of (21.9.14), we may rewrite this in the form of canonical Gibbs distribution:

$$P_i = \exp\left[\beta(F(\beta) - b_i)\right] \ , \quad F(\beta) = -\frac{1}{\beta} \ln \sum_i \exp(-\beta b_i) \ ,$$

(21.9.16)

where b_i acts as the effective Hamilton function, $F(\beta)$ is the relevant free energy, and $1/\beta$ is the effective temperature.

From the above formulas follows the linkage between the "free energy" $F(\beta)$ and the Renyi entropy:

$$F(\beta) = -\frac{1}{\beta} \ln \sum_i \exp(-\beta b_i) = -\frac{1}{\beta} \ln \sum_i (p_i)^\beta = \frac{1-\beta}{\beta} S_\beta[p] \ .$$

(21.9.17)

In particular, hence follows that the "free energy" is zero when $\beta = 1$ and Renyi entropy coincides with Shannon entropy:

$$F(\beta = 1) = -\ln \sum_i p_i = 0 \ ,$$

(21.9.18)

and the escort distribution is then

$$P_i = p_i \ .$$

(21.9.19)

Now it is easy to establish correspondence between distribution functions f_0, $\tilde{f}_0$, and the effective characteristics for the state of "physical chaos" on the one hand, and distributions p_i, P_i , and functions b_i, $F(\beta)$, β:

$$p_i \Leftrightarrow f_0(X_i, a_0) \, , \quad P_i \Leftrightarrow \tilde{f}_0(X_i, a_0) \, , \quad b_i \Leftrightarrow H_{\text{eff}} \, ,$$

$$F(\beta) \Leftrightarrow F_{\text{eff}} \, , \quad \frac{1}{\beta} \Leftrightarrow T_{\text{eff}} \, . \tag{21.9.20}$$

Given this rule of correspondence, the relation (21.9.17) between the effective free energy and the Renyi entropy takes on the following form:

$$F_{\text{eff}}(T_{\text{eff}}) = (T_{\text{eff}} - 1) S_{T_{\text{eff}}}[f_0(X_i, a_0)] \, . \tag{21.9.21}$$

Hence it follows that at $T_{\text{eff}} = 1$ the effective free energy is zero, $F_{\text{eff}} = 0$, and Renyi entropy coincides with Shannon entropy for the initial state of "physical chaos":

$$S_{\beta=1} = S_0 = -\sum_i f_0(X_i, a_0) \ln f_0(X_i, a_0) \, . \tag{21.9.22}$$

We have dwelled at length on the concepts of Renyi entropy and escort distribution, which are widely used in the analysis of complex motion, but whose physical meaning often remains not quite clear.

Now we see that these concepts have a sound statistical background, and characterize the nonequilibrium state which has been taken for the state of "physical chaos" in the formulation of S–theorem. Moreover, we saw that in the formulation of S–theorem the parameter of distribution $\beta = (T_{\text{eff}})^{-1}$ becomes "active". Indeed, from condition of constancy of $\langle H_{\text{eff}} \rangle$ follows the dependence of the effective temperature (and hence of the parameter β of Renyi entropy) on the change in the controlling parameter Δa. What is more, it is possible to find, using (21.9.10), the dispersion of Renyi entropy as function of Δa. This function carries additional useful information about complex motion in the state of "physical chaos".

The above is just a fragment of Renyi's theory. In particular, we did not touch upon such an important concept as the so-called *Renyi's dimensionality*, which is necessary for the analysis of local properties of open systems. The interested reader is referred to (Beck – Schlögl 1993), which also contains numerous practical examples.

Finally, let us mention some applications of the criterion of S–theorem (see also Klimontovich V, 1994a).

The first paper concerned with medico-biological diagnostics based of the criterion of S–theorem (Andreev – Gridina – Kramar – Klimontovich 1989) dealt with the study of the relative degree of order in a living organism (in experiments with rats) by electrical activity of the stomach. Compared were the states before and after injection of carcinogenic substance; the entropy was found to increase by as much as 40 per cent.

In (Vasilyev – Klimontovich 1994) the criterion of S–theorem was used for assessing the relative degree of order for a chain of diffusion-coupled oscillators, simulating the peristaltic action of the straight intestine. The coefficient of diffusion served as the controlling parameter; the entropy increased as the controlling parameter fell below the value corresponding to normal functioning of the rectum.

S–theorem was used in (Landa – Rosenblum 1989) in preliminary investigations of the change in the relative degree of order from time realizations associated with the location of the patient's center of mass. This paper also contains intercomparison of different criteria.

In (Anishchenko – Saparin – Anishchenko 1993; Kurts 1993) the criterion of S–theorem was applied to cardiograms, with sonic alarm serving as stress factor. It was demonstrated that the reaction of men and women to the same stress is opposite: the degree of chaoticity increased in females, and decreased in males. The same trend had been observed earlier in experiments with animals on the basis of biochemical tests.

After this digression into medicine, let us return to the realm of physics and discuss a number of problems whose solution which may benefit from the availability of additional information concerning the relative degree of order of different states.

We shall start with the analysis of fluctuations of the volume of fluid, such as the fluctuations of the column of liquid in a thermometer. Fluctuations are described by the distribution function of values of the volume $f(V, p, T)$. The pressure and the temperature serve as controlling parameters, since they can be used for changing the state of matter.

There are two ways for establishing the form of this distribution function. The first consists in obtaining the time realizations $V(t, p, T)$ at different values of the controlling parameters. Then the comparison of the relative degree of order may be based on the criterion of S–theorem as described in the present chapter.

In the alternative approach the form of the distribution is established for the physical model under consideration on the basis of Boltzmann's principle, formulated in Ch. 4. This method is discussed in the next section as applied to the Van der Waals model.

21.10. Self-Organization upon Transition Across Gas-Liquid Critical Point

Let us calculate the fluctuations of the volume of gas and liquid for different values of pressure and temperature — in particular, for the critical region. To find the distribution function of values of the volume we turn to Boltzmann's principle, formulated in

Sect. 4.10. Numerous examples convinced us that the more natural of the two expressions (4.10.4), (4.10.6) is the representation of the desired function in the form of canonical Gibbs distribution (4.10.6) with a certain effective Hamilton function. Now we take volume V for the variable A, and pressure p and temperature T for the external parameters. With this choice of the external parameters it is natural to use thermodynamic potential $\Phi(p,T)$, which is related to the free energy by

$$\Phi(p,T) = F + pV_{\text{m.p.}} \; . \tag{21.10.1}$$

Here in place of the commonly used mean volume we introduce the most probable value with respect to the distribution

$$f(V,p,T) = \exp\frac{\Phi(p,T) - H_{\text{eff}}(V,p,t)}{\kappa T} \; , \quad \int f dV = 1 \; . \tag{21.10.2}$$

Recall that in the statistical justification of the laws of thermodynamics the thermodynamic variables are represented by the mean values of the appropriate functions of dynamic variables. Defined in this way were the internal energy (3.3.3), the thermodynamic forces (3.3.6), the entropy (3.4.5). The mean values (that is, the first moments) act as the main characteristics in the statistical foundation of thermodynamics.

In Sect. 4.3 we wondered whether this choice is always justified. After all, instead of first moments one could choose to use the most probable values with respect to the distribution function of the macroscopic characteristic in question. The resulting values are very close (within $1/N$) as long as equation (4.3.3) holds good. As indicated in Sect. 4.3, the situation is much different when we come to dealing with phase transitions. Now the time has come for the gun to go off.

As a matter of fact, the gun went off when we dealt with nonequilibrium phase transitions in oscillators (Sect. 16.8), and phase transition in the system of nonlinear atoms oscillators (Sect. 18.6). Now we are going to consider the classical problem of transition across the critical point in a system of liquid and gas. The critical point is the origin of the two-phase gas-liquid state. In the statistical description this is manifested by the appearance (at temperatures above critical) of two maxima of the distribution function of the values of the volume of the system in place of one. Then small microscopic fluctuations of the volume, which are proportional to $1/N$, give rise to macroscopic deviations from the mean value. The two maxima of the distribution correspond to the two phases. According to the criterion of S–theorem, the transition across the critical point, which results in the two-phase liquid-gas state, is an example of self-organization.

Let us particularize the distribution (21.10.2) for the case of Van der Waals model. The effective Hamilton function is then defined by

$$H_{\text{eff}}(V,p,T) = -N\kappa T \ln(V - Nb) - \frac{N^2}{V}a + pV .$$ (21.10.3)

Some function of the temperature may be added to this, which will not, however, affect the nature of distribution with respect to V.

From condition of minimum of the effective Hamilton function

$$-\left(\frac{\partial H_{\text{eff}}}{\partial V}\right) = \frac{N\kappa T}{V - Nb} - \frac{N^2}{V^2}a + p = 0$$ (21.10.4)

we find the most probable value of the volume. This value satisfies equation

$$p = \frac{N\kappa T}{V_{\text{m.p.}} - Nb} - \frac{N^2}{V_{\text{m.p.}}^2}a ,$$ (21.10.5)

which formally coincides with the Van der Waals equation.

If the mean value is close to the most probable value $V_{\text{m.p.}}$, equation (21.10.5) differs little (within $1/N$) from the conventional Van der Waals equation. The situation, however, becomes much different when we deal with transition across the critical point.

To get a better understanding of what happens when the critical point is approached, let us define the range of parameters p, T for ideal gas. Then a and b vanish, and

$$H_{\text{eff}}(V,p,T) = -N\kappa T \ln V + pV , \quad p = \frac{N\kappa T}{V_{\text{m.p.}}} .$$ (21.10.6)

Now we carry out expansion in $V - V_{\text{m.p.}}$ and consider Gaussian approximation

$$f(V) = \sqrt{\frac{1}{2\pi\langle(\delta V)^2\rangle}} \exp\left[-\frac{(\delta V)^2}{2\langle(\delta V)^2\rangle}\right] , \quad \delta V = V - V_{\text{m.p.}} .$$ (21.10.7)

For the ideal gas the dispersion is given by

$$\langle(\delta V)^2\rangle = \frac{V_{\text{m.p.}}^2}{N} , \quad \langle(\delta V)^2\rangle = -\kappa T \left(\frac{\partial p}{\partial V_{\text{m.p.}}}\right)^{-1} ,$$ (21.10.8)

and the relative dispersion is therefore proportional to $1/N$. Since in the Gaussian approximation the most probable and the mean values are the same, expression (21.10.8) coincides with (4.9.3).

Now our task is to show how the transition across the critical point affects the relative degree of order. Since there are two free parameters, p and T, there are different paths of transition. For definiteness, let us choose one of them. Recall that the critical values of volume, pressure and temperature in the Van der Waals model are expressed in terms of constants a and b as follows:

$$V_c = 3Nb , \quad p_c = \frac{a}{27b^2} , \quad \kappa T_c = \frac{8a}{27b} . \tag{21.10.9}$$

We choose that path to the critical point on which the most probable value of the volume at every point coincides with V_c:

$$V_{\mathrm{m.p.}} = V_c = 3Nb . \tag{21.10.10}$$

Then from (21.10.5) follows the relationship between p and T,

$$\frac{p - p_c}{p_c} = 4\frac{T - T_c}{T_c} , \tag{21.10.11}$$

and we are left with just one independent controlling parameter (say, the temperature). We move down the temperature scale. For the Van der Waals model (with H_{eff} given by (21.10.3)) the dispersion in Gaussian approximation is given by

$$\frac{\langle (\delta V)^2 \rangle}{V_{\mathrm{m.p.}}^2} = \frac{1}{N}\frac{4}{9}\frac{T}{T - T_c} . \tag{21.10.12}$$

We see that for temperatures much above critical the result differs from the relevant expression for ideal gas only by a coefficient. On approaching the critical point, however, the dispersion increases according to Curie's law, and goes to infinity at the critical point. This implies that Gaussian approximation is inadequate, and one has to retain higher terms in the expansion in δV.

To obtain a result which would hold also for the critical region, let us continue expansion of the effective Hamilton function in $V - V_{\mathrm{m.p.}}$. From the outset we define the critical region by inequality

$$|T - T_c| << T_c . \tag{21.10.13}$$

Then, retaining the terms of up to the fourth order, for the critical region we get the following expression:

$$H_{\text{eff}} = N\kappa T_{\text{c}} \frac{9}{8}\left[\frac{T-T_{\text{c}}}{T_{\text{c}}}\frac{(\delta V)^2}{V_{\text{m.p.}}^2} + \frac{(\delta V)^4}{8V_{\text{m.p.}}^4}\right], \quad \delta V = V - V_{\text{m.p.}} \cdot \qquad (21.10.14)$$

Substituting this expression into the distribution (21.10.2) and reducing it to the form of Gaussian distribution with respect to square of deviation,

$$f(V) = C\,\exp\left[-N\frac{9}{64}\left(\frac{(\delta V)^2}{V_{\text{m.p.}}^2} + 4\frac{T-T_{\text{c}}}{T_{\text{c}}}\right)^2\right], \qquad (21.10.15)$$

we see that the nature of the distribution changes dramatically upon transition across the critical point. Indeed, for temperatures much above critical we have Gaussian distribution with the dispersion (21.10.12). As the critical point is approached, the distribution ceases to be Gaussian, and at $T = T_{\text{c}}$ becomes

$$f(V) = C\,\exp\left[-N\frac{9}{64}\frac{(\delta V)^4}{V_{\text{m.p.}}^4}\right]. \qquad (21.10.16)$$

The relative dispersion is now proportional to $\sqrt{1/N}$, rather than to $1/N$ as in (21.10.12). Upon transition across the critical point (at $T < T_{\text{c}}$), a bistable state sets in with the maxima at

$$\delta V = \pm\,2V_{\text{m.p.}}\sqrt{\frac{T_{\text{c}}-T}{T_{\text{c}}}} \propto N\,. \qquad (21.10.17)$$

The presence of two signs of macroscopic deviations signals the birth of a new two-phase state.

According to (21.10.17), the relative dispersion for the region of coexistence of two phases (gas and liquid) does not depend on N:

$$\frac{\left\langle(\delta V)^2\right\rangle}{V_{\text{m.p.}}^2} = 4\frac{T_{\text{c}}-T}{T_{\text{c}}}\,. \qquad (21.10.18)$$

Distribution (21.10.15) can be used for calculating the dispersion for arbitrary temperatures. The result is expressed in terms of special (parabolic cylindrical) functions. It is possible, however, to obtain a simpler approximate (but common) expression for the dispersion (Klimontovich IV, V):

$$\left(\frac{9}{4}\frac{T-T_{\mathrm{c}}}{T_{\mathrm{c}}}+\frac{9}{16}\frac{\left\langle(\delta V)^2\right\rangle}{V_{\mathrm{m.p.}}^2}\right)\frac{\left\langle(\delta V)^2\right\rangle}{V_{\mathrm{m.p.}}^2}=\frac{1}{N}\,. \tag{21.10.19}$$

For temperatures above critical hence follows (21.10.12), and (21.10.18) for temperatures below critical. For the critical point this formula yields a result which differs from the exact solution only by a numerical coefficient.

So we see that macroscopic (coherent) fluctuations of either sign, proportional to the number of particles N, arise below the critical point. This means that we are dealing with a transition to a more structured, and therefore more ordered, state. A quantitative assessment of the change in the degree of order can only me made on the basis of the criterion of S–theorem. The technique is similar to that used in Ch. 16 for the analysis of fluctuations in Van der Pol oscillator. Recall also the results of Sect. 18.6: expressions (21.10.12), and (21.10.18) correspond to formula (18.6.12) for the medium comprised of nonlinear atoms oscillators. The relative degree of order has been calculated in Sect. 18.6.2.

The above example is based on the Van der Waals model. It is also possible to subject the main results to direct experimental verification — for instance, by recording the fluctuations of the piston plugging a cylinder filled with liquid or gas as functions of time $V(t, p, T)$ for different pressures and temperatures. If the time of observation is greater than the characteristic relaxation times, the random process observed will be stationary and homogeneous. Then the time realizations may be used for finding the relevant distribution functions, which will contain all the necessary statistical information about the system under consideration.

21.11. Some Unsolved Problems

In the present chapter we have only discussed the major issues associated with the development of criteria of self-organization. A lot had to be left out. Let us now try to fill some of the gaps.

The problem of predictability. The current state of this problem is reflected in a recent collection of papers (Kravtsov 1993). The book leaves the impression that even the statement of this very sophisticated problem is not quite clear. In this situation it would be interesting to consider new approaches, which may be based, in particular, on the criteria of self-organization discussed above.

Of course, one should not overestimate the capability of the new approaches in trying to solve the general problem of self-organization. Some scholars argue that a long-term forecast for systems with complex behavior is altogether impossible. This scepticism is not unjustified, and actually one may only speak of extending the horizon of forecast within the specified limits of accuracy.

Any prediction is based on the information available about the process within certain time limits — say, $0 \leq t \leq \tau_{obs}$. The task consists in using this information for predicting this process at $\tau_{predict} > \tau_{obs}$.

Let us confine ourselves to the case when two time realizations of the observed quantity X can be obtained: the short one ($0 \leq t \leq \tau_{obs}$), and the long one ($0 \leq t \leq \tau_{predict}$). We can use these for constructing the appropriate distribution functions

$$f_{(0)}(X, \tau_{obs}) , \ f_{(p)}(X, \tau_{predict}) , \tag{21.11.1}$$

which, owing to the process being nonstationary, will, in general, depend on the time intervals for which the realizations have been obtained.

Let us be optimistic and assume (against the possible odds) that the longer realization corresponds to a more ordered state. Then we may take state "0" for the state of "physical chaos", and define the effective Hamilton function on the basis of $f_{(0)}$. Using the already familiar procedure — renormalize to the specified value of the effective Hamilton function, check the correct choice of the more chaotic state and, if necessary, revert the choice, find the difference in entropies

$$\tilde{S}_{(0)}(\tau_{obs}) - S_{(p)}(\tau_{predict}) = \int \ln \frac{f_{(p)}}{f_{(0)}} f_{(p)} dX \geq 0 . \tag{21.11.2}$$

Now if we specify a certain "margin of chaoticity" ΔS for the class of systems under consideration, the result of (21.11.2) may be interpreted as follows.

The forecast is feasible if the difference in entropies (21.11.2) falls within ΔS, and not feasible otherwise. Naturally, this conclusion only applies to the class of systems which includes the one for which the calculation of entropies has been performed.

Principle of maximum entropy. The criterion of S–theorem requires information about the structure of distribution functions of stationary states corresponding to different values of controlling parameters. We have considered two possible sources of such information: equations for the distribution functions, and experimental time realizations.

There is yet another possibility. Assume that we know not the distribution functions themselves, but a limited number of moments of these distributions. The problem of reconstruction of distribution functions required for calculating the relative degree of order may be solved with the aid of the so-called "principle of maximum entropy" (Jaynes 1979; Haken 1988; Bevensee 1993). The controlling parameters may be included into the number of the known moments of distribution functions. This method of finding the distribution functions opens additional opportunities for applying the criterion of self-organization.

Criteria of self-organization for non-Markovian processes. When the criterion of S–theorem is applied to mathematical models, the problem of the structure of equations for distributions functions is important. In particular, of interest is the question of the structure of Lyapunov functionals for non-Markovian kinetic equations (equations with time lag).

The question of the role of non-Markovian behavior arises also in the analysis of statistical properties based on the time realizations. In this connection it would be useful to recall the results of the preceding chapter, concerned with anomalous Brownian motion. These results demonstrate the practical necessity of studying the relative degree of order as connected with the "extent of anomaly" of the Brownian motion — or, in other words, with the "extent of non-Markovianity" of processes under consideration.

Now we shall consider some problems of the statistical theory of turbulent motion. In particular, it will be demonstrated that the transition from laminar to turbulent flow is an example of transition from a less ordered to a more ordered state, which allows us to regard it as a process of self-organization.

Traditionally, the theory of turbulence is based on equations of hydrodynamics. The linkage between the kinetic and the hydrodynamic description of nonequilibrium processes, established in Ch. 13, stimulates the quest for a unified kinetic and hydrodynamic description of turbulent motion.

The new view on the nature of turbulent motion makes the theory of turbulence one of the major issues of the statistical theory of open systems.

CHAPTER 22

TURBULENT MOTION.
KINETIC DESCRIPTION OF TURBULENCE

> *I recall that von Karman, in his opening address,*
> *said that, when he finally came face to face with*
> *his Creator, the first revelation he would*
> *supplicate would be an unfolding of the*
> *mysteries of turbulence.*
>
> H. Moffatt

Systematic experimental and theoretical studies of turbulent motion in liquid started in the first half of the 19th century (Hagen 1839). First fundamental results, however, were only obtained half a century later by British physicist and engineer Osborne Reynolds. In 1883 he established that the transition from laminar to turbulent flow in a pipe occurs at a certain critical value of dimensionless parameter (Reynolds number). He also formulated the main hypothesis of the theory of turbulence, which essentially consists in the following.

Complex turbulent motion in incompressible fluid may be described on the basis of Navier – Stokes equation, assuming, however, that velocity is a nondeterministic (random) function of coordinates and time. After averaging over a certain statistical ensemble, or over small enough space-time scales, the Navier – Stokes equation generates an infinite chain of equations in the moments of random velocity. In particular, the equation for the first moment involves, by virtue of the initial equation being nonlinear, the second moments of velocity fluctuations — the so-called Reynolds stress tensor. Because of this, one has to deal with the problem of closure, so as to obtain an approximate closed set of equations in the lower moments. The zero approximation with respect to velocity fluctuations corresponds to the conventional Navier – Stokes equation.

In defiance of earnest efforts of many outstanding scholars (L.F. Richardson, G.K. Batchelor, Z.A. Prandtl, T. Karman, A.N. Kolmogorov, L.D. Landau, A.M. Obukhov, G.L. Taylor, to name but a few), turbulent motion remains one of the most enigmatic physical phenomena. The important results obtained so far are mostly concerned with the special case of well-developed turbulence. In recent years considerable progress has been made in the experimental and theoretical analysis of the onset of turbulence. Practical calculations, however, are based mostly on the semi-empirical theory of turbulence, which definitely has many useful applications. Still, the question of the development of theory of turbulence as part of the general statistical theory of open systems remains on the agenda.

Until recently the theory of turbulence was regarded as a chapter of mechanics of continuous medium, and was included into the university course of statistical physics not long ago (Klimontovich IV). This feat was facilitated by the growing recognition of the fact that turbulent motion is not a "privilege of hydrodynamics", but rather a most universal type of motion in Nature. Today turbulent motion is considered definitive for many phenomena in the atmosphere, in plasma, in chemically reacting media, in nonlinear optics, etc.

For a detailed study of the theory of turbulence and its numerous practical applications the reader is referred to specialized literature (see, for example, Landau – Lifshitz 1951, 1986; Batchelor 1953; Monin – Yaglom 1965, 1967, 1992; Schlichting 1968; Frost – Moulden 1977; Lesieur 1990).

The main purpose of the present chapter consists in elucidating the general properties of turbulence as a physical phenomenon, which will allow us to treat the transition from laminar to turbulent flow as a chain of nonequilibrium phase transitions, a self-organization sequence. This conclusion is conceptually based on the analysis of the main properties of turbulent motion, and corroborated quantitatively by calculations based on the criterion of S–theorem. In this way, the transition from laminar to turbulent flow in open systems is a transition "from chaos to order".

Of course, in a closed system a reverse transition from order to chaos is also possible. For example, turbulent motion initially created in a closed system will degrade to the most chaotic equilibrium state with the corresponding increase in the temperature. This simple example shows that the process of self-organization is only feasible in open systems.

22.1. Is Turbulent Motion Chaos or Order?

Although the concept of turbulent motion was introduced into physical usage more than a hundred years ago, even today it is hard to tell what exactly is turbulence. Symbolically, one of the recent books on turbulence (Lesieur 1990) opens with the chapter entitled *Is it possible to define turbulence?* In the same chapter the author formulates three properties of turbulence.

Briefly, they are as follows:

(a) Turbulent motion is unpredictable.

(b) Internal mixing, as a result of dynamic instability of motion, is inherent in turbulent motion. Because of this, turbulent viscosity and temperature conductivity are much higher than in case of laminar motion.

(c) Turbulent motion is characterized by a large number of spatial scales.

An essentially similar definition of turbulence can be found in (Monin – Yaglom 1992, p. 125).

As a matter of fact, such definition does not distinguish "turbulence" from "chaos" (or, more precisely, from "space-time chaos").

The above properties indubitably are characteristic of turbulent motion. Professor Lesieur, however, is not quite satisfied with this definition, and in the third chapter entitled *Turbulence, order and chaos* he returns to this issue, this time taken in philosophical and historical perspective.

In particular, he quotes Lucretius, who in his famous poem *De Nature Rerum* described the Universe as "turbulent order" which emerged from "primitive chaos". This is a rather provocative statement.

Contrary to the ancient authority, professor Lesieur still concludes that "in the reality of the fluid dynamics, it seems nevertheless difficult to accept blindly statements like this in order to explain turbulence" (Lesieur 1990, p. 87).

The view on turbulence as chaotic motion is recognized almost universally. For instance, in (Frost – Moulden 1977) we read:

"'Chaotic' in this context is almost synonym to 'turbulent'. Chaoticity constitutes the main property of such motion".

This view was challenged by the author in (Klimontovich IV, 1984a, V), and in (Klimontovich – Ebeling 1984). We argue that the transition from laminar to turbulent flow is a process of self-organization. This is one of the examples for which the criterion of S–theorem had been initially formulated. In a way, this is a retreat to Lucretius. Unfortunately, the author came to this view independently, and in his discussions with colleagues could not draw support from the great Roman.

A similar opinion is very clearly expressed in (Prigogine – Stengers 1984):

"...the transition from laminar flow to turbulent is a process of self-organization. Part of the energy of the system, which in laminar flow was in the thermal motion of molecules, is being transformed in macroscopic organized motion".

Naturally, it is not sufficient to proclaim one's faith. The professed view must be proved within the framework of the general statistical theory of open systems. This is one of the tasks of the present chapter.

22.2. Characteristic Features of Turbulent Motion

What then is turbulent motion?

It seems that this query can only be answered by enumerating the main characteristic features of turbulence.

1. Turbulent motion is essentially nonequilibrium, and is characterized by a very large number of macroscopic degrees of freedom.

2. Depending on the level of description (kinetic or hydrodynamic), the macroscopic characteristics are the moments of nondeterministic (random) distribution functions in the phase space, random hydrodynamic functions, electromagnetic field strengths, etc. In accordance with the Reynolds hypothesis, the equations for these random functions formally coincide with the relevant dissipative kinetic, hydrodynamic or Maxwell equations for the deterministic macroscopic functions.

3. The approximation of continuous medium as used in the kinetic theory, in hydrodynamics and electrodynamics, is not adequate for the description of turbulent motion. The onset and development of turbulence are governed by fluctuations, the higher moments which reflect the existence of the structure of "continuous medium". The role of fluctuations increases as the system departs further and further from the laminar regime. In this way, the collective macroscopic degrees of freedom of turbulent motion are due to the conversion of disordered molecular motion into the higher organized motion of collective macroscopic variables.

4. The transition from the equations of motion of "atoms" to the more coarse description based on kinetic and hydrodynamic equations is made possible, as we already know, by the constructive role of dynamic instability of motion of "atoms". In case of turbulent motion, the dynamic instability of motion — of macroscopic characteristics this time — also plays a constructive role. It is owing to the dynamic instability of motion that equations for a large number of macroscopic variables can be used for describing far-from-equilibrium states. The number of variables, although very large, is still much smaller than the number of microscopic degrees of freedom. This facilitates a much simpler (as compared with the microscopic level) description of complicated turbulent motion.

5. Owing to the dynamic instability of macroscopic characteristics of laminar motion (high sensitivity to small variations), turbulent motion is practically unpredictable.

6. The transition from laminar to turbulent flow is a nonequilibrium phase transition (a sequence of nonequilibrium phase transitions), resulting in a more organized motion. In other words, we are dealing with a process of self-organization.

We see that Lucretius's statement to the effect that "order (that is, turbulence) emerges from chaos", is quite consistent with the concept of turbulent motion as characterized by properties 1 – 6.

On the strength of arguments developed above, the main attention in the present chapter is paid to the following issues:

(1) The classical example of turbulent motion in incompressible fluid is used for elucidating the role of nonequilibrium fluctuations for both the onset and the development of turbulent motion. The "fluctuation definition" of turbulent viscosity is given in terms of entropy production.

(2) Certain aspects of the semi-empirical theory of turbulence are considered on this basis. The example of Poiseuille flow is used for investigating the possible extension of this theory to the transition region.

(3) Turbulent motion in incompressible liquid is analyzed with the aim of justifying the view on the transition from laminar flow to turbulent as a process of self-organization. Reynolds stresses play the role of the parameter of order. The relative degree of order of laminar and turbulent flows is assessed with the aid of criterion of S–theorem. The conclusion that turbulent motion is higher organized is supported also by calculations of entropy production. These calculations are used for formulating the "principle of minimum entropy production at processes of self-organization".

(4) Discussed are the necessity and feasibility of unified kinetic and hydrodynamic description of turbulent motion. The general expression for entropy production is used for defining the dissipative characteristics of turbulent motion which are common for the kinetic and the hydrodynamic description. The dissipative characteristics in the description of hydrodynamic turbulence are the coefficients of turbulent self-diffusion, turbulent viscosity, and turbulent temperature conductivity.

22.3. Incompressible Liquid. Reynolds Equations and Stresses

Let us return to equations of hydrodynamics of incompressible liquid (14.4.8), (14.4.6). This model assumes that the temperature and the pressure are constant; the problem is reduced to finding the field of the curl velocity $u^{\perp}(r,t)$ (superscript "$\perp$" hereinafter is dropped).

Following Reynolds hypothesis, we assume that these equations hold also for the random (pulsating) velocity $\tilde{u}(r,t)$. On the strength of this assumption, we take the following equations for the initial ones in the theory of turbulent motion:

$$\frac{\partial \tilde{u}_i}{\partial t} + \frac{\partial \tilde{u}_i \tilde{u}_j}{\partial r_j} = -\frac{1}{\rho} \frac{\partial \tilde{p}}{\partial r_i} - \frac{1}{\rho} \frac{\partial \tilde{\pi}_{ij}}{\partial r_j} \ , \quad \mathrm{div} \, \tilde{u} = 0 \ , \tag{22.3.1}$$

$$\Delta \tilde{p} = -\rho \frac{\partial \tilde{u}_i \partial \tilde{u}_j}{\partial r_j \partial r_i} \ , \tag{22.3.2}$$

$$\tilde{\pi}_{ij} = -\eta \left(\frac{\partial \tilde{u}_i}{\partial r_j} + \frac{\partial \tilde{u}_j}{\partial r_i} \right) \equiv -\eta \tilde{u}_{ij} \ . \tag{22.3.3}$$

These equations will serve as initial ones for the description of both laminar and turbulent flows in incompressible liquid.

Laminar flows are characterized by a limited number of parameters of length and time. By contrast, a well-developed turbulent flow has a large number of macroscopic degrees of freedom, and is therefore characterized by a whole spectrum of characteristic scales. The largest (the main) scales of length and time for turbulent flow we denote by L and T, and the smallest by L_0 and T_0. Of course, for the approximation of continuous medium to hold good, all the characteristic scales of turbulent motion must be much greater than the relevant physically infinitesimal scales — that is,

$$L \gg L_0 \gg l_{\mathrm{ph}}^{(\mathrm{H})} \ , \quad T \gg T_0 \gg \tau_{\mathrm{ph}}^{(\mathrm{H})} \ , \tag{22.3.4}$$

where superscript "H" stands for "hydrodynamics". Recall that we have used superscript "G" in case of gas dynamics. The relevant scales are defined by (1.1.21). Smoothing over physically infinitesimal volume gives the smoothed velocity $\tilde{u}$ and the smoothed pressure $\tilde{p}$.

As before, we introduce the Gibbs ensemble for nonequilibrium processes. By averaging over the ensemble we define the mean velocity $u = \langle \tilde{u} \rangle$. We also introduce fluctuations of velocity, pressure, and tensor of velocity derivatives:

$$\tilde{u}_i = u_i + \delta u_i \ , \quad \tilde{p} = p + \delta p \ , \quad \delta \tilde{u}_{ij} = u_{ij} + \delta u_{ij} \ . \tag{22.3.5}$$

Finally, we carry out averaging of equations (22.3.1) – (22.3.3) in random functions. The equation for the mean velocity is

$$\frac{\partial u_i}{\partial t} + \frac{\partial u_i u_j}{\partial r_j} = -\frac{1}{\rho} \frac{\partial p}{\partial r_i} - \frac{1}{\rho} \frac{\partial \pi_{ij}}{\partial r_j} - \frac{\partial}{\partial r_j} \left\langle \delta u_i \delta u_j \right\rangle \ , \tag{22.3.6}$$

and includes, as compared with the conventional Navier – Stokes equation, the additional term

$$\langle \delta u_i(r,t) \delta u_j(r,t) \rangle, \tag{22.3.7}$$

the so-called *Reynolds stress tensor*. The equation for the mean pressure, obtained by averaging (22.3.2), contains the second moment of derivatives of velocity fluctuations.

In this way, the equations for velocity and pressure u, p (the first moments of the corresponding random functions $\tilde{u}$, $\tilde{p}$) are not closed, since they involve the second moments. The equations for the second moments contain, in turn, the higher moments. As a result, we arrive at the infinite chain of meshing equations in the moments of hydrodynamic functions. As before, we are faced with the *problem of closure*, so as to obtain an approximate closed finite set of equations.

A similar problem was encountered when Boltzmann equation was being derived from the chain of equations in the distribution functions (BBGKY equations; see Ch. 7). Then the problem was solved with the aid of perturbation theory in small density parameter for the Boltzmann gas $\varepsilon = n r_0^3$.

In Ch. 9, 10 we once again met with the problem of closure in connection with the kinetic theory of fluctuations. In Ch. 9 the kinetic theory of fluctuations was based on the Leontovich equation, which was used for obtaining a sequence of dissipative equations for the distribution functions. In Ch. 10 a chain of equations was obtained for the moments of random (pulsating) distribution function $\tilde{f}(r,p,t)$ which satisfies equation (10.1.1). Because of nonlinearity of the latter, averaging over the Gibbs ensemble again resulted in an infinite chain of meshing equations, this time for the moments of random distribution function.

Here we readily discern the analogy with the theory of turbulence. Indeed, the initial equation in Ch. 10 is (10.1.1) for the random distribution function. Neglecting fluctuations δf, it coincides with the Boltzmann equation for the deterministic (nonrandom) distribution function $f(r,p,t)$. So, equation (10.1.1) may be compared to the hydrodynamic equation (22.3.1) for the random (pulsating) velocity $\tilde{u}(r,t)$, and the Boltzmann equation for the distribution function $f(r,p,t)$ to the Navier – Stokes equation for the deterministic velocity $u(r,t)$. Can this analogy be of any help in solving the problem of closure?

Recall that in Ch. 9, 10 this problem was dealt with in the following manner. We used the approximation of the second moments, or, more precisely, of the second correlation functions. Both are based on the smallness of fluctuations $\delta f = \tilde{f} - f$. Fluctuations are small because the number of particles within a "point" of continuous medium is large, since large is the number of particles N_{ph} contained within physically infinitesimal volume. It is important that "continuous" was the medium in the six-dimensional space of coordinates and momenta!

This approximation turned out to be efficient, and allowed us to calculate the nonequilibrium fluctuations on both kinetic and hydrodynamic scales. This expanded the domain of traditional (neglecting fluctuations) kinetic theory and hydrodynamics.

Hydrodynamics also employs the model of continuous medium, in three-dimensional rather than in six-dimensional space. Each "point" of continuous medium also contains many particles. It would seem that the approximation of small fluctuations, used in the kinetic theory, ought to work in the hydrodynamic theory of turbulence. Unfortunately, this is not so. The problem of closure in the hydrodynamic theory of turbulence is immensely more complicated, and has so far defied the efforts of many prominent scientists.

The reasons behind this are not yet clear. In the forthcoming sections we shall try to throw some light on the issue. At this point we just mark the following.

Both the kinetic theory of fluctuations and the hydrodynamic theory of turbulence are actually based on the same assumption that the equations for deterministic functions hold good also for random functions. There is, however, an important dissimilarity in the choice of the initial equations. In the former case it is the kinetic equation, and in the latter the Navier – Stokes equation for incompressible liquid. Since we know that equations of hydrodynamics, and, in particular, the Navier – Stokes equation can be derived from the kinetic equation, there is a certain "bridgehead". To wit, one may start not with the Navier – Stokes equation, but rather with the kinetic equation. It would be more natural to use the generalized kinetic equation (13.4.1) instead of Boltzmann equation, which will allow us to give a unified description of the kinetic and hydrodynamic processes. Accordingly, possible is a unified kinetic and hydrodynamic description of turbulent motion. We shall discuss such approach at the end of this chapter.

Another important question concerns the foundations of the Reynolds hypothesis. The fact is that at certain values of the controlling parameters (for instance, of the Reynolds number), the solution of the Navier – Stokes equation for laminar flow (nonrandom velocity) becomes unstable. We have seen this in case of thermal convection. The onset of instability is accompanied by the growth of fluctuations. These developments indicate that the description in terms of the mean velocity (in the approximation of first moments) becomes inadequate, and call for transition to a more general equation in random functions, which carries additional information about the higher moments.

As we know, this transition brings up the problem of closure, which so far remains unsolved for the case of hydrodynamic turbulence. Nevertheless, the hydrodynamic theory of turbulence has yielded many important fundamental and practical results. A detailed review can be found in a recent book entitled *Statistical Hydromechanics* (Monin – Yaglom 1992). Here we shall only touch upon those matters which are important for further discussion. It will be natural to start with the problem of stability of laminar flow.

22.4. Hydrodynamic Instability and Onset of Turbulence

Instability of laminar flow in a pipe, resulting in turbulence, was first discovered in 1839 (Hagen 1839). Since then, instabilities of many other flows have been observed. Analytical calculation of stability of laminar flows is, as a rule, an extremely complicated task. The relatively simple problems include the calculation of instability of convective flow in liquid layer heated from below, and the flow of liquid between two rotating coaxial cylinders. The temperature gradient or the difference in angular velocities of the cylinders first give rise to new steady flows, Benard cells in the convective flow or Taylor vortices between the cylinders. Turbulence sets in only when the temperature gradient or the differential velocity are increased further.

Even more complicated is the description of transition from laminar to turbulent flow. Currently there are four tentative scenarios of such transition; their detailed description can be found in (Berge – Pomeau – Vidal 1984; Landau – Lifshitz 1986; Monin – Yaglom 1992). Here we just note that according to Landau and Hopf the transition to turbulence takes place through a sequence of quasi-periodical flows. Ruelle and Takens associate the onset of turbulence with the appearance of strange attractor; the same view is held by Feigenbaum who, however, believes that strange attractor appears after a cascade of period-doubling bifurcations. Finally, the scenario of Pomeau and Manneville assumes that the transition to turbulence occurs through a sequence of alternating laminar and turbulent flows. The experiment reveals that different scenarios are realized for different kinds of flows.

One may say that the onset of turbulence is a transition to the state of dynamic chaos. It must be remembered, however, that the above processes only characterize the initial stage of turbulence, when the number of collective degrees of freedom is not large. Shortly we shall prove that well-developed turbulent motion is associated with a very large number of degrees of freedom, so that the question of its nature remains open. The description of such motion seems to require new metaphors, which will be the object of our immediate concern. Further on, however, we shall return to the problem of incipient turbulence, armed with the results of semi-empirical theory.

22.5. Well-Developed Turbulence. Number of Degrees of Freedom

Assume that the transition to turbulent regime is accomplished, and consider Reynolds numbers $Re \equiv R = uL/v$ so large that turbulence is well developed. This means that inequalities (22.3.4) hold good, and the motion is therefore characterized by a large number of scales. In the definition of Reynolds number L is is the main scale, and u is the corresponding characteristic velocity. We also define the characteristic Reynolds number via the minimum scale of well-developed turbulent motion L_0:

$$R^0 = \frac{u_0 L_0}{\nu} . \tag{22.5.1}$$

We shall see that this characteristic parameter is linked with one of the two constants in the semi-empirical theory of well-developed turbulence, the Karman constant $K = 0.4$.

Well-developed turbulence is *locally homogeneous and isotropic*. Let us consider an extreme case of completely homogeneous and isotropic turbulence. For this state the mean velocity is zero, $u = 0$, and therefore the fluctuation δu coincides with the pulsating velocity $\tilde{u}$. For completely homogeneous and isotropic turbulence the third moments of velocity vanish. Given this, from (22.3.1) we find the following equation of balance for the density of the mean kinetic energy of turbulent motion:

$$\frac{\partial}{\partial t}\left(\rho \frac{\langle \tilde{u}^2 \rangle}{2}\right) = -\rho\varepsilon , \quad \varepsilon = \frac{\nu}{2}\left\langle \left(\frac{\partial \tilde{u}_i}{\partial r_j} + \frac{\partial \tilde{u}_j}{\partial r_i}\right)^2 \right\rangle , \tag{22.5.2}$$

where ε determines the mean dissipation of kinetic energy of well-developed turbulent motion per unit time per unit mass. In the theory of turbulence this quantity is known as the Kolmogorov parameter. In the next section we shall establish the linkage between ε and the production of entropy. We shall also see that Kolmogorov parameter defines the intensity of Langevin source in the equations of hydrodynamics of incompressible liquid.

By L', T', u' we denote the current scales of well-developed turbulence, and define, according to Kolmogorov, two characteristic intervals: the *inertial interval*

$$L > L' > L_0 , \quad T > T' > T_0 ; \tag{22.5.3}$$

and the *viscous interval*

$$L' \sim L_0 , \quad T' \sim T_0 . \tag{22.5.4}$$

The left-hand inequalities in (22.5.3) define the region of well-developed turbulence where the dependence on the main scales is not explicit. The right-hand inequalities allow disregarding the effects of viscosity on the inertial interval.

The selection of inertial interval for the case of well-developed turbulence is justified by the following arguments.

From definition of dissipative function in equation (22.5.2) follows the linkage between spatial spectral densities of the dissipative function and the mean kinetic energy

$$\mathbf{E}_k = 2\nu k^2 (\bar{u})_k^2 . \tag{22.5.5}$$

It would be natural to assume that the maximum of the mean kinetic energy falls on the scales close to L — that is, on the wave number $k \approx 1/L$. This assumption is supported by experimental results.

Then it follows that, owing to the factor of k^2, the dissipative function in the region of the main scale is small, and is at a maximum on the verge of the viscous interval. This allows us to write the following estimate for ε:

$$\varepsilon \sim \nu \frac{u_0^2}{L_0^2} . \tag{22.5.6}$$

Using (22.5.1), (22.5.6), we find three relations between L_0, T_0, u_0, which involve two parameters ε, R^0:

$$u_0 \sim \left(\varepsilon L_0 R^0 \right)^{1/3} , \quad u_0 \sim \left(\varepsilon T_0 R^0 \right)^{1/2} , \quad T_0 \sim \left(\frac{L_0^2}{\varepsilon R^0} \right)^{1/3} . \tag{22.5.7}$$

The *Kolmogorov hypothesis* states that the statistical characteristics of turbulence on the inertial interval are completely determined by ε and R^0, and do not depend on the viscosity ν. On the strength of this hypothesis, we may use relations (22.5.7) also for the current scales of the inertial interval:

$$u' \sim \left(\varepsilon L' R^0 \right)^{1/3} , \quad u' \sim \left(\varepsilon T' R^0 \right)^{1/2} , \quad T' \sim \left(\frac{L'^2}{\varepsilon R^0} \right)^{1/3} . \tag{22.5.8}$$

The first of these expresses the "two-thirds law" of Kolmogorov – Obukhov. As we shall see later on, the factor of R^0, which for well-developed turbulence is of the order of unity, is important in the semi-empirical theory of turbulence.

The above relations allows us to estimate the number of degrees of freedom in case of well-developed turbulence. This number may be defined as the ratio of volumes corresponding to the length scales L and L_0:

$$N_{\text{turb}} \sim \frac{L^3}{L_0^3} . \tag{22.5.9}$$

Making use of the definition of Reynolds number $R = uL/v$, and the above relations with $L' \Rightarrow L$, we find that

$$\frac{u}{u_0} \sim \left(\frac{L}{L_0}\right)^{1/3}, \quad \frac{L}{L_0} \sim \left(\frac{R}{R^0}\right)^{3/4}. \tag{22.5.10}$$

From the last two relations follows the desired estimate of the number of degrees of freedom in terms of Reynolds number R and parameter R^0:

$$N_{\text{turb}} \sim \left(\frac{R}{R^0}\right)^{9/4}. \tag{22.5.11}$$

We see that the number of degrees of freedom N_{turb} is very large for the region of well-developed turbulence, when $R \gg R^0$. It is, however, much smaller than the number of the microscopic degrees of freedom.

22.6. Intensity of Langevin Source, Entropy Production and Turbulent Viscosity for Well-Developed Turbulence

Intensity of Langevin source. Let us show how Langevin source can be defined for well-developed turbulence — for a state very far from equilibrium.

In Sect. 19.3.2 we introduced Langevin sources into equations of hydrodynamics for calculating the equilibrium fluctuations of hydrodynamic functions. Fluctuations of tensor of viscous stress and vector of thermal flux were given by (19.3.8), (19.3.9). In those expressions we separated the "induced contributions", proportional to the derivatives of hydrodynamic functions u, T, and the proper sources of fluctuations π_{ij}^{source} and q_i^{source}. Correlators of the sources are given by (19.3.10) and (19.3.11).

Along with these sources, it is convenient to use the relevant hydrodynamic random forces (19.3.12). Correlators of these forces are given by (19.3.13) – (19.3.16). Let us now generalize these results so as to find the appropriate Langevin sources for the description of fluctuations in case of turbulent motion in incompressible liquid.

Let us turn to formula (5.6.9) for the correlator of fluctuations of phase density $\delta N(r,p,t)$. To calculate the source of fluctuations caused by the atomic structure of the medium it is sufficient to take care of the contribution not associated with the correlation function g_2. This contribution defines the correlator of the source of fluctuations of phase density

$$\langle \delta N \delta N \rangle^{\text{source}}_{x,x',t} = n \left[\delta(x - x') f(x,t) - \frac{1}{V} f(x,t) f(x',t) \right] . \tag{22.6.1}$$

Now we can find the correlators of the sources of fluctuations of hydrodynamic functions. Here we shall need the correlator of velocity fluctuations

$$\langle \delta u_i \delta u_j \rangle^{\text{source}}_{r,r',t} = \frac{\kappa T(r,t)}{\rho} \delta_{ij} (r - r') . \tag{22.6.2}$$

With the aid of this expression, formula (19.3.15) for the intensity of fluctuations may be reduced to the equivalent form

$$A^{(\eta)}(r,r',t) = \frac{\rho \eta}{2} \left\langle \left(\frac{\partial \delta u_i}{\partial r_j} + \frac{\partial \delta u_j}{\partial r_i} - \frac{2}{3} \delta_{ij} \frac{\partial \delta u_k}{\partial r_k} \right)^2 \right\rangle^{\text{source}}_{r,r',t} . \tag{22.6.3}$$

On the strength of Reynolds hypothesis we may substitute $\delta u_i \Rightarrow \delta \tilde{u}_i$. For isotropic and homogeneous turbulence we have $u = 0$, and the replacement $\delta u_i \Rightarrow \tilde{u}_i$ is possible. Hence follows the desired expression for the intensity of the source of fluctuations in case of well-developed turbulent motion of incompressible liquid:

$$A^{(\eta)}_{\text{turb}}(r,r',t) = \frac{\rho \eta}{2} \left\langle \left(\frac{\partial \tilde{u}_i}{\partial r_j} + \frac{\partial \tilde{u}_j}{\partial r_i} \right)^2 \right\rangle^{\text{source}}_{r,r',t} \equiv \frac{\rho \eta}{2} \left\langle (\tilde{u}_{ij})^2 \right\rangle . \tag{22.6.4}$$

Substituting this into (19.3.13) for the correlator of the relevant Langevin force and using (22.5.2) for the Kolmogorov parameter, we get

$$\left\langle f^{(\eta)}_{\text{turb}} f^{(\eta)}_{\text{turb}} \right\rangle^{\text{source}}_{r,t,r',t'} = 2\rho^2 \varepsilon \delta(t - t') , \quad \varepsilon = \frac{\nu}{2} \left\langle (\tilde{u}_{ij})^2 \right\rangle . \tag{22.6.5}$$

We see that for the case in question the correlator of the source if determined by the Kolmogorov parameter, and hence by the rate of dissipation of the mean kinetic energy of well-developed turbulent motion of incompressible liquid. Interestingly, this result coincides with one obtained earlier on the level of physical intuition (Novikov 1964).

Entropy production. To find the expression for the production of entropy at well-developed turbulent motion of incompressible liquid we turn to formula (11.4.4), which defines entropy production in case of laminar flows. By virtue of Reynolds hypothesis, we replace $u \Rightarrow \tilde{u}$, and carry out averaging over the ensemble. This brings us to the following expression for the mean entropy production at turbulent motion:

$$\langle \tilde{\sigma} \rangle = \frac{\eta}{2T} \left\langle \left(\frac{\partial \tilde{u}_i}{\partial r_j} + \frac{\partial \tilde{u}_j}{\partial r_i} \right)^2 \right\rangle \equiv \frac{\eta}{2T} \left\langle \left(\tilde{u}_{ij} \right)^2 \right\rangle , \quad \tilde{u} = u + \delta \tilde{u} . \tag{22.6.6}$$

So far we did not assume turbulence to be homogeneous and isotropic, and the mean velocity may be nonzero. If $u = 0$, the production of entropy can also be expressed in terms of Kolmogorov parameter — that is, in terms of the rate of dissipation of the mean kinetic energy:

$$\langle \tilde{\sigma} \rangle = \rho \varepsilon , \quad \varepsilon = \frac{v}{2} \left\langle \left(\tilde{u}_{ij} \right)^2 \right\rangle . \tag{22.6.7}$$

Turbulent viscosity. In general, the mean entropy production is a sum of two contributions. The first coincides with σ for the laminar flow, whereas the second depends on the turbulent pulsations of velocity. By defining turbulent viscosity as

$$\eta_{\text{turb}} = \eta \left[1 + \frac{\left\langle \left(\delta \tilde{u}_{ij} \right)^2 \right\rangle}{u_{ij}^2} \right] , \tag{22.6.8}$$

we may reduce (22.6.6) to a form similar to that for the laminar flow:

$$\langle \tilde{\sigma} \rangle = \frac{\eta_{\text{turb}}}{2T} u_{ij}^2 . \tag{22.6.9}$$

So we have defined entropy production and turbulent viscosity for well-developed turbulent flow. Let us prove that our expression for turbulent viscosity agrees with its counterpart in Kolmogorov's theory.

Evaluating the term defined by pulsations $\delta \tilde{u}$ in terms of the minimum scales L_0, u_0, and the term with u_{ij} in terms of the current scales of the inertial interval L', u', and making use of the relations established for the inertial interval of well-developed turbulence, we get the following order-of-magnitude estimate for the second term on the right-hand side of (22.6.8):

$$\frac{\left\langle \left(\delta \bar{u}_{ij} \right)^2 \right\rangle}{u_{ij}^2} \sim \frac{u_0^2}{L_0^2} \frac{L^2}{u'^2} = \frac{u'L'}{u_0 L_0} \frac{u_0^3 L'}{u'^3 L_0} \ .$$

By $R' = u'L'/v$ we denote Reynolds number for the current scales. Then the first fraction on the right-hand side may be expressed as the ratio of Reynolds numbers R'/R^0. The second fraction, by virtue of (22.5.7), (22.5.8), is independent of ε, R^0, and is of the order of unity. Then we arrive at the following estimate for the turbulent viscosity in Kolmogorov's theory:

$$\eta_{\text{turb}} = \eta \left(1 + \frac{R'}{R^0} \right) , \quad R' = \frac{u'L'}{v} \ . \tag{22.6.10}$$

We see that the additional (turbulent) contribution is proportional to the relevant Reynolds number.

This result is useful in the semi-empirical theory of turbulence. It is possible to further particularize the expression (22.6.10). Note that turbulent viscosity, like entropy production, depends on the derivatives of velocity rather than on the velocities themselves. Because of this, the dependence is actually not on the Reynolds number R, but on the so-called *dynamic Reynolds number* R_* instead. These numbers are associated with the *drag law*

$$R = R(R_*) , \quad R_* = \frac{v_* L}{v} \ , \tag{22.6.11}$$

where v_* is the dynamic velocity, associated with the stresses on the wall of, say, the channel. Given all this, for the flow in a flat channel one may use the following expression for turbulent viscosity (Klimontovich – Engel-Herbert 1984; Monin – Yaglom 1992):

$$\eta_{\text{turb}}(y) = \eta \left[1 + f(y) \frac{R_*}{R^0} \right] . \tag{22.6.12}$$

Here we have introduced the function $f(y)$ (with the y–axis directed normally to the wall of the channel of width $2h$), which is derived from experimental data (Schlichting 1958). On the walls of the channel $f(y = \pm h) = 0$, which ensures that $\eta_{\text{turb}} = \eta$ on the channel boundaries.

22.7. Semi-Empirical Prandtl – Karman Theory

Consider two-dimensional steady Couette and Poiseuille flows in a flat channel with the distance of $2h$ between the walls. The x–axis is directed along the channel, and the y–axis is perpendicular to the walls. For the steady two-dimensional flow only the longitudinal velocity component $u_x \equiv u(y)$ is nonzero, and therefore (22.3.6) takes on the form

$$-\frac{d}{dy}\langle \delta u_x \delta u_y \rangle + v\frac{d^2u}{dy^2} = 0; \quad -\frac{\Delta p}{\rho l} , \tag{22.7.1}$$

where Δp is the pressure difference, and l is the length of the channel. Here and further the first value on the right-hand side (before the semicolon) corresponds to the Couette flow, and the second to the Poiseuille flow. The boundary conditions are

$$u(y = \pm h) = \pm u_0; \quad 0 . \tag{22.7.2}$$

The dynamic velocity v_* and the Reynolds numbers are given by

$$\tau_h = \eta \left|\frac{du}{dy}\right|_{y=h} = \rho v_*^2 , \quad R = \frac{u2h}{v} , \quad R_* = \frac{v_* h}{v} , \tag{22.7.3}$$

where τ_h is the stress on the wall.

Integration with respect to y with due account for (22.7.2) and (22.7.3) yields:

$$-\langle \delta u_x \delta u_y \rangle + v\frac{du}{dy} = v_*^2; \quad -v_*^2 \frac{y}{h} , \quad (-h \le y \le h) . \tag{22.7.4}$$

This equation is not closed, since is includes, along with the velocity u, a component of Reynolds tensor. According to the condition of closure proposed by von Karman, this component is proportional to the velocity derivative:

$$\langle \delta u_x \delta u_y \rangle = -\left[v_{\text{turb}}(y) - v \right]\frac{du}{dy} . \tag{22.7.5}$$

For the laminar flow the right-hand side of (22.7.5) is obviously equal to zero. The we get the following equation for the velocity:

$$v_{\text{turb}}(y)\frac{du}{dy} = v_*^2; \ -v_*^2\frac{y}{h}, \ (-h \le y \le h) \ . \tag{22.7.6}$$

This equation is also not closed, because the turbulent viscosity $v_{\text{turb}}(y)$ is as yet not known. To solve the problem of closure, one may take advantage of definition (22.6.12). The form of the function $f(y)$ is found experimentally.

Now let us return to the Karman relation (22.7.5). It contains three functions. Making use of the last equation, we eliminate the function du/dy and find the relationship between the Reynolds stress and the turbulent viscosity:

$$\langle \delta u_x \delta u_y \rangle = -\frac{v_{\text{turb}}(y) - v}{v_{\text{turb}}(y)}v_*^2; \ \frac{v_{\text{turb}}(y) - v}{v_{\text{turb}}(y)}v_*^2\frac{y}{h} \ . \tag{22.7.7}$$

The interesting feature of this relation is that it does not contain the velocity. From (22.6.12) it follows that the dependence of the Reynolds stress on y is completely defined by the form of the function $f(y)$.

A detailed calculation of velocity profile and the relevant drag law $R(R_*)$ can be found in papers referred to above. Here we shall only touch upon those results which, apart from being of interest by themselves, are useful for the forthcoming discussion.

The drag law for the laminar Couette flow is

$$R = 2R_*^2 \ , \text{ or } u_0 = \frac{v_*^2}{v}h \ . \tag{22.7.8}$$

Let us compare this with the drag law in case of well-developed turbulence ($R_* \gg R^0$):

$$R = 2R^0 R_* \ln\left(2\frac{R_*}{R^0}\right) + 2\delta R_* \ , \tag{22.7.9}$$

where δ is a dimensionless parameter. Unlike (22.7.8), this formula contains two "constants" (R^0 and δ), which exhibit a very weak dependence on the Reynolds number. In order to find their numerical values, we use the drag law for turbulent flow as established experimentally by Reichardt:

$$R = 5R_* \ln R_* + 14.5R_* \ . \tag{22.7.10}$$

One of the two contributions here is proportional to $R_* \ln R_*$, and the other to R_*. Expression (22.7.10) already includes numerical constants. Juxtaposition of (22.7.9) and (22.7.10) reveals that

$$R^0 \equiv \frac{1}{K} = 2.5 , \quad \delta = 7.8 , \tag{22.7.11}$$

where K is the Karman constant.

Finally, let us quote the expressions for velocity profile of Couette flow. For the laminar flow the dependence is linear,

$$u(y) = \frac{v_*^2}{v} y = u_0 \frac{y}{h} , \quad (-h \le y \le h) . \tag{22.7.12}$$

For the turbulent flow the profile near the wall is logarithmic,

$$u(y) = v_* R_{cr}^0 \ln\left(\frac{R_*}{R_{cr}^0} \frac{y}{h}\right) + \delta v_* . \tag{22.7.13}$$

At $y = 2h$ hence follows the expression (22.7.9) for drag law in case of turbulent Couette flow.

The results of semi-empirical theory only apply to the regime of well-developed turbulence, when Reynolds number R is large enough. It is possible, however, to extend this theory to the transition region. Let us briefly consider this possibility (Klimontovich V).

22.8. Onset of Steady Turbulent Flow. Evaluation of Critical Reynolds Number

In Sect. 22.4 we considered several scenarios of the early stage of transition from laminar to turbulent flow. Let us now change the statement of the problem, and evaluate the critical Reynolds number from condition of appearance of steady flow with nonzero Reynolds stresses, using the so-called quasilinear approximation for Couette and Poiseuille flows.

As it turns out, there is no stationary solution for Couette flow in this approximation. This agrees with the linear theory of stability, according to which the laminar Couette flow is stable (Monin – Yaglom 1992). Let us show that the situation is different for the steady Poiseuille flow.

Consider equations of quasilinear approximation. One of these is equation (22.7.1) for Poiseuille flow:

$$\frac{d}{dy}\langle \delta u_x \delta u_y \rangle - v\frac{d^2 u}{dy^2} = \frac{\Delta p}{\rho l} \ . \tag{22.8.1}$$

The second equation is approximate and has the following form:

$$-\frac{d\langle \delta u_x \delta u_y \rangle}{dy} = \frac{vk}{\left[\dfrac{\omega}{k} - u(y)\right]^2 + v^2 k^2}\frac{d^2 u}{dy^2}\langle \delta u_y \delta u_y \rangle \ . \tag{22.8.2}$$

This set of two equations contains three unknown functions (two components of Reynolds tensor and the mean velocity u), and can be supplemented by one more equation. For our purposes, however, this is not necessary.

In (22.8.2) we carry out integration with respect to y from 0 to h. With due account for (22.8.1) and boundary condition

$$\langle \delta u_x \delta u_y \rangle\big|_{y=0,h} = 0 \ ,$$

we come to the following integral equation:

$$\int_0^h \frac{vk}{\left[\dfrac{\omega}{k} - u(y)\right]^2 + v^2 k^2}\left[\frac{d\langle \delta u_x \delta u_y \rangle}{dy} - \frac{\Delta p}{\rho l}\right]\langle \delta u_y \delta u_y \rangle dy = 0 \ . \tag{22.8.3}$$

Naturally, these equations are satisfied for the laminar flow, when $\delta \tilde{u} = 0$. It is important that they also hold with good accuracy (in terms of parameter $1/R$) when Reynolds stresses are nonzero. For the transition region, when $R \sim R_{cr}$, the parameter $1/R$ is small, since $R_{cr} \sim 10^3$ in case of Poiseuille flow.

The first term in the integrand in (22.8.3) indicates that the wave resonance $\omega/k = u(y_0)$ occurs at a certain point y_0. For the main mode, when $kh \approx 1$, its width is determined by the only small parameter $1/R$. In the zero approximation it is therefore possible to replace the resonance curve by delta-function:

$$\frac{vk}{\left[\dfrac{\omega}{k} - u(y)\right]^2 + v^2 k^2} \Rightarrow \pi\delta\left[\frac{\omega}{k} - u(y_0)\right] \ . \tag{22.8.4}$$

In this approximation, equation (22.8.2) becomes

$$\left.\frac{d^2u}{dy^2}\right|_{y_0} = 0 \text{ if } u(y_0) = \frac{\omega}{k} .$$

(22.8.5)

We see that the onset of turbulence is signalled by the appearance of inflection point at a certain value of $y = y_0$ which satisfies the condition of wave resonance. The condition of wave resonance at $kh \approx 1$, however, does not yet define the value of y_0. Unambiguous definition of y_0 requires additional information, which can be gained in the following way.

The profile of steady Poiseuille flow is characterized by two scales of length: the width of the channel h and the thickness of the so-called laminar sublayer $l_{\text{lam}} = (\delta/R_*)h$ (where δ is one of the constants of semi-empirical theory of well-developed turbulence). The scale l_{lam} defines the thickness of boundary layer in which the viscosity differs little from that of the laminar flow. It constitutes a few hundredths of the width of the channel h. Accordingly, the steady turbulent flow starts at a distance from the wall about equal to the thickness of laminar sublayer.

So, one may assume that condition (22.8.4) is satisfied at the point

$$y_0 \approx h - l_{\text{lam}} .$$

(22.8.6)

Of course, equation (22.8.5) need not be satisfied exactly; it will do if the second derivative is of the order of $1/R$.

Experimental investigations of changes in the Poiseuille flow profile upon transition across the critical point were carried out in (Dubnishchev – Rynkevichus 1982) using the method of laser anemometry. An inflection point was detected on the velocity profile at $R = R_{\text{cr}}$ in the region adjacent to the wall of the channel.

Now let us evaluate the critical Reynolds number. From (22.8.1), (22.8.2) we eliminate the derivative $d\langle \delta u_x \delta u_y \rangle/dy$ and get the following equation:

$$\left\{ \frac{\langle \delta u_y \delta u_y \rangle}{\left[\frac{\omega}{k} - u(y) \right]^2 + v^2 k^2} + 1 \right\} v \frac{d^2 u}{dy^2} = -\frac{\Delta p}{\rho l} .$$

(22.8.7)

This equation is not closed, since it includes, along with function $u(y)$, also the Reynolds stress $\langle \delta u_y \delta u_y \rangle$. Observe that the first term in braces defines the "turbulent correction" to v. Because the resonance is narrow (of the order of $1/R$), it is sufficient to evaluate the Reynolds stress $\langle \delta u_y \delta u_y \rangle$ only at point y_0.

Let us do this taking advantage of the concept of the number of macroscopic degrees of freedom N_{turb} for the steady turbulent Poiseuille flow. The existence of two characteristic scales (the main scale h, and the smallest scale l_{lam}), allows us to define the number of degrees of freedom, by analogy with (22.5.9), as the ratio of two characteristic volumes:

$$N_{turb} = \frac{h^3}{l_{lam}^3}. \qquad (22.8.8)$$

In case of laminar flow the "sublayer" occupies the entire channel, and $N_{turb} = 1$.

The results of physical and numerical experiments indicate that the excitation of turbulence in Poiseuille flow is *hard*. This implies that the steady flow in question can only exist at finite values of the Reynolds stress $\langle \delta u_y \delta u_y \rangle$. To assess this value, we take advantage of the expression for the relative dispersion

$$\frac{\langle \delta u_y \delta u_y \rangle}{\langle u \rangle^2} = \frac{1}{N_{turb}}, \qquad (22.8.9)$$

defined by the number of degrees of freedom of turbulent flow.

Here we have introduced the mean velocity $\langle u \rangle$, which is related to the maximum velocity $u_{max} \equiv u$, in terms of which the Reynolds number is expressed, as $\langle u \rangle = 2u/3$. Given this, equation (22.8.9) may be rewritten as

$$\langle \delta u_y \delta u_y \rangle = \frac{4}{9} \frac{u^2}{N_{turb}}. \qquad (22.8.10)$$

Substituting this into (22.8.7), we obtain a closed equation for the velocity. Now we introduce dimensionless variables

$$u' = \frac{u}{u_{max}}, \quad y' = \frac{y}{h}, \quad \frac{\omega}{ku} = c; \quad R = \frac{uh}{v}, \quad R_* = \frac{v_* h}{v}, \qquad (22.8.11)$$

make use of the drag law for laminar Poiseuille flow

$$2R = R_*^2,$$

which holds approximately in the transition region, and the condition of resonance $c = u(y_0)$. Finally, dropping the prime at dimensionless variables, we arrive at the following equation:

$$\left[\frac{1}{\frac{9}{4}N_{turb}\left[u(y_0)-u(y)\right]^2+\frac{1}{R^2}}+1\right]\frac{d^2u}{dy^2}=-\frac{R_*^2}{R}=-2 \ . \tag{22.8.12}$$

To find the linkage between the critical Reynolds number R_{cr} with the number of degrees of freedom N_{turb}, we set $y=y_0$ and get

$$\left[\frac{R_{cr}^2}{\frac{9}{4}N_{turb}}+1\right]\frac{d^2u}{dy^2}=-2 \ . \tag{22.8.13}$$

The term in brackets defines (in dimensionless variables) the coefficient of turbulent viscosity. Since we are considering the regime of hard excitation, the Reynolds stresses are finite. This gives grounds to assume that turbulent viscosity is much greater than unity. Then the first term in brackets is much greater than the second, and (22.8.13) may be rewritten as

$$\left|\frac{d^2u}{dy^2}\right|_{y=y_0}=\frac{2}{R}\frac{\frac{9}{4}N_{turb}}{R} \ . \tag{22.8.14}$$

Let us compare this with condition (22.8.5), which is necessary for hard excitation of steady turbulence in Poiseuille flow. As indicated above, this condition must only be interpreted as condition of smallness of the second derivative at the point of resonance with respect to parameter $1/R_{cr}$. With due account for the factor of two, which appears in (22.8.4) because of the drag law in the form of (22.8.11), the condition of onset of turbulence may be written as

$$\left|\frac{d^2u}{dy^2}\right|_{y=y_0}=\frac{2}{R_{cr}} \ , \tag{22.8.15}$$

Juxtaposition of the last two results leads us to the definition of the critical Reynolds number,

$$R_{cr}=\frac{9}{4}N_{turb}=\frac{9}{4}\left(\frac{h}{l_{lam}}\right)^3 \ , \tag{22.8.16}$$

where we have used definition (22.8.8) of the number of turbulent degrees of freedom. Finally, using definition of $l_{lam} = (\delta/R_*)h$ and the drag law (22.8.11) at $R = R_{cr}$, the critical Reynolds number can be expressed in terms of the constant δ:

$$R_{cr} = \frac{8}{5}\left(\frac{\delta}{2}\right)^6 . \tag{22.8.17}$$

For evaluating the critical Reynolds number we identify δ with the respective constant (22.7.11) of well-developed turbulence. Then we get $R_{cr} = 5623$, which is close to the numerical value obtained from the approximate solution of the problem of stability of the laminar Poiseuille flow.

Of course, one should not exaggerate the significance of this numerical result. What matters more is that we have established the linkage between the critical Reynolds number and the number of degrees of freedom of steady turbulent flow in the channel. This explains why the critical Reynolds number is much greater than unity. Moreover, we have linked the critical Reynolds number to one of the constants in the semi-empirical theory of turbulence δ. This allows us to extend the semi-empirical theory of turbulence to the transition region without increasing the number of constants (Klimontovich V).

Because of the importance of this constant it should be noted that δ can be regarded as a new characteristic Reynolds number

$$R_{**} = \frac{v_* l_{lam}}{v} \equiv \delta . \tag{22.8.18}$$

In this way, both constants of the semi-empirical theory of Prandtl – Karman K and δ may be expressed in terms of the relevant Reynolds numbers R_{**} and R^0.

On the strength of arguments developed above one may regard the transition from laminar to turbulent flow as a nonequilibrium phase transition. It is natural to accept Reynolds stress for the parameter of order; for the flow in a channel the Reynolds stress is linked with turbulent viscosity by (22.7.7). At the transition point the parameter of order goes to zero. This standpoint is currently shared by other authors (see, for example, Zubarev – Morozov 1994).

Earlier we discussed some other examples of nonequilibrium phase transitions, and showed that the states with nonzero parameter of order are also more ordered according to the criterion of S–theorem. Calculations for the transition from laminar to turbulent flow will be carried out in Sect. 22.10. Another important criterion of the relative degree of order is based on comparison of values entropy production. As demonstrated in Ch. 21, entropy production is associated with the metric Krylov – Kolmogorov – Sinai entropy.

22.9. Entropy Production at Laminar and Turbulent Flows

In (Klimontovich – Engel-Herbert 1984) compared were the values of entropy production for two types of motion: (1) steady turbulent Couette and Poiseuille flows, and (2) unsteady laminar motion at Reynolds numbers above critical $(\text{Re} > R_{cr})$. It was demonstrated that, under additional condition of constant stress on the walls of the channel, the production of entropy at turbulent motion (stable at $\text{Re} > R_{cr}$) is less than that in case of laminar motion (unstable at $\text{Re} > R_{cr}$) — that is,

$$\sigma_{\text{lam}} - \sigma_{\text{turb}} > 0 \; . \tag{22.9.1}$$

This result can be obtained in the following way.

We turn to expression (22.6.9) for the mean production of entropy at turbulent motion,

$$\langle \tilde{\sigma} \rangle \equiv \sigma_{\text{turb}} = \frac{\rho}{2T} \, v_{\text{turb}} \left(\frac{\partial u_i}{\partial r_j} + \frac{\partial u_j}{\partial r_i} \right)^2 \; . \tag{22.9.2}$$

For steady turbulent flow in a channel hence follows that

$$\sigma_{\text{turb}}(y) = \frac{\rho v_{\text{turb}}(y)}{T} \left(\frac{du}{dy} \right)^2 \; . \tag{22.9.3}$$

Further calculations will be done for the Couette flow.

We eliminate the derivative of velocity from (22.9.3) with the aid of (22.7.6), getting as a result

$$\sigma_{\text{turb}}(y) = \frac{\rho}{T} v_*^4 \frac{1}{v_T(y)} \; , \quad \text{whence} \quad \sigma_{\text{turb}}(y) = \frac{1}{h} \int_0^h \sigma_{\text{turb}}(y) dy \; . \tag{22.9.4}$$

The last formula will be used for calculating the production of entropy at laminar and turbulent flows. The following is important.

Recall that the criterion of S–theorem is applied under condition of constant mean effective energy of the open system in question. The values of entropy production for comparison of the relative degree of order of different states must also be calculated under a certain additional condition. This condition may be formulated from the following considerations.

The characteristics of motion in hydrodynamics may be divided into two classes, depending on whether they are defined by the velocity of motion or by the velocity derivative. Accordingly, two different Reynolds numbers are introduced (R and R_*). Reynolds number R_* is expressed in terms of the dynamic velocity v_*, which is related to the velocity derivative on the wall, and hence to the stress on the wall (see (22.7.3)). Since the production of entropy is determined by the velocity derivatives, the values of entropy production for different types of flows ought to be compared at one and the same value of the stress on the wall — that is,

$$v \left| \frac{du}{dy} \right|_{y=h} = v_*^2 = \text{const} . \tag{22.9.5}$$

Then in calculating the production of entropy by (22.9.4) we may assume that the dynamic velocity is constant.

Let us consider two values of y–averaged entropy production at at $Re > R_{cr}$:

(1) Entropy production σ_{turb} at steady turbulent flow, which is stable at $Re > R_{cr}$.
(2) Entropy production σ_{lam} for an imaginary laminar flow, which is not stable at $Re > R_{cr}$ and therefore cannot exist as a steady flow. Turbulent viscosity v_{turb} in the calculations of entropy production is replaced by v.

Let us just quote the results of calculations carried out in (Klimontovich – Engel-Herbert 1984):

$$\frac{\sigma_{turb}}{\sigma_{lam}} = \frac{R}{2R_*^2} \leq 1 . \tag{22.9.6}$$

We see that the ratio of values of entropy production for stable turbulent and unstable laminar flows is defined by the drag law $R = R(R_*)$. The "equals" sign corresponds to the transition point, where the drag law for Couette flow has the form of (22.7.8).

The result for the Poiseuille flow is

$$\frac{\sigma_{turb}}{\sigma_{lam}} = 2 \frac{R}{R_*^2} \leq 1 . \tag{22.9.7}$$

In this case the ratio of values of entropy production is also defined by the drag law, this time in the form of (22.8.11) rather than (22.7.8).

These results allow us to conclude that the real (stable) process at the transition point takes the path of the lesser entropy production. This is an indication that the steady turbulent flow is more ordered than the imaginary laminar flow. On these grounds one may regard the transition from laminar to turbulent flow as a process of self-organization. The analysis based on the entropy criterion leads to the same conclusion.

So, for the exemplary steady Couette and Poiseuille flows we have established inequality (22.9.1). This particular result is very important, since it allows us to formulate a tentative general "principle of least entropy production at processes of self-organization", which has been already discussed in the Introduction. We should add that this hypothesis may form the basis of a variation method for calculating hydrodynamic and other nonequilibrium processes in open systems.

Let us now analyze the relative degree of order of laminar and turbulent flows on the basis of S–theorem.

22.10. Entropy Decrease upon Transition from Laminar to Turbulent Flow

Let us quote the results of calculations of the relative degree of order of laminar and turbulent flows on the basis of S–theorem. We have briefly touched upon this topic in the Introduction; now we shall start with some important details.

Assume that local Maxwell distribution holds for steady laminar flow of incompressible liquid, and has the form

$$f_{\text{lam}}(r,v) = \left(\frac{m}{2\pi\kappa T}\right)^{3/2} \exp\left[-\frac{m(v - u_{\text{lam}}(r))^2}{2\kappa T}\right], \tag{22.10.1}$$

where $u_{\text{lam}}(r)$ is the local velocity of laminar flow. Using this distribution, we find the local entropy

$$S_{\text{lam}}(r) = -\kappa n \int \ln n f_{\text{lam}}(r,v) f_{\text{lam}}(r,v) dv , \quad n = \frac{N}{V} . \tag{22.10.2}$$

In case of turbulent flow, the local velocity is a random function. Random (pulsating) is also the relevant local Maxwell distribution:

$$\tilde{f}_{\text{turb}}(r,v,t) = \left(\frac{m}{2\pi\kappa T}\right)^{3/2} \exp\left[-\frac{m(v - \tilde{u}(r,t))^2}{2\kappa T}\right] . \tag{22.10.3}$$

For the measure of uncertainty one may take the averaged entropy

$$S_{\text{turb}} = \left\langle \tilde{S}_{\text{turb}} \right\rangle = -\kappa n \int \left\langle \ln n \tilde{f}_{\text{turb}}(r,v,t) \tilde{f}_{\text{turb}}(r,v,t) \right\rangle dv . \tag{22.10.4}$$

Now let us formulate the additional condition of constancy of the mean effective energy, which is required for comparing the degree of order of the two states in question. For the state of physical chaos we choose the laminar flow; this choice will be later justified by calculations of the effective temperature. From (22.10.1) we see that the kinetic energy with respect to the mean flow must be taken for the effective energy; then the condition of constancy of the mean energy is

$$\int \frac{m\left(v - u_{\text{lam}}(r)\right)^2}{2} f_{\text{lam}} dv = \int \frac{m\left(v - u_{\text{turb}}(r)\right)^2}{2} \left\langle \tilde{f}_{\text{turb}} \right\rangle dv , \tag{22.10.5}$$

where $u_{\text{turb}} = \langle \bar{u}(r,t) \rangle$ is the mean velocity of turbulent flow. By solving this equation we find the effective temperature of the laminar flow, which we have taken for the state of "physical chaos":

$$\kappa T_{\text{lam}} = \kappa T_{\text{turb}} + \frac{1}{3} m \left\langle (\delta u)^2 \right\rangle , \quad (T \equiv T_{\text{turb}}) . \tag{22.10.6}$$

We see that in order to satisfy the additional condition (22.10.5) we have to raise the temperature of the laminar flow. This proves that our choice of the state of physical chaos was correct. The additional thermal (disordered) motion is converted in the turbulent flow into the higher-ordered motion of the collective degrees of freedom.

Taking advantage of this result, we find the difference in entropies of the laminar flow and the steady turbulent flow, which gives us a quantitative measure of the relative degree of chaoticity of the two flows:

$$T\left(S_{\text{lam}} - S_{\text{turb}}\right) = \frac{3}{2} \kappa T_{\text{lam}} n \ln \frac{\kappa T_{\text{lam}}}{\kappa T_{\text{lam}} - \frac{1}{3} m \left\langle (\delta u)^2 \right\rangle} \approx \frac{mn}{2} \left\langle (\delta u)^2 \right\rangle \geq 0 . \tag{22.10.7}$$

So, under condition (22.10.5) the entropy of averaged turbulent flow is less than that of the laminar flow. The decrease in entropy is due to the appearance of collective degrees of freedom which are characterized by Reynolds stresses. This is why turbulent motion is more ordered, and another reason to believe that the transition from laminar to turbulent flow is a process of self-organization.

As said in the Introduction, the considerable increase in viscosity upon transition from laminar to turbulent flow signals the switchover from "individual" transfer of momentum between adjacent layer to a "collective" process. In other words, individual unorganized resistance gives way to organized collective resistance.

The change in internal friction causes the change in the velocity profile, which also becomes more organized (more uniform across the channel).

In this way, the transition to turbulent regime is associated with the emergence of new macroscopic (collective) couplings. The structure of motion from macroscopic point of view becomes so complicated that we see it as "chaos". Because of this, the main task of the theory of turbulence may be formulated as elucidation and description of the *structure of chaos*.

The higher orderedness of turbulent motion has been established here on the basis of S–theorem, by comparing the degree of order of two stationary states corresponding to two different values of the controlling parameter (Reynolds number). The higher order of turbulent motion can, of course, be proved by considering the time evolution in a closed system.

Consider the transition from nonequilibrium initial state to equilibrium. According to H–theorem, the entropy then increases, and we are dealing with degradation rather than self-organization. Degradation can be detected through the increase in temperature. Let us illustrate this with an appropriate example.

Assume that well-developed turbulent motion is created by mixing in the system. This motion is characterized by distribution (22.10.3). Then the system is isolated, and evolves towards equilibrium, characterized by distribution (22.10.1) with zero mean velocity and entropy S_0. Since the mean energy of a closed system does not change in the course of evolution, the relative degree of order of the equilibrium state and the nonequilibrium initial state can be defined in terms of the difference in the corresponding entropies S_0 and $\langle \tilde{S}_{\text{turb}} \rangle$.

From condition of constancy of the mean kinetic energy at the initial instant and at the state of equilibrium we find the relationship between the corresponding temperatures,

$$\kappa T_{\text{turb}} + \frac{1}{3} m \langle (\delta \tilde{u})^2 \rangle = \kappa T \ , \quad \delta \tilde{u} = \tilde{u} - \langle \tilde{u} \rangle = \tilde{u} \ . \tag{22.10.8}$$

We assume that the initial turbulent motion is homogeneous. In the course of time evolution the temperature increases.

This result is similar to (22.8.6). This time, however, the state of equilibrium plays the role of laminar flow, accepted earlier for the state of "physical chaos". The corresponding entropy increase is given by expression

$$S_0 - S_{\text{turb}} = \frac{\rho}{2} \frac{\langle (\delta \tilde{u})^2 \rangle}{T_0} \ , \tag{22.10.9}$$

which is similar to (22.10.7). The equilibrium entropy S_0 now replaces the entropy of "heated laminar flow".

So, we again come to the conclusion that turbulent motion is more ordered. The order of two different turbulent motions can be compared by comparing the temperatures of the two corresponding equilibrium states.

To end this section, let us mark the following.

Almost all of the characteristic features of turbulent motion, listed in Sect. 22.2, have been confirmed by now for the example of hydrodynamic description of turbulent motion in incompressible liquid. It seems that what is left is just a few little clouds in the blue sky. We shall see, however, that the attempt to dispel the remaining dark spots calls for a thorough revision of the entire concept of description of turbulent motion.

22.11. Arguments in Favor of Kinetic Description of Turbulent Motion

Paragraph 3 in the list in Sect. 22.2 says that the approximation of continuous medium as used in the kinetic theory and in hydrodynamics is not adequate for the description of turbulent motion. This statement makes no distinction between different levels of description. At the same time, however, we know that the kinetic description is more powerful, because the "point" in the kinetic theory is much smaller than in hydrodynamics. Because of this, the inadequacy of the hydrodynamic model of continuous medium should not prevent us from trying to make good use of a similar model on the kinetic level. This is what we are going to do now.

Let us give some reasons which necessitate the transition to the kinetic description of turbulent motion.

22.11.1. *Maximum Values of Reynolds Number in Kolmogorov Theory*

The number of degrees of freedom for well-developed homogeneous isotropic turbulence in Kolmogorov's theory is defined in terms of the main L and the smallest L_0 scales by (22.5.9). Relations (22.5.10) establish the linkage (22.5.11) between N_{turb} and Reynolds number R.

The hydrodynamic theory of turbulence is based on the model of continuous medium. Because of this, the minimum scale L_0 must be greater than the size of the "point" of the medium. The latter is defined by the physically infinitesimal length $l_{ph}^{(H)}$, as given by (1.1.21) (where we have used superscript G in place of H). The physically infinitesimal length itself depends on the main scale. Because of this, the following inequality must be satisfied:

$$L_0 \gg l_{\text{ph}}^{(\text{H})} \sim \frac{L}{N^{1/5}} \sim L\left(\frac{r_{\text{av}}}{L}\right)^{3/5} , \quad N = nL^3 , \tag{22.11.1}$$

where r_{av} is the average distance between particles of the medium, defined in terms of the mean density of the number of particles n.

We see that the acknowledgment of atomic structure of "continuous medium" imposes a considerable constraint (with the given L, n) on the minimum scale L_0. With the aid of the second relation in (22.5.10), this constraint can be interpreted as a limitation of Reynolds number from above within the framework of the model of continuous medium:

$$\frac{R_{\text{max}}}{R^0} \ll N^{4/15} \sim \left(\frac{L}{r_{\text{av}}}\right)^{4/5} . \tag{22.11.2}$$

Finally, let us quote the estimate for the maximum number of turbulent degrees of freedom:

$$\left(N_{\text{turb}}\right)_{\text{max}} \sim N^{3/5} \sim \frac{N}{N_{\text{ph}}^{(\text{H})}} . \tag{22.11.3}$$

This relation quite naturally states that the largest number of turbulent degrees of freedom for a given scale L is defined by the number of "points" which can be packed into the volume L^3. Accordingly, the kinetic description is more powerful because the volume of the point is smaller.

22.11.2. *Problem of Closure in Kinetic and Hydrodynamic Description*

Let us return to Sect. 22.3, where we discussed the problem of closure of infinite chain of equations in the moments of velocity in the theory of hydrodynamic turbulence of incompressible liquid. We noted that the problem of closure had been encountered earlier in the calculations of both kinetic and hydrodynamic fluctuations. In those cases the problem admitted a satisfactory solution, since the fluctuations could be considered small because of the large number of particles within the relevant physically infinitesimal volumes. This was the reason why the Gaussian approximation worked well enough, with the statistical properties of the system being described by the first two moments.

In the theory of well-developed turbulence the situation is quite different. Despite the fact that this theory is also based on the approximation of continuous medium, and the "point" contains many particles, deviations from the Gaussian approximation are large. Because of this, there is actually no small parameter which could be used for limiting the number of moments. As a matter of fact, all the moments (all the collective degrees of freedom) are important for the state of well-developed turbulence.

Formally, the fluctuations may be not small. In such case, however, one might argue that the choice of the averaged state, which respect to which the fluctuations are being considered, has been made in not the best possible way.

How can one improve the selection of the initial state? It seems that we already know the answer: it is necessary to go over to the kinetic description based on the equation in the distribution function $f(r,v,t)$. Observe that the kinetic description on the basis of, for instance, the Boltzmann equation, also corresponds to the approximation of continuous medium, but now in the six-dimensional space of coordinates and momenta. Because of this, v in the kinetic theory is by no means the velocity of an individual particle. This velocity applies to the collective of particles within a "point", whose size in the kinetic theory is much smaller than in hydrodynamics.

In the hydrodynamic theory of turbulence we considered fluctuations

$$\delta u(r,t) = \bar{u}(r,t) - u(r,t) ,$$

that is, the deviations of the dynamic velocity

$$u^{(d)}(r,t) = \sum v_i \delta(r - r_i),$$

smoothed over the hydrodynamic physically infinitesimal volume $V_{ph}^{(H)}$, from the mean hydrodynamic velocity $u(r,t)$.

On the kinetic level of description we are dealing with fluctuations of velocity v, smoothed over the smaller volume of the "point". Fluctuations are again considered with respect to the mean velocity, which is defined via the distribution function as

$$u(r,t) = \int v f(r,v,t) dv .$$

Further simplifications are required for going over to the hydrodynamic definition of the mean velocity — for example, the use of the perturbation theory in Knudsen number.

So, the kinetic theory is potentially capable of giving a more comprehensive description of turbulent motion, when the conventional hydrodynamic equations are inadequate, and there is no efficient way of closing the chain of equations in moments of hydrodynamic random velocity $\bar{u}(r,t)$. Let us try to take advantage of this potential. First, however, we shall discuss yet another snag in the traditional hydrodynamic description of turbulent motion: equations of hydrodynamics of incompressible liquid are not consistent with the equation of entropy balance.

22.11.3. *Equation of Entropy Balance for Turbulent Motion in Incompressible Liquid*

In Sect. 14.5 we have noted that equation of entropy balance does not hold for the model of incompressible fluid. Indeed, the temperature and pressure for incompressible fluid are constant, and therefore the entropy is constant too. At the same time, the production of entropy is given by (14.5.1), and so the hydrodynamic motion is accompanied by the change in entropy.

This inconsistency remains in equations of turbulent motion, based on the Reynolds hypothesis. To wit, the mean value of local entropy of turbulent motion $\langle \tilde{S} \rangle$ is constant, whereas the production of entropy depends on the process. We shall see that this contradiction is removed in the kinetic theory of turbulence.

In this connection it should be noted that the situation is different when the criterion of S–theorem is used. Namely, the mean entropy of steady turbulent flow (22.4.10) is constant, but the entropy of laminar motion, renormalized to the given value of the mean effective energy, depends on the Reynolds number and thus reflects the nature of turbulent process. This entropy is not a locally equilibrium characteristic, being a function of Reynolds stress.

22.12. Two Possible Kinetic Descriptions of Turbulent Motion

We start with the generalized kinetic equation (13.4.1), with the replacement $R \Rightarrow r$:

$$\frac{\partial f}{\partial t} + v \frac{\partial f}{\partial r} + \frac{F}{m} \frac{\partial f}{\partial v} = \frac{\partial}{\partial v} \left[D_{(v)}(r,t) \frac{\partial f}{\partial v} \right] + \frac{\partial}{\partial v} \left[v - u(r,t) f \right]$$

$$+ \frac{\partial}{\partial r} \left[D_{(r)} \frac{\partial}{\partial r} - \frac{1}{\tau} \frac{F}{m} f \right]. \tag{22.12.1}$$

The coefficient of diffusion in the space of velocities is given by (13.3.12); the properties (13.3.13) hold for the corresponding "collision integral" (13.3.11); τ is the free path time.

The problem of turbulent motion in a rarefied gas (for example, in the atmosphere) is also very important. Because of this, the description of turbulent motion on the basis of kinetic equation (22.12.1) is of practical interest.

The kinetic equation (22.12.1) contains two time scales which characterize the dissipative processes: the free path time τ, and the diffusion time τ_D. In the zero approximation with respect to the ratio of these two time scales

$$\frac{\tau}{\tau_D} \sim \left(\frac{l}{L}\right)^2 \sim (\text{Kn})^2$$

(that is, in the square of Knudsen number), the solution of the kinetic equation is the local Maxwell distribution. In this approximation one gets a complete set of gas dynamics (14.1.2) – (14.1.4) in functions ρ, u, T, which can be used for describing laminar flows.

So, the locally equilibrium solution of the kinetic equation

$$f_{1.\text{eq.}}(r, |v - u(r,t)|, t) , \tag{22.12.2}$$

which pertains to the class of distributions (14.1.1), is sufficient for the description of laminar flows.

To obtain a less general solution, corresponding to the approximation of incompressible liquid, two additional conditions must be imposed on the distribution function (cf. (14.4.1), (14.4.2)):

$$\int f(r,v,t)dv = \text{const} , \quad \int (v - u(r,t))^2 f(r,v,t)dv = \text{const} . \tag{22.12.3}$$

The first of these ensures conservation of the density, and the second makes sure that the temperature is constant. The continuity equation reduces to (14.4.3), whence it follows that the velocity field is vortical, and satisfies the Navier – Stokes equations (14.4.8), (14.4.6). These equations have been used for the transition (based on the Reynolds hypothesis) to the initial equations of the theory of turbulence of incompressible liquid (22.3.1), (22.3.2).

Now we are ready to attack the problem of description of turbulent motion on the basis of the kinetic equation.

Local equilibrium. Reynolds hypothesis in the kinetic theory. Let us extend the Reynolds hypothesis so as to apply it to the kinetic equation (22.12.1). This feat consists in replacing the deterministic distribution function with the pulsating (random) distribution,

$$f(r,v,t) \Rightarrow \tilde{f}(r,v,t) . \tag{22.12.4}$$

As a result, we get the following equation:

$$\frac{\partial \tilde{f}}{\partial t} + v \frac{\partial \tilde{f}}{\partial r} + \frac{F}{m} \frac{\partial \tilde{f}}{\partial v} = \frac{\partial}{\partial v} \left[\tilde{D}_{(v)}(r,t) \frac{\partial \tilde{f}}{\partial v} \right] + \frac{\partial}{\partial v} \left[\frac{1}{\tau} (v - \tilde{u}(r,t)) \tilde{f} \right]$$

$$+ \frac{\partial}{\partial r} \left[D_{(r)} \frac{\partial}{\partial r} - \frac{1}{\tau} \frac{F}{m} \tilde{f} \right] . \tag{22.12.5}$$

Of course, this equation is much more general than equations (22.3.1), (22.3.2), which lie in the basis of the theory of turbulence in incompressible fluid. To accomplish the transition to the latter, we must single out the class of pulsating functions (cf. (22.12.2))

$$\tilde{f}(r,v,t) = \tilde{f}_{1.\text{eq.}} \left(r, |v - \tilde{u}(r,t)| \right) . \tag{22.12.6}$$

In this approximation the kinetic equation (22.12.5) leads to the equations (22.3.1), (22.3.2) in the pulsating velocity for incompressible liquid.

We see that all information about turbulent motion in incompressible liquid corresponds to the particular solution of the kinetic equation (22.12.5). Because of this, the kinetic equation has many resources for the development of the theory of turbulent motion. Such resources, however, are inherent also in the kinetic equation (22.12.1) in the distribution function $f(r,v,t)$.

Before discussing this point, let us make two remarks.

The class of distributions (22.12.6) has already been introduced in Sect. 22.10 in connection with application of S–theorem for comparing the degree of order of laminar and turbulent flows. Moreover, in Ch. 10 and 19 we used the kinetic Boltzmann equation and the generalized kinetic equation for the pulsating distribution $\tilde{f}(r,v,t)$.

Accordingly, there are two reasons for going over from the generalized kinetic equation (22.12.1) in the deterministic distribution $f(r,v,t)$ to the corresponding equation (22.12.5) in the pulsating distribution.

The first reason is connected with the atomic structure of the "continuous medium" under consideration. The kinetic equations in $\tilde{f}$, used in Ch. 10 and 19, allowed us to take care of the kinetic fluctuations in both passive (Ch. 10) and active (Ch. 19) media. The approximation of second moments proved to be efficient enough for the calculation of fluctuations, assuming that the fluctuations are small. The smallness of fluctuations was ensured by the factor of $1/N_{\text{ph}}$, where N_{ph} is the number of particles within the "point" of continuous medium.

In this chapter the transition from the Navier – Stokes equation to the corresponding equation (22.3.1) in the pulsating velocity was necessitated by the instability of laminar flow. On the kinetic level this corresponds to the transition from the particular solution (22.12.2) to the particular solution (22.12.6) of equation (22.12.5) in the pulsating distribution function.

In the hydrodynamic theory the problem of atomic structure of continuous medium does not arise — at least, not in an explicit form. Both laminar and turbulent flows are described on the basis of equations of fluid mechanics. Because of this, the small parameter $1/N_{ph}$ does not come up, and there are no grounds for using the approximation of small fluctuations.

Now, from arguments developed above it follows that both the theory of kinetic fluctuations and the theory of turbulent motion are based on the same kinetic equation (22.12.5). This means that the theory of kinetic fluctuations, developed in Ch. 10, 19, and the theory of turbulent motion, discussed in the present chapter, represent particular cases of the general kinetic theory of laminar and turbulent motion. Of course, this theory is very complicated; at the same time, it is capable of elucidating the general features of turbulent motion and breaking out of the traditional domain of the theory of turbulent motion.

22.13. Analogy with Gas-Liquid Transition in Van der Waals System

Let us return to Sect. 22.10, in which we considered the critical point of gas-to-liquid transition. There are two levels of description of this transition, thermodynamic and statistical.

The first is based on the Van der Waals equation (21.10.4), which may be rewritten as

$$p = \frac{n\kappa T}{1 - nb} - n^2 a . \qquad (22.13.1)$$

This equation shows no dependence on N, which implies that this is an equation of continuous medium.

Let us assimilate the state above the critical point with the laminar motion. There are no peculiarities in this range of thermodynamic parameters. When the critical point is crossed, the state of the system from thermodynamic point of view becomes unstable. Instability is associated with the emergence of the two-phase state, which may me likened to the state of turbulence.

In the statistical theory the states at any vales of parameters are described by distribution (21.10.2) with the effective Hamilton function (21.10.3), which may be approximately (!) replaced by the Gaussian distribution above the critical point. As the critical point is approached, the dispersion (22.10.12) increases according to the Curie's law. This points to the developing instability of the "laminar flow", or, in other words, to the incipient restructuring of the one-phase state of the system.

It is important that the onset of instability is only observed when the general distribution

(21.10.2) is replaced by the approximate Gaussian distribution, which means that we are using the approximation of small fluctuations. The initial distribution itself amounts to taking all moments into account, and does not exhibit any peculiarities.

In this way, the entire calculation is based on the deterministic distribution, which may be regarded as the equilibrium solution of some kinetic equation. The Van der Waals equation itself corresponds to the most probable value of the volume of the system under consideration, the fluctuations being regarded as deviations of the volume from the most probable value. The relative dispersion is small (proportional to $1/N$) only at the "laminar" state; as the system moves from the critical point further into the domain of turbulent motion, the fluctuations become of the order of unity. This means that the two-phase state has set in, and each component is a macroscopic subsystem, the number of particles in which constitutes a finite share of the total number of particles N.

As follows from this analogy, by analyzing the deterministic distribution one can not only describe the two extremes of laminar motion and well-developed turbulence, but also trace deviations from the initial laminar motion. These deviations, which increase as the critical point is approached, are definitive for the onset and development of turbulence. Importantly, the equations of laminar motion hold not for the first moments of the relevant variables, but rather for their most probable values.

The distinction between these values is only small for the laminar state, far from the critical point. For well-developed turbulence the relative deviations are of the order of unity, which implies that new macroscopic characteristics of the medium come on the scene.

Let us try to use this analogy for demonstrating the feasibility of unified kinetic description of laminar and turbulent motion in incompressible liquid. We shall start with the kinetic equation (22.12.1) with additional conditions (22.12.3). The more general kinetic equation (14.6.8), which describes convective motion in rarefied gas, is also used for practical applications.

22.14. Unified Kinetic Description of Laminar and Turbulent Motion

Recall that the approximate solution (the locally equilibrium distribution (22.12.2)) is sufficient for accomplishing the transition from the kinetic equation (22.12.1) to the Navier – Stokes equation. The distribution function for the description of laminar and turbulent motion we represent as a sum of two parts,

$$f(r,v,t) = f_{1.\text{eq.}}(r,v,t) + f_{\text{turb}}(r,v,t) \ . \tag{22.14.1}$$

The second term on the right-hand side describes deviation from the laminar flow. We assume that the condition of constant density in (22.12.3) holds for the general distribution, whereas the condition of constant temperature only holds for the laminar flow:

$$\int f(r,v,t)dv = \text{const} , \quad \int (v - u(r,t))^2 f_{\text{l.eq.}}(r,v,t)dv = \text{const} , \tag{22.14.2}$$

where $u(r,t)$ is the mean velocity for the general distribution. Then

$$u(r,t) = \int v f_{\text{l.eq.}}(r,v,t)dv , \quad \int v f_{\text{turb}}(r,v,t)dv = 0 . \tag{22.14.3}$$

The local temperature is, as before, constant, but there is an additional contribution to the tensor of velocity fluctuations

$$\langle \delta v_i \delta v_j \rangle = \int (v - u)_i (v - u)_j f_{\text{turb}}(r,v,t)dv , \tag{22.14.4}$$

which plays the role of stress tensor for the turbulent flow, and vanishes for the laminar regime.

Equation for the mean velocity follows from the kinetic equation (22.12.1), with due account for properties (13.3.13) of the "collision integral" defined by the redistribution in the space of velocities. As a result, we come to the equation which formally coincides with the first one in the chain of Reynolds equations in moments. There are, however, two distinctions.

First, the tensor of viscous stress for the general kinetic equation is not symmetrical, and is given by (13.4.13).

Secondly, the additional stress tensor is expressed by (22.14.4) via the solution of the kinetic equation. This definition is more specific than the definition of Reynolds stress tensor

$$\langle \delta u_i \delta u_j \rangle \equiv \langle (\tilde{u} - u)_i (\tilde{u} - u)_j \rangle , \tag{22.14.5}$$

which leaves open the question of the nonequilibrium statistical ensemble used for carrying out the averaging.

We see that the calculation of turbulent flows may be based on the kinetic equation. Let us do this for Poiseuille flow in a flat channel.

22.15. Kinetic Description of Steady Turbulent Poiseuille Flow in Flat Channel

The problem of steady turbulent Poiseuille flow in a flat channel was discussed in Sect. 22.7, 22.8. The distribution function in this case only depends on three variables,

$$f(r,v,t) \Rightarrow f(y,v_x,v_y) \; ; \; u_x \equiv u(y) \; , \; u_y = 0 \; , \tag{22.15.1}$$

and the kinetic equation (22.12.1) reduces to

$$v_y \frac{\partial f}{\partial y} - \frac{\Delta p}{ml} \frac{\partial f}{\partial v_x} = \frac{\partial}{\partial v} \left[D_{(v)} \frac{\partial f}{\partial v} + \frac{1}{\tau}(v - u(y)f) \right] + v \frac{\partial^2 f}{\partial y^2} \; , \; v = (v_x, v_y) \; . \tag{22.15.2}$$

Here we have noted that the force is determined by the pressure drop, and the coefficient of spatial diffusion is identified with the coefficient of viscosity. The coefficient of diffusion in the space of velocities is now given by (cf. (13.3.12))

$$D_{(v)} = \frac{1}{\tau} \frac{1}{3} \int (v - u)^2 f dv = \frac{\kappa T}{\tau m} + \frac{1}{\tau} \frac{1}{3} \int (v - u)^2 f_{\text{turb}} dv \; . \tag{22.15.3}$$

The additional term on the right-hand side takes care of the turbulent contribution to the temperature. In this way, we have arrived at the generalized Einstein relation.

Nonlinearity enters the kinetic equation via the "collision integral" — via the dependence of the coefficient of diffusion $D_{(v)}$ and the mean velocity on the sought-for distribution function. To find the equation in the mean velocity $u(y)$, we multiply (22.15.2) by v and integrate with respect to v_y, v_z. As a result, with due account for the above conditions, we get

$$\frac{d}{dy} \langle \delta v_x \delta v_y \rangle = v \frac{d^2 u}{dy^2} + \frac{\Delta p}{ml} \; . \tag{22.15.4}$$

This equation formally coincides with (22.7.1) for the Poiseuille flow; the difference is that the tensor of turbulent stress in (22.7.1) is the Reynolds stress tensor (22.14.5), whereas now it is given by (22.14.4), and is therefore defined by the solution of the kinetic equation (22.15.2).

For laminar flow, when $f_{\text{turb}} = 0$, the equation of motion (22.15.4) is closed. Given the boundary conditions, it defines the profile of laminar Poiseuille flow in flat channel.

The kinetic equation also becomes much simpler in case of laminar flow: it assumes the linear form

$$v_y \frac{\partial f}{\partial y} - \frac{\Delta p}{ml} \frac{\partial f}{\partial v_x} = \frac{\partial}{\partial v} \left[\frac{\kappa T}{\tau m} \frac{\partial f}{\partial v} + \frac{1}{\tau}(v - u_{\text{lam}}(y)f) \right] + v \frac{\partial^2 f}{\partial y^2} \; . \tag{22.15.5}$$

At large enough pressures, when the free path time τ is much smaller than the velocity diffusion time $\tau_v = h^2/v$, the main contribution is defined by the local Maxwell distribution

$$f_{1.\text{eq.}} = \left(\frac{m}{2\pi\kappa T}\right)^2 \exp\left(-\frac{m(v - u_{\text{lam}})^2}{2\kappa T}\right), \quad v = (v_x, v_y), \qquad (22.15.6)$$

where $u_{\text{lam}}(y)$ is the profile of laminar Poiseuille flow.

Using the kinetic equation (22.15.5), one can find the correction to this distribution: it is proportional to the small parameter τ/τ_v. Of interest is also the opposite extreme of free molecular flow, when the free path length is much greater than h.

The analysis of kinetic equation (22.15.2) is much more complicated. The most interesting results can only be obtained by numerical calculations. Here we shall confine ourselves to the analysis of the relevant equation of entropy balance. In particular, we are going to show that beyond the limitations of the local Maxwell distribution the expression for entropy production includes additional dissipative terms due to the turbulent motion.

22.16. Equation of Entropy Balance for Turbulent Motion

Let us reproduce the expression for local entropy (6.4.3),

$$S(r,t) \equiv \frac{\rho(r,t)}{m} s(r,t) = -\kappa n \int \ln(nf(r,v,t))f(r,v,t)dv, \qquad (22.16.1)$$

where $s(r,t)$ is the entropy per one particle, which enters equation (11.4.2) of the second law of thermodynamics for the state of local equilibrium. Assume that $F(r,t) = 0$. For the sake of simplicity we further assume that the turbulent distribution functions in (22.14.1) also belong to the class (14.1.1), but do not coincide with the locally equilibrium distribution $f_{1.\text{eq.}}$. Then the equation of entropy balance can be represented in the form (cf. (13.4.6))

$$\frac{\partial \rho s}{\partial t} + \frac{\partial}{\partial r}\left[\left(\rho u - D\frac{\partial \rho}{\partial r}\right)s\right] = \frac{\partial}{\partial r}\left(D\rho\frac{\partial s}{\partial r}\right) + \sigma(r,t). \qquad (22.16.2)$$

The flux of entropy is the sum of convection and diffusion terms, the latter being proportional to entropy gradient. As we know, this allows us to give a general (kinetic) definition of thermal flux (13.5.1). The production of entropy is given by

$$\sigma(r,t) = \kappa n \int \left[D_{(v)} f \left(\frac{\partial}{\partial v} \ln \frac{f}{f_{1.\text{eq.}}} \right)^2 + Df \left(\frac{\partial}{\partial r} \ln \frac{f}{f_0} \right)^2 \right] dv \geq 0 \ , \tag{22.16.3}$$

and is comprised of two positive contributions. One is determined by the redistribution of particles with respect to velocities, and is proportional to the coefficient of diffusion $D_{(v)}$; the other is determined by the redistribution of particles with respect to coordinates, and is proportional to the relevant coefficient of diffusion D.

This equation of entropy balance is general (for the class of distributions (14.1.1)), and holds for both laminar and turbulent flows. Of course, in case of laminar flows it can be much simplified, since the local Maxwell distribution can be used as the solution of the kinetic equation.

In this approximation the first term on the right-hand side of the expression for entropy production vanishes, and the second becomes

$$\sigma(r,t) = \frac{\kappa}{m} \left[D\rho \left(\frac{\text{grad}\,\rho}{\rho} \right)^2 + v\rho \frac{m}{\kappa T} \left(\frac{\partial u_i}{\partial r_j} \right)^2 + \frac{3}{2} \rho \chi \left(\frac{\text{grad}\,T}{T} \right)^2 \right] \geq 0 \ . \tag{22.16.4}$$

In case of incompressible liquid, the first and the third terms on the right-hand side vanish, and the production of entropy, as we already know, is defined by the derivatives of velocity.

At local equilibrium the entropy is given by

$$S(r,t) \equiv \rho s = \rho \left[\frac{3}{2} \frac{\kappa}{m} \left(\ln \frac{2\pi m \kappa T}{m} + 1 \right) - \frac{\kappa}{m} \ln \frac{\rho}{m} \right] , \tag{22.16.5}$$

and is constant for incompressible liquid. It turns out that the production of entropy is nonzero, while the entropy remains constant. This is the inconsistency we mentioned in Sect. 22.11.3.

Is there a way out?

In the description of laminar motion the condition of incompressibility must be at least partially renounced, as this is done in the theory of thermal convection in the Boussinesq approximation. The description is then based on the set of equations (14.6.10) and (14.6.11) for vortical velocity and temperature. From the second of these equations we see that the spatially inhomogeneous vortical field is a source of heat. Consequently, the nonzero production of entropy causes the entropy to change, and so the contradiction is removed.

Observe that when local equilibrium is violated (which, in accordance with the above discussion, implies the existence of turbulent motion), the equation of entropy balance is in agreement with the second law of thermodynamics even in the approximation of incompressible liquid.

Indeed, redefined is not only the production of entropy, which is given by the more general expression (22.16.3) instead of (22.16.4), but also the entropy itself. The entropy is now defined by the more general formula (22.16.1), which expresses the entropy in terms of the solution of the kinetic equation (22.12.1).

Observe that the weight of nonequilibrium "additions" to entropy production (22.16.4) increases dramatically as turbulence develops. The kinetic coefficients of viscosity, temperature conduction and self-diffusion also increase in sympathy. Reliable calculations can be made for the above two particular cases: well-developed spatially homogeneous and isotropic turbulence, and steady turbulent Poiseuille flow. In both cases the "turbulent additions" become predominant at large Reynolds numbers.

22.17. What Is Turbulent Motion? Final Remarks

Let us return to Sect. 22.2, which lists the main characteristic features of turbulent motion. The first item on the list states that turbulent motion is essentially nonequilibrium, and is characterized by a very large number of macroscopic degrees of freedom. Now we can elaborate on this statement.

We saw that the transition to equations of hydrodynamics, on which the description of laminar motions is based, corresponds in the kinetic theory to the approximation of local equilibrium. All the states, for which the distribution functions differ from the local Maxwell distribution, are turbulent. The extent of turbulence is thus defined by the number of moments which approximate the given nonequilibrium solution. For the state of full-fledged turbulence this approach is not efficient, and the problem of closure cannot be adequately solved.

The key concept in the transition from kinetic to hydrodynamic description was the local Maxwell distribution. This is justified if Knudsen number is small for the states close to equilibrium. The other extreme is also possible, which corresponds to the so-called free molecular flow. A typical example of the latter is the case of "collisionless plasma', when the free path length is much greater than the size of the system. This situation is encountered in installations designed for thermonuclear fusion (tokamaks), where turbulence is the predominant type of motion. The relevant theory is based on the use of kinetic equations (Tsytovich 1971; Kadomtsev 1988; Ichimaru 1973; Sitenko 1977; Balescu 1967; Klimontovich V; Nicholson 1983).

Boltzmann gas, simple liquids, completely ionized plasma are just a few examples of systems in which turbulent motion plays an important, and sometimes a crucial role. In recent years much attention is being paid to the study of turbulent motion in partially ionized plasmas and other chemically reacting media, in nonlinear optical distributed systems, in medico-biological systems — that is, in active media of all kinds. All in all, turbulence is one of the main types of motion in the statistical theory of open systems.

CHAPTER 23

BRIDGE FROM CLASSICAL STATISTICAL THEORY OF OPEN SYSTEMS TO QUANTUM THEORY

In the last chapter of this volume we are going to pave the way for the quantum statistics of open systems. So far we have been concerned with the one-component rarefied gas of structureless particles (the Boltzmann gas). In the next volume our expansion will proceed in two main directions, and we shall consider two more basic models.

The first is a rarefied electron-ion plasma, which is a three-component system comprised of gases of electrons and ions, and electromagnetic field.

The second is the system of atoms and field, which in the simplest case consists of two components. The gas component differs from Boltzmann gas in that the structure of atoms or molecules is taken into account. The second component is again the electromagnetic field.

A natural generalization of these two particular models is the so-called plasma-molecular system, which consists of at least four components (Klimontovich – Wilhelmsson – Yakimenko – Zagorodnii 1989). An example of such system is provided by partially ionized plasma. Foundations of the kinetic theory of systems of this kind have been laid in (Klimontovich 1967, 1968, II – IV; Klimontovich – Kremp – Kräft 1981, 1987).

A consistent description of plasma-molecular systems is only possible within the framework of quantum theory; hence the need of bridging the gap between the classical theory of open systems to the appropriate quantum theory. This is best done with the concrete examples of simple but real systems, such as the system of noninteracting hydrogen atoms or free electrons in electromagnetic field. We shall use these examples to illustrate the transition from the reversible microscopic operator equations to irreversible equations for deterministic quantum "distribution function" — the matrix of density and the moments of fluctuation electromagnetic field. Such transition may be interpreted as replacement of a system of particles and field oscillators by a continuous medium.

In particular, this implies that Schrödinger equation of quantum mechanics for conventional (deterministic, not operator) wave function also describes the evolution of continuous medium, but ignores the dissipative terms. In this sense, there is an analogy between Schrödinger equation in quantum mechanics and Euler equation in hydrodynamics (Klimontovich 1993e).

It would be natural to begin our study of quantum system of atoms and field with the microscopic equations, which give complete quantum electrodynamic description of atoms (or free electrons) and electromagnetic field.

Known are two formulations of the initial equations of quantum electrodynamics. The first is the Schrödinger equation for the wave function of all variables — for instance, of the atoms and all field oscillators. This can be used for constructing the equation for quantum distribution function in, say, Wigner representation. In the classical limit it will go over into the Liouville equation.

The second formulation is based on the set of reversible equations for operator wave function of electrons in atoms, and field operators. This is the method of secondary quantization. In the classical limit, the quantum operator equations correspond to equations in the microscopic phase density of atoms (or free electrons) and microscopic field strength. In the irreversible equations of statistical theory these functions are regarded as random.

Like in case of Boltzmann gas, we again are faced with important questions. What are the smallest scales on which the initial equations of quantum electrodynamics lose their reversibility? What is the cause of irreversibility?

Similar questions arise at different levels of description. We might ask, for instance, what are the smallest scales which allow going over from operator equations of quantum electrodynamics to the corresponding Schrödinger equation for the deterministic distribution function of, let us say, the atom of hydrogen.

The fact is that quantum mechanical Schrödinger equation is more coarse than the corresponding equations of quantum electrodynamics, since it only involves the mean field rather than the coordinates of individual field oscillators.

The problem consists therefore in finding the starting point of transition towards irreversible equations. This is mainly what the present chapter is about. In this connection we should mention stimulating influence of the work of Ilya Prigogine (Prigogine 1980; Prigogine – Stengers 1984), who for many years has been studying the possible generalization of the second law of thermodynamics to the microscopic level. The main role is assigned to the dynamic instability of motion of microscopic objects. Dynamic instability in quantum theory is manifested in significant changes of wave functions when the relevant initial conditions are varied even slightly.

Recall that in case of Boltzmann gas the procedure of smoothing over physically infinitesimal volume (the "point" of continuous medium), which is necessary for going over from the system of particles to the approximation of continuous medium, reflects the existence of dynamic instability of motion. To put it differently, the very feasibility of smoothing over physically infinitesimal scales in the construction of irreversible equations of statistical theory is based on the existence of dynamic instability of motion of microscopic elements of the system. In this respect the dynamic instability of motion plays a constructive role in the statistical theory, as we have repeatedly emphasized.

Since the theory of dynamic instability in quantum theory is still in its early stage, it will be expedient to start with the evaluation of the smallest scales on which smoothing is possible. This is necessary also for defining the limits of applicability of the initial microscopic equations of quantum electrodynamics.

This will provide us with a bridgehead for the second step, at which the conditions are clarified which allow going over from operator equations of quantum electrodynamics to equations in the relevant deterministic wave functions. This problem is closely associated with the structure of quantum kinetic equations. Examples of such equations will be analyzed in the second volume.

The question of completeness of description of physical phenomena on the basis of quantum mechanics was the subject of the famous debate between Einstein and Bohr at the 5th Solvay Conference in 1927 (see Mehra 1975). Later this problem was discussed in the paper entitled *Can quantum-mechanical description of physical reality be considered complete?* (Einstein – Podolski – Rosen 1935), and also in the papers by de Broglie (1953) and Bohm (1952). The issue of "deficiency of quantum-mechanical description" is closely related to the problem of the so-called *hidden parameters* in quantum mechanics.

Most physicists argued that quantum mechanical description is complete, and the problem of hidden parameters does not exist. This view was based for the most part on the book by John Neumann (1932), in which he proved that hidden parameters are incompatible with the foundations of quantum theory. It was not mentioned, however, that Schrödinger equation of quantum mechanics is itself approximate.

Further theoretical and experimental studies were stimulated by the results of Bell (1965, 1966), who formulated the condition of existence of hidden parameters as Bell's inequality. This was seen as a new possibility of experimental verification: if quantum mechanical description is complete, then Bell's inequality does not hold.

In a recent review (Belinskii – Klyshko 1993) we read:
"The problems formulated many years ago by Einstein, Podolski and Rosen, by Bohm and Bell, still excite the new generation of physicists. To a large extent this is due to the fact that the contradiction between the predictions of quantum theory and the theory of hidden parameters can be settled in a convincing way (in favor of quantum theory, of course) by *experimentum crucis*, unlike most other quantum paradoxes. The theory of hidden parameters is closely related to the ensemble-statistical interpretation of quantum theory, and this is the reason why such experiments (real or imaginary) add serious evidence to the eternal debate between the advocates of statistical (Einsteinian) and orthodox (Bohr's, or Copenhagen) interpretations, and their numerous modifications."

And, in the next paragraph:
"It is yet possible that in the future this debate will be settled (perhaps in favor of a third way), and historians of science will see it as a vivid example of fallacies which plagued even the brightest minds of the past."

So, the authors do not rule out the possibility of a third way! Let us take advantage of this option. Our treatment will be largely based on the ideas and methods described in detail in (Klimontovich 1992a,c, 1993b,c). In the preceding chapters we have already used some of these results, and this will make our task somewhat easier.

23.1. Microscopic and Macroscopic Schrödinger Equations

Imagine a quantum system which consists of an ideal gas of "particles" and fluctuation electromagnetic field. The role of particles is played by electrons which are either free or bound within hydrogen atoms. On the level of statistical description the electrons may be regarded as microscopic Brownian particles. The electromagnetic field will then act as the thermostat.

Microscopic description of quantum processes in such system starts with the reversible dynamic equations for particles and field. There are two possible approaches. One is based on the Schrödinger equation for the wave function of complete set of variables of particles and field. The other relies on on the equation for quantum (operator) wave functions of electron and electromagnetic field. In both cases the initial dynamic equations give, in principle, an exhaustive quantum mechanical description of time evolution of the system under consideration.

For the sake of simplicity, we shall describe interaction with the electromagnetic field in the dipole approximation. By $\hat{\Psi}(r,R,t)$ we denote the operator wave function of electron in hydrogen atom, and by $\hat{E}(R,t)$ the operator of strength of quantum electric field. In the dipole approximation the equation for the operator wave function may be written as

$$i\hbar\frac{\partial\hat{\Psi}}{\partial t} = -\frac{\hbar^2}{2m}\frac{\partial^2\hat{\Psi}}{\partial r^2} + U(r)\hat{\Psi} - d\hat{E}(R,t)\hat{\Psi} \ , \quad U = -\frac{e^2}{r} \ , \tag{23.1.1}$$

where $d = er$ is the dipole moment of electron. In the dipole approximation the vector R enters as a parameter.

We shall also use the appropriate equation for the "quantum distribution function", the operator density matrix in Wigner's form,

$$\hat{f}(r,p,R,t) = \frac{1}{(2\pi)^3}\int\hat{\Psi}\left(r+\frac{1}{2}\hbar\tau,R,t\right)\hat{\Psi}^*\left(r-\frac{1}{2}\hbar\tau,R,t\right)\exp(-i\tau p)d\tau \ , \tag{23.1.2}$$

$$\int\hat{f}(r,p,R,t)drdpdR = \int\left|\hat{\Psi}(r,R,t)\right|^2 drdR = 1 \ .$$

This equation follows from the Schrödinger operator equation and has the following form:

$$\frac{\partial \hat{f}}{\partial t} + v\frac{\partial \hat{f}}{\partial r} + \frac{i}{\hbar(2\pi)^3}\int \left[U\left(r + \frac{1}{2}\hbar\tau\right) - U\left(r - \frac{1}{2}\hbar\tau\right)\right]\hat{f}(r,p',R,t)$$

$$\times \exp\{i\tau(p'-p)\}d\tau dp + e\hat{E}(R,t)\frac{\partial \hat{f}}{\partial p} = 0 \ . \tag{23.1.3}$$

We also define the operator of polarization vector as

$$\hat{P}(R,t) = e\int r\hat{f}(r,p,R,t)drdp = e\int r\left|\hat{\Psi}(r,R,t)\right|^2 dr \ . \tag{23.1.4}$$

Then the equation for field operator can be written in the form

$$\frac{\partial^2 \hat{E}(R,t)}{\partial t^2} - c^2\frac{\partial^2 \hat{E}}{\partial R^2} = -4\pi\frac{\partial^2 \hat{P}^{\perp}}{\partial t^2} \ , \quad \mathrm{div}\,\hat{E}(R,t) = 0 \ . \tag{23.1.5}$$

Superscript $\perp$ denotes the curl component of the polarization vector.

So, we have obtained a closed set of equations for operator wave function and field operator in the dipole approximation. These equations can be used for various quantum electrodynamic calculations in the dipole approximation.

However, owing to the nonlinearity of operator equations, averaging over the Gibbs ensemble results in a very sophisticated system of meshing equations in the moments of different order. Accordingly, we once again have to deal with the problem of closure, so as to obtain a closed system of approximate equations.

There is an analogy with the two approaches of the classical theory. Schrödinger equation for the wave function of the system of particles and field corresponds to the Liouville equation for the distribution function of variables of particles and field. Operator equations of quantum electrodynamics correspond to the classical system of equations for the microscopic phase density and microscopic field $E^m(R,t)$. For the system of noninteracting particles, in place of phase density (2.4.1) one may use, like it is done in the theory of Brownian motion, the relevant dynamic (microscopic) distribution function $f^m(r,p,R,t)$. The equation for this function follows in the classical limit from (23.1.3), and is

$$\frac{\partial f^m}{\partial t} + v\frac{\partial f^m}{\partial r} - \frac{\partial U}{\partial r}\frac{\partial f^m}{\partial p} + E^m(R,t)\frac{\partial f^m}{\partial R} = 0 \ , \tag{23.1.6}$$

whereas the equation for the microscopic field formally coincides with (23.1.5). As a result, we have a closed system of classical dynamic equations.

In quantum mechanics, equation (23.1.1) is replaced by Schrödinger equation for deterministic wave function of the electron:

$$i\hbar\frac{\partial\Psi}{\partial t} = -\frac{\hbar^2}{2m}\frac{\partial^2\Psi}{\partial r^2} + U(r)\Psi - dE(R,t)\Psi \ , \ \int|\Psi(r,t)|^2\,dr = 1 \ . \tag{23.1.7}$$

As compared with Schrödinger equation for the wave function of the system of particles and field, this equation is approximate but still reversible. From (23.1.7) we may go over to the equation for deterministic "quantum distribution function" in Wigner's form,

$$f(r,p,R,t) = \frac{1}{(2\pi)^3}\int\Psi\left(r+\frac{1}{2}\hbar\tau,R,t\right)\Psi^*\left(r-\frac{1}{2}\hbar\tau,R,t\right)\exp(-i\tau p)d\tau \ ,$$

$$\tag{23.1.8}$$

$$\int f(r,p,R,t)drdpdR = \int|\Psi(r,R,t)|^2\,drdR = 1 \ .$$

Let us also write the corresponding reversible kinetic equation:

$$\frac{\partial f}{\partial t} + v\frac{\partial f}{\partial r} + \frac{i}{\hbar(2\pi)^3}\int\left[U\left(r+\frac{1}{2}\hbar\tau\right) - U\left(r-\frac{1}{2}\hbar\tau\right)\right]f(r,p',R,t)$$

$$\times \ \exp\{i\tau(p'-p)\}d\tau dp + eE(R,t)\frac{\partial f}{\partial p} = 0 \ , \tag{23.1.9}$$

which in the classical limit becomes

$$\frac{\partial f}{\partial t} + v\frac{\partial f}{\partial r} - \frac{\partial U}{\partial r}\frac{\partial f}{\partial p} + E(R,t)\frac{\partial f}{\partial R} = 0 \ . \tag{23.1.10}$$

We see that Schrödinger equation (23.1.7) for deterministic wave function corresponds in the classical limit to the kinetic equation for the one-particle distribution function $f(r,p,t)$. Like Schrödinger equation, this equation is reversible since it does not take into account the dissipation due to the interaction of electron with fluctuation electromagnetic field.

Now, to facilitate the transition to the irreversible equations, let us refresh some points from Boltzmann's kinetic theory.

The kinetic Boltzmann equation differs from (23.1.10) in that it includes the "collision integral" which takes care of the dissipation due to redistribution of the particles' velocities because of collisions between the particles. As before, by τ and l we denote the relaxation parameters — the free path time and length. We also introduce the characteristic parameters of the problem T and L.

As we know from Sect. 6.6, important are two extreme cases, corresponding to the approximations of gas dynamics and of free molecular flow. The gasdynamic approximation is used when $\tau \ll T$, $l \ll L$. Then the kinetic Boltzmann equation may be replaced by the simpler equations in the gasdynamic functions $\rho(R,t)$, $u(R,t)$, $T(R,t)$.

In the opposite extreme, when $\tau \gg T$, $l \gg L$, the dissipative term (the "collision integral") in the zero approximation may be dropped. This brings us to the reversible kinetic equation which formally coincides with (23.1.10).

Is it possible to exploit this analogy?

Recall first of all that the nature of description is changed dramatically when we go over from the Liouville equation, which carries all information about the motion of particles of Boltzmann gas, to the kinetic Boltzmann equation. To wit, from the system of particles whose motion is described by the reversible Hamilton equations, we go to a continuous medium in six-dimensional space of coordinates and momenta (or coordinates and velocities). Naturally, the transition to the approximation of continuous medium is associated with restrictions from the side of small scales which depend on the size of the "point". Since the information about the motion of particles within the "point" is lost, the equations of continuous medium must be dissipative and therefore irreversible.

Accordingly, the transition to the approximation of continuous medium changes the entire time symmetry of the original system. In this situation the initial reversible equations of Hamiltonian mechanics for systems with a large number of microscopic degrees of freedom (Boltzmann gas, or atoms and field) can only be regarded as the "starting point".

Of course, this transition should not be viewed as just a simplification. Rather than that, it is necessitated by dynamic instability of motion of microscopic elements of the system. It is the dynamic instability of motion, combined with uncontrollable small exertions from the surrounding world, that makes the transition to irreversible equations inevitable.

In this connection it is worthwhile to recall that dissipation is usually seen as damping of motion, scattering of energy, loss of information. At nonequilibrium phase transitions, however, which may result in the appearance of new structures (whose sequences form the processes of self-organization), the dissipation plays a constructive role. Persistent space-time dissipative structures are not feasible without dissipation.

This explains why the construction of dissipative equations is of crucial importance for the statistical theory of open systems. The first step in this direction consists in the revision of the concept of continuous medium. As we know, this requires giving a concrete definition of physically infinitesimal scales.

We know how to do this for Boltzmann gas. We have also defined physically infinitesimal scales for a plasma (Klimontovich II – IV, 1992a,c, 1993b,c). On this basis we defined Gibbs ensemble for nonequilibrium processes, and gave examples of kinetic equations for unified description of evolution on both kinetic and hydrodynamic scales.

How then do we apply this knowledge to the quantum system of atoms and electromagnetic field.

Let us return to the classical kinetic equation (23.1.10). This equation has many applications. For Boltzmann gas, for example, it can be used for describing free molecular flows, when the characteristic scales of the problem T and L are much less than the corresponding relaxation kinetic scales of rarefied gas. The approximation of continuous medium still holds, however, since equation (23.1.10) does not carry information about the motion of individual particles (atoms of the gas). This restricts the admissible values of T and L from below: they must be much greater than the relevant physically infinitesimal scales,

$$T \gg \tau_{\mathrm{ph}} \,, \quad L \gg L_{\mathrm{ph}} \,. \tag{23.1.11}$$

And yet, equation (23.1.10) is reversible. How does this comply with the above statement that the processes in continuous medium must be described by irreversible equations?

The answer is that equation (23.1.10) holds for Boltzmann gas only in the zero approximation with respect to a small parameter — for instance, to the ratio of the characteristic scale of the problem to the free path length. In the first approximation with respect to L/l the kinetic equation becomes irreversible. Moreover, the solution of (23.1.10) requires supplementing it with a boundary condition which will bring in irreversibility for any real system.

What is the situation in the quantum theory/

To take care of the dissipation, we must include into consideration the interaction of atom with fluctuation electromagnetic field. This will result in dissipative kinetic equations for the quantum distribution function (or density matrix) and the field. The appropriate "collision integrals" are determined by the small-scale fluctuations whose characteristic scales are much smaller than the characteristic scales of the kinetic equations. In this way, the problem of the structure of "continuous medium" is brought up explicitly (or implicitly, in case of formal application of perturbation theory) in the derivation of kinetic equations.

This problem is closely associated with the definition of quantum ensemble. The initial microscopic description corresponds to "pure ensemble", when the entire statistical information is incorporated in the "exact" wave function of the system.

In the pure ensemble, the operator density matrix (for example, in Wigner's representation; see (23.1.2)) is expressed via the product of operator wave functions $\hat{\Psi}(r,R,t)$. Upon transition to the continuous medium, when the kinetic equations become irreversible, the pure ensemble is replaced by "mixed" ensemble. Then there is no representation in which the density matrix can be expressed via the product of wave functions.

It might seem that the above formulas defy this statement. Indeed, the quantum distribution function (23.1.8) for hydrogen atom is defined as the product of wave functions which satisfy Schrödinger equation (23.1.7). It is as if we returned to the "pure ensemble" again.

Naturally, there is a fundamental difference between these two definitions of pure ensemble. The first exactly renders the statistical properties of quantum mechanical description, whereas the second definition of pure ensemble corresponds to the "coarse" approximation of continuous medium. The approximation amounts to neglecting the dissipation altogether. For the system of N particles this approximation corresponds to Hartree equation in quantum mechanics, or to self-consistent Vlasov approximation in plasma theory, or to Euler approximation in hydrodynamics which disregards the dissipation due to viscous friction and heat conduction.

In order to define the structure of "continuous medium" in quantum theory, we must first of all introduce the characteristic scales of the system in question. The characteristic length and time (frequency) for hydrogen atom are given by

$$r_0 = \frac{\hbar^2}{me^2}, \quad \omega_0 = \frac{|E_0|}{\hbar} = \frac{me^4}{2\hbar^3}, \quad v_0 \sim \frac{e^2}{\hbar}. \tag{23.1.12}$$

The first of these defines the characteristic length scale of distribution $|\Psi(r)|^2$ in the ground state (Bohr's radius); frequency ω_0 determines the energy of the ground state. Additional parameters arise when we go over to the dissipative kinetic equations; we divide them into two classes.

The first class includes those parameters which characterize the process of relaxation towards equilibrium. They are similar to the relaxation scales τ, l of Boltzmann gas. In quantum theory, when the relaxation towards equilibrium proceeds from one energy level to another, we may introduce the time of relaxation τ_{nm} for each transition. This time is defined by Einstein's coefficient for spontaneous transitions,

$$\tau_{nm} = \frac{1}{A_m^n}, \quad A_m^n = \frac{4|d_{nm}|^2 \omega_{nm}^3}{3\hbar c^3} \sim \gamma(\omega_{nm}) = \frac{2e^2 \omega_{nm}^2}{3mc^3}. \tag{23.1.13}$$

Einstein's coefficient A_m^n is proportional to the coefficient of radiation friction $\gamma(\omega_{nm})$ at the transition frequency ω_{nm}. The following estimate then holds good:

$$A_m^n \sim \mu^3 \omega_{nm} << \omega_{nm}, \quad \text{where} \quad \mu = \frac{e^2}{\hbar c} \tag{23.1.14}$$

is the "constant of fine structure".

We see that the relaxation for the system of hydrogen atoms, governed by interaction with the fluctuation electromagnetic field, is a slow process compared to the characteristic time interval $1/\omega_{nm}$. At equilibrium, the population of levels is described by Boltzmann's formula. At zero temperature the atoms are in the ground state, and the density of distribution of electron's positions in the atom is isotropic. The most probable distance from the nucleus is characterized by Bohr's radius r_0.

The displacement of spectral lines due to the interaction between hydrogen atom and the fluctuation electromagnetic field is also determined by the same kinetic equations — such as, for instance, Lamb's shift of levels. By order of magnitude, the shift of energy levels is $\Delta E_n \sim \mu^3 |E_0|/n^3$, where n is the principal quantum number. Accordingly, for the lower levels this effect is of the same order of smallness as the broadening of spectral lines because of spontaneous transitions.

Now let us try to answer the following question: does the distribution of electron's positions in the ground state display a finer structure than this would follow from Schrödinger equation (23.1.7)? What are the smallest times which the isotropic distribution takes to become established in the ground state?

It is obvious that the structure of "continuous medium" is defined by the scales smaller than those of (23.1.12). In classical and quantum electrodynamics there are two scales which are smaller than those of the theory of hydrogen atom:

$$r_e = \frac{e^2}{mc^2} \ , \quad \lambda_C = \frac{\hbar}{mc} \sim \mu r_0 \ll r_0 \ , \quad \text{where} \quad \mu = \frac{e^2}{\hbar c} \ . \tag{23.1.15}$$

The first is the "classical electron radius". In quantum electrodynamics, the main contribution to the effective cross section of scattering of photons by free electrons is proportional to r_e^2 (see Sect. 23.2). The second parameter is Compton's length, which defines the shift of wavelength in case of scattering of x-rays by free electrons. Which of these characterizes the commencement of irreversibility, and thus defines the finest structure of "continuous medium" in quantum mechanics?

Observe that the study of the structure of "continuous medium" in quantum mechanics is stimulated, in particular, by those difficulties which are associated with the attempts to calculate quantum fluctuations, like fluctuations of velocity of free electron moving in a fluctuation electromagnetic field. The solution of this problem will allow us to define the smallest relaxation times over which the irreversibility sets in. This will also help us to define the new scales which are pertinent to the structure of "continuous medium" of atoms and field.

23.2. Electron in Equilibrium Electromagnetic Field

We use the equation of motion of electron with radiation friction, external field, and Langevin force. The latter, as usual in the theory of Brownian motion, reflects the structure of the medium. In the dipole approximation the forces are assumed to depend only on the time. Then

$$\frac{dv}{dt} - \Gamma \frac{d^2v}{dt^2} = \frac{1}{m}F(t) + \frac{1}{m}F_L(t) , \text{ where } \frac{1}{\Gamma} = \frac{2e^2}{3mc^3} \sim \frac{r_e}{c} \qquad (23.2.1)$$

is the time parameter which by order of magnitude is the time taken by photon to travel over the distance equal to the classical electron radius r_e. Traditionally, the dissipative term is defined by the radiation friction and is proportional to the third derivative. This leads to certain grave complications, and even to a contradiction with the second law of thermodynamics, as we shall see in the next section.

On the basis of equation (23.2.1) we are going to treat two problems: (1) scattering disregarding fluctuations, and (2) Brownian motion of free electron.

Recall that equation (23.2.1) only holds in the dipole approximation, when the wavelength λ is much greater than the electron radius:

$$\frac{\lambda}{2\pi} = \frac{c}{\omega} \gg r_e \sim 10^{-13} \text{ cm} , \text{ or } \omega \ll \frac{c}{r_e} \sim \Gamma . \qquad (23.2.2)$$

Observe that these conditions do not involve Planck's constant: quantum effects become manifest on larger space-time scales. For Compton's effect, for instance, such scales are the Compton's length and the corresponding frequency:

$$\lambda_C = \frac{\hbar}{mc} \sim \frac{1}{\mu} r_e \gg r_e , \quad \omega_C = \frac{mc^2}{\hbar} . \qquad (23.2.3)$$

Assume now that the Langevin force is absent, and the external force is harmonic, $F(t) = eE\cos(\omega t)$. Then the total effective cross section is given by the known formula

$$\sigma = \sigma_T \frac{1}{1 + \dfrac{\omega^2}{\Gamma^2}} \equiv \sigma_T \frac{1}{1 + \dfrac{2\pi\sigma_T}{3\lambda^2}} , \quad \sigma_T = \frac{8\pi}{3} r_e^2 . \qquad (23.2.4)$$

Under conditions (23.2.2) this expression reduces to Thomson's formula for σ_T.

The effective cross section is defined by the classical electron radius, which is the smallest of the scales (23.1.15) of quantum electrodynamics and does not depend on Planck's constant. Thomson's cross section delimits the region where the quantum theory ("wave mechanics") does not work. Indeed, the uncertainty relation does not allow for the existence of de Broglie waves with $\lambda_B \sim r_e$.

It follows that the scales smaller than or about equal to r_e must be excluded from quantum-theoretical consideration. This gives us grounds for accepting the classical electron radius for the minimum point size when using Schrödinger equation for deterministic wave function — or, in other words, when going over from the system of particles to the approximation of "continuous medium". Accordingly, the smallest in the hierarchy of physically infinitesimal scales may be defined as

$$l_{ph} \sim r_e \sim \mu \lambda_C \ll \lambda_C \ , \quad \tau_{ph} \sim \frac{r_e}{c} \sim \frac{1}{\Gamma} \ , \quad V_{ph} \sim r_e^3 \ . \tag{23.2.5}$$

We see that they are μ times as small as the relevant minimum scales in quantum theory for the system in question. These classical scales must be eliminated from the initial microscopic dynamic equations. It is the corresponding smoothing procedure that first brings irreversibility into the quantum theory.

Since the structure of effective cross section of photon scattering by electrons is so important, it will be worthwhile to consider another method of deriving the expression for the effective cross section. For this we take advantage of the linkage between the effective cross section and the imaginary part of the refraction index k, or the coefficient of extinction h. The values of k and h are linked by

$$h = 2k\frac{\omega}{c} = \frac{4\pi}{\lambda_0}k \ , \quad \lambda_0 = \frac{2\pi c}{\omega} \ . \tag{23.2.6}$$

We also use the linkage between h and the effective cross section,

$$h = n_{eff}\sigma \ , \quad \text{where} \quad n_{eff} = \frac{N}{V_{eff}} \tag{23.2.7}$$

is the number of particles in unit volume. In case of scattering by one electron, $N = 1$. In experimental investigations of photon scattering by electrons of, for example, rarefied plasma, N is the number of electrons in the effective scattering volume V_{eff}.

We can express the coefficient of extinction and the effective cross section from dielectric permittivity of the medium. Let us begin with assessing the relevant time scales.

We turn to formula (23.1.13) for the coefficient of radiation friction γ, and consider its value at the maximum frequency ω_{max}, as defined by the proper energy of the electron

$(\hbar \omega_{max} = mc^2)$:

$$\gamma(\omega_{max}) \sim \frac{e^2}{mc^3}\left(\frac{mc^2}{\hbar}\right)^2 \sim \frac{mce^2}{\hbar^2} = \frac{c}{r_0} \,, \quad \omega_{max} = \frac{mc^2}{\hbar} \,, \tag{23.2.8}$$

where r_0 is Bohr's radius, which this time is interpreted as the length of light train emitted by electron within the minimum time compatible with the uncertainty relation. The relevant velocity is $v_0 \sim \omega_0 r_0$.

Useful is also another interpretation of relations (23.2.8).

Let us move one step further in the hierarchy of relaxation times, and introduce a time interval which is a combination of τ_e and τ_C,

$$\tau_B = \frac{\tau_C^2}{\tau_e} \equiv \frac{\Gamma}{\omega_{max}^2} \sim \frac{r_0}{c} \,, \quad \tau_B = \frac{1}{\gamma(\omega_{max})} \,, \tag{23.2.9}$$

and which will be referred to as Bohr's relaxation time, since it is the time taken by photon to travel to the distance equal to Bohr's radius.

The minimum time corresponding to the problem of photon scattering by free electrons (Thomson's problem) is of the order of the time over which a photon travels to the distance equal to the classical electron radius,

$$\tau_e = \frac{1}{\Gamma} \sim \frac{r_e}{c} \sim \mu \frac{r_0}{c} \,. \tag{23.2.10}$$

Like the classical electron radius r_e, this time interval does not involve Planck's constant, and falls beyond the domain of wave mechanics. The times as small as that do not comply with the Heisenberg time-energy uncertainty relation.

Return now to the problem of defining the effective cross section in terms of the "dielectric properties" of the system of electrons and field. Using the equation of motion (23.2.1) for free electron we arrive at the following expression for the permittivity:

$$\varepsilon = 1 - \frac{4\pi e^2}{mV_{eff}} \frac{1}{\omega^2 + i\omega\gamma(\omega)} \,. \tag{23.2.11}$$

Now we write the appropriate equations for the real (n) and the imaginary (k) parts of the refraction index:

$$n^2 - k^2 = 1 - \frac{4\pi e^2 N}{mV_{\text{eff}}} \frac{1}{\omega^2 + \gamma^2(\omega)} \ , \quad nk = \frac{2\pi e^2 N}{mV_{\text{eff}}} \frac{\gamma(\omega)}{\omega(\omega^2 + \gamma^2(\omega))} \ , \qquad (23.2.12)$$

and use (23.2.6) for k and h. Since the refraction index is close to one in the pertinent range of frequencies for all realistic values of V_{eff}, and the coefficient of damping k is small, the coefficient of extinction may be written as

$$h = \frac{4\pi e^2 N}{mV_{\text{eff}}} \frac{\gamma(\omega)}{\omega^2 + \gamma^2(\omega)} \frac{1}{c} \ . \qquad (23.2.13)$$

Finally, using (23.2.7) and the definition of Γ (23.2.1), we find the desired expression for the effective cross section σ:

$$\sigma = \frac{4\pi e^2}{mc} \frac{\gamma(\omega)}{\omega^2 + \gamma^2(\omega)} \equiv \frac{4\pi e^2}{m} \frac{\Gamma}{\omega^2 + \Gamma^2} \equiv \sigma_T \frac{1}{1 + \dfrac{\omega^2}{\Gamma^2}} \ , \qquad (23.2.14)$$

which coincides with the expression obtained for the scattering of photons.

The first equation in (23.2.14) implies that the effective cross section in the dipole approximation is proportional to the coefficient of radiation friction. When the dissipation is completely disregarded, the effective cross section goes to zero.

So, we have calculated the effective cross section of scattering by free electrons, and the results seem to be quite reasonable. Fundamental difficulties arise, however, when the same equation (23.2.1) is used for calculating the fluctuation characteristics: the condition of statistical equilibrium in the system of electrons and field is violated.

23.3. Equilibrium State of System of Electrons and Field. Fluctuation-Dissipation Relation

Return to the equation of motion of electron (23.2.1) with radiation friction, assuming now that the external field is absent, and the fluctuations of the field are taken into account by introducing the Langevin force. Let us write the appropriate equation for the Fourier components:

$$(-i\omega + \gamma(\omega))v_\omega = \frac{f(\omega)}{m} = \frac{e}{m}E_\omega \ . \qquad (23.3.1)$$

The spectral density of fluctuation field is given by Planck's formula

$$(EE)_\omega = 4\pi^2 \rho_\omega = \frac{4\pi\omega^2}{c^3}\kappa T_\omega , \quad \kappa T_\omega = \frac{1}{2}\hbar\omega \coth\frac{\hbar\omega}{2\kappa T} . \tag{23.3.2}$$

Hence, given the definition of the coefficient of radiation friction, we obtain the following expression for the spectral density of Langevin force:

$$(ff)_\omega = e^2(EE)_\omega = 3\cdot 2m\gamma(\omega)\kappa T_\omega , \quad \gamma(\omega) = \frac{3e^2\omega^2}{3mc^3} . \tag{23.3.3}$$

Then the frequency distribution of the mean kinetic energy of electron is

$$\frac{m(v^2)_\omega}{2} = \frac{\gamma(\omega)}{\omega^2 + \gamma^2(\omega)}3\kappa T_\omega . \tag{23.3.4}$$

For the system of electrons and field we have obtained the quantum Nyquist's formula; the dissipation is characterized by the coefficient of radiation friction.

However, relations (23.3.3), (23.3.4) lead to a contradiction with the second law of thermodynamics (Klimontovich 1987, V), since the condition of equilibrium of "Brownian particle" and field is violated. Indeed, at zero temperature, for instance, we get the following for the mean kinetic energy of electron at equilibrium:

$$\frac{m\langle v^2\rangle}{2} = \frac{1}{2\pi}\frac{3}{2}\int \frac{2\gamma(\omega)}{\omega^2 + \gamma^2(\omega)}\frac{1}{2}\hbar\omega d\omega = \frac{3}{4\pi}\hbar\Gamma \ln\left(1 + \frac{\omega_{max}^2}{\Gamma^2}\right) . \tag{23.3.5}$$

We see that at the state of equilibrium the mean kinetic energy is proportional to the dissipative factor Γ. This is a contradiction with the second law of thermodynamics, which states that the mean energy characteristics are functions of state and therefore cannot depend on dissipative parameters.

This calls for a revision of the structure of the coefficient of radiation friction, and the coefficient of friction in the field equation.

Traditionally, the coefficient of radiation friction is found from the lagging solution of the field equation in dipole approximation. The first nonvanishing term with odd time derivative of the polarization vector in the expansion with respect to lagging defines the dissipative term in the equation of motion of electron (23.2.1).

Observe that by this method the dissipation is found from solution of the reversible field equation. The mechanism of dissipation remains unknown. In particular, it is not clear what are the scales on which the irreversibility sets in. As a matter of fact, the above-mentioned fundamental difficulties result from the loose definition of dissipation in the field equations (and hence in the equation of motion of electron).

Let us try to introduce dissipation into the equation of motion of electron from a somewhat different standpoint. The main assumption will consist in the following.

Dissipation arises when small scales are eliminated — the scales within r_e and τ_e, which cannot be included in the equations of quantum mechanics. Recall that the characteristic non-quantum time interval τ_e is linked with the dissipative factor $1/\Gamma$ by (23.2.10). Smoothing, as will be shown in the next section, brings the dissipative term Γv into the equation of motion of electron in place of the traditional term with radiation friction. So, instead of (23.2.1), we get the following equation of motion:

$$\frac{dv}{dt} + \Gamma v = \frac{1}{m}F(t) + \frac{1}{m}F_L(t) \,, \quad v = \frac{dr}{dt} \,, \quad \Gamma = \frac{3mc^3}{2e^2} \sim \frac{c}{r_e} \,. \tag{23.3.6}$$

The effective cross section is now given by

$$\sigma = \frac{4\pi e^2}{mc} \frac{\Gamma}{\omega^2 + \Gamma^2} \equiv \sigma_T \frac{1}{1 + \frac{\omega^2}{\Gamma^2}} \,. \tag{23.3.7}$$

Note that this expression for σ coincides with (23.2.16) established on the basis of equation (23.2.1) with radiation friction. The advantages of the new approach become obvious in the calculation of fluctuations.

Let us quote the relevant results for the fluctuations of velocity.

Using the Langevin equation (23.3.6) instead of (23.3.4), we get the following expression for spectral density of fluctuations of velocity:

$$\frac{m\langle v^2 \rangle_\omega}{2} = \frac{\Gamma}{\omega^2 + \Gamma^2} 3\kappa T \,, \quad \frac{1}{2\pi}\int \frac{m\langle v^2 \rangle_\omega}{2} d\omega = \frac{3}{2}\kappa T \,. \tag{23.3.8}$$

Now there is no contradiction with the second law of thermodynamics, since these equations express the existence of equilibrium between the "Brownian particle" (electron) and the electromagnetic field. The mean kinetic energy is then determined only by the temperature, and does not depend on the dissipative parameters. This result agrees with the fact that the equilibrium solution of the corresponding quantum kinetic equation is the Maxwell distribution.

Observe that for a free electron the result (23.3.8) has the classical form. Because of the existence of the so-called relict radiation (background fluctuation electromagnetic radiation with the temperature $T \approx 2.7$ K), the intensity of the electron velocity fluctuations is always nonzero.

The above-introduced characteristic frequencies form a chain of inequalities

$$\Gamma \gg \omega_{max} \gg \gamma_{max} \sim \frac{c}{r} \gg \omega_0 \gg \gamma(\omega_0) \, , \tag{23.3.9}$$

and the relevant inequalities for the characteristic lengths are

$$r_e \ll \lambda_C \ll r_0 \ll \lambda \ll \frac{c}{\gamma(\omega_0)} \, . \tag{23.3.10}$$

These inequalities allow us to distinguish between the fast (small-scale, or fine-grained) and the slow (large-scale, or coarse-grained) fluctuations. The line of demarcation may be defined, for instance, by the time interval $1/\omega_{max}$ and the Compton length λ_C. Of course, the separation is to a large extent a matter of convention, and will depend on the particular problem under consideration. In any case, however, $V_C = \lambda_C^3$ is the minimum volume which may be regarded as the "point" of "continuous medium" in quantum theory. This volume is much larger than the "classical volume of electron" V_e. Their ratio defines the corresponding small "density parameter"

$$\varepsilon_e = \frac{V_e}{V_C} \sim \mu^3 \, . \tag{23.3.11}$$

In the parlance of the kinetic theory of gases and plasmas, the small-scale fluctuations may be referred to as "collisionless", since their characteristic scales are much less than the characteristic scales of "collision integrals".

So, using the example of calculation of the effective cross section of scattering of photons by free electrons, and the electron velocity fluctuations, we have demonstrated the efficiency of dissipation introduced by means of smoothing with respect to "non-quantum" small scales τ_e, r_e. Accordingly, these scales act as "hidden parameters" in quantum theory.

As indicated above, the feasibility and the expedience of hidden parameters for many years have been a matter of debate. Let us reproduce here some passages from de Broglie's paper mentioned in the beginning of the present chapter.

"Bohr — a Rembrandt of contemporary physics, who sometimes displays a kind of taste for chiaroscuro — said that a particle is a 'loosely defined entity within finite space-time limits'. [...]

"In classical theories, such as the kinetic theory of gases, the probability laws are regarded as the result of our ignorance of the perfectly deterministic though erratic and sophisticated movements of innumerable gas particles. Had we known the positions and velocities of each molecule, we could have made a precise calculation of evolution of the gas. In practice, however, we have to rely on probabilities because we do not know these hidden parameters.

"The purely probabilistic formulation of quantum mechanics, however, rejects such interpretation of probability laws: these laws are due not to the fact that we do not know the hidden parameters, which evidently are the coordinates and velocities of particles, but rather to the non-existence of hidden parameters, since a particle may only be detected with definite coordinates and definite velocity at the instant of observation or measurement. Thus, in quantum mechanics the probability is no longer the result of ignorance. [...]

"A couple of decades ago, Neumann has proved that the experimentally tested probability laws in the form accepted in wave mechanics are incompatible with the existence of hidden parameters. For a long time I regarded Neumann's theorem as irrefutable. Now I have some reservations. [...] His proof is mainly based on the assumption that all probability distributions allowed by wave mechanics are physically realized irrespective of whether the experiment which realizes one of such states is carried out or not."

Then de Broglie turns to analyzing Bohm's paper:

"Like Bohm, one can assume that while the probabilistic interpretation is good for predicting the events on the atomic scale (from 10^{-8} to 10^{-11} cm), it may become inadequate on the nuclear scale (10^{-13} cm). [...]

"Einstein and Schrödinger have flatly refused to accept the probabilistic interpretation of quantum mechanics. As far as I know, there has been no satisfactory answer to Einstein's objections. [...]

"Bohr's answers are characterized by that same 'chiaroscuro' of which I have spoken above.

"It seems that physics urgently needs the possibility to introduce the electron "radius", like in the old Lorentz theory. [...] One must avoid the danger of too much faith in the purely probabilistic interpretation of quantum mechanics eventually making this possibility futile. [...] Wave mechanics, the way it is currently taught, has to a large extent exhausted its ability to explain things. [...]

"The progress of science is continually harassed by the tyrannic influence of certain concepts which in the course of time have become dogmas. Because of this, the principles which have been recognized as final must be subjected to most thorough revision."

After this digression into history, let us return to formulas (23.2.14), (23.3.7) for the effective cross section of scattering of photons by free electrons. From these expressions it follows that the time dispersion is only observable in the ultrarelativistic domain, since the ratio

$$\frac{\omega^2}{\Gamma^2} \sim \left(\frac{\hbar\omega}{mc^2}\right)^2 \mu^2 \, , \ \mu = \frac{e^2}{\hbar c} \tag{23.3.12}$$

is very small.

We see that even in the relativistic region, when the ratio $\hbar\omega/mc^2$ is of the order of unity, the cross section is practically independent of the frequency and is determined by the Thomson cross section. Quantum electrodynamic calculation of the effective cross section leads to the known Klein – Nishina formula. In the linear approximation with respect to dimensionless parameter $\hbar\omega/mc^2$ we get

$$\sigma = \sigma_T\left(1 - 2\frac{\hbar\omega}{mc^2}\right). \tag{23.3.13}$$

The correction is proportional to the ratio of the Compton length to the wavelength, λ_C/λ. In the non-quantum approximation, when Planck's constant is zero, we return to the classical Thomson formula.

So, the "non-quantum" Thomson formula for the effective cross section holds not only for the optical, but also for the x-ray frequency range — that is, for the domain of quantum mechanics. This seemingly paradoxical result reveals certain limitations of quantum mechanics (or, more precisely, wave mechanics) on the scales of the order of the classical free electron radius $r_e \sim 10^{-13}$ cm, in harmony with de Broglie's statement quoted above.

Now, however, there is a possibility of giving a more concrete definition of "hidden parameters": they are those small scales which fall beyond the scope of applicability of quantum mechanical Schrödinger equation regarded as an equation of continuous medium neglecting the dissipation. Elimination of these small scales results in dissipative equations for the density matrix with appropriate "collision integrals".

We shall make just one step on the way towards construction of dissipative equations of quantum theory. Namely, we shall see how the structure of the initial microscopic equations of quantum theory is affected by smoothing with respect to small but non-quantum scales.

23.4. Transition from Reversible Dynamic Equations to Dissipative Equations of "Continuous Medium"

When the initial equations in operator functions $\hat{\Psi}$, $\hat{E}$ are used in quantum electrodynamics, the question of limitations from the side of small scales remains open. On the strength of arguments developed above, we may assume that these equations do not work on the scales about equal to or smaller than τ_e, r_e. To eliminate such scales, let us introduce the simplest relaxation term

$$-\frac{1}{\tau_e}\left(\hat{f}(r,p,R,t) - \tilde{\hat{f}}(r,p,R,t)\right), \quad \tau_e = \frac{1}{\Gamma} \tag{23.4.1}$$

into equation (23.1.3) for the quantum operator function. This term ensures adjustment of the dynamic distribution to the smoothed distribution over the smallest relaxation time τ_e.

Smoothing allows us to define the Gibbs ensemble for nonequilibrium processes in our system of particles and field. The incompleteness of description of individual systems within the Gibbs ensemble is due to the lack of information about the processes taking place inside the "point" with the volume $V_e \sim r_e^3$.

Now there is a natural question of the relationship between operations of smoothing and averaging with respect to the Gibbs ensemble.

Here we are only considering the initial phase of transition towards irreversible equations, associated with elimination of small non-quantum scales. It would be natural to assume that at this stage the procedure of averaging with respect to the Gibbs ensemble is practically equivalent to averaging over the volume V_e, which is regarded as the physically infinitesimal volume:

$$\left\langle \hat{f}(r,p,R,t) \right\rangle = \tilde{\hat{f}}(r,p,R,t) . \tag{23.4.2}$$

After the relaxation of the initial dynamic distribution to the smoothed distribution, the dynamic equation (23.1.3) retains its form, but holds now for the dynamic distribution smoothed with respect to the volume of the "point".

So we see that the form of operator equations of quantum electrodynamics remains the same. Now it is clear, however, that they do not carry information about the behavior of the system on the scales smaller than τ_e, r_e — that is, inside the "point" whose volume is V_e. Accordingly, from the standpoint of quantum electrodynamics the motion within these scales is "hidden".

Is it yet possible to gain any additional knowledge on top of the information provided by equations of quantum electrodynamics for smoothed operator functions? The answer is yes, since we can calculate the deviation of the initial dynamic operator distribution from the smoothed distribution,

$$\delta f(r,p,R,t) = \hat{f}(r,p,R,t) - \tilde{\hat{f}}(r,p,R,t) , \quad \delta E(R,t) = \hat{E}(R,t) - \tilde{\hat{E}}(R,t) . \tag{23.4.3}$$

An example is the expression (23.3.8) for the spectrum of velocity fluctuations of electron, which describes spectral distribution of fluctuations with the correlation time of the order of τ_e.

We have used just one option from the hierarchy of scales available for smoothing in the dynamic equations. Now we can move one step further, and carry out smoothing with respect to the volume defined by Compton length, the smallest of the quantum scales. In this fashion we pass on to a coarser definition of "continuous medium" with the volume of the "point" of the order of $V_C \sim \lambda_C^3$. The Gibbs ensemble is also redefined in a matching way.

By angular brackets $\langle ... \rangle$ we denote averaging over the ensemble. Then, under condition of local ergodicity, the "deterministic" quantum distribution function is

$$f(r,p,R,t) = \langle \hat{f}(r,p,R,t) \rangle = \tilde{\tilde{f}}(r,p,R,t) . \tag{23.4.4}$$

Into the equation for operator distribution function $\hat{f}$ we introduce a relaxation term, which now has the form (dropping the tilde "~")

$$-\frac{1}{\tau_C} \left(\hat{f}(r,p,R,t) - f(r,p,R,t) \right) . \tag{23.4.5}$$

In order to obtain the equation in function f, we carry out averaging of the equation for operator distribution function $\hat{f}$ over the Gibbs ensemble after introducing the relaxation term. As a result, we come to the equation for the deterministic distribution:

$$\frac{\partial f}{\partial t} + v \frac{\partial f}{\partial r} + \frac{i}{\hbar(2\pi)^3} \int \left[U\left(r + \frac{1}{2}\hbar\tau\right) - U\left(r - \frac{1}{2}\hbar\tau\right) \right]$$

$$\times f(r,p',R,t) \exp\{i\tau(p'-p)\} d\tau dp + eE(R,t) \frac{\partial f}{\partial p} = I(r,p,R,t) . \tag{23.4.6}$$

The "collision integral" here is determined by the correlator of fluctuations of the field and the distribution function,

$$I(r,p,R,t) = -e\frac{\partial}{\partial p} \langle \delta E(R,t) \delta f(r,p,R,t) \rangle , \tag{23.4.7}$$

where the fluctuations are defined by

$$\delta f(r,p,R,t) = \hat{f}(r,p,R,t) - f(r,p,R,t) ,$$

$$\delta E(R,t) = \hat{E}(R,t) - E(R,t) . \tag{23.4.8}$$

To obtain a closed kinetic equation we must calculate the fluctuations and show that the collision integral can be expressed in terms of the distribution function f. Since the equation for fluctuations is, in general, nonlinear, we get an infinite meshing chain of equations in moments, and are once again facing the problem of closure of the infinite chain of equations in the moments of fluctuations δf and δE. To find the condition of closure, we take advantage of the following.

The next one after λ_C in the hierarchy of quantum scales is "Bohr's radius" r_0. Accordingly, we have two characteristic volumes,

$$V_C \sim \lambda_C^3, \quad V_0 \sim r_0^3, \quad \text{and} \quad \varepsilon_0 \sim \frac{V_C}{V_0} \sim \mu^3 << 1 . \tag{23.4.9}$$

This allows us to use the approximation of small fluctuations. When the structure of "continuous medium" is taken into account, this approximation, like in the kinetic theory of gases and plasmas, corresponds to the approximation of second correlation functions, which is often referred to as the polarization approximation.

In this approximation the equation for fluctuations is linear, but contains the Langevin source which reflects the existence of the structure of "continuous medium":

$$\frac{\partial \delta f}{\partial t} + v \frac{\partial \delta f}{\partial r} + \gamma_{max} \delta f + \frac{i}{\hbar (2\pi)^3} \int \left[U\left(r + \frac{1}{2}\hbar\tau\right) - U\left(r - \frac{1}{2}\hbar\tau\right) \right]$$

$$\times \; \delta f(r,p,R,t) \exp\{i\tau(p'-p)\} d\tau dp + e\delta E(R,t) \frac{\partial f}{\partial p} = y(r,p,R,t) . \tag{23.4.10}$$

The distribution function f is defined by the sought-for kinetic equation.

The existence of solution of this equation is also based on inequalities (23.4.9): since the relaxation time of fluctuations and that of the distribution function f are much different, the distribution function may be regarded as constant in calculating the fluctuations on the time scale of the order of τ_C.

As a result, we get the expression for the collision integral. For a free electron, the kinetic equation describes, in particular, evolution towards the state of equilibrium, when the Maxwell distribution results from the balance between the radiation friction and the field fluctuations. For the hydrogen atom, the evolution leads to the Boltzmann distribution with respect to energy levels, and the Planck distribution at the transition frequencies ω_{nm} for the spectral density of field fluctuations. Examples of quantum kinetic equations can be found in numerous books (see, for example, Fain 1972; Haken 1981; Klimontovich II – IV).

Let us return to the Schrödinger equation for deterministic wave function. Now we are able to define more clearly its position and role in the description of processes in the system of atoms and field.

For this purpose we consider the kinetic equation (23.4.6) in the zero approximation with respect to dissipation. In this approximation the collision integral may be dropped; then equation (23.4.6) coincides with (23.1.9), and the quantum distribution function is linked with the wave function by (24.1.8). Substituting this into (23.1.9) we come to the Schrödinger equation (23.1.7). This confirms our earlier assumption that Schrödinger equation corresponds to the approximation of continuous medium for that range of scales where the dissipation is negligibly small. As indicated above, in this respect we have an analogy with the Vlasov approximation for plasma, and the Euler approximation in hydrodynamics.

The capabilities of the quantum theory are much enhanced when one goes over from the reversible Schrödinger equation for the deterministic distribution function to the appropriate irreversible kinetic equations. The latter can be used for describing the far-from-equilibrium processes in open systems — for instance, in lasers.

Like in the classical theory, it is possible to construct the generalized kinetic equations for unified description of kinetic and hydrodynamic processes. This allows one to proceed to the hydrodynamic description without using perturbation theory. As far as the question of self-diffusion in quantum systems is concerned, the generalized kinetic equations permit handling the problem of time generalization of equations of Ginzburg – Landau type in a more rational fashion, giving thus a more consistent treatment of such phenomena as, for example, the effect of superconductivity.

23.5. Example of Ground State Structure

So, by eliminating the small-scale fluctuations one can obtain the closed dissipative kinetic equation for the quantum distribution function $f(r, p, t)$ (Wigner function).

To avoid confusion, let us recall that Wigner function is one of the possible representations of the density matrix (Klimontovich – Silin 1960; Klimontovich II – V; Tatarskii 1983). It may assume negative values, and is therefore not a distribution function in the ordinary sense. Such are the simpler distribution functions with respect to coordinates $f(r, t)$ and momenta $f(p, t)$.

There is, however, an important case when the combined quantum distribution is positive and may be regarded as a conventional distribution function in the six-dimensional phase space (r, p). This is the case of quantum harmonic oscillator.

For the one-dimensional oscillator with proper frequency ω_0, the equilibrium distribution $f(x, p)$ has the form of canonical distribution, with the "quantum temperature" as defined by Planck's formula:

$$f(x,p) = C \exp\left(-\frac{\frac{p^2}{2m} + \frac{m\omega_0^2 x^2}{2}}{\kappa T_{\omega_0}}\right), \quad \kappa T_{\omega_0} = \frac{1}{2}\hbar\omega_0 \coth\frac{\hbar\omega_0}{2\kappa T_{\omega_0}}. \tag{23.5.1}$$

Recall that we have used the kinetic equation for distribution function f_n with the "collision integral" (15.7.2) to obtain, by expanding in $1/n$, the appropriate Fokker – Planck equation (15.7.9) for the distribution of energy values of oscillator. Formally it coincides with the classical distribution; the difference consists in that κT is now replaced by the quantum expression for the mean energy of oscillator.

For the system of quantum oscillators in the field of equilibrium radiation, one can also write the kinetic Fokker – Planck equation for the distribution function $f(x,v,t)$.

Now we return to the Fokker – Planck equation (15.7.9) for the distribution function of energy of vibrations of quantum atom oscillator. In the same approximation we may write the corresponding equation for the distribution function $f(x,p,t)$:

$$\frac{\partial f}{\partial t} + v\frac{\partial f}{\partial x} - m\omega_0^2 x\frac{\partial f}{\partial p} = D_{(p)}\frac{\partial^2 f}{\partial p^2} + \frac{\partial}{\partial p}(\gamma p f), \tag{23.5.2}$$

where

$$D_{(p)} = \gamma m \kappa T_{\omega_0}$$

is the coefficient of diffusion in the space of momenta. The derivation of this equation can be found in (Klimontovich IV, Ch. 16).

The structure of the kinetic equation is, in general, not symmetrical with respect to coordinates and momenta. It can be reduced to a symmetrical form only on condition that the coefficient of friction is $\gamma \ll \omega_0$, which allows one to carry out averaging over the period of oscillations. Then in place of (23.5.2) we get

$$\frac{\partial f}{\partial t} = D_{(p)}\frac{\partial^2 f}{\partial p^2} + D_{(x)}\frac{\partial^2 f}{\partial x^2} + \frac{\partial}{\partial p}(\gamma p f) + \frac{\partial}{\partial x}(\gamma x f) \equiv I_{F-P}, \tag{23.5.3}$$

where γ is the coefficient of radiation friction as given by (15.7.3). The coefficients of diffusion are given by

$$D_{(p)} = \frac{1}{2}\gamma m\kappa T\omega_0 \ , \quad D_{(x)} = \frac{1}{2}\frac{\kappa T\omega_0}{m\Gamma} \ , \quad \Gamma = \frac{\omega_0^2}{\gamma} \ . \tag{23.5.4}$$

These kinetic equations describe the process of time evolution towards the state of equilibrium with the distribution (23.5.1). Without any additional assumptions we may return from the last equation to equation (15.7.9), taking advantage of the relationship between the distribution functions:

$$f(E,t) = \int \delta\left(E - \frac{p^2 + m\omega_0^2 x^2}{2}\right) f(x,p,t)dxdp \ . \tag{23.5.5}$$

Note that the coefficient of radiation friction in the above equations is given by (15.7.3). The proper frequency ω_0 so far remains a free parameter. In order to proceed further we must use a concrete model of the atom.

Like in Sect. 17.4, we shall use Thomson's model of atom oscillator, in which the electron performs vibrations with respect to the center of a sphere carrying a distributed electric charge equal to minus the charge of electron. The radius of the sphere is r_0, the same as in Bohr's model.

For the sake of simplicity we consider a one-dimensional oscillator. At equilibrium, when the quantum distribution (23.5.1) holds good, the amplitude of oscillations is given by

$$x_0 = \sqrt{\frac{\kappa T\omega_0}{m\omega_0^2}} \ ; \quad x_0 = \sqrt{\frac{\hbar}{m\omega_0}} \quad \text{if } T = 0 \ . \tag{23.5.6}$$

For the one-dimensional model, the amplitude x_0 coincides with the size of the atom. The frequency of oscillations ω_0 in Thomson's model is linked with the amplitude x_0 by

$$\omega_0 = \sqrt{\frac{e^2}{mx_0^3}} \ . \tag{23.5.7}$$

Now we can express the amplitude and the frequency in terms of charge and mass of electron e, m, and Planck's constant $\hbar$ (disregarding the numerical coefficients):

$$x_0 \sim \frac{\hbar^2}{me^2} \ , \quad \omega_0 \sim \frac{me^4}{\hbar^3} \ , \quad v_0 = x_0\omega_0 \sim \frac{e^2}{\hbar} \ . \tag{23.5.8}$$

In this way, for the quantum atom oscillator we have obtained Bohr's scales (23.1.12) of the quantum theory of hydrogen atom. From the kinetic equation (23.5.3) it follows that, given the characteristic scales (23.5.8), the times of diffusion with respect to momenta and coordinates are of the same order of magnitude,

$$\tau_{D_{(p)}} \sim \tau_{D_{(x)}} \sim \frac{1}{\gamma(\omega_0)} , \qquad (23.5.9)$$

and are defined by the coefficient of radiation friction at Bohr's characteristic frequency ω_0.

In quantum theory of hydrogen atom and atom oscillator based on the Schrödinger equation, the length scales (23.1.12), (23.5.8) are the smallest. It is they that define the size of the region corresponding to the ground state. Schrödinger equation does not carry any additional information about the structure of the ground state and the process of its formation. Nor is any such information contained in the corresponding quantum kinetic equations (23.4.6), (23.5.3): they only describe the relaxation processes on the time scales much larger than the relevant oscillation periods.

It follows that in order to elucidate the structure of the ground state and to explain its feasibility we have to consider those processes which are characterized by the smaller scales.

In the hierarchies of time and space scales (23.3.9) and (23.3.10), the smallest are the "classical electron radius" r_e and the corresponding transit time τ_e, which is the smallest of the relaxation times. It is this time that defines, in accordance with (23.4.1), the procedure of smoothing at the first step of transition from the dynamic equations to the dissipative equations of quantum statistical theory.

This conclusion is supported by the calculation of photon scattering by free electrons on the basis of equation (23.3.6) with the characteristic relaxation time $\tau_e = 1/\Gamma$, which results in a correct expression for the effective cross section.

On the strength of these findings, the description of fast fluctuations in the region of ground state with Bohr's parameters (23.5.8) may be based on the quantum kinetic model equation with pulsating (smoothed over the volume V_e) distribution function $\tilde{f}$. This equation differs from (23.5.3) by an additional dissipative term (cf. (17.4.5)), which describes the relaxation of pulsating distribution towards the deterministic distribution $f(x,p,t)$ satisfying equation (23.5.3). Let us use the simplest form of such equation,

$$\frac{\partial \tilde{f}}{\partial t} = \hat{I}_{F-P} \tilde{f} - \frac{1}{\Gamma}\left(\tilde{f} - f\right) . \qquad (23.5.10)$$

The "collision integral" is given by (15.5.3). Given that

$$\langle \tilde{f} \rangle = f , \tag{23.5.11}$$

this equation after averaging over the Gibbs ensemble coincides with (23.5.3). The adjustment to the deterministic distribution takes place over the relaxation time

$$\tau_e = \frac{1}{\Gamma} . \tag{23.5.12}$$

From equation (23.5.10) we may go over to equation for the distribution function of energy values $\tilde{f}(E,t)$ (23.5.5). The corresponding equation for the mean energy $\tilde{E} = \int E\tilde{f}(E,t)dE$ now has the form

$$\frac{d\tilde{E}}{dt} = -\Gamma\left(\tilde{E} - \langle E \rangle\right) - \gamma(\omega_0)\left(\langle E \rangle - \kappa T_{\omega_0}\right) . \tag{23.5.13}$$

We see that the relaxation towards the equilibrium energy value occurs in two stages: the fast relaxation over the time τ_e towards $\langle E \rangle$, and the slow relaxation over the time $1/\gamma(\omega_0)$ towards the equilibrium value of mean energy. The fast process is "hidden" when the quantum kinetic equation (23.5.3) is used.

Recall that smoothing in the transition from the dynamic operator equations to the kinetic equation (23.4.6) in Sect. 23.4 has been carried out in two steps: first over the time τ_e, and then over the larger time interval τ_C. Now the smoothing with respect to τ_C was not needed, since we have used the explicit form of the kinetic equation (23.5.3) for the deterministic distribution. Smoothing, however, must be taken into account in the calculation of the kinetic fluctuations of distribution function f.

Now it is time to summarize the results of this chapter, which we present in the form of answers to three important questions. As a matter of fact, these questions (or answers) are not independent of one another. They all are actually concerned with the same problem: the limitations of quantum mechanical description of real systems based on the corresponding Schrödinger equation. The three questions just emphasize certain aspects of this many-sided problem.

23.6. Is the Concept of "Pure Ensemble" Justified in Quantum Mechanics? Is Quantum Mechanical Description Complete? Are There "Hidden Parameters" in Quantum Theory?

We saw that the concept of "pure ensemble" is not that simple. For our system of atoms and electromagnetic field it is clearly defined only in the case of complete quantum mechanical description. As indicated in the beginning of this chapter, this can be done in two ways. The first is based on Schrödinger equation for the wave function of complete quantum set of variables. Its counterpart in the classical theory is the Liouville equation for distribution function of the complete set of positions and momenta of particles and field. The second is based on the system of equations in operators $\hat{\Psi}$, $\hat{E}$ (or operators $\hat{f}$, $\hat{E}$).

In quantum mechanics, however, the term "pure ensemble" is used also in those cases when the description is based on Schrödinger equation (23.1.7) for the deterministic wave function $\psi(r,t)$ of only part of the variables of the system in question. On this basis, however, complete description of the system of atoms and field is not possible. This approach corresponds to a coarse approximation, when the dissipation is not taken into account, and is not capable of describing transitions between stationary levels accompanied by emission of radiation.

In this situation the meaning of "pure ensemble" becomes quite different. To wit, the inclusion of arbitrarily small dissipation (which is inevitable, for instance, in the neighborhood of resonances) leads to a "mixed ensemble", which actually has no real alternative in quantum theory.

Since the description based on Schrödinger equation is not complete, there exist "hidden parameters" which are revealed through the use of a more realistic approach to phenomena under consideration.

Earlier we have illustrated this possibility with a concrete example of the system of atoms and field. By introducing the scales of "continuous medium" we were able to describe in greater detail the small-scale fluctuations which define the structure of "collision integrals" in quantum kinetic equations. The calculation of small-scale fluctuations is also of interest by itself: these fluctuations, for instance, are definitive for the effective cross section of scattering of photons by free electrons.

It is important that the effective cross section in the quantum domain (visible and x-ray ranges) does not depend on Planck's constant, and can therefore be found from classical calculations. The relevant scales r_e, τ_e are so small that Heisenberg's uncertainty relation does not hold.

In this way, there are two "exits" from quantum theory. One of these corresponds to the domain of large scales and slow spatial variations, whereas the other is associated with the transition to the scales which are much smaller than any of the quantum scales of the system in question, which brings us into the realm of "hidden parameters".

This, of course, is just a rather preliminary attempt at solving a whole class of interrelated problems of statistical description of quantum phenomena. One of the main issues here is the problem of inclusion of the procedure of measurement into the quantum theory. This problem has been on the agenda for a long time; despite certain progress in this direction, the understanding is far from being complete.

The second question in the heading of this section was the subject of the famous debate between Einstein and Bohr. In the light of arguments developed above, the answer to this question is negative. The fact is that quantum mechanical description based on the reversible Schrödinger equation can always be supplemented by the inclusion of fluctuations. Then, owing to the existence of fluctuation dissipation relations, the dissipation is inevitable. Consequently, any kind of quantum mechanical description is in practice incomplete, and the concept of "pure ensemble" is just an abstraction.

The above arguments score in favor of Einstein's standpoint. In this connection it would be interesting to note that the problem of incompleteness of quantum mechanical description had been actually solved by Einstein long before the emergence of quantum mechanics as such.

As early as 1916, Einstein formulated the concept of *spontaneous emission* for the exemplary case of a two-level atom interacting with the equilibrium electromagnetic field. This term emphasizes the inevitability of energy loss by radiation, the dissipation thus being inseparable from real processes. This reason is good enough to assume beforehand that a theory of atomic processes of interaction between atoms and field cannot be based on equations in the variables of atoms alone. Such equations — in particular, the Schrödinger equations for hydrogen atom, atom oscillator, and other mechanical systems — can only be regarded as a useful idealization.

At this point we would like to quote Ilya Prigogine, who said (Prigogine 1980, p. 70):

"...Or, on the contrary, should we argue that nobody has ever seen an atom that would not decay when brought into an excited level? The physical "reality" then corresponds to systems with continuous spectra, whereas standard quantum mechanics appears only as a useful idealization, as a simplified limiting case. This is much more in line with the view that elementary particles are expressions of basic fields (such as photons with respect to the electromagnetic field) and fields are in essence not local because they extend over macroscopic regions of space and time."

Prigogine does not belong to the founding fathers of quantum mechanics. His statement, however, closely echoes the words of Louis de Broglie.

De Broglie expressed doubt in the completeness of quantum mechanical description even at the heyday of the quantum theory. His scepticism was not, however, shared by his contemporaries, and so he also abandoned this attitude, becoming one of the most brilliant advocates of the "Copenhagen" formulation of quantum mechanics. He recalled this period in the following words (de Broglie 1952):

"Some people, remembering that I abandoned my first attempts and used the interpretation of Bohr and Heisenberg in all my works for twenty-five years thereafter, will accuse me of being inconsistent when they see that I am once again doubtful and ask myself whether my initial orientation had been right after all. Should I feel like joking, I could reply in Voltaire's words, 'Only foolish people never change their minds.'

"The answer, however, can be more prudent.

"The progress of science is continually harassed by the tyrannic influence of certain concepts which in the course of time have become dogmas. Because of this, the principles which have been recognized as final must be subjected to most thorough revision."

At that time these words of de Broglie were the voice of one crying in the wilderness. This can be illustrated with a quotation from an article published in an influential American newspaper:

"The principle of uncertainty has eventually made all contemporary physicists (with the exception of Dr. Einstein) recognize that there is no causality or determinism in nature. Dr. Einstein in majestic solitude has held out against all these concepts of quantum theory" (The New York Times, 30 March 1952).

Yet another pertinent passage is taken from Dirac's paper published shortly before his death (Dirac 1978):

"I do not exclude the possibility that Einstein's standpoint may eventually prove to be correct, because the contemporary status of quantum mechanics cannot be regarded as final. The modern quantum mechanics is a great achievement, but it will hardly exist for ever. It seems very likely to me that some time in the future quantum mechanics will be modified and will return to causality, which will justify Einstein's point of view. This return to causality, however, will be only possible at the price of rejection of some fundamental idea which is currently accepted without any reservations. If we are going to revive causality, we shall have to pay, and today we can only speculate about the ideas which may have to be sacrificed."

We do not wish to comment on these statements: the clarity and boldness of the classics can only be admired. Their words encourage further studies in the quantum statistical theory of open systems, which will be continued in the second volume.

In this chapter we have tried to define the author's position with regard to the fundamental problems of quantum theory. The Conclusion gives an outline of the problems treated in the second volume.

CONCLUSION

In the Introduction we briefly discussed some of the ideas and results of the statistical theory of open systems. By the end of the book we have made considerable progress in solving a number of problems of this theory, illustrating the general results with concrete examples. Although many things had to be left out, we did our best to give a consistent presentation of the capability of the theory.

The main restriction consisted in the choice of models of macroscopic open systems. On the microscopic level the elements of the system were either structureless small particles— as, for instance, atoms in case of Boltzmann gas, or quantum atom oscillators interacting with equilibrium electromagnetic field. Macroscopic elements served as structural units in phenomenological description of active media comprised of bistable elements or oscillators, as well as in the treatment of Brownian motion.

On of the possible and promising generalizations of the theory consists in the extension of the basic macroscopic model — the transition to the statistical theory of plasma-molecular open systems (Klimontovich – Wilhelmsson – Yakimenko – Zagorodnii 1989). Models of this kind include the completely ionized plasma and the gas of neutral particles as two extreme cases. The unified model of partially ionized plasma consists of at least four components: electrons, ions, atoms, and electromagnetic field. The analysis of such sophisticated systems calls for the development of quantum theory of open systems. Some elements of this theory were discussed in the last chapter of the present book.

Of course, individual branches of the theory call for being developed further. In the first place this applies to criteria of self-organization and the theory of turbulence. One of important problems is concerned with the feasibility study of long-term forecast based on the criteria of statistical theory of open systems. Then, there is the problem of diagnostics of most diverse open systems on the basis of the same criteria.

The solution of these problems depends on the progress in computer simulations of open systems. In particular, it is necessary to obtain numerical solutions of the generalized kinetic equations. Hopefully, this will facilitate the development of the theory of turbulent motion. So far, as one will gather from the epigraph to Ch. 22, the unfolding of the mysteries of turbulence has to a large extent depended on the good will of the Creator.

Throughout the book we tried to take advantage of the ideas and concepts of the new interdisciplinary science of synergetics, associated first and foremost with the name of

Hermann Haken, founder and director of the Institute of theoretical physics and synergetics at Stuttgart university. In recent years synergetics has much extended its sphere of influence. Its methods are, for instance, applied for studying such intricate processes as the transition from perception to thinking (Caglioti 1992; see also Klimontovich 1993). According to Caglioti, the disordered sensations gradually correlate and are organized in the brain into coherent structures, which turn into a thought. In other words, the transition from perception to thinking is the transition from the less ordered to the more ordered state of the mind.

This is a very beautiful scheme. Of course, it is still not clear whether this picture complies with the reality. This can only be checked with the aid of criteria of the relative degree of order, which may be applied to, for instance, electroencephalograms. The main difficulty here is associated with the finite length of the experimental time realizations of the process.

Time will show whether this will help us to see how a thought is born. It is clear, however, that such problems are highly stimulating for the development of the statistical theory of open systems.

REFERENCES

Akhromeeva T.S., Kurdyumov S.P., Malinetskii G.G., Samarskii A.A. 1992:
Nonstationary Structures and Diffusion Chaos[*], (in Russian) (Nauka, Moscow).

Andreev A.V., Emel'yanov V.I., Il'inskii Yu.A. 1993: *Cooperative Effects in Optics*
(Institute of Physics Publishing, Bristol, Philadelphia).

Andreev E.A., Gridina N.Ya., Kramar V.M., Klimontovich Yu.L. 1989: *S–Theorem-
Based Criterion for Relative Evaluation of Order Degree of Biological Objects*, Z.
Naturforschung C, **38**, 447.

Anishchenko V.S. 1990: *Complex Oscillations in Simple Systems*, (in Russian) (Nauka,
Moscow).

Anishchenko V.S., Astakhov V.V. 1983: *Experimental Investigations of the Mechanism
of Genesis and Structure of Strange Attractor with Inertial Nonlinearity*, (in Russian)
Radiotekhnika i Elektronika, **28**, 1109.

Anishchenko V.S., Klimontovich Yu.L. 1984: *Evolution of Entropy in Generator with
Inertial Nonlinearity in the Course of Transition to Stochasticity Through a Cascade of
Bifurcations of Period Doubling*, (in Russian) Pis'ma v ZhTF, **10**, 816.

Anishchenko V.S., Neiman A.B., Chua L.O. 1993: *Chaos-Chaos Intermittency and $1/\omega$
Noise in Chua's Circuit*, Int. J. of Bifurcation and Chaos, **3**.

Anishchenko V.S., Saparin P., Anishchenko T.G. 1993: *On the Criterion for a Relative
Order Degree of Self-Oscillating Regimes. Illustration of Klimontovich's S–Theorem*,
Pis'ma v ZhTP, **24**; (English transl.: Proc. SPIE, Bellingham, USA, in press).

Arnold V.I. 1990: *Theory of Catastrophes*, (in Russian) (Nauka, Moscow).

Bakai A.S., Sigov Yu.S. 1989: *Multifaced Turbulence*, (in Russian) (Znanie, Moscow).

Balescu R. 1975: *Equilibrium and Nonequilibrium Statistical Mechanics* (Wiley, New
York, London, Sydney, Toronto).

Belinskii A.V., Klyshko D.N. 1993: *The Interference of Light and the Bell's Theorem*,
(in Russian) Uspekhi Fiz. Nauk, **163**, 1.

Bell J.S. 1965: *On the Einstein – Podolski – Rosen Paradox*, Physics, **1**, 165.

Bell J.S. 1966: *On the Problem of Hidden Variables in Quantum Mechanics*, Rev. Mod.
Phys. **38**, 447.

---

[*] Please note that many original Russian titles have been translated here for the reader's convenience.

Belyi V.V., Klimontovich Yu.L. 1978: *Kinetic Fluctuations in Partially Ionized Plasma and Chemically Reacting Gases*, (in Russian) ZhETF, **74**, 1160.

Bethe H. 1947: *The Electromagnetic Shift of Energy Levels*, Phys. Rev. **72**, 339.

Bevensee R.M. 1993: *Maximum Entropy Solutions to Scientific Problems* (PTR Prentice Hall, Englewood Cliffs, New Jersey).

Bogolyubov N.N. 1946: *Problems of Dynamic Theory in Statistical Physics*, (in Russian) (Gostekhizdat, Moscow).

Beck C., Schlögl F. 1993: *Thermodynamics of Chaotic Systems* (Cambridge University Press, Cambridge).

Berge P., Piomeau Y., Vidal Ch. 984: *L'Order dans le Chaos* (Hermann, Paris).

Bohm D. 1952: *On the Possible Interpretation of Quantum Mechanics on the Basis of Concept of "Hidden Parameters"*, Phys.Rev. **85**, 166, 180.

Born M., Green H.S., 1949: *A General Kinetic Theory of Liquids* (Cambridge University Press, Cambridge).

Bunkin F.V., Kirichenko N.A., Luk'yanchuk B.S. 1985: *Bifurcations, Catastrophes and Structures in Laser Thermochemistry*, (in Russian) Izvestiya AN SSSR, **49**, 1054.

Casati G. (Ed.) 1985: *Chaotic Behavior in Quantum Systems* (Plenum Press, New York).

Chandrasekhar S. 1943: *Stochastic Problems in Physics and Astronomy*, Rev. Mod. Phys. **15**, 823.

Chandrasekhar S. 1961: *Hydrodynamic and Hydromagnetic Stability* (Clarendon Press, Oxford).

Chapman S., Cowling T.G. 1939: *The Mathematical Theory of Non-Uniform Gases* (Cambridge University Press, Cambridge).

Chua L. 1993: *Genesis of Chua's Circuit*, Applied Nonlinear Dynamics, **1**, 4.

Davydov B.I. (Ed.) 1936: *Brownian Motion* (in Russian) (ONTI, Moscow).

De Broglie L. 1953: *La physique quantique resterat-elle indeterministe?* (Paris).

De Groot S.R, Mazur P. 1962: *Nonequilibrium Thermodynamics* (North Holland, Amsterdam).

Dirac P.A.M. 1978: *Directions in Physics*, edited by H.Hora, J.R. Shepanski (Wiley, New York, London, Sydney, Toronto).

Dmitriev F.S., Kislov I.Ya. 1989: *Stochastic Oscillations in Radio Engineering and Electronics* (in Russian) (Nauka, Moscow).

Ebeling W., Engel A., Feistel R. 1990: *Physik der Evolutionsprozesse* (Akademie-Verlag, Berlin).

Ebeling W., Klimontovich Yu.L. 1984: *Self Organization and Turbulence in Liquids* (Teubner, Leipzig).

Ebeling W., Klimontovich Yu.L., Neiman A. 1994: *Random Walk Models and Long-Range Correlations in Human Writings and in DNA Sequences*, Chaos, Solitons and Fractals (in press).

Eckmann J., Ruelle D. 1985: *Ergodic Theory of Chaos and Strange Attractors*, Rev. Mod. Phys. **57**, 617.

Einstein A. 1916: *Strahlung Emission und Absorption nach der Quantentheorie*, Verhandl. Dtsch. Phys. Ges. **18**, 318.

Einstein A., Podolski B., Rosen N. 1935: *Can Quantum-Mechanical Description of Physical Reality Be Considered Complete?* Phys. Rev. **47**, 777.

Elyutin P.V. 1988: *Problems of Quantum Chaos*, Sov. Phys. — Uspekhi, **155**, 397.

Engel-Herbert H., Ebeling W. 1988: *The Behavior of the Entropy During Transitions Far From Thermodynamic Equilibrium*, Physica, **149A**, 182-194, 195-205.

Fain V.M. 1972: *Photons and Nonlinear Media*, (in Russian) (Sovetskoe Radio, Moscow).

Feigenbaum M. 1983: *Universality in the Behavior of a Nonlinear System*, (in Russian) Uspekhi Fiz. Nauk, **141**, 343.

Ferziger J.H, Kaper H.G. 1972: *Mathematical Theory of Transport Processes in Gases* (North Holland, Amsterdam).

Fox R., Uhlenbeck G. 1970: *Contribution in Nonequilibrium Thermodynamics II. Fluctuation Theory for the Boltzmann Equation*, Phys. Fluids, **13**, 881.

Friedrich R., Wunderlin A. (Eds.) 1992: *Evolution of Dynamic Structures in Complex Systems* (Springer, Berlin, Heidelberg, New York).

Frost W., Moulden T. 1977: *Handbook of Turbulence*, v.1 (Plenum Press, New York, London).

Gantsevich S.V., Gurevich V.L., Katilus R. 1979: *Theory of Fluctuations in Non Equilibrium Electron Gas*, Rivista del Nuovo Cimento, **2**, 1.

Gaponov-Grekhov A.V., Rabinovich M.I. 1987: *Equations of Ginzburg – Landau and Nonlinear Dynamics of Non Equilibrium Media*, (in Russian) Radiofizika, **30**, 131.

Gardiner C.V. 1983: *Handbook of Stochastic Methods* (Springer, Berlin, Heidelberg, New York).

Gibbs H. 1985: *Optical Bistability: Controlling Light With Light* (Academic Press, New York).

Gibbs J.W. 1902: *Elementary Principles in Statistical Mechanics* (Yale University Press, New Haven).

Gihman I.I., Skorokhod A.V. 1972: *Stochastic Differential Equations* (Springer, Berlin, Heidelberg, New York).

Ginzburg V.L. 1975; 1981: *Theoretical Physics and Astrophysics*, (in Russian) (Nauka, Moscow).

Gladyshev G.P. 1988: *Thermodynamics and Macrokinetics of Natural Hierarchical Processes*, (in Russian) (Nauka, Moscow).

Gledzer E.B., Dolzhanskii F.V., Obukhov A.M. 1981: *Systems of Hydrodynamic Type and Their Applications*, (in Russian) (Nauka, Moscow).

Goldshtik M.A. Shtern V.N. (Eds.) 1985: *Stability and Turbulence*, (in Russian) (IT AN SSSR, Novosibirsk).

Grad H. 1963: *Asymptotic Theory of the Boltzmann Equation II*, in: *Proc. 3rd Int. Symp. on Gas Dynamics* (Academic Press, New York).

Grandy W.T. 1988: *Foundations of Statistical Mechanics*, v. II: *Nonequilibrium Phenomena* (D. Reidel, Dordrecht).

Grechanny O.A. 1988: *The Stochastic Theory of Irreversible Processes*, (in Russian) (Naukova Dumka, Kiev).

Grossmann S., Thomae S. 1977: *Invariant Distribution and Stationary Correlation Function in One Dimensional Discrete Processes*, Z. Naturforschung, **32A**, 1353.

Haken H. 1978: *Synergetics* (Springer, Berlin, Heidelberg, New York).

Haken H. 1981: *Light*, v.1 (North Holland, Amsterdam, New York, Oxford).

Haken H. 1983: *Advanced Synergetics* (Springer, Berlin, Heidelberg, New York).

Haken H. 1988: *Information and Self-Organization* (Springer, Berlin, Heidelberg, New York).

Hanggi P., Talkner P., Borkovec M. 1990: *Reaction Rate Theory: Fifty Years After Kramers* Rev. Mod. Phys. **62**, 251.

Holden A. (Ed.) 1986: *Chaos* (Manchester University Press).

Horstemke W., Lefever R. 1984: *Noise Induced Transitions* (Springer, Berlin, Heidelberg, New York).

Ivanitskii G.R., Krinskii V.I., Sel'kov E.E. 1978: *Mathematical Biophysics of the Cell*, (in Russian) (Nauka, Moscow).

Izrailev F.M., Chirikov B.V., Shepelianskii D.L. 1986: *Quantum Chaos*, (in Russian) (IYaF SO AN SSSR, Novosibirsk).

Jaynes E.T. 1979: *Where do we stand on Maximum Entropy?* in: *Maximum Entropy Formalism*, edited by R.D. Levine and M. Tribus, pp. 15-118, (M.I.T. Press, Cambridge MA).

Jaynes E.T. 1980: *The Minimum Entropy Production Principle*, in: *Annual Review of Physical Chemistry 31*, pp. 579-601 (Annual Review Inc. Palo Alto CA).

Joseph D. 1976: *Stability of Fluid Motion* (Springer, Berlin, Heidelberg, New York).

Jumarie G. 1994: *Further Results on the Modelling and the Control of Stochastic Dynamics Subject to Fractional Brownian Motion Inputs*, Syst. Analysis Modelling Simulation, **14**, 163.

Kadomtsev B.B. 1957: *On Fluctuations in a Gas*, (in Russian) ZhETP, **32**, 943.

Kadomtsev B.B. (Ed.) 1984: *Synergetics*, (in Russian) (Mir, Moscow).

Kapur J.N. 1989: *Maximum-Entropy Models in Science and Engineering* (Wiley, New York, London, Sydney, Toronto).

Karlov N.W., Kirichenko N.A., Luk'yanchuk B.S. 1992: *Laser Thermochemistry*, (in Russian) (Nauka, Moscow).

Keizer J. 1987: *Statistical Thermodynamics of Nonequilibrium Processes* (Springer, Berlin, Heidelberg, New York).

Kerner B.S., Osipov V.V. 1989: *Autosolitons*, Sov. Phys. — Uspekhi, **157**, 201.

Kerner B.S., Osipov V.V. 1994: *Autosolitons* (Kluwer Academic Publishers, Dordrecht).

Kirkwood J.G. 1947: *The Statistical Mechanical Theory of Transport Processes*, J. Chem. Phys. **14**, 180, 347 (1946); **15**, 72 (1947).

Klimontovich Yu.L. (I): *The Statistical Theory Of Non Equilibrium Processes in a Plasma* (Izd. MGU, Moscow, 1964; English transl.: Pergamon Press, Oxford 1967).

Klimontovich Yu.L. (II): *Kinetic Theory of Non Ideal Gases and Non Ideal Plasmas* (Nauka, Moscow 1975; English transl.: Pergamon Press, Oxford 1982).

Klimontovich Yu.L. (III): *The Kinetic Theory of Electromagnetic Processes* (Nauka, Moscow 1980; English transl.: Springer, Berlin, Heidelberg, New York 1983).

Klimontovich Yu.L. (IV): *Statistical Physics* (Nauka, Moscow 1982; English transl.: Harwood Academic Publishers, New York 1986).

Klimontovich Yu.L. (V): *Turbulent Motion and the Structure of Chaos* (Nauka, Moscow 1990; English transl.: Kluwer Academic Publishers, Dordrecht 1991).

Klimontovich Yu.L. 1956: *Definition of Eigenvalues of Physical Quantities by the Method of Quantum Distribution Function*, (in Russian) Doklady AN SSSR, **108**, 1033.

Klimontovich Yu.L. 1957: *Second Quantization in Phase Space*, (in Russian) ZhETP, **33**, 982.

Klimontovich Yu.L. 1967, 1968: *Statistical Theory of Inelastic Processes in a Plasma*, (in Russian) ZhETP, **52**, 1233 (1967); ZhETP, **54**, 136 (1968).

Klimontovich Yu.L. 1971a: *On Nonequilibrium Fluctuations in a Gas*, TMF, **8**, 109.

Klimontovich Yu.L. 1971b: *On the Kinetic Boltzmann Equation for Nonideal Gas*, (in Russian) ZhETP, **60**, 1352.

Klimontovich Yu.L. 1973: *Kinetic Equations for Nonideal Gas and Nonideal Plasma*, (in Russian) Uspekhi Fiz. Nauk, **110**, 537.

Klimontovich Yu.L. 1974: *Boltzmann's H-Theorem for Nonideal Gas*, (in Russian) ZhETP, **63**, 150.

Klimontovich Yu.L. 1978: *Callen – Welton and Kubo Formulas for Nonequilibrium States*, (in Russian) ZhETP, **75**, 361.

Klimontovich Yu.L. 1982a: *Fluctuation Dissipation Relations in the Problem of Scattering and Method of Fluctuations in the Kinetic Theory of Gases*, (in Russian) Pis'ma v ZhTP, **8**, 1005.

Klimontovich Yu.L. 1982b: *Nonequilibrium Sources of Hydrodynamic Fluctuations. Mutual Influence of Hydrodynamic Motion and Nonequilibrium Fluctuations in the Kinetic Coefficients*, Ann. der Physik, **39**, 1417.

Klimontovich Yu.L. 1983a: *Dissipative Equations for Multiparticle Distribution Functions*, (in Russian) Uspekhi Fiz. Nauk, **139**, 689.

Klimontovich Yu.L. 1983b: *Natural Flicker Noise*, (in Russian) Pisma v ZhTP, **9**, 406.

Klimontovich Yu.L. 1983c: *Entropy Decrease in the Process of Self Organization. S–Theorem*, (in Russian) Pis'ma v ZhTP, **9**, 1089.

Klimontovich Yu.L. 1984a: *Entropy and Entropy Production in the Laminar and the Turbulent Flows*, (in Russian) Pis'ma v ZhTP, **10**, 80.

Klimontovich Yu.L. 1984b: *Linkage Between the Critical Reynolds Number and Thickness of Laminar Sublayer in the Poiseuille Flow*, (in Russian) Pis'ma v ZhTP, **10**, 326.

Klimontovich Yu.L. 1985: *Dynamic and Statistical Theory of Generalized Van der Pol Generators with the Number of Stable Limiting Cycles 12 for k = 0, 1, 2,...*, (in Russian) Pis'ma v ZhTP, **11**, 21.

Klimontovich Yu.L. 1987a: *S–Theorem*, Z. Phys. B, **66**, 125.

Klimontovich Yu.L. 1987b: *Entropy Evolution in Self Organization Processes. H–Theorem and S–Theorem*, Physica, **142A**, 390.

Klimontovich Yu.L. 1987c: *Sequences of Bifurcations of Limiting Cycle Energy and Period of Oscillations in Generators with Two Controlling Parameters: Amount of Feedback and Scale of Discrete Time*, (in Russian) Pis'ma v ZhTP, **13**, 175.

Klimontovich Yu.L. 1987d: *Irreversibility and Fluctuation Dissipation Relations*, Physica, **142A**, 1374.

Klimontovich Yu.L. 1987e: *Fluctuation Dissipation Relations. Quantum Generalization of Nyquist's Formula*, Sov. Phys. – Uspekhi, **151**, 309.

Klimontovich Yu.L. 1988: *Definition of Relative Degree of Order in Open Systems on the Basis of S–Theorem Using Experimental Data*, (in Russian) Pis'ma v ZhTP, **14**, 631.

Klimontovich Yu.L. 1990a: *Ito, Stratonovich and Kinetic Forms of Stochastic Equations*, Physica, **163A**, 515.

Klimontovich Yu.L. 1990b: *Natural Flicker Noise (1/f-Noise) and Superconductivity*, (in Russian) Pis'ma v ZhETF, **51 (1)** 43.

Klimontovich Yu.L. 1990c: *Effects of Self Diffusion on the Spectra of Hydrodynamic Fluctuations*, (in Russian) Pis'ma v ZhTF, **16**.

Klimontovich Yu.L. 1990d: *The Kinetic Description of Hydrodynamic Motion. Entropy Balance*, (in Russian) Pis'ma v ZhETF, **51 (11)** 569.

Klimontovich Yu.L. 1991: *Some Problems of The Statistical Description of Hydrodynamic Motion and Autowave Processes*, Physica, **179A**, 471.

Klimontovich Yu.L. 1992a: *The Unified Description of Kinetic and Hydrodynamic Processes*, Physica, **197A**, 434.

Klimontovich Yu.L. 1992b: *Alternative Description of Stochastic Processes in Nonlinear Systems. "Kinetic Form" of Master and Fokker – Planck Equations*, Physica, **182A**, 121.

Klimontovich Yu.L. 1992c: *On The Need for and the Possibility of a Unified Description of Kinetic and Hydrodynamic Processes*, TMF, **92**, 312.

Klimontovich Yu.L. 1993a: *The Reconstruction of the Fokker – Planck and Master Equations on the Basis of Experimental Data. H–Theorem and S–Theorem*, Bifurcations and Chaos, **1**, 119.

Klimontovich Yu.L. 1993b: *From the Hamiltonian Mechanics to a Continuous Media. Dissipative Structures. Criteria of Self-Organization*, TMF, **96**, 3.

Klimontovich Yu.L. 1993c: *Unified Description of Kinetic, Hydrodynamic and Reaction Diffusion Processes in Gases and Plasmas*, Contributions to Plasma Physics, **33** (5-6) 421.

Klimontovich Yu.L. 1993d: *Criteria of the Relative Degree of Order in Self-Organization Processes*, Nanobiology, **2**, 229.

Klimontovich Yu.L. 1993e: *To the Statistical Ground of Schrödinger Equation*, TMF, **97** (**1**), 3.

Klimontovich Yu.L. 1994a: *Criteria of Self-Organization*, Chaos, Solitons & Fractals (in press).

Klimontovich Yu.L. 1994b: *Nonlinear Brownian Motion*, Sov. Phys. – Uspekhi, **164**, 8.

Klimontovich Yu.L. 1994c: *What is Turbulence?*, Applied Nonlinear Dynamics, **2**, (in press).

Klimontovich Yu.L., Bonitz M. 1988: *Evolution of the Entropy of Stationary States in Self-Organization Processes in the Control Parameter Space*, Z. Phys. B, Condensed Matter, **70**, 241.

Klimontovich Yu.L., Boon J.P. 1987: *Natural Flicker Noise (1/f-Noise) in Music*, Europhys. Lett. **3(4)** 395.

Klimontovich Yu.L., Chetverikov V.I. 1987: *Bifurcations and Distributions of Energy in the Generalized Van der Pol Generator Associated with the Change of Feedback and Discrete Time Scale*, (in Russian) Pis'ma v ZhTP, **13**, 977.

Klimontovich Yu.L., Engel-Herbert H. 1984: *Averaged Stationary Turbulent Couette and Poiseuille Flows in an Incompressible Fluid*, (in Russian) ZhTP, **54**, 440.

Klimontovich Yu.L., Kovalev A.S., Landa P.S. 1972: *Natural Fluctuations in Lasers*, (in Russian) Uspekhi Fiz. Nauk, **106**, 279.

Klimontovich Yu.L., Kremp D. 1981: *Quantum Kinetic Equation in Systems with Bound States*, Physica, **109A**, 512.

Klimontovich Yu.L., Kremp D., Kräft W.D. 1987: *Kinetic Theory for Chemically Reacting Gases and Partially Ionized Plasmas*, Adv. Chem. Phys. **58**, 175.

Klimontovich Yu.L., Shevchenko A.Yu., Yakimenko I.P. 1989: *Molecular Systems*, Contributions to Plasma Physics, **28**.

Klimontovich Yu.L., Shevchenko A.Yu., Yakimenko I.P., Zagorodnii A.G. 1990: *Theory of Bremsstrahlung in Plasma Molecular Systems*, Contributions to Plasma Physics, **29**, 1.

Klimontovich Yu.L., Slin'ko E.F. 1974: *Kinetic Equations for Nonideal Gases with Account for Triple Collisions*, Theor. Math. Phys. **19**, 137.

Klimontovich Yu.L., Wilhelmsson H., Yakimenko I.P., Zagorodnii A.G. 1989: *Statistical Theory of Plasma Molecular Systems*, Phys. Rep. **175 (55-56)** 264 (1989); Izd. MGU, Moscow, 1990.

Klyatskin V.I., Tatarskii V.I. 1973: *Approximation of Diffusive Random Process in Some Nonstationary Statistical Problems of Physics*, (in Russian) Uspekhi Fiz. Nauk, **110**, 1980.

Kogan M.N. 1967: *Rarefied Gas Dynamics*, (in Russian) (Nauka, Moscow).

Kogan Sh.M. 1985: *Low-Frequency Current Noise with $1/f$ Spectrum in Solids*, (in Russian) Uspekhi Fiz. Nauk, **145**, 286.

Kogan Sh.M., Shul'man A.Ya. 1969: *To the Theory of Fluctuations in a Nonequilibrium Gas*, (in Russian) ZhETP, **56**, 862.

Kolmogorov A.N., Petrovsky I.G., Piskunov N.S. 1937: *Study of Equation of Diffusion Combined with the Increasing Quantity of Matter and Its Application to a Particular Biological Problem*, (in Russian) MGU Bulletin, Matematika i Mekhanika, **1**, 1-26.

Kramers H.A. 1940: *Brownian Motion in Field of Force and Diffusion Model of Chemical Reactions*, Physica, **7(4)**, 284.

Kravtsov Yu.A. 1993: *Limits of Predictability* (Springer, Berlin, Heidelberg, New York).

Krinsky V.I. (Ed.) 1984: *Self Organization. Autowaves and Structures Far from Equilibrium* (Springer, Berlin, Heidelberg, New York).

Krylov N.M. 1950: *Works for the Foundations of Statistical Physics*, (in Russian) (Nauka, Moscow).

Krylov N.M., Bogolyubov N.N. 1934: *Application of Methods of Nonlinear Mechanics to Theory of Stationary Oscillations*, (in Russian) (Izd. AN SSSR, Moscow).

Krylov N.M., Bogolyubov N.N. 1937: *Introduction to Nonlinear Mechanics*, (in Russian) (Izd. AN SSSR, Moscow).

Kullback S. 1959: *Information Theory and Statistics* (Wiley, New York, London, Sydney, Toronto).

Kuramoto Y. 1984: *Chemical Oscillations, Waves and Turbulence* (Springer, Berlin, Heidelberg, New York) ,

Kuroda S., Kubota K. 1976: *Dye Laser Action in a Liquid Crystal*, Appl. Phys. Lett. **29**, 737.

Kurts J., Saparin P., Vossing H.J., Witt A., Voss A., Dietz R., Fiehring H., Kleiner H.J. 1993: *Some Methods of Nonlinear Dynamics in ECG-Rhythm Analysis*, in: *Proc. Int. Soc. Optical Eng.* (IEEE Comp. Soc. Press, Los Alamos).

Lamb W., Rutherford R. 1947: *Fine Structure of the Hydrogen Atom by a Microwave Method*, Phys. Rev. **72**, 241.

Landa P.S. 1980: *Self-Oscillations in Systems with Finite Number of Degrees of Freedom*, (in Russian) (Nauka, Moscow).

Landa P.S., Rosenblum M.G. 1991: *Time Series Analysis for System Identification and Diagnostics*, Physica, **48**, 232.

Landau L.D. 1937: *Kinetic Equations for the Case of Coulombian Interaction*, (in Russian) ZhETP, **7**, 203.

Landau L.D., Lifshitz E.M. 1957: *On Hydrodynamic Fluctuations*, (in Russian) ZhETP, **32**, 618.

Landau L.D., Lifshitz E.M. 1976: *Statistical Physics*, (in Russian) (Nauka, Moscow).

Landau L.D., Lifshitz E.M. 1986: *Hydromechanics*, (in Russian) (Nauka, Moscow).

Lax M. 1968: *Fluctuation and Coherence Phenomena in Classical and Quantum Physics* (Gordon and Breach, New York).

Lebowitz J.L. 1986: *It is no secret that there does not exist at present anything resembling a rigorous derivation of the hydrodynamic equations governing the time evolution of macroscopic variables from the laws governing their microscopic constituents*, in: *Proceedings of the 16th JUPAP Conference on Statistical Mechanics*, edited by H.E. Stanley (North-Holland, Amsterdam).

Leontovich M.A. 1935: *Basic Equations of the Kinetic Theory of Gases from the Standpoint of the Theory of Random Processes*, (in Russian) ZhETP, **5**, 211.

Lerner A.Ya. (Ed.) 1966: *Principles of Self Organization*, (in Russian) (Mir, Moscow).

Lichtenberg A., Liebermann M. 1982: *Regular and Stochastic Dynamics* (Springer, Berlin, Heidelberg, New York).

Lifshitz E.M., Pitayevsky L.P. 1979: *Physical Kinetics*, (in Russian) (Nauka, Moscow).

Lin S.S. 1955: *Theory of Hydrodynamic Stability* (Cambridge University Press, Cambridge).

Lorenz E. 1963: *Deterministic Nonperiodic Flow*, J. Atm. Sci. **20**, 167.

Maistrenko Yu.L., Romanenko E.Yu, Sharkovskii A.N. 1986: *Dry Turbulence*, in: *Mathematical Mechanisms of Turbulence*, (in Russian) (IM AN USSR, Kiev).

Malakhov A.N. 1968: *Fluctuations in Self-Oscillatory Systems*, (in Russian) (Nauka, Moscow).

Mandel P., Smith S.D., Wherrett B.S. 1987: *From Optical Bistability Towards Optical Computing* (North Holland, Amsterdam).

Mandelbrot B. 1982: *The Fractal Geometry of Nature* (Freeman, San Francisco).

Markin V.S., Pastushenko V.F., Chizmadzhev Yu.A. 1981: *The Theory of Excitable Media*, (in Russian) (Nauka, Moscow).

Martynov G.A. 1992: *Fundamental Theory of Liquids* (Adam Hilger, Bristol, Philadelphia, New York).

Mikhailov A.S. 1989: *Selected Topics in Fluctuational Kinetics of Reactions*, Phys. Rep. **184** (5-6), 309.

Mikhailov A.S. 1990, 1994: *Foundations of Synergetics I* (Springer, Berlin, Heidelberg, New York).

Maslov V.P. 1986: *Coherent Structures, Resonances and Asymptotic Non Single Valuedness for the Navier Stokes Equations at High Reynolds Numbers*, (in Russian) Uspekhi Mat. Nauk, **41**, 93.

Mehra J. 1975: *The Solvay Conferences on Physics* (D. Reidel, Dordrecht, Boston).

Moffatt G. 1984: *Some Directions of the Development of the Theory of Turbulence*, in: *Contemporary Hydrodynamics. Achievements and Problems*, (in Russian) (Mir, Moscow).

Monin A.S., Yaglom A.M. 1965, 1967, 1992: *Statistical Hydromechanics*, (in Russian) (Nauka, Moscow).

Moss F. 1992: *Stochastic Resonance from the Ice Ages to the Monkey's Ear* (University of Missouri, St. Louis).

Murray G. 1977: *Lectures on Nonlinear Differential Equations. Models in Biology* (Clarendon Press, Oxford).

Neimark Yu.I., Landa P.S. 1992: *Stochastic and Chaotic Oscillations* (Kluwer Academic Publishers, Dordrecht).

Neumann J. 1932: *Mathematische Grundlagen der Quantummechanik* (Verlag von Julius Springer, Berlin).

Nicolis G., Prigogine I. 1977: *Self Organization in Non Equilibrium Systems* (Wiley, New York, London, Sydney, Toronto).

Novikov E.A. 1964: *Functionals and Method of Random Forces in the Theory of Turbulence*, (in Russian) ZhETP, **47**, 1919.

Orayevskii A.N. 1986: *Dynamic Stochasticity and Lasers*, (in Russian) Trudy FIAN, **171**, 3.

Patashinskii A.Z., Pokrovskii V.L. 1982: *Fluctuation Theory of Phase Transitions*, (in Russian) (Nauka, Moscow).

Peng C.K., Mietus J., Hausdoeff J.M., Havlin S., Stanley H.E., Goldberger A.L. 1993: *Long-Range Anticorrelations and Non-Gaussian Behavior of the Heartbeat*, Phys. Rev. Lett. **70**, 1343.

Petrov W.M., Yablonskii A.I. 1993: *Mathematics and Social Processes. Hyperbolic Distributions and Their Applications*, (in Russian) (Moscow).

Polak L.S., Mikhailov A.S. 1983: *Self Organization in Nonequilibrium Physico-Chemical Systems*, (in Russian) (Nauka, Moscow).

Poincaré, H., 1893: *Le mécanisme et l'expérience*, Rev. Metaphys. **1**, 537.

Pontryagin L.S., Andronov A.A., Vitt A.A. 1933: *On Statistical Treatment of Dynamic Systems*, (in Russian) ZhETP, **3**, 3.

Prigogine I. 1980: *From Being to Becoming* (Freeman, San Francisco).

Prigogine I. 1993: *From Classical to Quantum Chaos*, (in Russian) Priroda, **12**, 13.

Prigogine I., Stengers I. 1984: *Order out of Chaos* (Heinemann, London).

Rabinovich M.I.,Trubetskov D.I. 1984: *Introduction into the Theory of Oscillations and Waves*, (in Russian) (Nauka, Moscow).

Risken H. 1984: *The Fokker – Planck Equation* (Springer, Berlin, Heidelberg, New York).

Romanovsky Yu.M., Stepanova N.V., Chernavsky D.S. 1984: *Mathematical Biology*, (in Russian) (Nauka, Moscow).

Ruelle D., Takens F. 1971: *On the Nature of Turbulence*, Commun. Math. Phys. **20**, 167.

Rytov S.M. 1976: *Introduction to Statistical Radiophysics*, (in Russian) (Nauka, Moscow).

Samarskii A.A., Galaktionov V.A., Kurdyumov S.P., Mikhailov A.P. 1987: *Peaking Regimes in Problems for Quasilinear Equations*, (in Russian) (Nauka, Moscow).

Schlichting H. 1958: *Grenzschicht - Theorie* (Karlsruhe).

Schram P. 1991: *Kinetic Theory of Gases and Plasmas* (Kluwer Academic Publishers, Dordrecht)

Schuster H.G. 1988: *Deterministic Chaos* (VCH, Weinheim).

Sheffield J. 1975: *Plasma Scattering of Electromagnetic Radiation* (Academic Press, New York, London).

Silin V.P. 1971: *Introduction into the Kinetic Theory of Gases*, (in Russian) (Nauka, Moscow).

Sinai Ya.G. 1979: *Stochasticity of Dynamic Systems*, in: *Nonlinear Waves*, (in Russian) (Nauka, Moscow).

Sinai Ya.G., Shilnikov L.P. (Eds.) 1981: *Strange Attractors*, (in Russian) (Mir, Moscow).

Sitenko A.G. 1975: *Fluctuations in Turbulent Plasma*, (in Russian) Fizika Plazmy, **1**, 45-49.

Sitenko A.G. 1977: *Fluctuations and Nonlinear Interaction of Waves in Plasma*, (in Russian) (Naukova Dumka, Kiev).

Sokolov A.A., Ternov I.M., Vukovskii W., Borisov A.W. 1983: *Quantum Electrodynamics*, (in Russian) (Izd. MGU, Moscow).

Spohn Herbert 1991: *Large Scale Dynamics of Interacting Particles* (Springer, Berlin, Heidelberg, New York).

Stanley H.E. 1971: *Introduction to Phase Transitions and Critical Phenomena* (Clarendon Press, Oxford).

Stratonovich R.L. 1967: *Topics in the Theory of Random Noise* (Gordon and Breach, New York).

Stratonovich R.L. 1975: *Theory of Information*, (in Russian) (Sovetskoe Radio, Moscow).

Stratonovich R.L. 1985: *Nonlinear Nonequilibrium Thermodynamics*, (in Russian) (Nauka, Moscow; English transl.: Springer, Berlin, Heidelberg, New York 1992).

Tatarskii V.I. 1983: *Wigner Representation of Quantum Mechanics*, (in Russian) Uspekhi Fiz. Nauk, **139**, 587.

Teodorchik K.F. 1945: *Generators with Inertial Nonlinearity*, (in Russian) Doklady AN SSSR, **50**, 191.

Terletskii Ya.P., Gusev A.A. (Eds.) 1955: *Problems of Causality in Quantum Mechanics*, (in Russian) (Inostrannaya Literatura, Moscow)

Turing A.M. 1952: *The Chemical Basis of the Morphogenesis*, Proc. Roy. Soc. B. **237**, 37.

Uhlenbeck G.E., Ford G.W. 1963: *Lectures in Statistical Mechanics* (Am. Math. Soc., Providence).

Van Kampen N.G. 1981: *Stochastic Processes in Physics and Chemistry* (North Holland, Amsterdam, New York, Oxford).

Vasilyev V.A., Klimontovich Yu.L. 1994: *Definition of Relative Degree of Order in a Chain of Diffusion-Coupled Oscillators Simulating the Dynamics of the Rectum*, Nanobiology, **3** (in press).

Vasilyev V.A., Romanovsky Yu.M., Yakhno V.G. 1987: *Autowaves*, (in Russian) (Nauka, Moscow).

Vedenov A.A. 1988: *Simulation of Elements of Thinking*, (in Russian) (Nauka, Moscow).

Velarde M. (Ed.) 1984: *Nonequilibrium Cooperative Phenomena in Physics and Related Fields* (Plenum Press, New York, London).

Velarde M. (Ed.) 1988: *Synergetics, Order and Chaos* (World Scientific, Singapore).

Vlasov A.A. 1938: *On the Vibrative Properties of Electron Gas*, (in Russian) ZhETF, **8**, 231.

Volkenshtein M.V. 1984: *Essence of Biological Evolution*, Sov. Phys. – Uspekhi, **143**, 429.

Voss R. 1992: *Evolution of Long-Range Fractal Correlations and 1/f Noise in DNA Base Sequences,* Phys. Rev. Lett. **68**, 3807.

Voss R., Clarke J. 1978: *1/f-Noise in Music. Music from 1/f Noise,* J. Acoustic Soc. Am. **63(1)**, 258.

Weissmann M.B. 1988: *1/f-Noise and Other Slow Nonexponential Kinetics in Condensed Matter*, Rev. Mod. Phys. **60**, 537.

Wiener N. 1958: *Nonlinear Problems in Random Theory* (Wiley, New York, London, Sydney, Toronto).

Zel'dovich Ya.B. 1984: *Chemical Physics and Hydrodynamics*, (in Russian) (Nauka, Moscow).

Zel'dovich Ya.B., Mikhailov A.S. 1987: *Fluctuation Kinetics of Reactions*, (in Russian) Uspekhi Fiz. Nauk, **153**, 469.

Zhabotinsky A.M. 1974: *Concentration Waves*, (in Russian) (Nauka, Moscow).

Zhdanov V.M. 1982: *Transfer Phenomena in Many-Component Plasma*, (in Russian) (Nauka, Moscow).

Zubarev D.N. 1971: *Nonequilibrium Statistical Thermodynamics*, (in Russian) (Nauka, Moscow).

Zubarev D.N., Morozov V.G. 1994: *Statistical Mechanics of Nonequilibrium Processes*, (Springer, Berlin, Heidelberg, New York (in press)).

Zubarev D.N., Morozov V.G., Troshkin O.V. 1992: *Turbulence as Nonequilibrium Phase Transition*, (in Russian) TMF, **92**, 293.

Fundamental Theories of Physics

Series Editor: Alwyn van der Merwe, *University of Denver, USA*

1. M. Sachs: *General Relativity and Matter*. A Spinor Field Theory from Fermis to Light-Years. With a Foreword by C. Kilmister. 1982   ISBN 90-277-1381-2
2. G.H. Duffey: *A Development of Quantum Mechanics*. Based on Symmetry Considerations. 1985   ISBN 90-277-1587-4
3. S. Diner, D. Fargue, G. Lochak and F. Selleri (eds.): *The Wave-Particle Dualism*. A Tribute to Louis de Broglie on his 90th Birthday. 1984   ISBN 90-277-1664-1
4. E. Prugovečki: *Stochastic Quantum Mechanics and Quantum Spacetime*. A Consistent Unification of Relativity and Quantum Theory based on Stochastic Spaces. 1984; 2nd printing 1986   ISBN 90-277-1617-X
5. D. Hestenes and G. Sobczyk: *Clifford Algebra to Geometric Calculus*. A Unified Language for Mathematics and Physics. 1984
          ISBN 90-277-1673-0; Pb (1987) 90-277-2561-6
6. P. Exner: *Open Quantum Systems and Feynman Integrals*. 1985   ISBN 90-277-1678-1
7. L. Mayants: *The Enigma of Probability and Physics*. 1984   ISBN 90-277-1674-9
8. E. Tocaci: *Relativistic Mechanics, Time and Inertia*. Translated from Romanian. Edited and with a Foreword by C.W. Kilmister. 1985   ISBN 90-277-1769-9
9. B. Bertotti, F. de Felice and A. Pascolini (eds.): *General Relativity and Gravitation*. Proceedings of the 10th International Conference (Padova, Italy, 1983). 1984
          ISBN 90-277-1819-9
10. G. Tarozzi and A. van der Merwe (eds.): *Open Questions in Quantum Physics*. 1985
          ISBN 90-277-1853-9
11. J.V. Narlikar and T. Padmanabhan: *Gravity, Gauge Theories and Quantum Cosmology*. 1986   ISBN 90-277-1948-9
12. G.S. Asanov: *Finsler Geometry, Relativity and Gauge Theories*. 1985
          ISBN 90-277-1960-8
13. K. Namsrai: *Nonlocal Quantum Field Theory and Stochastic Quantum Mechanics*. 1986   ISBN 90-277-2001-0
14. C. Ray Smith and W.T. Grandy, Jr. (eds.): *Maximum-Entropy and Bayesian Methods in Inverse Problems*. Proceedings of the 1st and 2nd International Workshop (Laramie, Wyoming, USA). 1985   ISBN 90-277-2074-6
15. D. Hestenes: *New Foundations for Classical Mechanics*. 1986
          ISBN 90-277-2090-8; Pb (1987) 90-277-2526-8
16. S.J. Prokhovnik: *Light in Einstein's Universe*. The Role of Energy in Cosmology and Relativity. 1985   ISBN 90-277-2093-2
17. Y.S. Kim and M.E. Noz: *Theory and Applications of the Poincaré Group*. 1986
          ISBN 90-277-2141-6
18. M. Sachs: *Quantum Mechanics from General Relativity*. An Approximation for a Theory of Inertia. 1986   ISBN 90-277-2247-1
19. W.T. Grandy, Jr.: *Foundations of Statistical Mechanics*.
   Vol. I: *Equilibrium Theory*. 1987   ISBN 90-277-2489-X
20. H.-H von Borzeszkowski and H.-J. Treder: *The Meaning of Quantum Gravity*. 1988
          ISBN 90-277-2518-7
21. C. Ray Smith and G.J. Erickson (eds.): *Maximum-Entropy and Bayesian Spectral Analysis and Estimation Problems*. Proceedings of the 3rd International Workshop (Laramie, Wyoming, USA, 1983). 1987   ISBN 90-277-2579-9

Fundamental Theories of Physics

22. A.O. Barut and A. van der Merwe (eds.): *Selected Scientific Papers of Alfred Landé.*
    [*1888-1975*]. 1988   ISBN 90-277-2594-2
23. W.T. Grandy, Jr.: *Foundations of Statistical Mechanics.*
    Vol. II: *Nonequilibrium Phenomena.* 1988   ISBN 90-277-2649-3
24. E.I. Bitsakis and C.A. Nicolaides (eds.): *The Concept of Probability.* Proceedings of the
    Delphi Conference (Delphi, Greece, 1987). 1989   ISBN 90-277-2679-5
25. A. van der Merwe, F. Selleri and G. Tarozzi (eds.): *Microphysical Reality and Quantum
    Formalism, Vol. 1.* Proceedings of the International Conference (Urbino, Italy, 1985).
    1988   ISBN 90-277-2683-3
26. A. van der Merwe, F. Selleri and G. Tarozzi (eds.): *Microphysical Reality and Quantum
    Formalism, Vol. 2.* Proceedings of the International Conference (Urbino, Italy, 1985).
    1988   ISBN 90-277-2684-1
27. I.D. Novikov and V.P. Frolov: *Physics of Black Holes.* 1989   ISBN 90-277-2685-X
28. G. Tarozzi and A. van der Merwe (eds.): *The Nature of Quantum Paradoxes.* Italian
    Studies in the Foundations and Philosophy of Modern Physics. 1988
                                                              ISBN 90-277-2703-1
29. B.R. Iyer, N. Mukunda and C.V. Vishveshwara (eds.): *Gravitation, Gauge Theories
    and the Early Universe.* 1989   ISBN 90-277-2710-4
30. H. Mark and L. Wood (eds.): *Energy in Physics, War and Peace.* A Festschrift
    celebrating Edward Teller's 80th Birthday. 1988   ISBN 90-277-2775-9
31. G.J. Erickson and C.R. Smith (eds.): *Maximum-Entropy and Bayesian Methods in
    Science and Engineering.*
    Vol. I: *Foundations.* 1988   ISBN 90-277-2793-7
32. G.J. Erickson and C.R. Smith (eds.): *Maximum-Entropy and Bayesian Methods in
    Science and Engineering.*
    Vol. II: *Applications.* 1988   ISBN 90-277-2794-5
33. M.E. Noz and Y.S. Kim (eds.): *Special Relativity and Quantum Theory.* A Collection of
    Papers on the Poincaré Group. 1988   ISBN 90-277-2799-6
34. I.Yu. Kobzarev and Yu.I. Manin: *Elementary Particles. Mathematics, Physics and
    Philosophy.* 1989   ISBN 0-7923-0098-X
35. F. Selleri: *Quantum Paradoxes and Physical Reality.* 1990   ISBN 0-7923-0253-2
36. J. Skilling (ed.): *Maximum-Entropy and Bayesian Methods.* Proceedings of the 8th
    International Workshop (Cambridge, UK, 1988). 1989   ISBN 0-7923-0224-9
37. M. Kafatos (ed.): *Bell's Theorem, Quantum Theory and Conceptions of the Universe.*
    1989   ISBN 0-7923-0496-9
38. Yu.A. Izyumov and V.N. Syromyatnikov: *Phase Transitions and Crystal Symmetry.*
    1990   ISBN 0-7923-0542-6
39. P.F. Fougère (ed.): *Maximum-Entropy and Bayesian Methods.* Proceedings of the 9th
    International Workshop (Dartmouth, Massachusetts, USA, 1989). 1990
                                                              ISBN 0-7923-0928-6
40. L. de Broglie: *Heisenberg's Uncertainties and the Probabilistic Interpretation of Wave
    Mechanics.* With Critical Notes of the Author. 1990   ISBN 0-7923-0929-4
41. W.T. Grandy, Jr.: *Relativistic Quantum Mechanics of Leptons and Fields.* 1991
                                                              ISBN 0-7923-1049-7
42. Yu.L. Klimontovich: *Turbulent Motion and the Structure of Chaos.* A New Approach
    to the Statistical Theory of Open Systems. 1991   ISBN 0-7923-1114-0

Fundamental Theories of Physics

Fundamental Theories of Physics

64. M. Evans and J.-P. Vigier: *The Enigmatic Photon.* Volume 1: The Field $B^{(3)}$. 1994
ISBN 0-7923-3049-8
65. C.K. Raju: *Time: Towards a Constistent Theory.* 1994   ISBN 0-7923-3103-6
66. A.K.T. Assis: *Weber's Electrodynamics.* 1994   ISBN 0-7923-3137-0
67. Yu. L. Klimontovich: *Statistical Theory of Open Systems.* Volume 1: A Unified
Approach to Kinetic Description of Processes in Active Systems. 1995
ISBN 0-7923-3199-0; Pb: ISBN 0-7923-3242-3

KLUWER ACADEMIC PUBLISHERS – DORDRECHT / BOSTON / LONDON